INTRODUCTION TO ECOLOGY

INTRODUCTION TO ECOLOGY
Paul A. Colinvaux

JOHN WILEY & SONS,INC.
New York London Sydney Toronto

This book was set in Optima by Progressive
Typographers, Inc., and printed and bound by
Universal-Jenkins, Inc. The designer was
Jerome B. Wilke. The drawings were designed
and executed by the Wiley Illustration De-
partment. The editor was Deborah P. Herbert.
Joan E. Rosenberg supervised production.

For ecological reasons, 40 percent of the
content of the paper used in this book consists of recycled, post consumer, waste paper.

Library of Congress Cataloging in Publication Data

Colinvaux, Paul A   1930–
  Introduction to ecology.

  Bibliography: p.
  1. Ecology.   I. Title.
QH541.C63   574.5   72-3788
ISBN 0-471-16498-4

Printed in the United States of America

10 9 8 7 6 5 4 3 2 1

PREFACE

Ecology is a pleasant science. Those who follow it pass their time in trying to understand functions of the natural world, which are as mystifying as anything in physics but which also appeal to the animal cravings in humans. An ecologist can savor the life of a naturalist while using the methods of chemistry or the philosophy of mathematics. A textbook on this subject should be easy to write, and my aim has been to capture enough of ecology's enjoyment to give a little pleasure in the reading. I hope the book might be read, not just consulted. Its many figures and tables are but second statements of information embodied in the prose.

There are four main themes or ways of looking at nature which ecologists have followed, and these I have used to organize the book into four parts. Part 1 traces the way ecology emerged from biogeography, from the attempt to explain the odd fact that animals and plants of different parts of the world lived in communities so distinct that they even looked different. This attempt led not only to the basic ecological understandings but also to the great unifying concept of the ecosystem. Part 2 describes how the ecosystem concept was developed to show how living things share space and raw materials while dwelling in complex systems driven by the energy flowing from the sun. Part 3 follows the long intellectual struggle which ecologists have waged in their attempts to explain the balance of nature. The themes of this part are the Darwinian themes of competition for limited resources and the battles between predators and prey. Part 4 is an account of how modern ecology seems to be capping the ecological efforts of a century by restating the Darwinian principle of fitness and using it to explain the qualities of individual animals and plants. These qualities give rise to such classical ecological phenomena as communities, succession, ecosystems, and the impression of a balance in nature. Part 4 is thus a review and synthesis of the earlier parts of the book.

I have written in short chapters intending that each should be a self-sufficient piece of writing with its own introduction and fully developing its own arguments. It should be possible to use the book for a course which organizes ecology into a different sequence of subjects by assigning chapters in a different order. Each chapter has a preview rather than a summary, for this conforms with my main principle that the book should be readable while serving a summary's purpose of succinctly stating the theme of a chapter.

The book evolved from the experience of lecturing on ecology to large classes of students majoring in the sciences, second-year students mostly but also including all ranks from freshmen to graduates, and from lecturing to a complete cross section of the university in courses on the human environment. The whole book could be assigned reading for a one-semester course for second-year science students, though some chapters might have to be deleted for a course of only one quarter. There is, however, nothing in the book which should be beyond the comprehension of first-year science students, though for use by freshmen it might be best to delete all of Part 4 except Chapter 40, which is a review of succession theory.

Part 3 might well stand alone for an introductory course on population studies. In Part 3 a few mathematical models are discussed, but nowhere is an understanding of the mathematics necessary in order to follow the argument. It is my hope that mathematics and jargon could be completely circumvented by those wishing to set them aside and follow the arguments in prose, that arts students and even artists would be able to glean from the book some of the great fun that ecologists find in their own peculiar way of looking at nature.

Parts of the book, particularly Part 2, could be used for general courses trying to acquaint nonscientists with something of the scientific method. The implications of ecological findings for human affairs are everywhere discussed, but with an attempt to keep the discussion within the bounds of realism and to what ecology truly has to say. While as worried as anybody about the sort of earth and living we will leave our descendants, I have yet been at pains to show where many of the hypotheses of ecological catastrophe which have been put forward are false. The book, however, is about ecology and not the human condition. References to men are thus scattered throughout the book, though, for convenience, they will be found listed in the index under "Human Ecology."

Every writer who tackles so wide a collection of subjects as this must know that many of the ideas which he develops as if they were his own have come to him from others in sundry conversations. This has surely been the way in the making of this book. Some of my thought, of course, had its roots in my formal training, at Cambridge, Duke, Belfast, and Yale. More dawned on me during wanderings in many lands, in Alaskan tundras, Nigerian forests, and the Galapagos Archipelago, on soil survey in the boreal forest of Canada, and assisting my wife in studies of coral reefs, under the Caribbean with a tank of air on my back. But the real nursery of the book has been The Ohio State University where for seven years I have had the high privilege of lecturing about ecology to classes several hundred strong, and where perennial arguments about ecology have gone round the lunch table in my laboratory. My many teachers, whether formal or as companions, will I hope, excuse my not listing them by name, but two of them, D. A. Livingstone and R. H. Whittaker, I must acknowledge, for they read early drafts of this book to such effect that their comments opened whole new understandings to me and have in places changed the book into something much better than it was.

Paul A. Colinvaux
Columbus, Ohio, 1972

CONTENTS

INTRODUCTION TO ECOLOGY

The business, and the fun, of science is to notice that phenomena of the natural world require explanation; then to find the explanations. In these twin tasks the physical sciences have advanced further than biology. It is centuries now since men saw that they must provide explanations, based on logic rather than superstition, for such familiar things as the alternation of night and day, the phases of the moon, and the falling of dropped rocks. The first realization that these phenomena needed explaining has been followed by the beautiful flowering of the physical sciences, which has developed as a steady quest for better explanations. But biology has moved more slowly, both in identifying the great problems and in moving toward their solution.

A medieval man (or a modern child), who was willing to accept with simple gladness the changing of the seasons and the falling of things down rather than up, took with the same simple acceptance some remarkable biological phenomena. When he went about his business in the early morning, he was serenaded by the songs of birds all singing together, the dawn chorus, evidently a God-given joy to start the day. And yet reason should think this singing of the birds in the morning to be decidedly odd. Why do the birds sing at dawn — indeed, why do they sing at all? And then, in the garden, there were many different kinds of plants; in the meadow as well, and in the forest. Why should there be all these different kinds of plants growing side by side in similar places? And the animals too; there were even more kinds of them than of plants, particularly the insects. Why should this be? And, for that matter, why should there not be more kinds? Questions such as these were not seriously asked even as science advanced toward explanations of the no more dra-

INTRODUCTION

matic phenomena of the physical world, and the most interesting of these questions have only been posed in our century. They provide the substance of the discipline of ecology.

Questions about the numbers of living things, of where they are found and what they do, may be neatly referred to as problems of habit and habitat. The word **"Ecology"** is invented to convey this idea of **"The study of animals and plants in relation to their habit and habitats."** It comes from the Greek word "oikos" which means "house" or "home" or "household" or something like that, and "logos" which is the word taken to mean "science" and which is the root of all the "-ologies" with which our language is cluttered. We take, then, this word from the Greek "ecology" to mean "the study of the household of nature." It is a useful word, one that has come to convey the idea of the study of all the related phenomena of the ways and environments of living things.

An ecologist must often be out in the field studying animals and plants in nature, but many other people do this without being ecologists. The distinction is best shown by an example. It is common English usage to talk of larks; of singing like a lark, being happy as a lark, or larking about; and this usage comes from poetic musings about the habits of the North-European skylark, *Alauda arvensis*. In the early summer skylarks trill beautifully, high in the sky over meadows and wheat fields. They start from the ground with a swiftly rising, fluttering flight, singing the while, and climbing up and up until they almost vanish against the blue, then they stop singing and plummet down to earth before repeating the whole performance. You may lie on your back in the sun for hours lulled by this pleasant serenade. Many poets have done so, and for many centuries. Some came to know the birds well, to sense on what days the larks sang, to know where to find larks, to see their nests and eggs and, in short, to be good field naturalists. And yet, for centuries there was no attempt to look at the lark's beautiful performance with the eye of reason, to realize that here was something odd that required explanation, and to ask the question: "Why does the skylark behave in this fetching but peculiar manner?" When that question is asked, the field study of the skylark becomes ecology. But reflect on the myriads who have watched skylarks without asking that question: naturalists all, but ecologists none.

A hundred years after the birth of modern science, the time when modern astronomy offered its reasoned explanations to the phenomena of night and day and when Newtonian physics set out on the long road of inference that was to lead more than two centuries later to relativity, came the birth of evolutionary biology. Linnaeus, and those who came after him, established that the animals and plants of the world were distinct enough to be classified. There truly were many different kinds of plants and animals. World travel became possible, and biologists became aware that different parts of the world held different collections of plants and animals. "Why was this so?" the question began to be asked. Some more practical questions began to be asked also, agricultural questions like, "Why does a crop do better on one field than another"—questions on which man had long pondered but which he now began to ask in a more reasoned manner. These questions of agriculture were ecological questions, too, but until the statement of the theory of evolution by Darwin none were asked with much conviction. If the number of species, their distribution, and how well they "did," was the result of the whim of a creator, then there was no point in asking "why" and expecting reason to provide an answer. Ecology was made possible by the theory of evolution.

The theory of evolution by natural selection showed *how* it could be that there were many different kinds of living things; species were changed into other species by selection or "the struggle for existence"; but the theory *assumed* that this struggle must be taking place. It declared in some instances the results of the struggle, the fleetness of foot of hares, the long legs of wading birds, and so on, but it could not see all the dimensions of the struggle nor did it show why there should be so very many different kinds of organisms rather than a more generalized few. But there are in the *Origin of Species,* clear clues to where the details of the struggle may be found. They lie in the quest for food or for living space, in the hostility of winter or other lean seasons of the year and, above all, in the drive to find resources for the young. The reason that there are so many species must lie in some interaction of the animals and plants with their physical world and with each other. It was sought, and is still being sought, by the new discipline of ecology.

The ecological quest has advanced slowly, from the more obvious problems to those more refined and subtle. Some of the first leads were geographical, "Why did the vegetation of different parts of the world look different?" and "Why did some countries produce more crops than others?" There were two approaches to these problems. Some people concentrated on single species and the conditions that controlled their lives (**autecology**) while others studied the aggregates of mixed organisms called communities (**synecology**). Then the problem of population size became, and remains, of central interest, "Why do we get the impression of a balance in nature with populations of animals and plants seemingly roughly constant, since they are all presumably breeding away as fast as they can go?" Why, on the other hand, do we see such noted exceptions to this conceived balance as plagues and less disastrous upsets? Ecology has worked steadily to answer questions such as these, but the problems are complex and movement toward the solution of each major problem has only been achieved by uncovering and solving many lesser problems that lay in the way. We are now beginning to offer answers for the major questions, but the end is not with us, for the answers given by all ecologists are not yet the same.

Now, while we are still reaching for some of the major answers, the answers have suddenly become vital to mankind, for our own increasing population and changing habits are dramatically altering the conditions of life for all plants and animals, including ourselves. Our population is outstripping its energy supply, and the excrement of our bodies and machines has become so copious that the world system is changing. We take crude and uninformed measures to improve our lot, such as inventing potent insecticides, only to find fearful and uncalled-for results. Some birds like the skylarks of which the poets sung may be vanishing now, poisoned out of existence by DDT. Why does DDT, used in tiny doses calculated to kill only insects, end up by killing off our birds? — And what will be the effect on us of the slaughter of the birds? Problems that were once fascinating only to the intellects of gifted men have taken on a grim significance.

PART 1

THE
ECOLOGICAL
SIDE OF
GEOGRAPHY

Perhaps the first great phenomenon to interest plant ecologists was that the vegetation of different parts of the earth looked different. There were rain forests, deciduous forests, coniferous forests, and tundra. Geographers drew maps showing the aerial extent of each of these formations, and their maps seemed to say that the world was neatly parceled out between different formations of plants. Ecologists were invited to explain why this should be true. A working hypothesis was that parameters of climate, particularly the availability of water and temperature, set the boundaries for plants; vegetation maps could thus be used to draw climatic maps. In a very general way the climatic maps were useful, suggesting that the general hypothesis that vegetation belts were correlated with climate was sound, but this was only a very general conclusion. Real vegetation types usually grade one into another so that it is impossible to tell where one formation ends and another begins. The boundaries drawn by map-makers were usually arbitrary. This was for long not realized, however, the boundaries being thought to be real. The experience of botanists working on high mountains encouraged this error for the vegetation of mountainsides seemed truly to be set out in discrete blocks, but the distinctness of belts of mountain vegetation has now been shown to be a myth. Plants are distributed up and down mountains according to their individual tolerances, producing a spectrum of distribution. The apparent distinctness of the vegetation belts, when viewed from afar, is largely an optical illusion as the eye picks out bands where individual species are concentrated. The same is true of distributions on continents where mixtures of individual kinds of plants grade one into another, but across whose gradients the map-maker has

7

drawn a line. Yet there are natural barriers such as mountain ranges which do divide formation from formation, and it has recently become apparent that climatic fronts may also serve as discrete physical boundaries that are recognized by plants. The circulation of the atmosphere divides the earth into more or less discrete climatic regions providing more or less discrete areas for plant and animal life.

CHAPTER 2

THE BROAD
DIVISIONS OF THE
EARTH

The vegetation of different parts of the world looks different. In the equatorial basins of the major continents, Africa, Asia, and South America, are forests of enormous trees whose thick trunks rise like columns from the gloom of the forest floor until the first branches are reached perhaps a hundred feet from the ground. Full-grown American elms or English oaks could be fitted under the lowest branches of these trees. They are evergreens whose interlocking canopy of leaves blocks out the sun and allows only a dim green light to penetrate, producing an effect below that the African traveler Stanley (1890) likened to the interior of a great cathedral, the rising Gothic columns of which were illuminated only by windows of green stained glass. Lianas trail from the upper branches where their leaves are in the light, and large epiphytes (literally "on-plants") grow attached to the branches of the forest giants. There are layers upon layers of plants between the canopy and the ground, but the ground itself is bare, being but sparsely covered with small plants. These forests are called "tropical rain forests" (Figure 2.1).

North of the tropics, in the temperate regions of Europe and eastern North America, where most ecologists have worked, there are still forests but they look very different. The trees are not the same shape, and they are deciduous. Small creepers and epiphytes such as lichens and mosses may still be found, but they are not so conspicuous as in the rain forests. There is a richer vegetation on the ground below, which is often covered by a carpet of herbs. The general appearance and feel of the forest is something quite different. Such a forest may be called "temperate deciduous forest" (Figure 2.2).

Blocks of rain forest or deciduous forest may spread for hundreds of miles without notable changes in their appearance. It is true that in a big area there is much local variation, either because of the works of man or as the pattern of the land is broken up by swamps, mountains, and river valleys. But rain forest or deciduous forest may be so much the norm over whole countries that the shapes of trees and forest may seem almost to be part of the national character.

Where the territory of the temperate deciduous forest approaches that of the rain forest are signs of one forest type tending toward the other; the trees of the Carolinas, for instance, are larger and more draped with

Tropical rain forest in the Congo. Tall, broad-leaved evergreen trees, often with
spreading buttress roots like that in the center of the photograph, form the high
canopy of the forest. These trees have long straight trunks with few lateral branches,
but they are often festooned with lianas and epiphytes. Smaller trees and shrubs
make successive layers of patchy canopy under the canopies of the tallest trees. The
forest floor, although carpeted with herbs, may be surprisingly open. Such forests are
hard to photograph from inside because so little light penetrates the successive layers
of canopy, and photographs like this one have only usually been taken where people
have been clearing or thinning the forest.

Figure 2.1

Figure 2.2   Temperate deciduous forest; a beech (*Fagus silvatica*) wood in France. The trees are broad-leaved, but deciduous, and there are less layers of canopy than in the tropical rain forest. The only epiphytes apparent are lichens and mosses on the trunks of the trees, and the forest floor is well-covered with herbs and shrubs. Such a forest in Europe is certain to be second growth and to have been managed by foresters. It is likely that a virgin forest in such a place would have much larger trees and perhaps more undergrowth. But the ''feel'' of this forest is yet quite different from the more closed evergreen forests of the wet tropics. Various intermediate forms can be found in intervening latitudes.

epiphytes than those of New Jersey. But we are denied looking to see what really happens where the forests meet in America by the insertion of the Isthmus of Panama between them. The Sahara Desert and the Mediterranean Sea likewise separate the forests of Europe from the rain forests of Africa, again denying us the chance of looking for a vegetation boundary. Only in Southeast Asia does a continental landmass connect temperate northern lands to the rainy tropics, and there the forests may be seen to blend over a gradient of hundreds of miles. But over most of the world these main forest types are isolated by geography, letting you map the bounds of each type with ease.

Some of the places where other blocks of vegetation meet are more satisfactory for a geographer, because the vegetation itself seems to provide you with nice boundaries that you can map. North of the deciduous forests of temperate New England lie the boreal forests of Canada. Instead of being bushy-topped, the trees are mostly of the triangular Christmas-

Boreal forest in the Pacific Northwest. The trees are evergreen, cone-shaped, needle-leaved, and branched over most of their height. They grow close together and there is little sign of layering in the canopy of the forest. The whole form of the forest is starkly distinct from that of temperate and deciduous forests.

Figure 2.3

tree shape, standing in dark-green brooding ranks (Figure 2.3). The change from the one kind of forest to the other may be startlingly swift. If you drive a fast car northeastward through New Hampshire and Maine, and on to New Brunswick, you may skip through the transition without being aware that you are upon it. There has been some merging of the forests, some mixing of outliers in the two blocks, some alternation of patches of one forest type with the other, but it has yet been a remarkably sudden transition. A few miles of road are all it takes. And yet to get this effect you must choose your crossing place well. Far into New Brunswick there are maple woods, and even walnuts and elms along the broad valley of the St. John River. There is a geographer's boundary in places but the generality is a grand merging of vegetation types as in Southeast Asia.

The trees of the boreal forest are evergreens, as are those of the tropical forests so many hundreds of miles of deciduous forest away. They branch low and are relatively close-packed. Climbers are absent and, apart from lichens, epiphytes are insignificant. The ground is carpeted brown with needles and has a different array of understory plants. The change to this from deciduous forest is remarkable, particularly for a traveler who has found one of those lucky places where the change is abrupt. But a more remarkable change still can be found a few hundred miles to the north.

The territory of the boreal forest is hundreds of miles wide, but eventually it meets the completely treeless vegetation of the tundra. There are many miles of transition, of ever scrubbier, ever more isolated, patches of trees. Even when the open tundra seems to occupy nearly all the land, there will be lines of spruce trees fingering out along the rivers, and woods of alders or poplars in hollows or sheltered places. The vegetations blend as if stippled together. But there is something dramatic about the changeover all the same, for one vegetation type is a forest and the other has no trees. To be with a tree or without a tree; it seems an uncompromising sort of choice. A map-maker must itch to draw the line between them, and this itch may be encouraged by the existence of places where the vegetation types do meet locally as the deciduous forest meets the boreal forest in parts of Maine. But the reality is a generous interfingering between the forest and the open tundra, an interfingering spread over many miles.

For all that these various blocks of vegetation generally merge in ways which must frustrate a map-maker, the distinctness of each general type seems clear. If you traveled the earth in bounds of several hundred miles at a time, you would be frequently set down in vegetation that was uncompromisingly different from what you saw at the last stop. You would get an impression of a world set out into different blocks of vegetation, as by a giant gardener who had yet been a bit careless about the edges of his plots so that the plants sometimes got mixed up. It is the difference in form of these different blocks of vegetation that are so obvious to us, which leads us to call them **formations.** The succession of formations between the equator and the Arctic is roughly mirrored in the Southern Hemisphere, although the absence of large landmasses makes the pattern less complete. There are also some other formations that are more irregularly scattered about the world, but which have distinctive form and, if not with territorial integrity, are yet typical of large areas. These are the formations of desert plants, prairie grasslands, savanna, and the sclerophylous bushlands variously called chaparral (California), maquis (Mediterranean), Mallee Heath (Australian), and so on (Figures 2.4, 2.5).

The naturalist travelers of the eighteenth and nineteenth centuries saw these strikingly different kinds of vegetation and brought back to their European universities the news that the vegetation of the earth was parceled out into "formations." They drew maps showing the extent of each formation. Now drawing a map involves drawing boundaries. Where there are real boundaries on the ground this is easy. But what if there are no real boundaries on the ground? A vegetation map-maker could plot the position of the tree line well enough, if he had a few reports from places with abrupt treelines which he could plot and link up. But what of the grand transition between the temperate woods and the tropical rain

Tropical savanna in East Africa. The form of this vegetation reflects the pressures of grazing mammals and fire. Plants of the turf are cropped low or burned down. Scattered trees preserve their leaves only where they are out of reach of most of the animals, and they are thorny and with fire-resistant bark.

Figure 2.4

forest? In America you could quickly draw a line somewhere through Mexico or Panama and have an answer that, at least, looked convincing. You could do the same sort of thing in Africa. But in Southeast Asia you faced a more serious problem. From the rain forests of Cambodia to the temperate woodlands of North China there were hundreds of miles of gradient of continuous change. To make your map, you must run your pencil through Asia, deciding a boundary that nature has not decided.

Separating tropical and temperate forests was not the only difficulty that map-makers faced. Many of the lesser formations such as deserts, prairies, and moist coniferous forests like those of western America (Figure 2.6) had no edge that you could see on the ground. There might be local edges set by landforms, or sometimes by where the fires of the last few years had burned themselves out, but on a continental scale it was more usual to find an interfingering or merging of the formations. Yet you had your map to make. So you drew your boundaries down what seemed to be the middle of the transitions. When you had finished, you had shown the world parceled out by formations of plants (Figure 2.7). Such a map reflected much underlying reality. There *were* blocks of vegetation of strikingly different form. These blocks *were* confined to broad geographic regions. And occasionally there were places where a pair had remarkable geographic integrity, confronting each other face to face, like nations at a frontier. Why should the vegetation of different parts of the world look so different? Why should enormous stretches of land have vegetation of a special form? And why should the plants of different formations respect each others territory so? These questions were among the first to be tackled by ecologists, and the answers given were to influence the thinking of ecologists for a century.

In 1855 Alphonse de Candolle made the first serious attempt to answer some of these questions. He was a taxonomist, an herbarium botanist who traveled little but who had at his disposal the great Paris collections representing all the known plants of the world. From this material he set out to compile his "Prodromus" (de Condolle, 1824–1873), the last attempt by one man to describe all the plant species known to science and,

Figure 2.5
Sclerophylous bushlands, chaparral of California (A), maquis of the Mediterannean (B) and mallee heath of Australia (C). The forms of these bushlands, on different continents and made up of different species of plants, have much in common. They are gnarled and twisted, rough and thorny. The leaves tend to be dark and are hairy, leathery, or with thick cuticles. All grow in places with hot dry summers and relatively cool moist winters; a climatic pattern which is both rare and always accompanied with vegetation of this form. The ability to be able to map this particular climate by the extent of this particular formation became, perhaps, one of the most persuasive arguments for the hypothesis that the forms of plants reflected climate.

A

B

C

Figure 2.6

Moist coniferous forest in California. The trees are tall, as tall or even taller than those of the tropical rain forest, but they are needle-leaved conifers whose individual shapes are like the trees of the boreal forest. There is often an understory of smaller deciduous trees up to 20 feet high and perhaps a dense undergrowth of bushes. Forest of this form once covered the coastal strip of western North America from northern California to southeastern Alaska. Climatologists used its extent to map the wet foggy coastal climate.

in doing so, he became aware of the ecological problem of the formations. Some aspect of the weather seemed the obvious general answer. Shortage of water, due to physiographic features, should certainly be the explanation of desert formations and, perhaps, of grasslands, but something else was needed to explain the rest. De Candolle sought the answer in temperature (de Candolle, 1874), supposing that there must be critical changes in the heat regimen at particular times of the year that accounted for the changes between one formation and another. The climatic data available to him, of course, were scanty, and his attempts to draw isotherms which should coincide with formation boundaries were not lastingly successful, but the attempt was on sound ecological lines (Figure 2.8) A general hypothesis of climatic control had been put forward and then an attempt had been made to narrow this down to limiting factors of the environment, water and heat.

No more detailed explanation for the existence of formations and their apparent boundaries was forthcoming for a long time, but de Candolle's work became, instead, the basis of climate maps. Mapping so fluid a thing as climate was a very hard task in the nineteenth-century world of few weather stations. By comparison, mapping vegetation was easy. If climate was the explanation for changes in vegetation, then vegetation was conversely the key to climate. So climatologists could use vegetation maps as a base for climatic maps.

More than half a century after the publication of de Candolle's treatise on plant geography, Vladimir Köppen (1918, 1844, 1900) used de Candolle's classification of vegetation to found the modern systems of classifying climate. De Candolle had ordered the formations according to their supposed heat-loving or drought-tolerating qualities; rain forests being formations of *megatherms,* deciduous forests of *mesotherms,* deserts of *xerophylls,* and the like (Figure 2.9). Köppen simply called the climate mapped by the boundaries of a megatherm formation an "A" climate (mercifully ignoring the opportunity to air his Greek), that of a xerophyllous formation a "B" climate, that of mesotherms a "C" climate, and so on. Suffixes that served to subdivide each type resulted in the Köppen climatic system which is the basis of all modern classifications (Figure 2.10). Unnumbered meteorologists have learned and used this system. It works. There can thus be no doubt that the extent and rough limits to formations are determined by climate.

This conclusion is satisfactory as far as it goes. The remarkable discovery that the earth had been parceled out between different sorts of plants ceased to be remarkable when we found that climate was parceled out into roughly the same-sized blocks. Our vegetation maps had become the necessary outcome of things such as Coriolli's force and the spread of mountain ranges. But "climate" is too general an explanation to give much intellectual comfort. Why do different climates require different forms of plants? Why do formations commonly grade together after being recognizable over continental expanses? And why, on the other hand, must there sometimes be natural boundaries between formations so that plants as different as maple trees and spruce trees face each other like rival armies at a disputed frontier?

Although the distinctness of form of any one formation is so obvious to the eye, the description and classification of form is a difficult matter rather akin to describing the uniqueness of human faces. An obvious tactic is to rely on notably different plant shapes and to talk of spruce-tree forms, palm-tree forms, and so on, but such classifications soon become confused with familial classifications; they classify evolutionary affinities

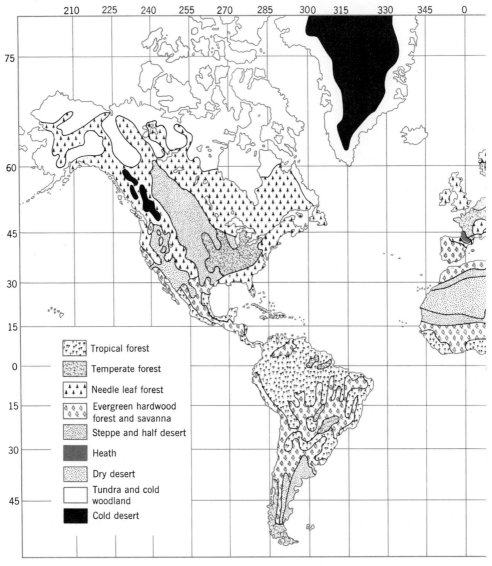

210   225   240   255   270   285   300   315   330   345   0

Tropical forest

Temperate forest

Needle leaf forest

Evergreen hardwood
forest and savanna

Steppe and half desert

Heath

Dry desert

Tundra and cold
woodland

Cold desert

Figure 2.7   A vegetation map of the earth. This is the sort of map that evolved from attempts to plot the boundaries of formations. The real distinct boundaries that geographers can use are those between land and sea, along deserts or beside mountain ranges, and the main shapes on this map are familiar enough, within continents as well as between them. Where geography has

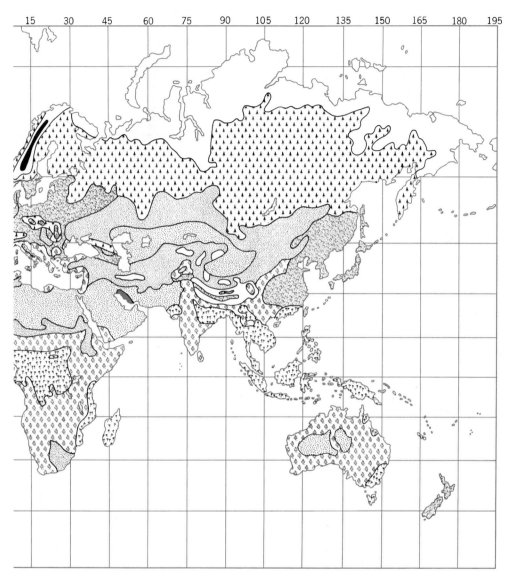

not helped, so that vegetation manifestly grades from one formation to another, the map-makers have yet had to draw a line between formations. A map like this shows the vegetation of the earth to be more neatly set into compartments than it really is. (Redrawn from Dansereau, 1957.)

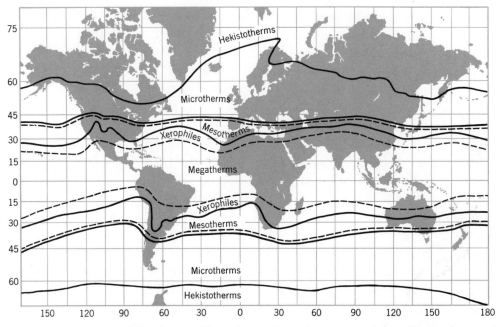

Figure 2.8 De Candolle's division of the earth according to the properties of plants. The broad extent of formations was thought to be set by temperature, except that shortage of water set the limits of plants of hot deserts (the xerophiles). De Candolle's ideas forshadowed the attempts of ecological biogeographers like Allen to show that the forms of life on earth were distributed in latitudinal bands. The hypothesis that formation boundaries could be correlated with temperature and water led to the use of vegetation maps as the basis for climatic maps. The isotherms shown here are those suggested by De Candolle but are plotted from modern meteorological data.

rather than functional shape, and they assign the majority of plants to a few common groups. Attempts of this sort have been made ever since the time of the Napoleonic wars when the great traveler von Humboldt (1807) made the first. But no such system has come into general use. Stylized sketches of plants in profile are often used as an aid to descriptive writing, and their construction is a useful aid to analytical thinking, rather like the police exercise of putting together an *identikit* drawing of a wanted fugitive. Figure 2.11 shows how sketches of this kind can be used to reveal the main features of formations, and Figure 2.12 shows the formalized version of such a system which has been put forward by Dansereau (1951). But to be a real use to ecology, the units of form chosen must clearly represent function. We know the forms are there. Now we want to know why they are there.

The Raunkiaer (1934) system for classifying life-forms has been the most illuminating, even though it approached the problem indirectly. Raunkiaer reasoned that the aspect of form which could be most surely correlated with function was that which set the position of the perennating organ. Since all weather is variable, plants must be equipped to overcome the most unfavorable season to which they are exposed, and

De Candolle's plant group	Postulated plant requirements	Formation	Köppen climatic division
Megatherms (most-heat)	Continuous high temperature and abundant moisture	Tropical rain forest	A (rainy with no winter)
Xerophiles (dry-loving)	Tolerate drought, need minimum hot season	Hot desert such as Sonoran	B (dry)
Mesotherms (middle-heat)	Moderate temperature and moderate moisture	Temperate deciduous forest	C (rainy with mild winters)
Microtherms (little-heat)	Less heat, less moisture. Tolerate long cold winters	Boreal forest	D (rainy climates with severe winters)
Hekistotherms (least-heat)	Tolerate polar regions "beyond tree-line"	Tundra	E (polar climates with no warm season).

Figure 2.9

De Candolle's plant groups as the basis of the Köppen classification of climate. De Candolle noted that plant formations occupied wide areas which seemed to be characterized by climate. He sought to describe the plants of each formation by their heat and water requirements. His formation tables and five major divisions were later used by Köppen to map and classify climate. The success of the resulting climatic classification as a working and predictive tool successfully tested the De Candolle hypothesis that the rough extents of formations were set by climate.

such seasons for many plants are, in fact, so unfavorable that they last it out in a state of dormancy. Deciduous trees lose their leaves in winter and grow new ones in the spring; for them the perennating bud (literally "the bud which helps them through the year") is the leaf bud. It is high and exposed. The perennating bud of a daffodil, on the other hand, is in the bulb. It is hidden in the ground and thus presumably is protected against the dangers of the unfavorable season. All the plants of an arctic tundra have their perennating buds buried or very near the ground where they are somewhat sheltered. On the other hand, most of the buds in a rain forest are high in the air and clearly exposed. Raunkiaer erected a number of classes, at first five and later more, defined by the degree of exposure of the perennating buds in the unfavorable season. Unfortunately he fell to the lure of bastard Greek and we must learn of *cryptophytes* (hidden plants), *hemicryptophytes*, and the like. The whole shape of a plant is so dependent on where it keeps its perennating bud that Raunkiaer had provided us with a realistic classification that both reflected form and catalogued function (Figure 2.13).

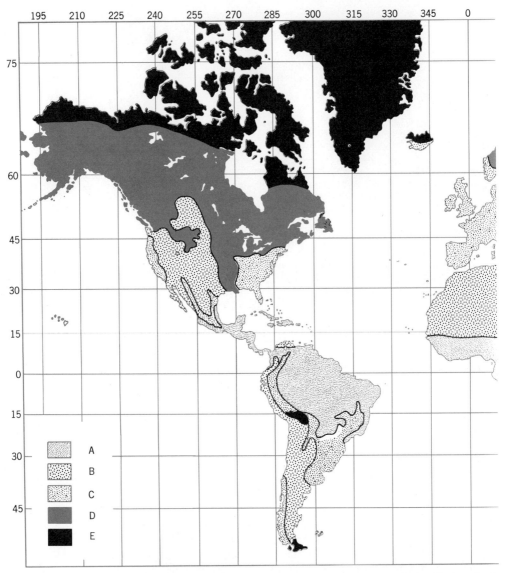

Figure 2.10   Climates of the earth mapped under the Köppen system. Köppen worked on the hypothesis that the boundaries between different kinds of vegetation were set by climate. His climatic maps were essentially vegetation maps, and the similarity between this figure and a formation map (Figure 2.7) is obvious. But climatologists found the Köppen system useful,

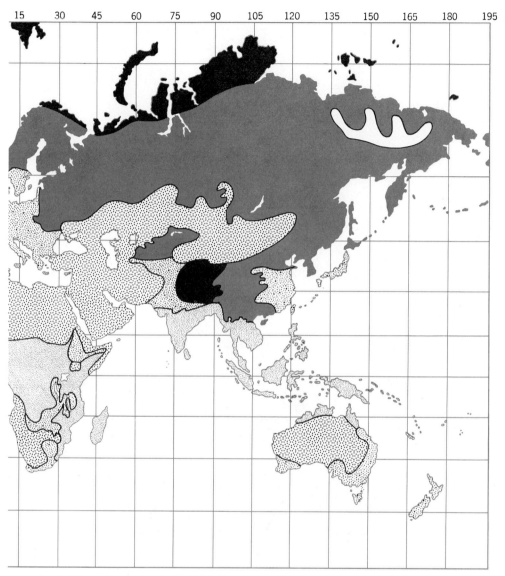

suggesting that the earth might really be divided into vague climatic regions each of which
resulted in a special kind of vegetation. Correlation of Köppen's units of climate with units of
vegetation is shown in Figure 2.9.

Tropical rain forest

Temporate deciduous forest,

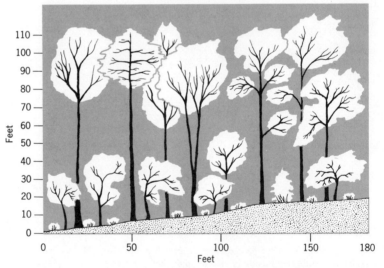

Figure 2.11

Elevation sketches of temperate deciduous forest and tropical rain forest. Notice the extra horizontal structure in the rain forest illustrated by the drawings. Such sketches are of practical use in field work because they may help in making analytical descriptions. There have been attempts to substitute formal systems of symbols for semi-realistic sketches like these, as shown in Figure 2.12. Upper diagram from Beard, 1966 lower diagram from Billings, 1968.

Figure 2.12

Dansereau's symbols for describing vegetation. A system to transform realistic profile diagrams of the kind shown in Figure 2.11 into a formal graphic language. Each descriptive term which a botanist normally uses is assigned a symbol and these are clustered on a shape drawn to scale on squared paper. Assuming that the scale used here is one division to 1 m, the plant at the left of the bottom row is a tree 8 m tall, deciduous (because the circle is not shaded) and with broad membranous leaves. Growing on it is a smallish liana, also deciduous. This graphic language could be used by field groups compiling data for a computer study. It has not come into wide use, however, partly for the reason that people are seldom ready to learn another's new language but more because it has been realized that refined description gives little help to answering grand ecological questions like why the plants are shaped as they are and why they live as they do. (Redrawn from Dansereau, 1957.)

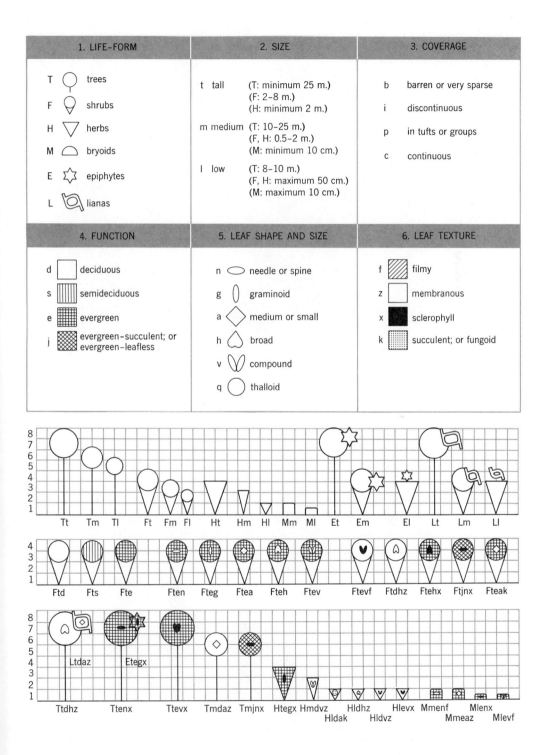

1. LIFE-FORM	2. SIZE	3. COVERAGE
T ⃝ trees	t tall (T: minimum 25 m.) (F: 2–8 m.) (H: minimum 2 m.)	b barren or very sparse
F shrubs		i discontinuous
H ▽ herbs	m medium (T: 10–25 m.) (F, H: 0.5–2 m.) (M: minimum 10 cm.)	p in tufts or groups
M bryoids		c continuous
E ☆ epiphytes	l low (T: 8–10 m.) (F, H: maximum 50 cm.) (M: maximum 10 cm.)	
L lianas		

4. FUNCTION	5. LEAF SHAPE AND SIZE	6. LEAF TEXTURE
d deciduous	n ⬭ needle or spine	f filmy
s semideciduous	g graminoid	z membranous
e evergreen	a ◇ medium or small	x sclerophyll
j evergreen-succulent; or evergreen-leafless	h ♡ broad	k succulent; or fungoid
	v compound	
	q thalloid	

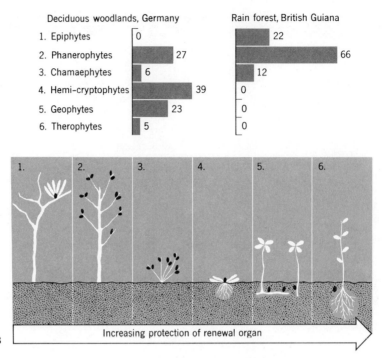

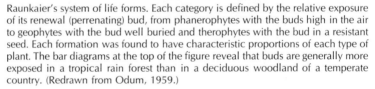

Figure 2.13

Raunkaier's system of life forms. Each category is defined by the relative exposure of its renewal (perrenating) bud, from phanerophytes with the buds high in the air to geophytes with the bud well buried and therophytes with the bud in a resistant seed. Each formation was found to have characteristic proportions of each type of plant. The bar diagrams at the top of the figure reveal that buds are generally more exposed in a tropical rain forest than in a deciduous woodland of a temperate country. (Redrawn from Odum, 1959.)

Plants of several of the Raunkiaer categories commonly live together in one formation, as is readily apparent from thinking of the temperate forests where there are trees with their buds high in the air, low shrubs with buds near the ground, and varieties of spring flowers with buds more or less hidden in the soil, so that one can only separate formations by the preponderance of one category over another. The tundra has a higher percentage of species with buried buds than does the deciduous forest, which has a higher percentage than does the rain forest.

Raunkiaer expected that the percentage composition of life-forms in each known formation would be characteristic, that each of the known blocks of vegetation would turn out to reveal its own **life-form spectrum.** And so it proved. The work of Raunkiaer himself and of many of the botanists of his period who followed in his footsteps showed this without a doubt (Cain, 1950). Each formation to which de Candolle had ascribed a heat-loving name and to which Köppen had ascribed a climate, could also be given one of Raunkiaer's Greek tonguetwisters to describe the prevailing life-form spectrum (Figure 2.14).

De Candolle	Raunkiaer phytoclimate	Köppen climatic type
Megatherms	Phanerophytic	A
Xerophiles	Therophytic	B
Mesotherms	Chamaephytic	C
Microtherms	Hemicryptophytic	D
Hekistotherms	Cryptophyic	E

Figure 2.14

Ecology seemed to be advancing nicely toward understanding the map of the earth. The bounds to formations were set by climate. The essential characteristics of form could be understood as adaptions to the harshest parameters of climate. And the limits to life-forms as measured by an objective statistical approach also turned out to correspond with Köppen's climatic maps. Very nice. But ecology had fallen into a trap of circular reasoning. The original vegetation boundaries had been drawn not by nature but by map-makers. These same boundaries were then used to draw the boundaries of climate, and then again were used to draw the limits of life-form classes. It should not be surprising to find that these three sets of boundaries were superimposed! What ecology had shown by this exercise was that there was, indeed, a broad correlation between the plants of a place and the weather of a place. On the grand-scale, maps of climate based on the plant geographer's boundaries were useful. And on the grand average, life-form groupings based on the adaptations of plants to adversity did reflect those same boundaries. But this was not to say that the world was really parceled out to grand nations of plants, the formations, each with territorial integrity like that of a nation-state. Ecologists had made the mistake of imagining real boundaries between collections of averages.

There were clear excuses to be made for ecologists in falling into this error. The first was that not all the boundaries drawn on vegetation maps were artificial. The boundaries separating the northern and southern edges of the boreal forest in North America, for instance, were real if diffuse. So were many of the boundaries made by the sweeps of prairie fires and the crests of mountain ranges. The mistake was to assume that all boundaries were as distinct as these; and, furthermore, to make light of steady changes in composition that took place across the length and breadth of all formations, even those with definite natural edges. As soon as the overseeing role of climate was properly understood, it should have been evident that vegetation changes must be gradual, because climate is a diffuse and changeable thing. But this diffuseness of climate and geography on a continental scale was further kept at arm's length by another related and perplexing phenomenon. This was the zonation of vegetation on high mountains, a phenomenon that seemed to provide powerful extra reasons for believing that vegetation did exist in discrete blocks.

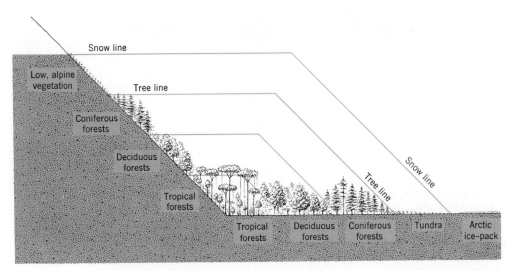

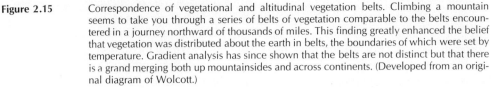

Figure 2.15 Correspondence of vegetational and altitudinal vegetation belts. Climbing a mountain seems to take you through a series of belts of vegetation comparable to the belts encountered in a journey northward of thousands of miles. This finding greatly enhanced the belief that vegetation was distributed about the earth in belts, the boundaries of which were set by temperature. Gradient analysis has since shown that the belts are not distinct but that there is a grand merging both up mountainsides and across continents. (Developed from an original diagram of Wolcott.)

THE ZONING OF
VEGETATION ON
HIGH MOUNTAINS

Many men have looked at mountain ranges from a distance, and remarked at the bands of color that their flanks display. Vegetation belts appear to be stacked one on another, all the way from, perhaps, a lowland forest out of which the mountains grow to the alpine meadows and snowcaps of the summits. As you look from a distance, you can trace with the mind's eye the lines where the belts of vegetation meet. Here is a phenomenon, well-known and apparently definite enough. Not only does vegetation change with altitude but it seems to exist in discrete belts, the edges of which can be sketched by a man who views them from across the valley. Why should this be? What accounts for those dividing lines on the mountain? And there is more to the problem than this. The successive belts encountered on the ascent of a high mountain run parallel to the succession of formations encountered when traversing a continent from south to north (Figure 2.15). If you can see the boundaries on the mountainside, why should similar boundaries not be present between the continental formations?

Zones on the flanks of high mountains became of particular interest to ecology as a result of the work in Arizona of one man, C. Hart Merriam. Merriam was charged with a biological survey of part of Arizona in 1889, in the pioneer days when the region was still biologically unknown. In one season, the young Merriam (1890) brought back 20 new species and subspecies of mammals. But he was particularly impressed with the vegetation belts he encountered on the ascent of the 13,000-foot San Fran-

Figure 2.16

The Sonoran Desert. The plants are spaced out, with bare ground showing between them. Parts of the plants are thick, spiny, and thorny, and the whole appearance of the formation is distinctive. Yet the San Francisco Mountain rises out of this desert formation to where boreal and arctic formations clothe its upper slopes. Attempting to separate the vegetation of this mountainside into distinct units led to the concept of life zones.

cisco Mountain. The mountain had its roots in the Sonoran Desert, a landscape studded with cacti, including the famous saguaro, and with leathery or thorny scrubs (Figure 2.16). This desert is left behind as you start to climb, and soon you are in oak scrub. At 6000 feet you are in pine forest, there are streams of water and rising fragrance from a carpet of needles underfoot. From the thirsty desert below you seem to have been transported to Canada, but all you have done is to climb a few thousand feet, to move yourself less than a mile. If you go higher still, the almost pure pine stands give way to a forest of Douglas Fir, then to spruce trees, and finally to a tree line at 12,000 feet. Above this is a rough flowering turf, a type of vegetation having a distinct resemblance to the tundras of the true Arctic and containing many of the same species of plants. Merriam described the main aspects of each piece of vegetation with care. He

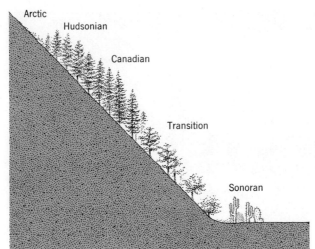

Figure 2.17

Merriam's life zones on San Francisco Mountain.

also collected animals, particularly mammals, within each vegetation belt, noting that there was a characteristic mammal assemblage for each (Figure 2.17 and 2.18). Each belt of plants and animals he called a **life zone.** Merriam knew that he had to be a bit arbitrary when he drew boundaries between life zones, particularly because many animals wandered up and down the mountain. But the vegetation was a guide, so that the task of mapping did not seem hard. When he had finished, he had a set of life zones, each separated by lines on his map, each with a characteristic flora and fauna, each comparable to the life of one of the great continental formations. Merriam mapped all of North America into life zones, each having plants and animals comparable to those of one of the zones ringing the mountains of the Southwest (Figure 2.19). Then he looked at his maps, saying to himself: "These boundaries I see before me are natural boundaries; what sets them?" As de Candolle had, he sought an answer in temperature.

Since Merriam's life zones were defined by animal collections as well as by plants, he could achieve a better intuitive idea of how the changes of temperature might operate than could de Candolle. It is easy to imagine how small changes in temperature might have dramatic effects on the lives of such things as short-lived rodents when the possible effect of similar changes on forest trees is obscure. Merriam suggested that an animal might be able to live in the coldward direction until the temperature of the breeding season was too low for reproductive success. Thus there might be a distinct northern or upward limit to the spread of the species. Life in the heatward direction should be possible until a place was reached where the heat of the hottest part of summer became intolerable. Thus each species of animal was sandwiched between the place where it was too cold in the breeding season and the place where the heat of

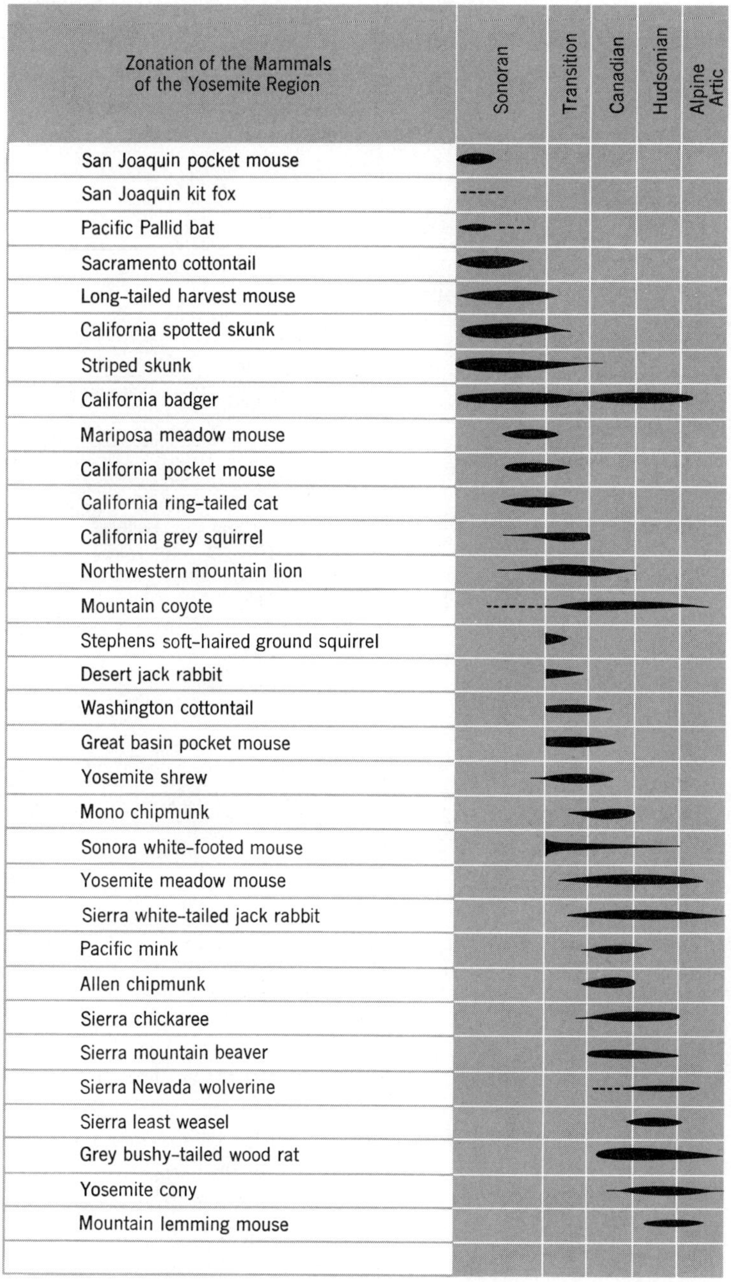

Zonation of the Mammals of the Yosemite Region	Sonoran	Transition	Canadian	Hudsonian	Alpine Artic
San Joaquin pocket mouse	●				
San Joaquin kit fox	-----				
Pacific Pallid bat	●-----				
Sacramento cottontail	●				
Long-tailed harvest mouse	●				
California spotted skunk	●				
Striped skunk	●				
California badger	●	●			
Mariposa meadow mouse		●			
California pocket mouse		●			
California ring-tailed cat		●			
California grey squirrel		●			
Northwestern mountain lion		●			
Mountain coyote	------	●			
Stephens soft-haired ground squirrel		●			
Desert jack rabbit		●			
Washington cottontail		●			
Great basin pocket mouse		●			
Yosemite shrew		●			
Mono chipmunk		●			
Sonora white-footed mouse		●			
Yosemite meadow mouse		●	●		
Sierra white-tailed jack rabbit		●			
Pacific mink		●			
Allen chipmunk		●			
Sierra chickaree		●	●		
Sierra mountain beaver		●			
Sierra Nevada wolverine		----	●		
Sierra least weasel			●		
Grey bushy-tailed wood rat			●		
Yosemite cony				●	
Mountain lemming mouse				●	

Figure 2.18

The ranges of mammals with altitude in the Yosemite region of California. (Redrawn from Odum 1959.) Notice that the ranges of the animals overlap. There is a gradient, not a series of disjunct populations as Merriam postulated.

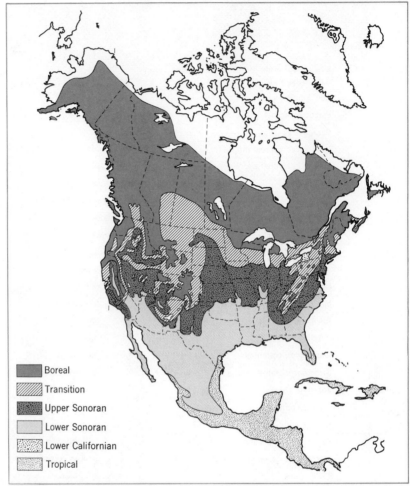

Boreal

Transition

Upper Sonoran

Lower Sonoran

Lower Californian

Tropical

Figure 2.19

Map of the life zones of North America as proposed by Merriam.

summer was too intense. This was a sound ecological hypothesis, embodying the idea of threshold requirements for critical activities in the lives of animals. Merriam followed up the idea by recording daily temperatures of the various zones, both in the breeding seasons and in the hottest weeks of summer, and by drawing isotherms that should define the places on the ground in which each collection of animals could live. He expected that these isotherms would coincide with his zone boundaries, when he would claim that the limits of his life zones were, indeed, set by temperature.

In the event Merriam never succeeded in his attempt to match isotherms to life zones, because mathematical errors were found in his published attempts, and he never announced the results of recalculation (Figure 2.20). But there were much more serious criticisms to be made of his methods than mere arithmetic error. That animals should have their distributions set by the air temperature during critical moments in their

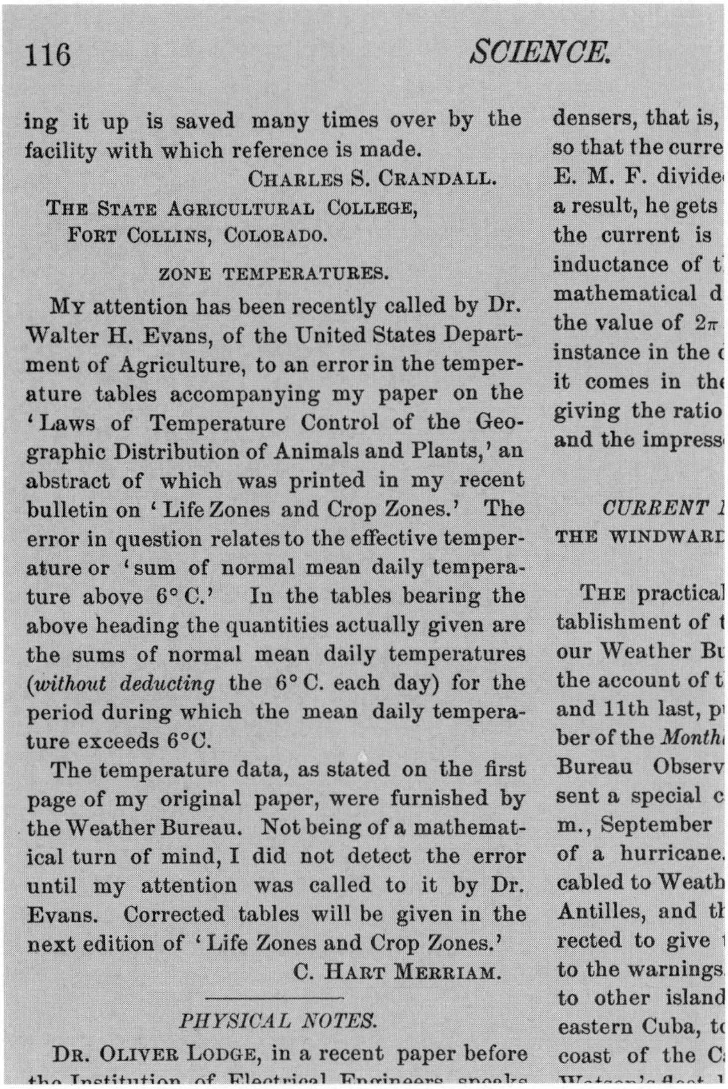

116   *SCIENCE.*

ing it up is saved many times over by the
facility with which reference is made.
                              CHARLES S. CRANDALL.
THE STATE AGRICULTURAL COLLEGE,
  FORT COLLINS, COLORADO.

ZONE TEMPERATURES.

MY attention has been recently called by Dr.
Walter H. Evans, of the United States Department of Agriculture, to an error in the temperature tables accompanying my paper on the
'Laws of Temperature Control of the Geographic Distribution of Animals and Plants,' an
abstract of which was printed in my recent
bulletin on 'Life Zones and Crop Zones.' The
error in question relates to the effective temperature or 'sum of normal mean daily temperature above 6° C.' In the tables bearing the
above heading the quantities actually given are
the sums of normal mean daily temperatures
(*without deducting* the 6° C. each day) for the
period during which the mean daily temperature exceeds 6°C.

The temperature data, as stated on the first
page of my original paper, were furnished by
the Weather Bureau. Not being of a mathematical turn of mind, I did not detect the error
until my attention was called to it by Dr.
Evans. Corrected tables will be given in the
next edition of 'Life Zones and Crop Zones.'
                              C. HART MERRIAM.

───────────

PHYSICAL NOTES.

DR. OLIVER LODGE, in a recent paper before
the Institution of Electrical Engineers speaks

densers, that is,
so that the curre
E. M. F. divide
a result, he gets
the current is
inductance of t
mathematical d
the value of 2π
instance in the
it comes in the
giving the ratio
and the impress

CURRENT
THE WINDWARD

THE practical
tablishment of t
our Weather Bu
the account of t
and 11th last, p
ber of the *Month*
Bureau Observ
sent a special c
m., September
of a hurricane.
cabled to Weath
Antilles, and th
rected to give
to the warnings
to other island
eastern Cuba, t
coast of the C
Watson's fleet l

Figure 2.20

Merriam's retraction of his original plot of isotherms correlating with life zone
boundaries. The promised recalculation of isotherms was never published.

lives was reasonable, but what possible reason was there for whole
collections of distantly related animals recognizing exactly the same temperature limits? Merriam never suggested a reason. Nor has anyone else.
The whole thrust of biological thinking is that different species have
evolved, by natural selection, different tolerances. This is why species
exist. But Merriam was supposing, as had de Candolle before him, that
the same tolerances were shared by all the animals and plants of large
areas. Such a conclusion should have been unacceptable to a student of
the theory of evolution.

Merriam and de Candolle have not been the only biologists to try and

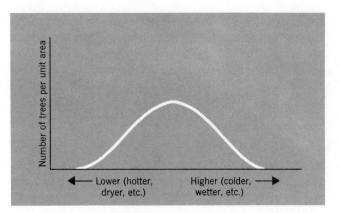

Figure 2.21

match isotherms to the boundaries on their biogeographic maps. Attempts of this sort are still sometimes made, for the compulsion is strong. There on the map in front of you lie the boundaries, between **life zones,** or **formations,** or **biotic provinces** (Dice, 1952), or whatever the classification might be. You must explain these boundaries somehow. The boundaries usually run along lines of latitude or round mountains, and it is common knolwedge that it gets colder as you go north or up. So you explain your boundaries with a temperature hypothesis, and try to test the hypothesis by plotting isotherms. But it is much more fruitful to question the existence of the boundaries themselves.

If you climb one of Merriam's Arizona mountains, you will learn that the edges of his life zones are very hard to find on the ground. The cactus desert (the **Sonoran life zone**) merges, interfingers, and grades into the oak scrub above it. You have to climb hundreds of feet before you are sure that you have crossed from one life zone to the next. Higher up the mountain it is worse, so that you might want to put a thousand feet behind you before you feel confident that the oak scrub of the **Transition life zone** is well behind you, and that you may now take out your camera and record for posterity a perfect example of a pine forest in the **Canadian life zone.** The zones merge together; it is only the map-maker who sets them so sharply apart (Figure 2.18).

And this is reasonable, because there should be gradients on the side of a high mountain. Climb it and you find that the environment changes gradually. The air gets steadily colder. The winds get steadily fiercer. You may encounter an abrupt change as you enter the base of clouds perched on the heights, but otherwise the climatic changes are gradual. So, surely, should the changes in vegetation and fauna be gradual. But why, if this is true, do we get the impression of discrete belts of vegetation, since we really are able to detect the different belts when we look at our mountainside from across the valley?

This question has been answered for us by a group of modern botanists generally referred to as the "continuum analysis school," notably by R. H. Whittaker (1967). Whittaker starts by noting that what a viewer from across the valley sees is mostly just the trees. On mountains with ap-

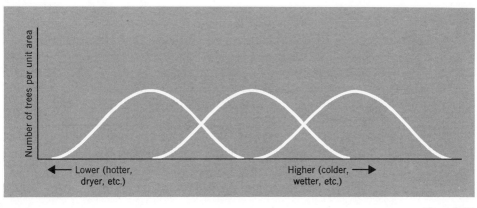

Figure 2.22

parently well-pronounced vegetation belts, like those of Arizona, there are not many species of trees present. Suppose each species were to be distributed up the mountainside according to its individual and specific tolerances, then a graph of its abundance plotted against altitude would probably look like Figure 2.21.

There would be individual trees that had extreme tolerances, but most would be grouped round the optimum conditions for the species. Other tree species would have different optimum requirements, although the needs of the extreme individuals of different species should certainly overlap. We might expect the distribution of three species of tree up a mountain to look like Figure 2.22.

The actual number of trees in any place does not change in this system, and there is much overlap in the distribution of neighboring species. But there are clear regions around the optimum for each species where there are almost pure stands. Someone looking at the mountainside from across the valley sees a succession of distinct lines denoting the optima. If the trees were of different colors, as for instance in New England in the fall, the effect would be of a row of colored bands. There would be a fuzzy region between the bands, where the populations overlapped, but the eye would make light of this. Whittaker likens the effect to that of the light spectrum thrown by a prism. There are not really distinct colored bands in a light spectrum, merely a continuous gradient of wavelengths. The eye breaks this spectrum up into distinct units. Whittaker put forward the hypothesis that mountain zones, like those of Merriam, were similar illusions in which the map-maker aided and abetted the original observer. He has tested this hypothesis with painstaking samples of plant abundance both in the mountains of Appalachia and the American Southwest (Whittaker, 1956; 1967; Whittaker and Niering, 1965). The technique is to measure the percentage composition of plants in the different vegetation layers at many sites up the side of the mountain and then to plot the relative importance of each along the gradient. In every instance the work shows that each plant species is separately distributed, and that the distributions of each overlap. No matter how distinct the veg-

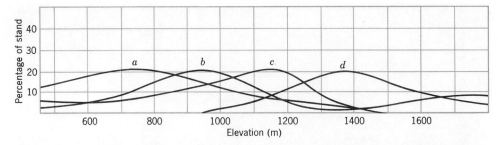

Figure 2.23   Distribution of tree species on a slope of the Great Smoky Mountains in Tennessee. The percentage compositions of the forest were calculated from sample counts of trees more than one centimeter in diameter 4 feet from the ground. The forest was sampled at 100 meter elevation intervals. Four tree species out of the sample are shown here. There are no distinct vegetation belts, for the distribution of each species grades into that of its neighbors. A distant observer might well resolve points a, b, c, d as the middles of separate vegetation belts: a, hemlock; b, *Halesia monticola* (the distribution of which is bimodal); c, lime; d, mountain maple. (Modified from Whittaker, 1956.)

etation "belts" may seem from across the valley, the supposed boundaries cannot be found by measurement on the ground. A typical result, for trees in the Great Smoky Mountains of Tennessee, is given as Figure 2.23.

Discrete zones of life on mountainsides are thus largely an illusion. There may be some real abrupt changes, as when different rocks or drainage patterns meet, but these are the result of changes in the physical habitat and are not a property of plants or isotherms. And the tree line near the summit may be distinct enough, perhaps being set by physical factors such as the cloud base or the sandblasting of flying ice crystals. But otherwise the vegetation and the animals it shelters are not collected into discrete zones.

Yet many maps are still based on Merriam's system of life zones. This may be partly because Merriam rose to be head of the Biological Survey and directed field work on lines of his own choosing so that all the pioneer studies used it. But it also reflects the fact that the system is deceptively convenient. The compiler of a field guide who believes in life zones need only find a few records of an animal to assign it to a life zone. He then maps its total distribution. The final map may well look pleasing in the guide. Unfortunately, it will be wrong.

**CLIMATE AND THE
DIVISIONS WHICH
ARE REAL**

If, as gradient analysis suggests, even the "banded" vegetation of mountainsides does not really exist in discrete blocks, what about the formations themselves? Are the divisions between these imaginary also? It is true that climatic maps based on botanical maps do provide a useful guide to climate; but we should expect grand climatic gradients across continents, not disjunct regions. If climates merge, then formations should merge also. Are formations mere map-maker's conventions, like the life zones on a mountain? Some of them certainly grade into one another, as do the forests of Southeast Asia. But what of more abrupt boundaries like that which you cross to enter the boreal forest of southern

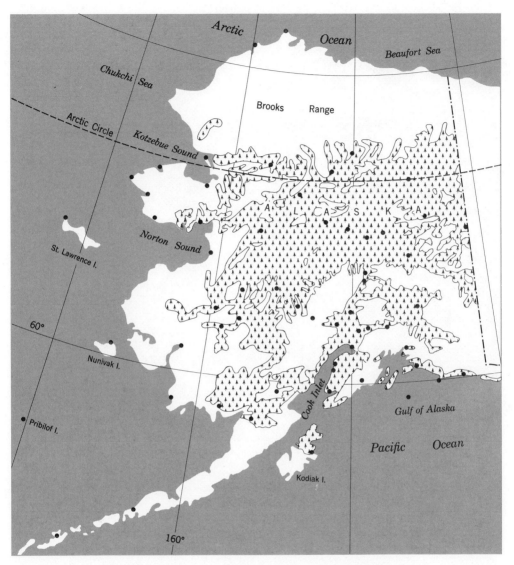

Distribution of the tree line in Alaska. Notice that the tree line runs parallel with the west coast of Alaska. This encouraged Hopkins' temperature hypothesis because coastal areas have maritime climates with summers that are cooler than those of the interior. Dots are the sites of weather stations whose data were used to calculate degree-days. (Redrawn from Hopkins, 1959.)

Figure 2.24

Canada, or that delight of map-makers, the tree line? These boundaries are not all illusion, if somewhat diffuse. How can we resolve the evidence for gradients with the existence of formation boundaries such as these? Some of the most recent insights into this problem have come from studies along portions of the arctic tree line where the boundaries between formations seem most distinct.

The tree line of northwestern Alaska is irregular, first running from the

Figure 2.25   The arctic tree line. The forest on the right, the tundra on the left, and a good enough separation between them to please any map maker. In such places formations almost seem to face each other like rival armies at a disputed frontier. This part of the tree line is in Alaska. The tall trees are black spruce (*Picea mariana*), but there are several kinds of broad-leaved bushes in the forest also. The lines of small bushes in the tundra will be growing along drainage channels, and the open tundra, which looks grassy, will be humpy, a mixture of grass and sedge tussocks with prostrate woody plants like dwarf birch and blueberry.

Canadian side westward, but then curving south to run parallel with the seacoast (Figure 2.24). It became the particular study of an Alaskan geologist, David Hopkins. Hopkins' interest in this biological question came from the fact that he found fossils of late Pleistocene trees several thousands of years old in places that are now in the tundra, and he wanted to know what he could infer about the climate when the fossil trees lived. So Hopkins addressed himself to the problem of the modern tree line, and he tried the now familiar hypothesis of temperature (Hopkins, 1959). He had many meteorological stations in Northern Alaska on which to call, some in the tundra, some in the forest. The winter temperature should surely be unimportant, since it was likely to be below zero at all sites, at which temperature a few degrees one way or the other are unlikely to make much difference. So he looked for an index reflecting the amount of warmth received by the different sites in summer and found it by calculating degree-days when the mean daily tempera-

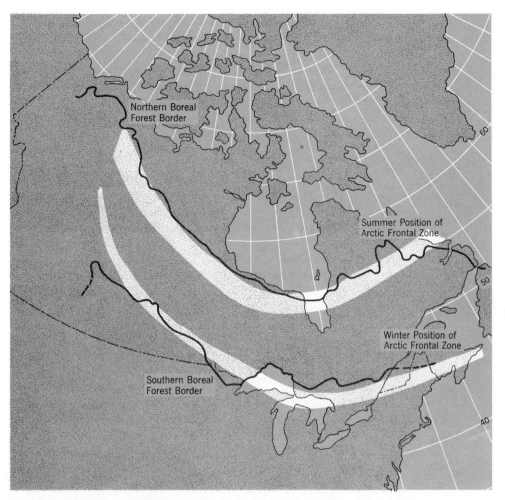

Forest boundaries and air-mass fronts in eastern Canada. The tree line and the southern edge of the boreal forest are perhaps the most distinct of continental vegetation boundaries. The plot shows that they closely follow the mean positions of the front of the arctic air mass at different seasons of the year. This study of Bryson's provides direct evidence that extents of the Canadian boreal forest and tundra are directly influenced by the sway of an air mass (Redrawn from Bryson, 1966.)

Figure 2.26

ture was more than 50°. If a site had only a week when the mean daily temperature was 55°F, Hopkins said it had 7 × 5 = 35 degree-days over 50°F. All the forest sites in his survey had more than 130 of these degree-days whereas all the tundra sites had less than 130 degree-days. It seemed as if temperature might have something definite to do with these boundaries after all. But how could temperature change abruptly, moving from one region to another in, say, less than a mile? This question seems to have been answered by the work of the Wisconsin meteorologist, Reid Bryson. Bryson (1966) devoted much thought to the problem of, perhaps, the most remarkable formation boundary of them all, that where the boreal forest of central Canada meets the tundra in a nice mappable boundary hundreds of miles long (Figure 2.25). This tree line lies somewhere in the path of the front of the arctic air mass as this sways north and south across the continent, now receding before the advance of the Pacific air mass from the southwest, now advancing as the Pacific air mass recedes. An air-mass front is scarcely a stationary thing like a tree line, but it is a disjunction, even though having appreciable thickness. By the arduous process of trajectory analysis, from many stations and the records of 10 years, Bryson established the average position of the arctic front, both in summer and in winter. The average summer position coincided beautifully with the tree line (Figure 2.26). In winter, of course, the arctic front lies further to the south. Bryson's 10-year average drew it neatly along the boundary between the boreal forest and the deciduous forest. Two formation boundaries had thus been shown to coincide with the mean position of an air-mass front during a major portion of the year. Such a conclusion offers comfort to those who have long thought that temperature has something to do with formation boundaries because the temperatures on either side of an air-mass front may differ widely.

But Bryson also used temperature as an indicator of the presence of an air mass that was independent of his trajectory data. The passage of fronts past his many stations would be recorded by changes in temperature so that Bryson could calculate the mean positions of fronts from his temperature data alone. And the positions calculated in this way coincided with those calculated by trajectory analysis and with vegetation boundaries.

There is nothing in Bryson's work to suggest that temperature rather than some other parameter of an air mass is the factor most affecting vegetation. He just shows us that a disjunction of climatic parameters exists where our formations change. The plotters of isotherms, and the calculators of degree-days, may have been doing no more than recording fronts for themselves, whereas the plants were more affected by other qualities of the air masses than temperature. But Bryson had shown us that there could sometimes be disjunctions in the air, in spite of its fluidity. Discrete formation boundaries might truly reflect discrete climatic boundaries. For the tree line itself, work on the heat balance of plants suggests that the temperature of the air masses may, indeed, be crucial (Chapter 20). Other growth forms than trees might be more concerned with different properties of air masses, such as moisture or wind speed.

But how does vegetation detect the 10-year mean position of a front that is continually moving? The answer to this riddle probably lies in the slow generation time and rate of spread of the major units of the formations, the trees and other perennial plants. Ten years is a short part of the generation time of a forest. No doubt the formation boundaries are not really static but are waxing and waning at a rate that produces changes too slow to be noted in a human lifetime. Fluctuating boundaries are in dynamic equilibrium with more swiftly fluctuating fronts. What we see as

The pattern of the earth's climate revealed by clouds. The photograph was taken by an earth satellite crossing the equator over Brazil in November 1967. Sets of such photographs may show directly the patterns of climate which were earlier mapped by plotting the approximate areas of pieces of vegetation. Note particularly the mass of cloud that hangs over the tropical rain forest of the Amazon basin. (NASA photograph.)

Figure 2.27

a discrete boundary is just one temporary position in an always-moving edge.

Other formation boundaries have not yet been investigated by meteorologists in so detailed a way, so it remains uncertain how many can be neatly correlated with the mean positions of fronts. It might be particularly worthwhile to examine some of the boundaries in drier areas where fire is known to be one of the agents in controlling the extent of particular formations. The American prairies are formations in which fire is a frequently recurring, and hence normal, part of the environment. Where the fires no longer often reach is the boundary with forest. It would be very nice to know if some such boundaries could be correlated with air-mass fronts, or whether perhaps the boundary is set by the random history of fire in a climatic continuum.

But most of the formation boundaries of the earth are probably not distinct. The vegetation of one region grades into that of another, letting us tentatively conclude that the sway of climate is usually less predictable than in the Arctic. There are few discrete air-mass boundaries on the earth, so that there are few discrete vegetation boundaries. But this does not invalidate the general hypothesis of there being large regional patches of vegetation, distinguished by form, and acquiring their distinctive appearance because of their adaptations to prevailing climates. Global vegetation maps, however arbitrary their divisions, are a useful base for climatic maps. It follows that formations, broadly defined, are real entities.

CONCLUSION:
THE TERRESTRIAL
PATCHWORK QUILT

Geographers present us with an earth which is a grand patchwork quilt of air masses. The circulation of the atmosphere effectively divides the surface of the earth into a number of separate compartments or environments for life. Climatologists first identified these compartments from vegetation maps, but now you can see some of them from satellite photographs; the cloud formations give them away (Figure 2.27). Usually there are broad gradients between compartments, and gradients of vegetation in consequence, but in parts of the Arctic the compartments may be more neatly divided.

But it is important to notice that the distinctive appearance of each of the resulting formations results from the adaptations of individual plants to conditions of the local climate. The formation looks as it does because the local plants share common adaptations to common environmental problems. A formation as a unit is no more than a conception, a human impression of the way plants look when they are all forced to adapt to a common environmental reality.

Vegetation and climate, so intimately linked, are collaborators in making the soil. Soils are formed by processes working from above. If the vegetation is coniferous forest, the ground will be covered by fallen needles which rot slowly, making the cold drainage waters of the region acid. The cold acid water percolates through the mineral soil bleaching it and giving the soil profile a characteristic appearance. Where the broad leaves of deciduous trees fall in warmer climes, the percolating water is less acid, there is less bleaching and earthworms live to dig and churn the soil. Where rainfall is very heavy, chemical changes go deeper. The warm rains of the tropics dissolve silica minerals and leave iron behind to stain the soil red. In the drier climates of prairies water does not penetrate far and leaves behind it layers of precipitate. The grand result of processes such as these is that each large formation of plants is roughly mirrored by the characteristic soil type covering a corresponding area. The way in which the climatic pattern of the earth results in a characteristic pattern of life is thus enhanced. Climate influences vegetation which influences soil which, in turn, influences vegetation. The study of the formation of soil also reveals that all agriculture, and hence the livelihood of our kind, depends on the top 2 feet of the earth surface. Some forms of agriculture proceed by mining the resources of that top 2 feet and thus cannot continue indefinitely.

SOILS AND
BIOGEOGRAPHY

43

CHAPTER 3

SOILS AND
BIOGEOGRAPHY

If you dig a trench in any well-vegetated place, you reveal a succession of layers in the soil. At the surface there is a litter of dead or rotting plant parts, and underneath are one or more distinctly different layers that separate the surface litter from the subsoil a few feet down. Sometimes these layers are sharply distinct, other times they merge gradually one into another, but always it is apparent that they have been formed within the subsoil by weathering processes working down from the surface. They are not strata in the geological sense, because they are not separate deposits. It is convenient to have a special word to describe these layers which have been differentiated at the top of the subsoil, and they have come to be called **"horizons"** by soil scientists. The horizons of the soil have formed as rotting plant parts were mixed with the upper layers of the mineral soil, and as drainage water percolated down through the litter to work a slow washing and chemistry on the lower horizons. The thickness of earth affected by these processes constitutes the soil. The subsoil underneath is the earth from which the soil was made and is, therefore, called the **"parent material."**

Underneath the litter of leaves, the top layer of the mineral soil is colored and structured by the organic particles mixed into it by soil animals such as earthworms, or by roots, and by various organic materials produced by decomposition and collectively called humic acids. The percolating water, which is a solution of various substances washed out of the litter, dissolves anything soluble in the surface of the mineral soil and carries it down to deeper horizons. Percolating water also removes finer particles from the top horizons, particles such as fragments of clay minerals, and carries them physically downward. The horizons at the top of a soil profile which are being continually denuded in this way are called the **"A" horizons** of the soil. Underneath them is a horizon or group of horizons that have caught things which were washed down from the top, such as the particles of clay which were trapped in the spaces of the deeper soil as in a filter, and solutes which are redeposited over the immense surfaces of that same filter bed. The detailed chemistry of this process of redeposition is not always known, but the fact that it occurs is clear enough. The result is a horizon in which there has been redeposition, and such horizons are called **"B" horizons.** Underneath them are unaltered parent materials, which are called the **"C" horizons.** Soils are thus in three parts: "A" horizons which include the organic matter of the surface but from which other material has been removed in the drainage water, "B" horizons which have collected material washed down from above, and "C" horizons of unaltered parent material (Figure 3.1). There are many subdivisions of these horizons, by which individual soils are described and defined.

Plants and animals living in the soil strongly influence the soil-forming process, and in doing so they directly affect the habitat in which they live. Much of the structure of the soil surface must reflect the sorts of plant pieces that are dropped there. The types of soil animals must also control structure and mixing, for such things as earthworms and soil arthropods

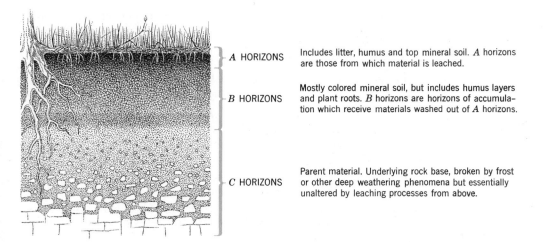

	A HORIZONS	Includes litter, humus and top mineral soil. A horizons are those from which material is leached.
	B HORIZONS	Mostly colored mineral soil, but includes humus layers and plant roots. B horizons are horizons of accumulation which receive materials washed out of A horizons.
	C HORIZONS	Parent material. Underlying rock base, broken by frost or other deep weathering phenomena but essentially unaltered by leaching processes from above.

Idealized soil profile. There may be several subhorizons in each of the main horizons and roots may penetrate them all, even deep into the parent material. There may be colored bands in each horizon, as some of the later figures in this chapter show. Finally the boundaries may be elusive. But the functional separations are clear enough: material is removed from A horizons, material is added to B horizons, and C horizons are essentially unaltered by processes acting from above.

Figure 3.1

do not dig and churn to the same extent, and the varieties of animals present, in turn, will be determined by the kinds of plant litter available to them as food. The mass of litter available will differ between different kinds of vegetation. And the type and quantity of solutes in the percolating water will also be dependent on the kinds of plant parts deposited and the manner of their rotting. In these various ways the appearance of a soil profile will be determined by the kind of vegetation that covers the ground.

This controlling influence of vegetation can, perhaps, best be seen in soils under boreal forest. Here a leaf litter is the fragrant carpet of brown needles, so familiar to people of north temperate lands and so different from the mulchy litter of the deciduous forest to the south. The needles rot slowly, and in such a way that the water draining out of them is notably acid, commonly with a pH of about 4. Earthworms and other large digging soil animals do not live in this acid litter, which is left to mites and other small animals that do not mix the surface layers of the soil. There is often a sharp discontinuity between the organic layer and the mineral soil beneath, and the influence of the vegetation on the mineral soil is almost entirely effected through the agency of percolating water. This effect is, however, profound. The acid water dissolves very many minerals from the lower part of the "A" horizon so that this may be bleached a whitish grey, and the colored minerals removed are then redeposited in the "B" horizon to such effect that this may be banded red and brown. The result is a striking soil profile of black rotted litter, whitish mineral soil, a brightly colored "B" horizon, and then the still different color of parent material. Soils of this kind were first critically examined in the continental boreal forests of European Russia by a man who laid the foundations of modern soil science, V. V. Dokuchaiev. He used the local peasant name for the soils, a name describing the ashy appearance of the ground if you pulled a plow through the bleached "A" horizon, and such soils are now universally known as **podzols,** which means "ash-earth" (Figure 3.2).

A trench dug in any well-drained place in the northern boreal forest will reveal the profile of a podzol, sometimes more strikingly developed than at others but always clearly recognizable for what it is. The hues and depths of the horizons will vary as the parent material changes from one geological region to another, but the essentials of the podzol profile will be expressed at well-drained sites no matter what kind of rock lies buried below. Only if the ground is waterlogged are you likely to find trees of the northern boreal forest over soils without the characteristic look of a podzol, but this is because the necessary flushing through with soil water is prevented. On the nearest ridge to a black-spruce bog, where soil is raised above the water table, you will find a podzol. Soil men say that the **Great Soil Group** characteristic of the northern boreal forest is the podzol, conceding only that local factors like impeded drainage may inhibit its expression.

A map of the northern boreal forests of Asia and North America is a rough map of the distribution of podzols. There can be podzols locally elsewhere, but not occupying very large areas. Some heath lands of northern Europe, with acid litter and leached soils, reveal podzolic profiles, and I have found a typical podzol profile under a shrub tundra growing on an old gravel beach ridge in the Alaskan Arctic. And south of the border of the boreal forest the podzols grade away into different soil types. But a crude map of the distribution of podzols can be made by drawing the boundaries of the boreal forest. No such approximation can be made by mapping geological formations. Apparently vegetation and climate combined influence soil formation more than does the rock from which the soil is made.

Under the mulchy leaf litter of deciduous forest, well to the south of the boreal forest boundary, a trench shows the soil profile to look quite different. Rotted leaves at the bottom of the surface litter are mixed with the mineral soil so that there is perhaps 10 or 20 centimeters of dark fertile-looking earth, quite like the surface of a ploughed field or a garden. Earthworms, which live on the nearly neutral leaf litter, have done the digging and mixing. There is often little obvious sign of a bleached layer under this dark upper horizon, although a reddening of the "B" horizon further down tells of materials brought down from above. The colored minerals of the "B" horizon have been washed down in the percolating water and redeposited but not nearly to the same extent as under the needle litter of the boreal forest, and this more gentle leaching leaves the profile looking unlike that of a podzol. Under deciduous forest there is a gentle gradation of horizons with subtle changes of hues: brown litter, a

Figure 3.2
SOIL PROFILES
REPRESENTATIVE
OF GREAT SOIL
GROUPS. SCALES
ARE IN FEET.

Photographs were provided by the Soil Conservation Service, U.S. Department of Agriculture, courtesy of Dr. Guy D. Smith, Director, Soil Survey Investigations, Soil Conservation Service.

Podzol from Northern France

A needle litter layer (A_0) and a layer of black humus (A_1) overlies a bleached A_2 horizon. The B horizon includes the reddish color of iron oxides brought down from above which make it more strongly colored than the parent material (C horizon) out of which it was made. The bleached A horizon is much paler than the parent material.

Gray-brown podzolic soil from Minnesota

There is a thicker humus-rich A_1 horizon than in the podzol, evidence that earthworms and other soil animals have mixed the surface litter more deeply into the soil. The bleached A_2 horizon is less pronounced, having been less completely leached than in a podzol, and the B horizon has received less iron and aluminum oxides from above which might color it.

Terra Rossa from Portugal

Formed under a warmer climate than a gray-brown podzolic soil, soils of the Mediterranean region are likely to have lost relatively more silica, reddening their overall color. It is likely that a few thousand years of agriculture have resulted in erosion and oxidation of their organic matter, revealing the red color of their iron oxides and resulting in a simple profile.

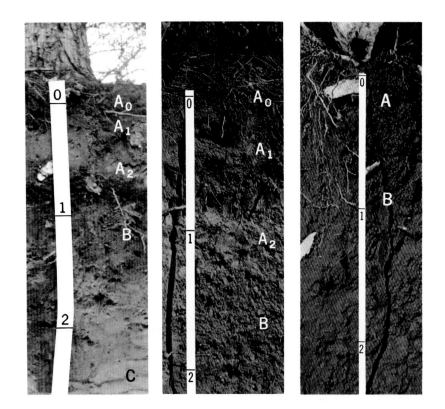

Latosol, or tropical lateritic soil, from Puerto Rico

Copious tropical rain selectively dissolves silica minerals and leaves iron and aluminum oxides behind in insoluble forms, giving the characteristic red color of tropical soils. Organic matter is worked deeply into the soil giving a thick but vague A horizon. The B horizon may be tens of feet thick in particularly well-drained sites.

Prairie soil from Iowa

Decomposition is slow in the dry prairie climate and a thick organic A_1 horizon collects. Leaching of minerals from the A horizons is slow, with a consequent poor development of the B horizon.

Chernozem from South Dakota

A pronounced development of a prairie soil in the drier parts of the prairies. Percolating waters that reach the top of the C horizon are usually drawn back to the surface as the soil dries, leaving their dissolved mineral load as deposits in the top of the C horizon. Commonly calcium salts are deposited in this way, resulting in a C_{Ca} horizon. Sometimes calcium carbonate may appear as a more pronounced layer or as larger deposits in rodent burrows.

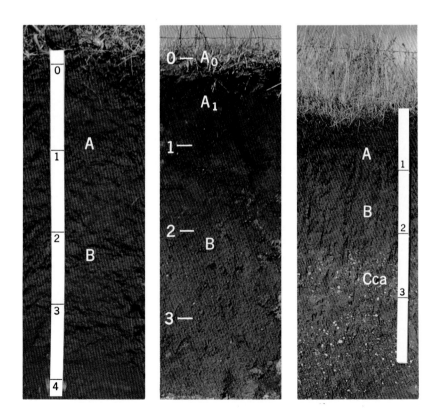

dark mixture of mineral soil and humus, the deep brown "B" horizon where minerals and clay have collected from above, and thence, gradually, the parent material. In the American mid-west, such a soil is known as **"grey-brown podzolic"** (Figure 3.2), at its finest development, as in parts of Europe where it was first described, it is known as a **"brown earth"**. As with the podzols, there are local variants but the essentials of the soil profile may be found over the vast areas occupied by the formation, and in many different geological regions.

Other formations are also associated with characteristic soils. Under tundra, the frozen earth impedes alike drainage, decomposition, and activities of soil animals. There is a characteristic simple profile; a layer of peaty plant parts, a thin layer of grey waterlogged mineral soil, and then the permanently frozen parent material. Under deserts the profiles are simpler still; a thin horizon of mixed organic and mineral fragments grading directly into the parent material.

Under grasslands, such as the Russian steppe or the American prairie, is a more striking soil. Here rainfall is limited, and there is rarely much water to percolate through the soil. In spite of the low rainfall, there may yet be heavy showers, providing enough water at one time to penetrate to considerable depths before evaporation at the surface draws it back. A profile forms within the depth reached by the percolating water, but with some peculiar characteristics. Dead grass parts decompose but slowly so that a thick largely organic layer builds up at the top of the soil, forming the famous black earth of the wheat lands. This black peaty layer grades gently into mineral soil below and is mixed with it by burrowing animals, although this mixing is not done so thoroughly as in the wetter soils under deciduous forest. The result is a deep black soil, getting grey toward the bottom, although still dark, and then an abrupt termination when the line of lowest water penetration is reached. Here there may be a mineral band, often white with carbonates, where the drying water has left its load of dissolved matter. Below this, and sharply distinct, is the parent material. Such a **prairie soil,** or **chernozem** to give it its Russian name, is quite as characteristic of prairie as is a podzol of boreal forest, and it too may be crudely mapped by mapping vegetation boundaries (Figure 3.2).

It must be obvious that vital soil properties are controlled by whether there is sufficient rainfall to flush through the soil all the way to the water table. If there is, a profile such as a podzol or a brown earth will be produced. Such a soil loses minerals by solution continuously, and must have them replaced, mostly by the weathering of parent material, if vegetation is to persist. But a soil in a place of low rainfall may never be flushed through. Solutes are merely carried to the base of the soil profile where they are redeposited as, most strikingly, in a chernozem. There must obviously be places where the rainfall is such that sometimes the soil is flushed through whereas at other times it isn't, so that a hard distinction between the two types of soil is not possible. But, for convenience, soil men talk of **pedalfers,** meaning soils that are flushed through,

and **pedocals,** meaning soils of dry (usually hot) places where the soil
water is usually evaporated from above. Actually soil men do not talk
about pedalfers and pedocals so much as they write about them in their
textbooks. Like other artificial distinctions, the terms are more attractive
to pedagogy than to practice.

The rough coincidence of soil boundaries with vegetation boundaries
can, in part, be explained as a direct consequence of vegetation itself, for
different plants leave different litter, with marked effect on the soil
animals, the pH of the percolating water, and the increment of organic
matter to the soil. But it is also clear that climate exerts a direct effect on
the soil. The cold of the Arctic shapes the soil by reason of the ice which
blocks the drainage, and the low rainfall of grasslands determines the
buildup of surface organic matter, just as it decrees that grass rather than
forest shall grow there.

But the direct effect of climate on soil formation is most clearly seen
when the soils of the humid tropics are compared with those of well-watered temperate regions. The leaf litter of a tropical rain forest does not
look very different from that of northern deciduous forest, except that
rapid decomposition makes it much thinner, but the dominant color of
the mineral soil underneath it is red instead of brown or grey. This red
color is the red of iron oxides, and mineralogical analysis quickly shows
that the mineral soil, in fact, largely consists of various oxides of iron and
aluminum without the silica minerals that are so prominent a feature of
most unweathered rocks and that give the greyer colors to northern soils.
The warm tropical water washing through the soil of a rain forest dissolves the silica minerals and leaves the iron and aluminum behind. The
cold percolating waters of the north do the reverse; they dissolve the iron
and aluminum and leave the silica minerals behind, and the extreme example of this cold region process is that bleached layer of a podzol,
which is sometimes almost pure silica. These solution processes are very
slow, as may be readily realized when you consider that the tropical
process implies the solution of something like glass in lukewarm water,
and the detailed chemical mechanisms are not yet completely known.
But there is no doubt that the temperature of the percolating water is
decisive; that the cool rain of the northern forest leaves silica but removes
iron whereas the warm rain of the tropics leaves the iron and removes the
silica. The deep red tropical soil so formed is called a **latosol** (Figure
3.2).

The oxides of iron and aluminum which are left behind by the
weathering process of tropical soils are very resistant to further weathering. Should the local climate change from warm to cool, as has often
happened during spans of geological time by processes such as continental drift, some of the thick red **laterites** may survive as fossil soils. It is
proper to think of this fossil red material as being the parent material of
the new soils forming at its top. In places, the superposition of modern
soil processes on fossil horizons weathered under different climates may
produce confusing color schemes. Some of the gradations of yellow and

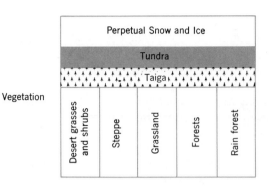

Vegetation

Dry cold   Wet cold

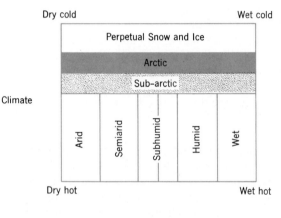

Climate

Dry hot   Wet hot

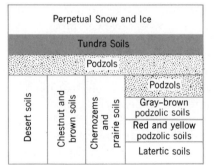

Soils

Figure 3.3

These diagrams summarize the hypothesis that mapping formations yields maps of climate, and that maps of formations
and climates combined yield maps of the great soil groups. As
a grand generalization the hypothesis provides a useful way
of thinking about the earth. But it is well that we draw the
earth as square, thus accentuating the fact that we are being
idealistic. Few of the boundaries shown in these diagrams can
actually be found on the ground. (*Taiga*, which appears in the
top diagram, describes the belt of more or less open forest that
commonly lies between the tundra and the closed boreal
forest; many would include it in the term "boreal forest"
when speaking in very general terms.)

red soils in the southeastern United States may reflect the imposition of the transition climate of the region on a medley of fossil soil horizons. But local confusions of color are minor compared with the global patterns.

It is possible to make a crude map of the **Great Soil Groups** of the world on the basis of a vegetation map. The distribution of podzols roughly reflects the distribution of boreal forest; of chernozems the prairies and steppes; of tropical red soils the tropical forests; and so on. Each area of such a map represents only a grand generalization, of course, because there is much local variation hidden by its scale. And the divisions between one soil type and another may be even harder to find on the ground than the division between some vegetation types. But the crude generalization remains. Soil men can map soils by mapping plant formations. Climatologists draw roughly the same maps from the same formations and call them maps of climate. Soil and vegetation between them affect greatly the lives of local animals; which, in turn, react on the vegetation and climate. And holding sway over all these interactions is the climatic pattern of the earth, a splendid if vague patchwork quilt of air masses stamping onto the geography of our globe a basic pattern of regions for life (Figure 3.3).

AGRICULTURE AND THE GREAT SOIL GROUPS

The impressions that climate and vegetation have made on the mineral surface of the globe have been slowly made. Just a foot or two under the primeval litter is the raw unproductive crust of rocks, in a medley of varieties and patterns, all obscured by the work of vegetation. A thin brown layer of litter covers the earth, and in this we farm. Where we have inherited a layer a few feet thick, as where chernozems and prairie soils can be made to yield fortunes in wheat, we can proceed by mining the organic layer, carrying off at harvest time the vegetation that in past years went to restore the litter. The inevitable future for such wheat lands is the restoration of the ancient bare parent material. But in brown earths, things almost as striking are possible. It seems likely that the Mediterranean lands credited with the birth of modern civilizations, were once the homes of lush forests growing in brown earth soils. Then we cleared the forests and ploughed the land. The hot sun dried the surface, speeding the oxidation of the soil organic matter. Farmers carried off produce, making sure that organic matter was thus not returned. If you remove the organic matter from the "A" and "B" horizons of a brown earth you take away the brown. What is left is the red color of the iron oxide left in the "B" horizon by the leaching water. The result is the famous **terra rossa** (Figure 3.2) of the Mediterranean, lovely to look at if you can stand the glare but not good for much else. A more extreme version of this history is the making of a desert by the ploughing up of land now supporting tropical forest, when you can lose the "A" horizon, the litter, and the hoard of soil nutrients almost completely (Chapter 14). In agriculture as in much else, we live in an ephemeral time of thriving on systems that must be short-lived.

Similar formations on different continents look alike but may be made up of quite unrelated species. The similar appearance of these unrelated plants is due to parallel or convergent evolution. There are thus two ways in which botanists may seek to classify broad regions of the earth. Regions may be collected together because their vegetation looks similiar or because their plants are related to each other. These two methods of classification beget quite different results. The ecological system of classifying together plants of similar appearance is a natural and obvious one, but it is not as simple to do this for animals that cannot be seen so easily in the mass. Furthermore, the pioneer zoogeographers were principally interested in demonstrating the family relationships of animals as crucial evidence for the theory of evolution. Mapping animal distributions in an ecological way thus did not become common until the theory of evolution had become generally accepted. Even then the greater ease of mapping plant distributions meant that animal maps would always be largely based on plant maps. Ecologists have found it useful to consider broad areas of the earth presenting common problems for life as units, referring, for instance, to the "tropical rain forest biome." Evolutionists and geneticists, on the other hand, refer to areas of common evolutionary history such as the "South American region." Both are valid ways of looking at nature.

THE CLASH WITH
EVOLUTIONISTS
OVER ANIMAL
GEOGRAPHY

51

CHAPTER 4

THE CLASH WITH
EVOLUTIONISTS
OVER ANIMAL
GEOGRAPHY

The great plant formations are so obvious a condition of our globe that they invite classification. Armed with a map of world vegetation, showing by its arbitrary boundaries the world parceled out into discrete blocks of vegetation, it is easy to order the blocks into formation types. The rain forests of Africa, South America, and Asia, for instance, are separated by thousands of miles of ocean, but they look so alike that it is easy to lump the plants of these distant places into one category called "the tropical rain forest." And yet the rain forests of different continents are made up of different species, often even of different genera and families. They are not closely related; they merely look the same. An evolutionist explains this as the result of parallel or convergent evolution. The conditions for life in these distant places were similar so that similar adaptations were evolved from different ancestral stocks. Classification of collections of plants along ecological lines into formations thus obscures the true family relationships, and a map drawn by an ecological plant geographer will be quite unlike that drawn by an evolutionary (or floristic) plant geographer.

These two attitudes to mapping are also possible for animal geographers but, unlike their botanist colleagues, zoologists have argued hotly over which was the better way. Collections of animals are much less easy to map than are pieces of vegetation because they move, they hide, and they are often small. Zoogeographers usually have to work with groups of animals only, particularly with those that are well-known because they are big and obvious or those on which the geographer is a specialist but, even so, clear ecological classifications of the principle areas of the earth are possible. J. A. Allen, working on the bird collections at the Harvard Museum of Comparative Zoology, published such a classification (Allen, 1871). He had noticed, as had many others, that you saw different birds in the different states of the eastern American seaboard; this was so even though many of the birds were migratory and flew north and south with the seasons. There were different lists of birds to be found at different latitudes and, furthermore, this was not just, a property of eastern America. The birds of the arctic parts of North America were similar to those of arctic Europe, Asia, and Greenland, so that a traveler across the arctic would find the birds always familiar. Then there seemed to be mirrors of the temperate American birds in the birds of Europe. And peculiar tropical birds like the toucans and hummingbirds of the South American rain forests had their counterparts in the hornbills and sunbirds of the Congo (Figure 4.1). Allen found it was possible to classify the birds of the world into **"realms"** that were roughly organized into latitudinal belts circling the globe (Figure 4.2).

His was a day when biologists felt compelled to express their findings in grand statements called "Laws," and Allen expressed his findings as, **"The law of the distribution of life in circumpolar zones."** He had done essentially what the botanists had done with their formations, although the mobility of birds had left him with much less definite boundaries. But he was using the very material, vertebrate distribution, out of which much of the theory of evolution was still being established, and his maps were very different from those of the evolutionists.

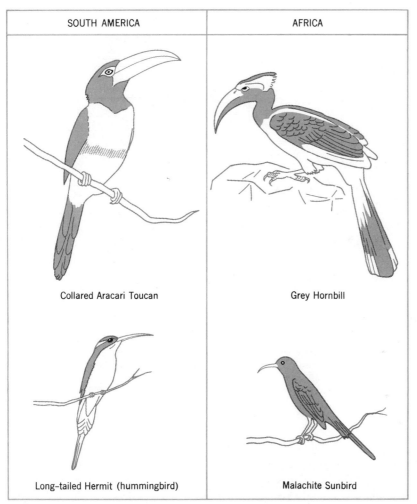

SOUTH AMERICA	AFRICA
Collared Aracari Toucan	Grey Hornbill
Long–tailed Hermit (hummingbird)	Malachite Sunbird

Figure 4.1

Ecological equivalents from South America and Africa. Toucans and hornbills look alike yet belong to different orders. So do hummingbirds and sunbirds. An ecological biogeographer will want to classify the parts of the earth so that the functional similarities between members of such pairs are revealed. He is content to call equatorial forests of both continents containing birds which look alike parts of the *tropical rain forest biome*. Evolutionary biogeographers have objected to such classifications, preferring to separate South America from Africa as different *regions* because there birds and other living things are not closely related even though they look alike.

Alfred Russel Wallace had deduced the theory of evolution independently of Darwin, and he had done it essentially from an insight into zoogeography developed during his own travels (Wallace, 1876). Because animals of isolated but similar environments were quite unrelated yet looked the same, it was evident that different stocks had evolved separately to the same end. The business of zoogeography must be to so map the world that the separate history of different landmasses should be

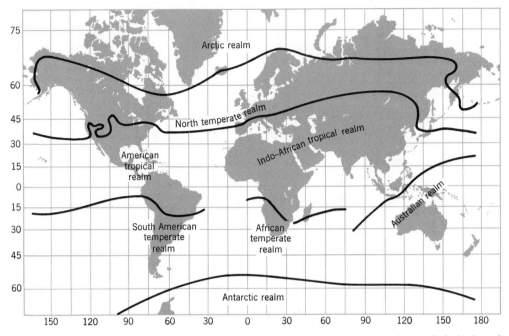

Figure 4.2   Allen's zoogeographic realms. Each of these realms was erected to include birds and mammals which shared common adaptations. Like a map of formations, the main divisions fall naturally along lines of latitude. But Allen, working as a museum taxonomist as well as a geographer, was yet aware of the great taxonomic distinctness of the tropical continents, hence the separation of the tropical and south temperate realms into continental units.

accentuated, and so all could see the truth of evolution. Wallace found his ideal system in the work of another bird man, this one working at the British Museum, P. L. Sclater (1858). The world's birds were mapped in six **regions,** roughly corresponding to the continents (Figure 4.3). All of South America was one region, ignoring the fact that this vast area included rain forest (with its ecologically distinct birds), pampas (with its ecologically distinct birds), temperate forest (with its), and so on. Nearly all South American birds had clearly evolved from a common stock. The Amazonian birds were much more closely related to the Patagonian birds than they were to the birds of Equatorial Africa. This was what counted. The other continents were treated in the same way. It was rotten ecology but good classical taxonomy.

When Wallace produced his great two-volume work on zoogeography in 1876, a work which is still the standard for the subject, he did so to reveal still more thoroughly the truth of evolution. He favored Sclater's classification as doing this most clearly, and he applied it not only to birds but to all animals. As a first step he found it necessary to demolish the "rightness" of Allen's ecological system and to do it in a way which should make the system seem ridiculous. There is a page of caustic writing on the lines of, "Mr. Allen's supposed law is merely . . . etc." (Fig-

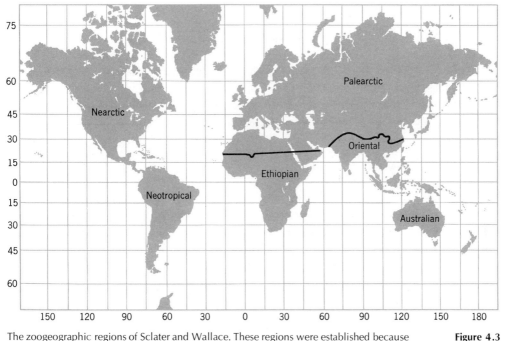

The zoogeographic regions of Sclater and Wallace. These regions were established because the birds and mammals of each continent or subcontinent were closely related to each other, even though they might have adapted to different ways of life. Use of this system by Wallace was one of the clues on which the theory of evolution was based.

Figure 4.3

ure 4.4). Wallace, of course, knew that the ecological classification was a logical possibility, that similarly adapted forms could be classified together, but this was in truth the reason he was so anxious to discredit the system. Classifying adaptations masked the evidence for evolution and was, indeed, the system long used by those who believed in special creation. For them the toucans of South America were, in effect, the same as the hornbills of Africa, and hummingbirds and sunbirds were also more or less the same. Wallace, and those like him, spent their lives struggling to make certain that the great truth of evolution penetrated mens' minds. There was a special creationist bishop hiding behind every page of an ecological classification, and the interest of these classifications was trivial compared with the importance of evolutionary truths which classification by family relationship revealed.

The best zoological brains of the time were engaged in the examination of evolutionary theory, and in exploring its ramifications in phylogeny and ontogeny, with the result that the study of mere adaptations, or ecology, received less attention. It still had its champions. Others besides Allen went on doing ecological geography, and among botanists it made the running. Allen responded to the Wallace strictures with more comprehensive and thorough documentation of his system, showing that it was suited to the distributions of mammals as well as of birds, but eventu-

Objections to the system of Circumpolar Zones.—Mr. Allen's system of "realsm" founded on climatic zones (given at p. 61), having recently appeared in an ornithological work of considerable detail and research, calls for a few remarks. The author continually refers to the *"law of the distribution of life in circumpolar zones,"* as if it were one generally accepted and that admits of no dispute. But this supposed "law" only applies to the smallest details of distribution—to the range and increasing or decreasing numbers of *species* as we pass from north to south, or the reverse; while it has little bearing on the great features of zoological geography—the limitation of groups of *genera* and *families* to certain areas. It is analogous to the *"law of adaptation"* in the organisation of animals, by which members of various groups are suited for an aerial, an aquatic, a desert, or an arboreal life; are herbivorous, carnivorous, or insectivorous; are fitted to live underground, or in fresh waters, or on polar ice. It was once thought that these adaptive peculiarities were suitable foundations for a classification,—that whales were fishes, and bats birds; and even to this day there are naturalists who cannot recognise the essential diversity of structure in such groups as swifts and swallows, sun-birds and humming-birds, under the superficial disguise caused by adaptation to a similar mode of life. The application of Mr. Allen's principle leads to equally erroneous results, as may be well seen by considering his separation of "the southern third of Australia" to unite it with New Zealand as one of his secondary zoological divisions. If there is one country in the world whose fauna is strictly homogeneous, that country is Australia; while New Guinea on the one hand, and New Zealand on the other, are as sharply differentiated from Australia as any adjacent parts of the same primary zoological division can possibly be. Yet the *"law of circumpolar distribution"* leads to the division of

F 2

Figure 4.4   Facsimile of a page from Wallace's *Geographical Distribution of Animals*, 1876, in which he attacks the Allen ecological classification for obscuring evolutionary truths.

ally his defense was taken up by a doughtier champion. C. Hart Merriam rested the general applicability of his life-zone system (Chapter 2) on Allen's circumpolar zones (Merriam, 1892). The arbitrary zones into which Merriam had divided the side of an Arizona mountain could be compared with equally arbitrary divisions of the North American continent into bands of latitude. Merriam used Allen's maps of "realms" to guide him in extending his mountain life zones across the lowlands of North America. The result was the familiar pattern that underlies many of the distribution maps found in American natural history books (Figure 2.17). Such maps, for all their shortcomings, are ecological maps that throw together animals adapted to common environmental problems, completely ignoring their familial relationships. There was no loud complaint from the evolutionists when Merriam's work had reached its ascendancy at the turn of the century because the theory of evolution was then solidly accepted. There was no longer any need to fear an ecological classification. The two ways of classifying could coexist.

Much of modern biogeography, both of plants and animals, has developed from the familial classifications of Sclater and Wallace. The standard works of biogeography, like those of Darlington (1957) and Good (1953), start their discussions with Sclater's six regions. Each region separates a flora and fauna of common ancestry, and the biogeographer then proceeds to reconstruct history from the detailed distributions and family relationships within a region. He sees in the mind's eye the common ancestors diverging with time to occupy all the vastly varied habitats of a region. Then he traces the family contacts between the regions themselves, working back further in time to develop a grand synthesis of life evolving and migrating throughout geologic time. In this grand synthesis the biogeographic region with which he started is but a temporary and imperfect grouping of animals and plants, physically permanent as long as the present pattern of continents persists but even then somewhat arbitrary. The continents do not really exist in isolation from each other, nor do their living things. Continents always share some families, perhaps even genera or species. The region as a clearly defined entity in evolutionary time is thus a conception, an abstraction, an artifice that simplifies the complex pattern of life on earth. But a great truth is revealed by this abstraction; that things living in the same place are usually related to each other, whatever they look like.

The "formations" of the ecological plant geographer are also abstractions, for they too commonly grade into one another and must be impermanent through geologic time. And a great truth is revealed by these abstractions also; that similar climates require similar forms of life, and that these will evolve no matter what the genetic constitution of the original inhabitants might be. The same could be said for the groups of animals collected in Allen's "realms" and Merriam's "life zones," but the great technical difficulty of plotting the distributions of animals has led ecologists to forbear developing maps of animal distributions very far. Instead they have conceded that the distribution of animals is so tied up with the distribution of the vegetation in which they live that it is best to accept the formation boundaries and to include the animals in the formation descriptions. The formations then become **biomes,** and people speak of the tropical rain forest biome instead of just the tropical rain forest.

The word **biome** is useful for impressing the layman with superior and mysterious knowledge, but otherwise it is unfortunate. The word is redundant, since "the tropical rain forest biome" conveys no more to the reader than does "the tropical rain forest." But the word is in current use,

PREFACE

DARWINIAN fitness is compounded of a mutual relationship between the organism and the environment. Of this, fitness of environment is quite as essential a component as the fitness which arises in the process of organic evolution; and in fundamental characteristics the actual environment is the fittest possible abode of life. Such is the thesis which the present volume seeks to establish. This is not a novel hypothesis. In rudimentary form it has already a long history behind it, and it was a familiar doctrine in the early nineteenth century. It presents itself anew as a result of the recent growth of the science of physical chemistry.

About fifteen years ago I first became interested in the connection between physical and chemical properties of simple substances and the organic functions which they serve. At that time the applications of the new physical chemistry to physiology were only just beginning, and the older speculations of natural theology upon such subjects had long since

v

Figure 4.5

Facsimile of the first page of Henderson's book, *The Fitness of the Environment*, 1913. Henderson's statement that the adaptations of animals could not be properly studied without examining the qualities of the environments in which they lived was central to the establishment of much of modern ecological thought.

having replaced the easily understood term "formation," and we shall have to live with it. The concept behind it is, however, useful, and central to much of ecological thought. There are great regions of the earth, roughly delimited by climate and sharing common environmental problems, in which life is lived in similar ways. "Regions" and "biomes" are complementary abstractions leading to different, but equally valid, truths.

Ecology was finally removed from the disrepute that the accent on evolutionary studies had brought it by the work of a Harvard chemist, Henderson. Henderson (1913) wrote a now classic book called, "The Fitness of the Environment" in which he pointed out how well suited was the terrestrial environment to terrestrial life, in particular how the remarkable and unique physical properties of water at terrestrial temperatures were suited to the life of carbon chemistry (Figure 4.5). Before Henderson's book appeared, studies of the goodness of the environment, of the life-supporting condition of the earth, had a biblical and nonscientific ring about them. Talking of the fitness of the environment was a bit like singing the lovely and rousing hymn:

> He sends the snow in winter,
>   The warmth to swell the grain,
> The breezes and the sunshine,
>   And soft refreshing rain.

This was acceptable in church, if you did not mean it literally, but not acceptable in the laboratory. Nevertheless, animals and plants are adapted to their environments, so the converse must be true; that the environments are suited to them. Ecology studies the interaction between both; it must study the fitness of both. Henderson made it respectable to look at the environment once more.

Plants grow in communities that commonly seem to be dominated by a few plants, the *ecological dominants*. Since naturalists were long aware that different kinds of communities existed, even within the territory of a formation, ecology attempted to classify communities. The result was the subdiscipline known as plant sociology or *phytosociology*. One method of classifying was to list all the plants usually found with common dominant species and to take this list as a unit, thus the *Beech-Maple Association*. This unit is an abstraction to which real pieces of vegetation more or less conform. The Braun-Blanquet school at Zurich-Montpelier made complete species lists of similar pieces of vegetation to find *character* species other than the dominants, then used these to define another generalized unit to which real vegetation would more or less conform, also an abstraction called an *association*. An alternative plan was followed in Scandinavia where vegetation was less varied and less easy to examine in an intuitive way. Census taking by quadrat samples was used to discover the approximately complete species list in a locality, and this list was then taken to be the unit of classification. Called a *sociation*, it was defined by dominant and *constant* species. All these methods failed in their real purpose of discovering a grand natural ordering of plant communities in nature. The reality was eventually more clearly shown by the methods of *gradient analysis* which demonstrated that the composition of vegetation changes gradually as you move from place to place, unless there are physical boundaries in the environment. Plants are distributed according to their individual requirements. They are adapted to the presence of each other, providing some of the familiar aspects of plant communities, but the communities themselves are not discrete units like taxonomic species.

61

CHAPTER 5

THE EXAMINATION
OF PLANT SOCIETIES

Plants grow in communities. An oakwood at the corner of an English farm is a community of plants, as is the woodlot of an American Midwest farm, with its prominent beeches and maples. The plants of a pasture are clearly another community, the member plants of which are mostly not those found in nearby woods. So much is common knowledge of botanist and layman alike. But how rigid is the constitution of these communities? Are all the oakwoods of a region alike, so that studies in one oak wood give results applicable to another oakwood? To what extent do the plants of one oakwood live there by accident, and to what extent are they compulsive oakwood dwellers, essential parts of the society of plants which is the oakwood community? May communities of plants be formally classified as the species of plants and animals can be classified? The approaches that have been developed to these problems can best be illustrated by example.

Woods left on good farmland in Ohio and neighboring states are likely to be beech-maple woods. There may well be a dozen or more tree species present, but the most obvious, because the most common, will be beeches and sugar maples. In a thorough study of 40 acres of beech-maple forest, A. B. Williams (1936) counted 3654 trees, 1920 being beeches and 1193 sugar maples. The remaining 541 trees were of 18 different species. The beeches and maples are so predominant that there is clearly sense in calling this a beech-maple wood. Ecologists say that the wood is dominated by beeches and maples, that these trees exhibit the phenomenon of **ecological dominance.** There can be little doubt that the other plants of the wood are adapted to the presence of the abundant beeches and maples so that to describe them as **dominants** in this way is good use of language. The beeches and maples determine the microclimate, and even the soil, to which all the other living things of the wood must adapt.

Although the species of trees are few in the beech-maple wood, the varieties of other plants are numerous. There is much shrubbery; there are creepers; there are tangled brambles and herbs; there are the small plants of the woodland floor whose flowers and scent make the place so pleasant during two weeks of the spring. Excluding mosses, Williams listed 95 species in his 40-acre lot. These plants are all part of the community that is dominated by the beeches and maples, and are thus in some way associated with them. We may know from experience that many of these plants are to be found in other woodlots dominated by beeches and maples, that perhaps they may usually be found associated with them. We thus have an idea of a Beech-Maple community-type, or association. This idea is an abstraction. The actual community which we (or Williams) studied is a concrete entity (it was real enough to keep Williams painstakingly occupied for several years), but the idea of lots of communities conforming to the same general pattern is a class-concept, an abstraction. Plant ecologists talk of this class or grouping of actual stands that are alike in important respects, as an **association** and identify it by the name of its dominant or most distinctive plants; thus the *Beech-Maple Association.*

The dominants need not be trees. A good sheep pasture of fescue grasses in Wales might lead to the idea of the *Festuca ovina-Festuca rubra Association*.

The idea of the unit of vegetation being the association was developed during the first quarter of this century by a group of botanists working in the Alps and the south of France from institutes at Zurich and Montpelier. The most prominent of them was J. Braun-Blanquet, director of the Montpelier Institute, who gave his and his institute's name to this "school" of **plant sociologists,** as they liked to call themselves.

The Braun-Blanquet method was to make detailed descriptions of a number of pieces of vegetation that seemed to conform to a general pattern, and then to compare the descriptions in search of common denominators (Braun-Blanquet, 1932; Whittaker, 1962). Each description took the form of an annotated species list. The investigator, a trained botanist of judgement, first chose a patch in a local piece of vegetation which he thought was typical and which should be large enough to be representative of all facets of the community. This was a purely subjective process. A rather large area would be sought in a forest, one big enough to be likely to include some of all the tree species present, but a much smaller patch should suffice for a meadow. In a recent American test by Benninghoff (1966), a 4-meter square was thought large enough to describe the association of a field of corn (*Zea mays*).

Once you have chosen your piece of vegetation, you proceed to list every species present in the patch. A species list alone is of only limited use; you need to know how important each species is. Its importance may be measured by its abundance, in the sense of the number of plants that are present per unit area, but also by the plant's size. One tree must mean more to the community than one lily. This extreme problem of size differences in a forest can be overcome by describing it layer by layer, first the trees, then the shrubs, then the ground cover, but the problem persists even in a relatively unlayered community like a pasture. One rosette-shaped dandelion spreads over much more ground than a thin wild onion, a fact that must surely be recorded. The problem is further compounded by the growth-form of such things as meadow grasses: how do you count grass plants? It is much easier to record how much ground they cover. The Braun-Blanquet answer to this problem is to consider cover and abundance together and to represent them by an arbitrary but carefully defined index. Against each plant name in the species list you place a **cover-abundance index,** a number between one and five, with "5" being the most important (covering more than three-fourths of the sampled area) and "4" next (covering one-half to three-fourths of the sampled area) and so down to one. For a whole series of descriptive indexes the Braun-Blanquet school set up five classes, a number chosen I believe because they had five fingers on each hand. For the cover-abundance rating, five turned out to be too few, and they had to add to the low end a "+" (for less important than "1") and an "r" (rare, usually only one example) (Figure 5.1).

cover-abundance ratings according to following scale:

5   covering more than 3/4 of the sampled area
4   covering 1/2 to 3/4 of the sampled area
3   covering 1/4 to 1/2 of the sampled area
2   with any number of individuals covering 1/20 to 1/4 of the sampled area, or very numerous individuals but covering less than 1/20 of the area
1   numerous, but covering less than 1/20 of the sampled area, or fairly sparse but with greater cover value
+   sparse and covering only a little of the sampled area
r   rare and covering only a very little of the sampled area (usually only 1 example)

(n.b. "+" is always spoken "cross")

The **sociability** is estimated for each species in terms of another scale:

5   in large solid stands; very dense populations
4   in small colonies or larger mats; rather dense populations
3   in small patches or polsters; distinct groups
2   in small groups or clusters or tufts
1   growing singly

Further symbols relating to the **condition** and **vitality** of the individual species, as those below, may be recorded beside the cover-abundance and sociability estimates.

oo - very poor and especially not fruiting (e.g., $+^{oo}$ or 2^{oo})

o - poor vitality (e.g. 1°)

g - germinating plant

Y - young plant
st - sterile
bu - budding

bl - blooming
fr - fruiting

no notation - normal growth

• - luxurious growth (e.g. 4•)
e - being driven out (by other plants)
d - dying
def - defoliated
dd - above-ground organs dead or dried out
s - present only as seed
- specimen collected

Figure 5.1   The descriptive indexes used by the Zurich-Montpelier school of phytosociology as interpreted by Benninghoff (1966). This is an arbitrary set of indexes, but one that has had such wide use in the literature that it should be used in preference to ad hoc systems. For the purpose of pure description it is the most expedient method, but it is well to remember that the labor involved in using it may be very great and that such description might only be worthwhile when the object is to record for posterity the present condition of changing vegetation.

A list of all the plants found in a good example of the community, with a cover-abundance rating for each, became a useful piece of descriptive data, but they also went further. Some plants grow in clumps, that is, they are aggregated, while others are more usually found growing singly or scattered. Braun-Blanquet called this quality the **sociability** of the species, and rated it on the usual scale of one to five. A plant received a rating of "5" if it grew in dense solid stands, but only of "1" if it was an evenly spaced loner. The sociability rating was always written after the cover-abundance rating, and separated from it by a stroke. In the American cornfield used by Benninghoff (1966) to test the system, the corn plant was, of course, dominant, but it also appeared in the species list as: *Zea mays* 3/1, that is, it had a cover abundance rating of only "3" since, although the commonest plant, it by no means covered nearly all the ground as it must to rate a "5," and a sociability of "1," earned because it was evenly spaced and thus the extreme loner.

A species list of a rich community annotated in this way is a formidable document, one that has required much labor. If compiled well it is a fine record for posterity; but the Braun-Blanquet school had more immediate hopes. The species list of one community might be compared to that from another community thought to be similar to see if there was validity in their abstraction of the "association." They compared many such lists by a routine sorting process, essentially by visual inspection (Figure 5.2). Some plants turned up in only a few lists, some were often there, but a few were always to be found. Among these few were the dominants, but there were also other faithful plants that were thought to be of more interest. These were humble members of the community whose fidelity would not have been discovered without the listing technique. The faithful species were called the **character** species of the association.

For the regions in which it was evolved, this system of describing and grouping plant communities into associations worked. The south of France is rich in plants, with many and varied plant communities of numerous species, and there is also the zoned vegetation of the Alps, rising like Merriam's Arizona mountains from warm sunny places to the arctic–alpine vegetation above the tree line. The area had been long settled so that many interesting patches of vegetation were left as discrete blocks. The problem of what piece of vegetation to choose for a description could be greatly simplified if there was but a small patch left, perhaps with a fence around it. Clearly defined pieces of vegetation could be comfortably organized into associations by this simple subjective approach: find a good piece, decide on the size plot you need to sample, make your species list. Then go away and find another plot. Compare lists to find character species and from then on you only had to look in a doubtful community for your character species and, once you found it, you could immediately docket the community in your mind with others like it as part of a specific association. The method was purely subjective, but it worked.

In a region with a less rich flora and with less patchy vegetation the method was not so attractive. In miles of unbroken boreal forest or peaty

Lists of species present in ten cornfield communities in Michigan to illustrate the Zurich-Montpelier method.

	75	90	95	75	70	40	10	60	40	25
Total vegetation cover (5)	75	90	95	75	70	40	10	60	40	25
Total number of species	9	12	5	8	9	5	7	7	9	11
Releve (list) number	1	2	3	4	5	6	7	8	9	10
Zea mays (planted)	3.1	3.1	3.1	3.1	3.1	3.1	3.1	3.1	3.1	3.1
Agropyron repens	r.1	+.3	+.2	r.3	r.1	•	+.1	2.2	+.1	+.1
Ambrosia artemisiifolia	+.1	1.2	•	r.2	+.2	r.1	•	+.1	1.1	+.1
Amaranthus retroflexus	r.1	•	2.2	r.1	1.2	r.2	r.1	•	1.1	r.1
Setaria glauca	r.2	+.2	•	•	•	1.2	r.1	2.3	+.1	+.1
Chenopodium leptophyllum	3.2	•	2.2	r.1	•	+.2	•	+.1	+.1	+.1
Daucus carota	•	2.3	•	•	r.1	•	r.1	•	+.1	+.2
Cirsium sp.	•	1.1	•	+.1	•	•	r.1	•	+.1	+.1
Chenopodium album	•	r.1	•	r.2	•	•	•	+.1	•	•
Portulaca oleracea	•	•	•	•	+.2	•	r.2	•	•	r.2
Digitaria sanguinalis	2.2	•	3.3	•	1.2	•	•	•	•	•
Panicum capillare	r.2	r.2	•	•	•	•	•	•	+.2	•
Plantago cf. major	•	r.1	•	•	•	•	•	•	•	r.1
Equisetum arvense	•	r.1	•	•	+.3	•	•	•	•	•
Xanthium chinense	•	•	•	•	•	•	•	2.2	•	•
Amaranthus sp.	•	r.1	•	•	•	•	•	•	•	•
Polygonum pensylvanicum v. laevigatum	•	r.2	•	•	•	•	•	•	•	•
Taraxacum officinale	•	•	•	r.1	•	•	•	•	•	•
Panicum dichotomiflorum (typical)	•	•	•	•	+.3	•	•	•	•	•
Echinochloa crusgalli	•	•	•	•	•	•	•	•	•	r.1
Echinochloa pungens v. microstachya	r.2	•	•	•	•	•	•	•	•	•

Figure 5.2   The lists were each prepared from 4 × 4 meter plots, a size chosen intuitively, after which they were compared in quest of character species. The lists are already long, in spite of the simplicity of a cornfield, suggesting how complex they might become for wild vegetation. (From Benninghoff, 1966.)

moor how could you make a value judgement on what was an association? The gradations were much more subtle, and yet you could be sure from botanical knowledge of the area that there were differences, that some communities of separate places were more like each other than their immediate neighbors. Such was the vegetation which confronted the botanists of Scandinavia at the time when Braun-Blanquet and the Zurich-Montpelier school were developing their lists a thousand miles to the south. The Scandinavians, too, wanted to see if unique vegetation types existed that could be classified into some natural order, but they had very different material with which to work. They solved this problem by taking an objective and quantitative approach to vegetation sampling. A school of plant sociology, a rival to that at Zurich-Montpelier, grew up at Uppsala in Sweden under the leadership of G. E. Du Rietz which sought to discover natural communities and their relationships from the data of random census (Du Rietz 1929, 1930; Whittaker, 1962).

The main instrument of the Uppsala school was the quadrat sample, and their use of it can best be illustrated by considering a simple unlayered piece of vegetation like a meadow or a bog. A number of points in the area to be investigated are chosen at random and around each is marked a square of standard size, commonly a one-meter square. Some men found their random points and marked them out at one step by throwing hoops, regarding the plants enclosed by each hoop as a quadrat sample of vegetation. It did not matter that the quadrat was not square as long as the area was known. The botanist would then go to the first quadrat and make a complete list of all the species enclosed in it, noting the number against the area. He would then go on to the next quadrat, repeat the process, and so on. With each successive quadrat sampled his total species list would grow, quickly at first but then more and more slowly as he began to record most of the plants present in the area. When he hardly found any more he would stop, assuming that his species list was virtually complete and that it comprised a full account of the plants associated in his community. The list was as complete as Braun-Blanquet's lists but it had been compiled entirely by objective means. And

Hypothetical species area curve as used by the Uppsala school for determining both the number of species in the *sociation* and the minimum area of that sociation. Such a curve never becomes asymptotic to the horizontal because indefinitely increasing the area sampled will always bring in a few more species, but the curve does noticeably flatten out. This one seems to flatten at the point indicated by the arrow. There is an extensive literature about just how this point of flattening-out should be determined. Once agreed upon the right point, the minimal area and the number of species in the sociation may be read off. In this hypothetical example, there are 14 species in the sociation, and they may be found in 1 and ⅛ square meters. Evidently, we are dealing with some sort of (hypothetical) pasture or other sociation of small plants.

Figure 5.3

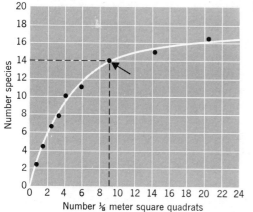

the Uppsala botanist had also found the right-sized piece of vegetation needed to contain almost the full species list; it was the sum of the area of all his quadrats. This area he called the **minimum area** (Figure 5.3).

The essence of the Braun-Blanquet system had been to say: "I am a good botanist, I see what is clearly a well-defined community, I place its boundaries so, now I will make a list of the plants contained therein." This method, for all that it worked, was decidedly subjective. Du Rietz and his men of Uppsala could pride themselves that they, on the contrary, were more carefully scientific, with their lists and areas found by means of random census. They could also identify dominant plants by their sheer numerical abundance in the samples or by physical measure of the area of the quadrats they covered, although in practice the dominance of plants is usually so obvious that even the most determinedly objective scientist will accept the evidence of his eyes for this.

Du Rietz could make many of these scientifically derived lists, just as Braun-Blanquet made many of his subjective lists, and compare them. At Uppsala, too, they looked for faithful species, but they did it on a percentage basis, and instead of calling them **character** species, they called them **constant** species. A species that was found in every list was 100 percent constant. In practice these were rare, and Du Rietz accepted 80 percent constancy as sufficient to qualify a plant for the rank of **constant species** (which was a subjective note creeping in). Du Rietz was then in a position to start calling his communities, defined by dominants, minimum area, and constant species, "associations" and to attempt to order them into a classification, just as Braun-Blanquet had done with his. The trouble was that they were quite different entities. Braun-Blanquet's "association" was an abstraction to which he sought to fit communities, but the Du Rietz unit was a concrete entity which he had found by census, and it was in practice a very much smaller piece of vegetation than were the communities that Braun-Blanquet assigned to his "associations." The two schools could not use the same terminology without confusion, so they used it anyway and bucked the confusion. The matter was resolved at the Tenth International Congress of Botany in Brussels in 1910 when, by formal motion of the delegates assembled, the Uppsala school was assigned the word **sociation** for their unit while the Zurich-Montpelier people kept **association** for theirs. I think the meeting must have been packed with Braun-Blanquet supporters. The temper of the congress may be indicated by a recorded reference of Braun-Blanquet's to the Uppsala school as "Die Herren Quadratica" (the *gentlemen* of the quadrat) (Whittaker, 1962).

It is to be remembered that many men were involved in this controversy. Dozens of men in various parts of Europe described vegetation according to the methods of one school or the other or, perhaps more often, by some hybrid system of their own. They invented much terminology and many other indexes, some by assigning ratings of one to five, others by deriving percentage classes from their census data. The botany journals are fat with their papers. In time, most of the more convenient aspects of both systems came into general use. In areas of patchy vegetation, particularly well-farmed areas, for instance, the subjective choice of a piece of vegetation is virtually unavoidable, but it is then often simple and reasonable to survey it by means of quadrat samples, with or without measuring minimum area. The descendants of the Braun-Blanquet school are not above using quadrats.

Out of this "plant sociology" have come recognized and simple methods for describing vegetation, and wherever the systems have been used a clear record has been left for posterity. As the rapid spread of the human population destroys the remaining patches of complex and interesting pieces of vegetation, this is an important achievement. I would like to see teams of students trained in each of the schools sent to the last interesting places, to the Galapagos Islands, of whose unique vegetation there is still no description, and to Amazonia, so that we may leave to our next generations a record of what they were like. But in other ways the labors of these botanists produced only a meager intellectual crop.

The most glittering possibility for the results of their work had been that the vegetation of the world could be ordered into a natural system, with the "associations" or "sociations" treated as the species of Linnaean taxonomy were treated. They were to be grouped into ever higher units of family relationship in the same way that species could be grouped into genera, families, and orders. Linnaean taxonomy had led to the theory of evolution; what new truths might lie beyond the rainbow of plant sociology? But experience was not to reward these hopes. Many systems of classification were proposed, but none stood, and the names used are known only to a few specialists. Often the units of real vegetation had no definite boundaries but rather merged into one another. Where boundaries did seem distinct, they could be ascribed to physical changes in the environment, to changes of rock, soil, or exposure, or to the sorts of climatic changes we find at the few distinct formation boundaries. Braun-Blanquet acknowledges this in his book summing up his life's work (Braun-Blanquet, 1932). The communities he studied occupied distinct habitats, but the parameters of the habitats usually defied measurement. He was reduced to using the plant communities themselves as a measure of the habitat, as indicators of environment rather as meteorologists used the formations. The distinctness of communities was limited to the distinctness of habitats.

The work of the plant sociologists may usefully be compared with that of the plant geographers who preceded them. If you devise a classification based on the general appearance of vegetation, on its physiognomy, the natural result is the *formation.* You have very large units in your classification because the vegetation of very large tracts of land, being subjected to a common climate, looks similar. But if you allow floristic uniformity to be your principal criterion, then your units are smaller. Requiring that your patches of vegetation have common dominant species results in a unit considerably smaller than a formation, but still large enough, a unit like the *Beech-Maple Association,* and other forest associations of the American botanists. Requiring common character species of lesser apparent importance than the dominants, as the Zurich-Montpelier school does, results in a still smaller piece of vegetation being called the critical unit or *association*. Finally comes the minimum area *sociation* chosen, stratum by stratum, by random quadrat census to produce yet another and finer classification. The exact criteria and qualifications of these various units for vegetational analysis are of little theoretical importance. What is important is to notice that the scale of each system resulted from the criteria chosen. The investigator made the choice. He was not in the position of Linnaeus describing discrete entities which nature had chosen for him. Plant sociologists were in reality confronted with an infinite variety of plant communities, no two exactly alike, none existing as discrete units in nature like biological species.

An echo of this experience is found in the modern American school of **continuum** or **gradient analysis** (Whittaker, 1967; 1967). The assumption is made that, where there is no obvious physical barrier, the composition of communities will gradually change and that this gradual change will reflect an environmental gradient. This assumption can be readily tested against gradients of real vegetation in the field, and can be shown to apply whatever the scale of vegetation mapping undertaken. Most of the great plant formations grade into one another (Chapter 2), there only being distinct boundaries where there are environmental boundaries as for the Canadian tree line. The vegetation of mountainsides was shown not to be separated into distinct bands as Merriam had thought (Chapter 2). And Braun-Blanquet's confession, after a lifetime of work, that all the associations which he could recognize were defined by habitats, is compatible with the same assumption. The composition of vegetation always changes gradually in space unless there is some discontinuity in the physical environment.

The grand lesson which plant sociology has for us is that plant distributions are set by the environment and that each plant finds its own optimum range. Familiar assemblages occur because many plants physically affect the places in which other plants must live, providing microhabitats to which other plants associated with them have become adapted. But the environment, acting on individuals, is the decisive factor in constructing a community.

Exciting to botanists were the changes that followed the colonization of fresh ground by plants. Such changes might take longer than a human lifetime, but their progress could be reconstructed from circumstantial evidence. Going back from the edge of a pond are concentric rings of vegetation which may be thought to be succeeding each other as the pond fills with mud and shrinks. A rather similar pattern of stages can be found round rocky outcrops where soil is being built up. Study of such sites suggested that the changes followed each other in orderly sequence. A typical sequence started with small herbs (the pioneer plants), progressed through small shrubs and ended with forest, after which changes were very slow or nonexistent. Botanists said a *climax* had been reached and that this *climax* was the end point of an *ecological succession*. The concept attracted very widespread interest when a dramatic succession was deduced to explain the development of a forest of beeches and maples on old sand dunes that are spread for many miles along the southern shore of Lake Michigan. Then it was noticed that similar successions took place on fields abandoned by agriculture. Many botanists accepted the idea that changes in vegetation were usually orderly, predictable ecological successions. For some, notably the followers of F. E. Clements in America, the idea became the dominant philosophy of their discipline. This idea seemed to require that plant communities were discrete taxonomic entities, just as had the work of the European phytosociologists. Many botanists, however, were convinced that all the phenomena of vegetation could be explained as the result of properties of individual plants. This view has been encouraged as it has become increasingly realized that the order of successions is by no means invariable. Even the succession of plant communities which grow on the infill of ponds, the so-called "hydrarch succession," seems to proceed in an almost random way, as has been shown by digging trenches in the mud beside English ponds and looking for fossil traces of past vegetation. Once ecological successions are seen to be the results of the dispersion and establishment of individual plants it becomes possible to make conceptual, and even mathematical, models of the development of vegetation freed from the shadow of some organizing principle.

CHAPTER 6

SUCCESSION TO
CLIMAX

A pond is usually ringed about with concentric belts of vegetation. In the shallow water at the edge are floating leaves of such things as water lilies and *Potamogeton,* plants that take root in the bottom mud where the water is shallow enough, the rooted aquatics (Figure 6.1). Next to them is a ring of reedy marsh in which cattails, rushes, or sedges are the usual dominants. The soil between the stems is waterlogged and squashy, but it becomes noticeably drier as the outer edge of the ring of this community is reached. The first shrubs and small trees are now encountered, and these may be called the dominants of another of the communities ringing the pond. Typical of such communities are those dominated by willows or alders or, in the boreal forest, black spruce (*Picea mariana*). The soil is noticeably drier, although obviously peaty. Beyond this ring of damp-site shrubs may be found one of the more widespread communities of the local forest formation, whatever this may be. A transect from the open water of the lake to the trees behind thus passes in successive order a series of distinct plant communities; rooted aquatics, reedy marsh, and a community dominated by water-loving shrubs or small trees. The exact details of the communities, and their number, differ from place to place, but a comparable succession of communities may be found around ponds in any formation of wetter climates.

Digging in the flat ground near any old pond unearths old swamp and pond mud, giving rise to the suspicion that the soil is the infill of an older and bigger pond. At the water's edge the filling process is obvious. Ponds fill with sediment, silt and debris washed into them by rains and the organic debris from life in the pond water. The pond shrinks and the shallows advance toward the center. As soon as each stretch of water becomes shallow enough, the rooted aquatics can be expected to grow. The stems of these plants must form a mesh in the water that should collect and concentrate settling particles, and the mass of their own root systems should further speed the process of building up the bottom mud. Soon the bottom is raised so close to the surface that the first reeds can take root. Their thicker lattice of stems, and the decomposing mass of

The vegetation at the edge of a pond. In the water can be seen floating leaves of water plants which are rooted in the bottom mud. In the wet mud at the edge is a line of reeds, and behind them bushes and small trees which grow in damp places. Well away from the pond, on dryer ground, are trees of the local woods. If you choose a good spot on the banks of such a pond you can trace a line which goes successively through rooted aquatics, reeds, water-loving bushes, and finally the local forest. This observation gave rise to the hypothesis that these successive kinds of vegetation were succeeding each other in time also, as the pond filled with sediment and the encircling bands of vegetation constricted towards the center.

Figure 6.1

their old stems and rhizomes, piles up the debris still more until the soil is firm enough to walk on. The plants of the shrub community can then grow. Deposition continues, so that the soil surface is raised still higher above the water table, affording a still drier site. The transpiration of the shrubs dissipates water to the air, lowering the water table under them. The soil may then be dry enough for plants of the surrounding forest community to grow. Examination of the successive stages has, therefore, allowed us to put forward the hypothesis that the pond is being filled and drained by the actions of a series of plant communities following each other in succession.

This hypothesis has been so important to the development of ecological thought that it is best to have a clear idea of what it claims. The observation is a succession of communities in space arranged in concentric rings around a pond. But the hypothesis requires that there be a succession in time also. The communities are supposed to succeed each other in orderly sequence, and furthermore each community prepares the ground for the next. Such a succession of plant communities through time, in which each successive stage is thought to be dependent on the one which preceded it, is given the special name of an **ecological succession.** The succession that is thus supposed to fill a pond commands the waters and so is called an **hydrarch succession** (Greek; literally "water commanding"). Each community in the succession is called a **seral stage** (from the same root as the English word "series").

Other apparent ecological successions are familiar in most regions, one of the most notable being that which develops on bare rock. Look for an outcrop of ancient hard rock, such as granite or metamorphosed sandstones. On the steeper parts of the sides there may be patches of bare polished rock which support no plants, not even lichens, but also there may be level places where a thin soil enables trees to grow. Digging through this thin soil, perhaps only 10 centimeters or 6 inches thick, may reveal rock as smooth and unfractured as that exposed on the steeper faces of the edges. The soil on which the trees grow has been built up on the surface of the rock with debris, and not made by erosion of the rock itself. Wandering about the outcrop will provide the evidence that shows how the soil had been built up. Next to the bare rock patches are patches of lichens and the low cushions of mosses called "moss polsters." Cutting through a moss polster shows that a thin layer of soil is building under it, perhaps 2 centimeters of black humus rotted from dead moss and lichen parts but including grains of sand and silt caught from the wind or from the runoff waters of a rainstorm. Further away may be heathy plants and grasses, growing in soil now a few centimeters deep. Searching through this soil may reveal parts of the mosses that had grown there until they had covered the rock with a sufficient thickness of soil for the larger plants to grow. We can imagine that, among the stems of the grasses and heaths, the soil continues to thicken as the plants supply their dead parts, and as silt and grains of sand continue to be trapped. At last, the first shrubs grow, and then the trees. On the bare rock, soil has been produced by the actions of a series of plant communities, each so modifying the environment that another community can take its place. A bare rock is a desert place for plants, a dry desert place, so this succession is called a **xerarch succession** (Greek for "dry commanding").

This idea that plant communites succeed each other in time is an old one. It has probably come to the mind of every naturalist who has pondered the filling of a pond or the covering of a rock with trees. The great French naturalist Buffon noted that forest trees may succeed each other in some sort of regular order, and wrote about it as early as 1742. But as botanists began their attempts to classify plant communities, the idea began to seem more interesting. The communities of these supposed successions were apparently distinct; at least, you could describe them well enough if you chose your place properly. And these apparently distinct communities were thought to be related to each other in a dynamic way. In many European botanical writings from about 1870 onward (Warming, 1909) the idea of classifying pieces of vegetation by successional relationship was inherent, but the most interesting attempts developed from later American work which sought to explain the existence of beech-maple forests on old sand dunes near the southern shore of Lake Michigan as resulting from a particularly dramatic xerarch succession.

A broad belt of sand fronts on the southern shore of Lake Michigan, a belt in places several miles wide. On sites furthest from the lake, forest covers the sand, in places beech-maple forest quite like that growing on the rich moist soils of other places. But under this forest there are only 10 centimeters of organic soil. Under this a probing spade finds only loose sand, a bit damp but otherwise similar to that of the modern beach fronting the lake. Once the site of the forest had been a sandy beach. It had been abandoned as currents in the lake built up the beaches of the southern side over the years, causing the shoreline to retreat steadily northward. But how had a beech-maple forest established itself on the shifting dunes that could be expected behind an abandoned beach?

This problem was tackled at the turn of the century by Henry S. Cowles (1899), a professor of the young University of Chicago. Careful field observations led him to infer that the beech-maple forest was the climax of a long succession of events, of an ecological succession. Beyond the storm line of the modern beach, the sand was being colonized by dune grasses, species able to tolerate very dry living, and spreading by means of stems that crept through the sands. Once these grasses were in place they stopped the wind from blowing the sand about. Amongst them seedlings of cottonwood (*Populus deltoides*) germinated and grew. These developed into big trees, providing shade and dropping their thick, slow-rotting leaves onto the sand. Soil animals mixed the leaves with sand grains, forming the first soil. Pine seedlings, sprung from seeds brought by chance of wind or animal carriage, grew under the cottonwood trees. Further back from the exposed dunes there were pine woods. There was little in the pine woods to show that large cottonwoods had once grown there, but the finding of the pine seedlings under patches of cottonwood elsewhere made it a valid inference that they had. Among and behind the pines was a forest of oaks. This was a simpler community than the beech-maple forest, but it was already deciduous woodland, a very different place to that of the pioneer grasses on the abandoned beaches. There was

Figure 6.2   Colonization of Lake Michigan sand dunes by creeping grasses and shrubs. The creeping
grasses and other herbs can be seen to hold the shifting sand; evidentally necessary for the
first shrubs to get established. With the sand fixed, leaf litter and other debris can collect,
making better soil for other plants. So much can be seen to happen. But further away from
the lake are woods of pines, oaks, and even beech-maple forest, all growing in soils built on
beach sand. It is deduced that the eventual forest was established as the result of a succession of plant communities, each making the site more suitable for the next.

no direct evidence that the whole complexity of a beech-maple forest
was developing in the shade of the oaks, since this must surely be a very
slow process of centuries or millennia so that it should be hard to see the
slow changes, but it was quite reasonable to suppose that oak woods had
existed where the beeches and maples now dominate the dozen other
trees and the hundred or so smaller plants of the beech-maple association. Cowles reasoned that it must have been so, that beech-maple forest
was the climax of a long xerarch succession which, through grasses, cottonwoods, pines, and oaks, had conquered the sand dunes and made
them a fit place for the richest forest of the region (Figure 6.2).

   This deductive work of Cowles was published in a paper of persuasive
logic which immediately led to the general acceptance of the idea of
ecological succession (Cowles, 1899). Men reminded themsleves of the
series of communities round ponds and rocks, and noted how change
and replacement were the normal fates of many of the plant communities

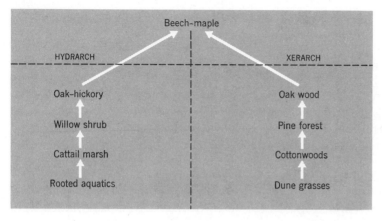

Figure 6.3

around them. Cowles' numerous graduate students described other successions in the terms of the master. Of particular note was the fact that successional series seemed to result in a community of plants that was apparently self-duplicating and perpetual. Only change of the environment by some outside agency such as new species, could supplant this community. It was called the **climax community.**

It is important to notice that the whole story so far rests almost entirely on deduction. No man had been alive long enough to witness the infilling of a lake, the covering of a rock with vegetation, or the planting of forests on the Lake Michigan sand dunes. But the circumstantial evidence that all these things had been so was decidedly strong. It seemed to suggest that plants typical of the formation widespread in any region could be established on the most unlikely sites, if given enough time and by the process of ecological succession. The **climax** of each succession looked as if it might be the same, the local forest or other formation. All the communities near Lake Michigan, for instance, had a dynamic relationship one to another and could be classified as shown in Figure 6.3. The communities are at once classified into a number of subordinate communities, the seral stages, and the generic taxon, the *Beech-maple climax community.*

The successions so far discussed are all **primary successions,** that is to say they are supposed to proceed by pioneering new sites. But there are other successions, those occurring on disturbed land and that climax in the replacement of the community which existed before the disturbance. The seral stages of these **secondary successions** can also be included in a classification as dependent taxa of the climax community. Secondary succession is best understood by considering what happens to a farm when it is abandoned.

Consider a stretch of country in the American Midwest where beech-maple woods are common. There is much cleared land laid out to plough or pasture but there are also many woodlots left in which there are likely to be representatives of nearly all the species that made up the primeval forest. Within perhaps half a mile of every field there are likely to be seeds or other propagules of every plant that grew on the farmed fields before the settlers came. If a farm is abandoned we can intuitively imagine it going back to the forest so that, if we left it for long enough, we should find it covered with a forest like that which the first settlers found. But experience shows that it does not go back to the primeval forest directly, for all that the seeds of the primeval forest plants are available and for all that we may believe that a good forester could plant the land to trees and start the forest coming back at once. Instead there is a long ecological succession on the abandoned land.

When the farmer quits he leaves ploughed fields behind. Next year these are covered with weeds, many of them annual plants which seem peculiarly adapted to getting their seeds into temporary spaces on the ground, but some being biennials or perennials such as the thistles and dandelions of the family Compositae whose parachute seeds also make good use of unoccupied sites. By the second summer the soil is completely covered with plants, and the perennial herbs have siezed all the available space. Grasses start to form a turf. In the next few years the turf of herbs thickens, but also woody plants, such as brambles and thorny shrubs, become established. The shrubbery grows until the field is blocked by an almost impenetrable tangle of thorny plants. Trees grow up through them, changing the field in a decade or two into a rough wood. Other communities of trees replace these trees in the succession (Figure 6.4, pages 79–85).

No farms of the Midwest have been abandoned so long that the primeval forest has returned, but the stages from old field to a forest of sorts have been witnessed many times and we can have little doubt that the original forest would be replaced in time. The exact details of the early seral stages vary from place to place, but in any one neighborhood they are quite predictable. A good botanist who knows his area can tell to within a year or two the time a farm was abandoned merely by seeing what community of plants occupies the old fields now. Such secondary successions, therefore, unlike primary successions, have been shown to occur by direct observation. Plant communities do succeed each other in a roughly predictable manner, and there is good reason to suggest that the climax should be typical of the formation of the region. The idea of ecological succession grew on these foundations until it acquired the status of a general theory.

THE HERESY OF THE SUPERORGANISM

Belief in ecological succession makes possible some heady ideas about the nature of plant communities. What is the result of secondary succession? It is to repair a wound in the old vegetation. A farmer had cut a great gash from the community and now, as soon as he had stopped his annual cutting and ploughing, the vegetation had closed the gash, had healed the wound by this intriguing process of succession. Healing wounds is a property of organisms.

A climax community does another organismic thing, too, it grows. By the xerarch succession, the climax community spreads over the inhospitable terrain of sand dunes or rocks; by the hydrarch succession it spreads where once there was open water. It is amusing to think of these growth processes as a sending out by the climax community of successive waves of colonizers, the communities of the seral stages. While Cowles was publishing his papers on the Lake Michigan sand dunes a young American botanist pondered the meaning of succession along these lines, with far-reaching consequences for ecology. His name was Frederick Clements.

Clements was born in the prairies just after the first settlement, in 1870, at Lincoln, Nebraska. The city of Lincoln had been founded only 10 years before his birth, and famous Indian fights took place while he was still a boy. He took advantage of the wilderness around him, wandering over much of the huge wild state with a team of pack mules, and writing a doctoral dissertation at the age of twenty-four on "The phytogeography of Nebraska." In his wanderings he had been particularly impressed with the idea of succession replacing the prairie turf over the ruts left by wagon trains and the migration roads of the vanished buffalo. Even though there was no forest to repair, a secondary succession was still obvious. The annual weeds and pioneer grasses formed a turf much different from that of the virgin prairie around, and only slowly gave way to it through the successive stages of the succession. Primary successions, too, seemed evident in the prairie, as sandy places or blowouts in the dry plains were colonized by pioneers which could be communities of early xerarch successions. On all sides, Clements could see the virgin prairie growing and repairing its wounds through succession. As an organism the prairie matured qnd grew (Figure 6.5, page 87).

For Clements (1916), the intellectual prize of a general classification of plant communities seemed to be won. There were a few great climax communities in the world, the formations. Climate set the boundaries to formations (as all agreed), but all other communities within the formation boundaries were but seral stages of various successions which should

Old field succession in Eastern North America. A corn field is abandoned (a) and a mixture of various weeds and crab grass (*Digitaria* sp) invade the bare ground between the stalks the same year. In the next two years a community of tall weeds (many of them in the Compositae, notice their branching and flowered tops, b) replaces the crab grass and annual weeds. Gardeners might be interested to note that crab grass is a delicate plant which will always be quickly replaced by other plants if left alone. By the third year (c) the tall weeds have been replaced by a nearly pure stand of broom-sedge (*Andropogon virginicus*, a true grass not a sedge). The broom-sedge may persist for some years as it is slowly invaded by pine seedlings (d) but eventually succumbs under the shade of the pines as they close their canopy (e). But even the pine wood is not permanent as we can tell by seeing young oaks develop among the litter of needles as in (f). In (f) notice the bulk of a large pine tree a few inches behind the oak seedling, but it is a baby oak and not a baby pine which grows there. Eventually there can be little doubt that an oak-hickory forest like those nearby would be established when this succession of events is complete. **Figure 6.4**

A

A cornfield is abandoned and crab grass moves in.

B

Compositae replace the crab grass.

C

An almost pure stand of broom-sedge follows.

D

Pine trees grow in the broom sedge.

E

A pine wood shades out the broom sedge.

F

Oak seedlings grow out of the pine litter.

lead to the areal climax or formation. And this satisfying classification apparently led to a new truth. The formation had many of the characteristics of an organism; the whole was more than the sum of the parts.

Clements was an eloquent prophet for his point of view, and he was also an accomplished field botanist, apparently much respected by foresters and farmers, so that his views on plants could be taken seriously. He wrote voluminously about succession, and invented a complex terminology to relate all the successional stages to the climax formation (Clements, 1916). He convinced a majority of American botanists to look at plant communities in his way, to treat all communities, however static they may appear, to be stages in the dynamic process of establishing the climax formation on the site. To an extraordinary degree the study of succession *was* ecology after Clements started writing about it.

The measure of the impact of Clement's writings may be seen in the philosophies of a famous contemporary, Jan Christian Smuts. Smuts was the warrior statesman of his age; intrepid leader of irregular cavalry against the British in the Boer War, prime minister of South Africa who placed his country alongside Britain in the Great War, unlistened-to voice for sanity at the Versailles peace conference, and strange religious mystic. Smuts (1926) produced a new philosophy called "Holism" in which the universe was seen as being built out of aggregations, each of which was more than the sum of its parts. The philosophy was built on the "evidence of ecology." Plant communities were more than the sums of their parts, so possibly the communities themselves were units of larger communities or holes, which were, in turn, more than the sum of *their* parts, and so on. The philosophy had some vogue. In the beginning of his book setting forth "Holism," the warrior statesman acknowledges the boy from Nebraska as supplying the original idea. This was perhaps the first abortive attempt of ecology to fasten on the popular mind as a code to live by.

Clements' views were very vulnerable. All could agree that successions led to a climax, but there seemed to be a number of different climaxes possible within the territory of one formation. Xerarch succession on Midwest sand dunes might lead to beech-maple (Clements' climatic climax for the region) but xerarch succession on old rock outcrops led to chestnut-oak or oak-hickory woods, with no indication that they would ever be anything different. On slopes of hills an oak-hickory association seemed always the end point of succession, even though beech-maple forest was only a short distance away on moist valley soils. Against these criticisms, Clements asserted that a long enough wait, perhaps thousands of years, might be required for the habitat to be sufficiently modified for the "true" climax to appear; other "climaxes" were just prolonged seral stages. This was not easy to accept and, indeed, most botanists came to believe that there could be many climaxes, (so-called **polyclimax theory**), and not just one as Clements claimed.

To objective scientists, the claim that the formation was in effect a superorganism was also unsubstantiated and disturbing. The chief critic on these lines was Gleason of the New York Botanical Garden, who

I. CONCEPT AND CAUSES OF SUCCESSION.

The formation an organism.—The developmental study of vegetation necessarily rests upon the assumption that the unit or climax formation is an organic entity (Research Methods, 199). As an organism the formation arises, grows, matures, and dies. Its response to the habitat is shown in processes or functions and in structures which are the record as well as the result of these functions. Furthermore, each climax formation is able to reproduce itself, repeating with essential fidelity the stages of its development. The life-history of a formation is a complex but definite process, comparable in its chief features with the life-history of an individual plant.

Universal occurrence of succession.—Succession is the universal process of formation development. It has occurred again and again in the history of every climax formation, and must recur whenever proper conditions arise. No climax area lacks frequent evidence of succession, and the greater number present it in bewildering abundance. The evidence is most obvious in active physiographic areas, dunes, strands, lakes, flood-plains, bad lands, etc., and in areas disturbed by man. But the most stable association is never in complete equilibrium, nor is it free from disturbed areas in which secondary succession is evident. An outcrop of rock, a projecting boulder, a change in soil or in exposure, an increase or decrease in the water-content or the light intensity, a rabbit-burrow, an ant-heap, the furrow of a plow, or the tracks worn by wheels, all these and many others initiate successions, often short and minute, but always significant. Even where the final community seems most homogeneous and its factors uniform, quantitative study by quadrat and instrument reveals a swing of population and a variation in the controlling factors. Invisible as these are to the ordinary observer, they are often very considerable, and in all cases are essentially materials for the study of succession. In consequence, a floristic or physiognomic study of an association, especially in a restricted area, can furnish no trustworthy conclusions as to the prevalence of succession. The latter can be determined only by investigation which is intensive in method and extensive in scope.

Viewpoints of succession.—A complete understanding of succession is possible only from the consideration of various viewpoints. Its most striking feature lies in the movement of populations, the waves of invasion, which rise and fall through the habitat from initiation to climax. These are marked by a corresponding progression of vegetation forms or phyads, from lichens and mosses to the final trees. On the physical side, the fundamental view is that which deals with the forces which initiate succession and the reactions which maintain it. This leads to the consideration of the responsive processes or functions which characterize the development, and the resulting structures, communities, zones, alternes, and layers. Finally, all of these viewpoints are summed up in that which regards succession as the growth or development

3

Facsimile of the first page of F. E. Clements' book on "Succession" (1916). His claim that the study of succession "necessarily rests . . . etc." though now seen to be false, and even silly, was for long taken seriously. It obstructed the objective study of plant distributions for many years.

Figure 6.5

pointed out in a series of papers (Gleason, 1917, 1926), quite as simply and forcefully written as Clements' own, that you could explain all successions and communities as being the result of the random spread and establishment of individual plants. Far from being an organism, the formation was not even ordered; it was merely the result of random process. This is essentially how a modern student of gradient analysis sees the vegetation. If the community changes gradually along an environmental gradient (Figure 2.21), then the vegetation (climax in Clements' thinking) of any one place is the sum of the distributions of all the individual species that live there.

In time, and particularly after his death, the influence of Clements' visionary thinking waned. But the idea of some organizing principle behind succession has held such a fascination for ecologists that new versions of the same theme have constantly appeared. Clements essentially considered only the establishment of a dominant climax community, but the same idea of organismic organization can be applied to diversity, biomass, activity, energy flow, and at last even information and negentropy. Even the Holism approach of Smuts seems to be resurrected in some of the less-informed ecology cults of the present day. The successive echoes of Clements appear frequently in the later parts of this book, and a reconsideration of the whole subject of succession in the light of them is reserved until near the end (Chapter 40).

We still must live with some of the more unfortunate legacies of Clements. Clementsian ecology was much more exciting than the descriptive work of the botanists of the Braun-Blanquet school or the census takers of Uppsala, and its flourishing in America during the time of the last settlement has meant that the more rigorously descriptive type of work was not attempted. The heritage of this was summed up neatly from the platform of a recent international conference by Margaret B. Davis, a student of the American vegetation of the past, when she said, "We do not know what the virgin vegetation of the pioneer days was like because all the ecologists were so busy looking for a non-existent climax that they forgot to record what was actually growing there." We should have been better off with Braun-Blanquet.

A DOUBT CAST ON
SUCCESSION THEORY

Secondary successions have often been witnessed. Vegetation on old fields really does develop from annuals to perennials to shrubs and back to forest. The stages may be somewhat erratic but within tolerable bounds the successions are predictable. They do seem to end with vegetation typical of the area, a local climax. But primary successions have not been witnessed. Their predictability, their determinism, even their very existence rests on deductive logic. Accordingly some direct historical test of a primary succession should be very interesting. Such an historical test is now available for classical hydrarch successions.

The record of infilling of an old pond should be preserved in its mud, and such mud usually contains fragments of plants that are preserved in its anoxic environment. It should thus be possible to trench or bore into

The Test of the Hypothesis of Hydrarch Succession in British Lakes   TABLE 6.1

		Succeeding vegetation											
		1	2	3	4	5	6	7	8	9	10	11	12	Σ
Antecedent vegetation	1	•	•	•	•	•	•	•	•	•	•	•	•	0
	2	•	•	•	1	1	•	•	•	•	•	•	•	2
	3	•	•	•	2	5	•	•	•	•	•	•	•	7
	4	•	•	1	•	3	•	1	•	2	1	•	•	8
	5	•	•	•	•	•	1	4	3	5	2	4	•	19
	6	•	•	•	•	•	•	•	1	•	•	•	•	1
	7	•	•	•	•	•	•	•	•	4	1	3	•	8
	8	•	•	•	•	•	•	1	•	1	1	•	•	3
	9	•	•	•	•	1	•	1	•	•	1	5	•	8
	10	•	•	•	•	•	•	•	•	•	•	4	•	4
	11	•	•	•	•	•	•	•	•	2	1	•	•	3
	12	•	•	•	•	2	•	1	•	•	•	•	•	3
	Σ	0	0	1	3	12	1	8	4	14	7	16	0	66

There were 12 possible successional communities ranging from open water (Community 1) to the driest association found near ponds (Community 12). Borings and trenches through the deposits at the edges of 66 British lakes show that almost any sequence of communities is possible. Classical hydrarch succession apparently does not occur round British lakes (from Walker, 1970).

the mud beside an old pond and reconstruct the sequence of the plant communities that occupied the site. Chose a good place, say among the willows and alders well back from the water, and your trench should pass down successively through remains of willows, reed swamp, rooted aquatics, and plankton, or whatever you expect the local communities to have been. Donald Walker (1970) has now collected together records of borings at the edges of lakes and ponds from various parts of Britain and has used them to test the hypothesis that infilling of ponds results from the actions of plant communities appearing in a predictable orderly sequence. A first part of his finding was that the fill in most of his ponds was material brought in from outside by erosion and runoff and was not autochthonous organic matter deposited by the plant communities themselves. The plants had passively responded to the infilling; they had not caused it. And his second finding was that there was no predictable sequence in which the plant communities appeared.

Walker had data from 66 sites. He identified 12 possible plant communities that can occur around ponds in the British Isles, and arranged them in what appeared to be their proper order, from 1 to 12 running from open water to a comparatively dry-site community. Then for all 66 sites he worked out the sequence in which the communities had actually appeared. The result is given in Table 6.1. A glance at it shows that there was no orderly sequence of succession. Armed only with the knowledge of what vegetation grew at any one time at the edge of a British pond it should not be possible to predict what would be growing there next. For these ponds, the hypothesis that hydrarch successions proceed in orderly predictable ways fails.

Walker came to another provocative conclusion; that the infilling of British ponds resulted not in the establishment of the local formation, but in peat bogs. Even when trees did grow, as when an alder or willow stage appeared, they did not last but were eventually replaced by peat bogs. If you could say that there was a climax to so irregular a process of infilling as Walker described, you must concede that it is an association of peat, not the local woodland.

Walker's work must shake the widespread belief in the generality of primary succession. Probably it is wiser to think that vegetation colonizes sites as the sites become modified to suit it. Plants may play some active role in the modifying, as when lichens trap soil on rocks or when dune grasses fix shifting sands, but more often the plants may passively benefit from changes caused by weathering and water transport. Discussions of successions are best reserved for those that can be shown to occur, such as the secondary successions of old fields.

Much of the distribution and abundance of life as we know it has been established in the last few tens of thousands of years of the last Ice Age and the time since it. A history of the vegetation of the earth through this period should tell us much of the climate and life of these times. Such histories were first found in the bogs of Scandinavia, thick layers of peat that collected on the gravels and pond muds left behind by retreating glaciers. Leaves and tree stumps among the peat provided a continuous history of the vegetation of the bog site throughout postglacial time, but these fossils might represent no more than the local vegetation of the bog itself as it changed in response to things such as changes in the drainage pattern. The bog peats also contained microscopic fossils of pollen grains which had been blown there from miles around, and these pollen fossils provided clues to the vegetation of a whole region. Pollen analysis from bogs all over northern Europe showed that changes in vegetation had been everywhere parallel. A climatic history for postglacial time was deduced, and pollen analysis became a stratagraphic tool for geologists, climatologists, and archaeologists. The invention of radio carbon dating, however, freed pollen analysis from this service, allowing ecologists to reconstruct the vegetation from samples of known age. The technique could be used to answer ecological questions about the past such as the true history of past vegetation or the environment in which various animals evolved. Pollen analysis has been used, for instance, to resolve the controversy over how the vegetation of the southeastern United States was affected by glacial advances in the north, and the conditions for life of the ancestors of American Indians who crossed into America by the Bering land bridge.

CHAPTER 7

HISTORY OF THE
FORMATION AS AN
ECOLOGICAL TOOL

The work of the plant sociologists established with great clarity that vegetation is a subtle and accurate indicator of the physical environment. Where boundaries between associations seemed distinct, it was always found that physical habitats changed. As vegetation graded from one kind to another, so the environment graded also. And the limits to the grandest vegetation units of all, the formations, were set by climate, the major arbiter of environment. It follows that reconstruction of the vegetation of the past from fossil traces allows reconstruction of the environment of the past.

The value of using fossils for reconstructing environmental histories has long been known to geologists, but knowledge of the climate of the past is also valuable to evolutionists and ecologists who must understand adaptation as an historical process. Records of past vegetation should also serve to settle some of the arguments of plant sociology itself: "Have climax formations really evolved over the centuries through the long succession we infer?" Reconstructions of fossil vegetations should provide answers to questions like this.

For the remoter periods of the earth's history, the knowledge of modern ecology is hard to apply, since the species that inhabited the earth are now all gone so that we cannot call on living specimens for an account of the environments to which they were suited. We are reduced to crude inferences such as "webbed feet mean water." But from the time of the Pleistocene Ice Ages we are on safer ground for we are then presented with fossil traces of species that still exist. And it was out of the conditions of the last Ice Age that much of the distribution and abundance of modern life was fixed, including, of course, the emergence and spread of human societies. Our knowledge of these crucial few tens of thousands of years of environmental history is largely based on reconstructions of the vegetation of these times. The story starts with observations in the late nineteenth century on the bogs of Norway and Sweden.

Peat bogs are common in northern Europe, where they have long been drained and mined for fuel. They usually rest in depressions in the bottoms of which are sand and gravels left behind by retreating glaciers. Ponds formed in those early depressions, leaving deposits of pond mud, then various forms of sedge and moss peats collected, and eventually perhaps a raised peat bog grew up and out of the original hollow. There are many places where a botanist can go and look at a face of drained peat, perhaps four times as high as himself, knowing that he looks at a fossil history of the vegetation of the hollow which spans all the thousands of years since the last glacier withered away. And in such bogs in Scandinavia could be seen layers of tree stumps; suggesting an ancient forest that later went away again. What did these traces of ancient forests mean? The obvious answer was "climatic change," that there was preserved in peat bogs the fossil evidence of vegetation changing in response to changing climate. But the habitat of a bog can change for purely local things, blocking or unblocking a drainage channel, for instance, so that it was necessary to compare the fossil evidence of many sites over large areas in order to substantiate an hypothesis of climatic change. This was done for Scandinavia by Axel Blytt and Rutger Sernander, who compared many bogs layer by layer, and constructed the celebrated Blytt-Sernander chronology for postglacial time (Sernander, 1908; Flint, 1971; West, 1968).

Under the bottom layer of stumps was marshy peat, as you would expect from a puddle in a hollow. But included among the leaves and seeds of sedges were leaves of the little arctic plant *Dryas,* a small white herbaceous rose with a yellow center, a plant exclusively of the arctic tundra. This bottom layer thus represented vegetation of what could be called an **arctic period.** Higher up was the first line of tree stumps, commonly birch but sometimes other trees too. It seemed certain that a bog must have been partially drained to allow trees to root, so we infer something that dries up the bogs of a whole continental isthmus as big as Scandinavia, climatic drought. The dry time becomes the **boreal period.** But there is peat between the *Dryas*-bearing bottom mud and the first line of stumps, a gradation probably, but one that could be used as a stratigraphic unit. It represented the **subarctic period.** Then more wet peat overlying the tree stumps, flooding doubtless, a wetter climate, the **Atlantic period.** Then another line of stumps, commonly pine-tree stumps this time, another dry time, the **subboreal period.** Then the modern peat which continued to the treeless top of the bogs, the modern climatic epoch, a wet time, the **subatlantic period.** In this way (Figure 7.1) Blytt and Sernander used plant fossils to infer climate, and then to separate postglacial time into a series of discrete climatic periods.

From the first moment this scheme was put forward there was room for doubt. What did we really have other than a record of water levels in bogs? When the bogs dried a little, trees grew on the tops of them; when they flooded again, the trees were drowned. A climatic explanation certainly seemed necessary to synchronize the water levels in bogs all over Scandinavia, but there was nothing in this to show that climate was divided into discrete epochs. All we really had was evidence that the bogs had dried out twice, slowly each time, and enough each time for trees to flourish for a generation or two. There was quite a leap from this to the impression of climatic succession given by Figure 7.1. This was forcibly pointed out to Blytt and Sernander by their friends, notably their fellow Scandinavian, Gunnar Andersson (Faegri and Iversen, 1964).

It was clear that the vegetation of a bog would always largely depend only on the water level of the bog itself. If you wanted to infer the climate of a whole region, you must surely have a record of the vegetation of that region, not just of bogs; and, if you wanted to test the hypothesis that there had been a succession of distinct climatic epochs in Scandinavia since the last ice age as Blytt and Sernander suggested, you must have a record that was continuous in time. Such a record must have seemed a very tall order until Lenard von Post, State Geologist of Sweden in 1916, invented pollen analysis. Von Post found that pollen grains of plants were preserved in peats by the hundreds of thousands, that they could be iden-

Climatic Periods
The Subatlantic Period.
Climate humid, and, especially at the beginning, cold.
The Subboreal Period.
Climate dry and warm, much as in central Russia.
The Atlantic Period.
Ciimate maritime and mild, probably with warm and long autumns.
The Boreal Period.
Climate dry and warm.
The Subarctic Periods of Blytt.
The climatic conditions more or less undetermined.
The Arctic Period.
In Scandinavia a climate like that of South Greenland.

Figure 7.1
The climatic sequence constructed by Blytt and Sernander from bog profiles in southern Scandinavia. Tree stumps define the boreal and sub-boreal periods, and leaves of arctic plants like *Dryas* the arctic period. The Atlantic and sub-Atlantic periods have peat of relatively moist and warm times. (From Sernander, 1908.)

tified, and that they came from miles around. Far from merely holding a record of their own water level, peat bogs actually preserved a quantitative sample of the vegetation of whole regions.

Over any plant community in spring and summer there hangs a cloud of pollen, undetected, save by sufferers from hay fever, yet composed of an almost unthinkably vast multitude of tiny drifting grains, gently settling as a pollen rain. Pollen grains from different plants do not look the same so that it is easy with a microscope to tell one genus from another (Figure 7.2), although, alas, not one species from another. Pollen that settles on waterlogged places is preserved, for the outer coat of pollen grains is made of one of the most resistant materials produced by living things. The pollen contents quickly rot but the outer shell may persist in a bog indefinitely, for thousands of years, and hundreds of thousands of years if the bog should last. We find them still preserved in sedimentary rocks millions of years old from the Tertiary epoch. The Swedish bogs are only about 10,000 or 12,000 years old at their oldest, and the pollen embedded in their peat looks as fresh as the pollen of a year ago. There may be many thousands of grains in one cubic centimeter of peat, a fine quantitative sample of ancient vegetation from which to reconstruct the community. It is a simple, although time-consuming, matter to sort a few thousand grains from the peat, destroying the organic matrix with sodium hydroxide and a sulphuric acid-acetic anhydride mixture, and dissolving minerals with hydrofluoric acid. Sometimes it helps to float pollen out from a mineral matrix on a heavy liquid such as a bromoform-acetone mixture, but with patience in applying routine methods it is always possible to make a pollen concentrate of thousands of grains which can be spread beneath the cover slip of a single microscope slide (Faegri and Iversen, 1968; Erdtman, 1969). There, on the microscope stage before you, is a quantitative sample of the vegetation of, perhaps, several square miles of country in some distant time. It is a comparatively simple task to traverse the microscope slide at a magnification of 400 diameters, and to identify and tabulate every pollen grain you come across until you have found enough to work out the percentage composition of that ancient pollen cloud.

Von Post and his many successors did this for the Swedish bogs, and then for the bogs of all northern Europe, taking small samples of peat at intervals of about 10 centimeters all up those imposing faces of peat, and producing pollen diagrams (Figure 7.3) to show how the composition of the pollen cloud changed with time. The first thing that seemed apparent was that you could use the record to defend the Blytt-Sernander hypothesis of a series of climatic episodes. In Figure 7.3, which is a synthetic composite diagram for a large part of Norway, the Blytt-Sernander episodes are labeled at the left, and their supposed boundaries (the lines of stumps, and the like) are drawn across the diagram. The pollen spectra do seem to be naturally collected into zones that roughly coincide with the climatic episodes, for all that there is gradation and overlap. The second obvious thing was that the vegetation description given by the pollen was very incomplete. There are many hundreds of species of plants in the vegetation of Norway, but only 20 names appear in Figure 7.3. Partly, this is because we identify pollen only to genus, but more important is the fact that a few of the Norwegian plants, most of them trees, produce great masses of pollen while others produce very little. Temperate forest trees are mostly wind pollinated, but many of the herbs are insect pollinated and thus need to produce much less pollen. Furthermore, the trees release their pollen high in the air, whence it can be blown about, whereas herb pollen has only to drop a few inches to be caught on the soil. The proportions of the pollen types in a pollen cloud are thus not the same as the proportions of plants in the parent vegetation. But for all this the suggestion of different vegetation types coinciding with the proposed climatic episodes is still strong.

The support that pollen analysis offered to the Blytt-Sernander scheme was, of course, of great interest to geologists and climatic historians. But it had perhaps a more profound effect on plant ecologists. Long-continued episodes of climate each had their characteristic vegetation. When there was climatic change, then there was swift vegetation change also. This

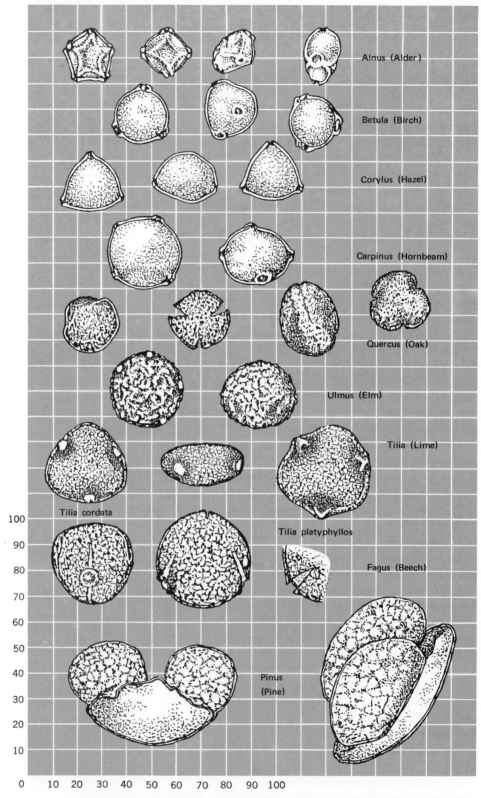

Figure 7.2   Pollen of European tree genera drawn to a common scale. The scale is in microns, which means that the 100 divisions shown represent one-tenth of a millimeter. The largest pollen grain in the figure, the pine, would be just visible to the naked eye as a minute speck. Notice how distinctive are the grains of different genera. Even when built on a common plan, as are the 3-pored grains of birch and hazel, there are distinctive differences (between these two

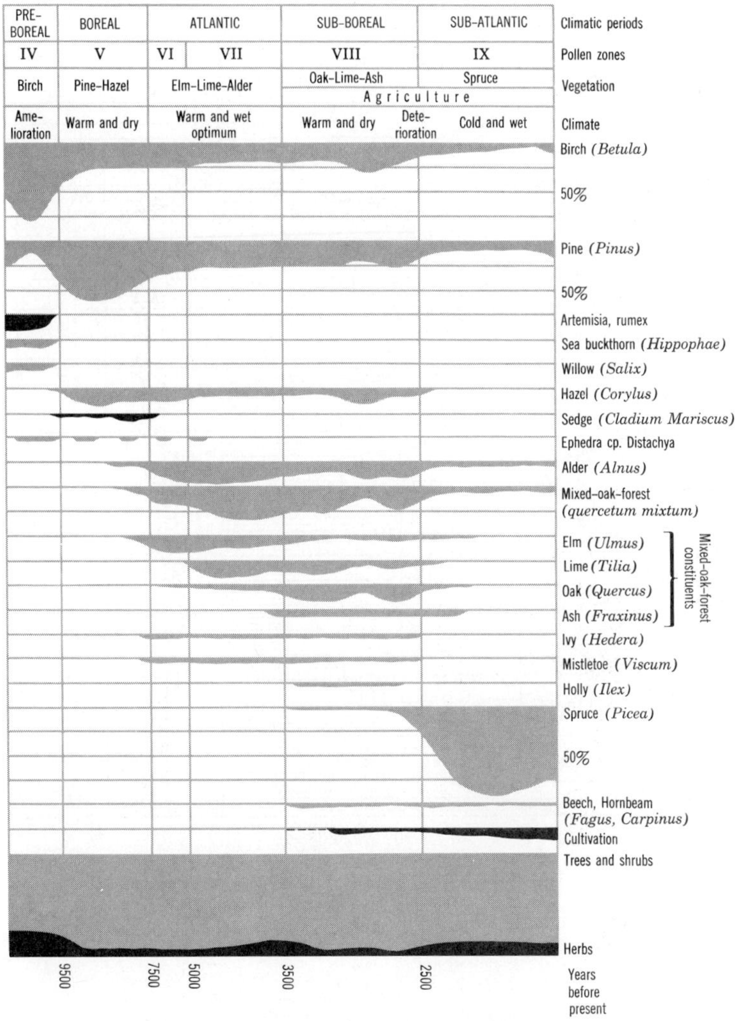

PRE-BOREAL	BOREAL	ATLANTIC		SUB-BOREAL	SUB-ATLANTIC	Climatic periods
IV	V	VI	VII	VIII	IX	Pollen zones
Birch	Pine-Hazel	Elm-Lime-Alder		Oak-Lime-Ash	Spruce	Vegetation
				Agriculture		
Ame-lioration	Warm and dry	Warm and wet optimum		Warm and dry	Dete-rioration Cold and wet	Climate

Figure 7.3

Schematic pollen diagram from bogs of Southern Norway showing the correlation between Blytt-Sernander climatic periods defined by bog stratigraphy and zones of pollen spectra. The pollen are shown as percent total arboreal (tree) pollen, and each division represents 25 percent. The few herbs in the total count (non-arboreal pollen or NAP) are shown in darker shading. It is common practice to show the relationship of herb to tree pollen in a separate part of the graph, as is done here at the right of the diagram. Although the diagram shows that the pollen zones grade across the arbitrary climatic boundaries, there is yet a considerable suggestion of a sequence of fairly well-defined intervals in the pollen diagram. The pollen data do encourage the impression of a series of alternating climatic episodes, even if they do grade one into another. (After Hafsten, redrawn from Faigri and Iversen, 1964.)

---

the shape of the pores) with which the analyst quickly becomes familiar. But it is seldom possible to distinguish between species within a genus. Alder grains, for instance, may have 4 or 5 pores (as shown) or even 6 or 7, yet there seems to be no correlation between the number of pores and the different species of alder. Sometimes you are lucky, however, and two species of lime can be told apart by the shape of their pores, as shown. (Redrawn from Godwin, 1956.)

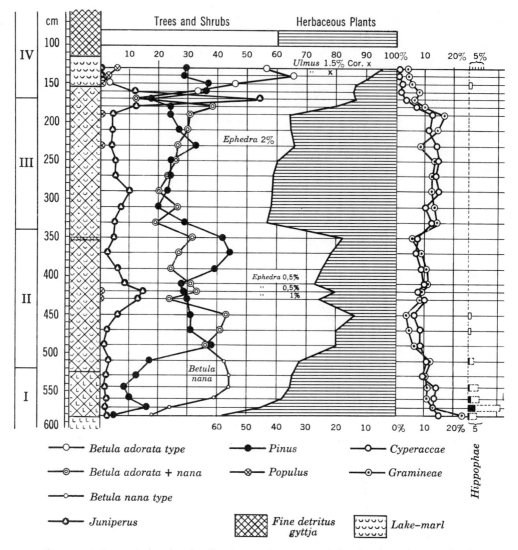

Figure 7.4 Composite lateglacial pollen diagram from Denmark showing the earlier periods added to the Blytt-Sernander sequence. Zone I is the Older Dryas, zone II the Allerød, and Zone III the Younger Dryas. The Allerød mild period shows as a blip in the juniper curve together with a loss of pollen of tundra plants such as dwarf birch *Betula nana*. The section ends with zone IV, the Preboreal period in the Blytt-Sernander scheme and the base of Figure 7.3. The composite diagram showing the general relationship between trees and shrubs is shown to the left of this figure and it reveals that herb pollen was more important than in the later zones of Figure 7.3. These were times just after the glaciers had gone when trees were beginning to colonize the tundra. The right hand part of the diagram shows herbs only except for willow (*Salix*) which is assumed to be a creeping tundra plant and not a heavy producer of pollen. Most of the herbs listed do not have familiar English names but Cyperaceae are sedges, Gramineae grasses, *Rumex* docks, and *Urtica* stinging nettles. (Redrawn from Faegri and Iversen, 1964.)

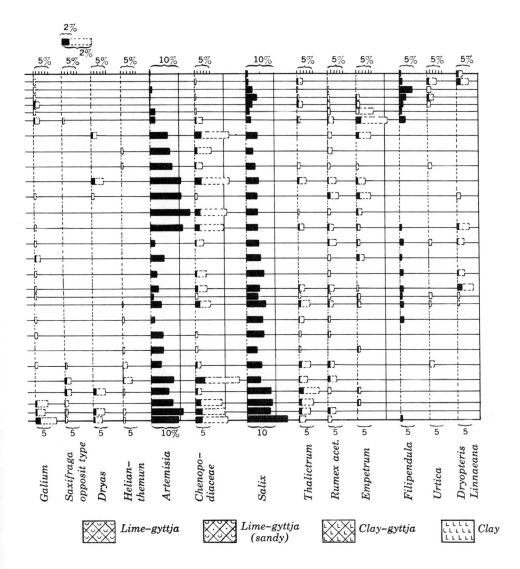

Lime-gyttja   Lime-gyttja (sandy)   Clay-gyttja   Clay

was what had been predicted by students of succession and climax. To them it seemed that the pollen analysts had clearly shown that a climax was established in response to climate, and that this climax remained unchanged unless there was further climatic change. In association with the geographic evidence, the pollen record left little room for further doubt that climate determined vegetation.

There were, it is true, some uncomfortable suspicions of circular reasoning in this conclusion. Climatic historians used the vegetation record to assert their claims of climatic episodes; then plant ecologists used those same claims for climatic episodes to say that the vegetation had been influenced by climate. But it was not actually quite as bad as this, since there were various independent clues to the past climates to

break the circle. The changing water levels of the bogs provided one independent clue, and stratigraphic correlations with the advance and retreat of glaciers provided another. Then the record was made more
striking by being extended further back in time: the ancient **arctic period**
of the Blytt-Sernander system was found to include yet another climatic
episode, the celebrated **Allerød oscillation** (Hartz and Milthers in
Godwin, 1956). Deep in a deposit near the village of Allerød in Denmark,
the **arctic period** with its included *Dryas* leaves was represented by clay,
but this clay was interrupted by a layer of organic mud. The local evidence was clear; arctic times had ended with warming, represented by
the organic mud, then the grip of the cold had returned resulting in the
burial of the organic mud by frost processes (so-called **solifluction**). An
extra two periods had been added to the bottom of the Blytt-Sernander
scheme thus:

> Subatlantic
> Subboreal
> Atlantic
> Boreal
> Preboreal
> Younger Dryas ⎫
> Allerød   ⎬ the old Arctic Period
> Older Dryas   ⎭

making eight episodes in all. The Allerød climatic episode was then
sought in pollen diagrams all over Europe, both from the bottoms of bogs
and from the sediments of lakes. And a pollen equivalent, commonly a
blip in pine or spruce spectra, was widely found (Figure 7.4). The episode
was also tied in to a glacial retreat, during the **Allerød,** and a readvance in
younger Dryas time. Here then was climate and vegetation determined
independently for the same event. And the climate had set the vegetation,
not just locally, but for all northern Europe. There seemed to be little further doubt that climate did indeed control vegetation, that in a very general sense, it determined a climax state.

  After this initial triumph for ecology of European pollen analysis, the
method fell very largely, and for many years, into the hands of geologists
and archaeologists as a stratigraphic tool, as a method of dating samples.
Any polliniferous sample from northern Europe could be aged or, at least,
put into its proper stratigraphic position, by looking at its pollen. At the
same time, a rough account of the climate could be made. Pollen men all
over northern Europe had identified all eight of the Blytt-Sernander episodes, and had even added an extra one and sometimes two of their own,
making a nine or ten pollen-zone sequence which they named in Roman
numerals (Figures 7.3 and 7.4). They could thus be asked by an archaeologist to say into which pollen zone the sample from a dig fitted, and the
archaeologist would then use the pollen zone given him as an "age" for
his sample, and as an account of environmental conditions to boot. This
approach was immensely useful to archaeologists and geologists, but it

weakened the usefulness of pollen analysis as an ecological tool. An ecologist looking at the past wants to be told what the environment is like by the students of some other discipline, and he wants the same sample independently aged. Then he can use his fossil traces to see how the vegetation of past times reacted to its environment. But the geologists and archaeologists wanted him to deduce both climate and time from his fossil data, and the two lines of reasoning, the ecological and the stratigraphic, are mutually exclusive. Fortunately this dilemma was ended with the invention of radiocarbon dating, which freed pollen analysis from its bondage as a stratigraphic tool.

The vegetation of much of Europe was decimated by the continental glaciers, ground down between ice sheets advancing southward from Scandinavia and northward from the Alps. As a result, the flora of Europe is impoverished. The vegetation episodes studied by the European pollen analysts result from long distance dispersal or were developed from the scant survivors of the ice which persisted in small refugia. But more interesting possibilities for plant ecologists are offered by the history of the vegetation of eastern North America.

DEVELOPMENT OF EASTERN AMERICAN VEGETATION, AND THE LIMITATIONS OF THE POLLEN METHOD

Eastern North America was beset by ice sheets only from the north and locally from mountaintops. South of the ice sheets the rich native vegetation, far richer in species than that of Europe, should have persisted subject only to climatic change. And climatic change there must have been, the continental ice sheets were evidence for that. A cold strip along the glacier front in which life was affected by cold winds off the palisade of ice was at least probable, surely an arctic sort of place? And the vegetation further south should have been affected by whatever worldwide climatic changes caused the glaciers themselves. Did the complex vegetation of America respond to these different climates of the past as associations, recognizable from their pollen in terms of modern associations? Or was past vegetation built up by immigration and the establishment of individuals to suit the unique conditions of the time so that the past associations were unique also? A long pollen record from eastern America should thus provide a further test of the claims of plant sociologists. But there was yet another possibility; that the vegetation of North America south of the ice margin was hardly affected by the glacial episode, that it existed much as it does now and presumably had done since the Tertiary epoch when it had evolved, in equilibrium with the conditions of its latitude, oblivious of climatic changes far to the north which sent the glaciers down to its very borders.

So there were three possible histories for the eastern American vegetation:

1 That it was essentially unaffected by the glacial events.
2 That it migrated south as complete associations or formations so that the latitudinal zoning of modern America was essentially intact but displaced to the south.
3 That plants responded to changing climate as individuals so that the

different climates of the last Ice Age resulted in plants being associated together in combinations not found in modern America.

Pollen analysts could tackle the first of these alternatives most easily. It was put forward by Lucy Braun (1950), who concluded from studies of the vegetation itself that the unique formations of eastern America, with their many species and their ancestries far in the Tertiary, were too ancient and too complex to have been shifted around as belts when the ice sheets came and went. Braun postulated that the glaciers bulldozed their way into the edge of the forest, causing local disturbances beyond the reach of the ice, but no more. This closeness between advancing ice and rich vegetation is not without precedent elsewhere, since it can be seen in places such as New Zealand where tropical plants like tree ferns grow within yards of an advancing valley glacier. The hypothesis was plausible, but any long pollen record from a critical region should test it. Such records are now available from, among other places, North Carolina, due to the work of Frey (1953) and Whitehead (1964, 1965). Figure 7.5, a composite pollen diagram of Frey's work, shows the completeness with which pollen analysis disposes of the hypothesis. Radiocarbon dating shows that Frey's history spans from sometime during the last Ice Age and on into postglacial time. It is not necessary to attempt an actual reconstruction of vegetation to see that the changes must have been very great during the time covered, even when so far removed from the ice as North Carolina always was. It is apparent that vegetation is not so hardily organized that it can resist climatic events, nor that it is difficult for seemingly ancient vegetation to be constructed in a time span of a few millennia at most. The Braun hypothesis fails.

It is more difficult to distinguish between the second and third hypotheses, to decide if vegetation moved in belts or if the associations of the past would seem entirely novel to a modern botanist. The reason for this difficulty lies in the limits of the pollen method itself.

A test of the hypothesis that vegetation has migrated intact, as zones of familiar vegetation keeping rough pace with the advance or retreat of climate, involves finding a polliniferous site in the path of the supposed vegetation advance, and then reconstructing the vegetation of the past time from a pollen diagram. Now it may be easy to show that vegetation differed in the past, but it is another matter to show exactly what is was like from pollen data. There are a series of difficulties with the method that tend to compound to obstruct you.

A first difficulty with the sorts of sites that were initially used for pollen analysis, sections in bogs, is that much of the pollen you find was produced on the bog itself. There may be what a pollen analyst calls **over representation** of pollen of bog plants. You want not a history of bog plants but a history of the vegetation of a broad area, yet the intense rain of local bog pollen may make you see the other pollen "through a glass darkly." The simplest way out of this difficulty is to abandon bogs in favor of lakes. A lake, say one-half mile across, traps pollen from the local vegetation just as well as a bog, but it does not obscure the record with terres-

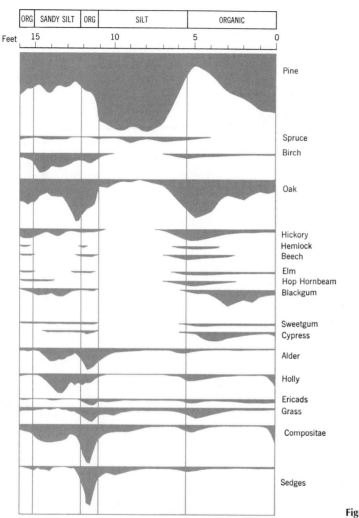

| ORG | SANDY SILT | ORG | SILT | ORGANIC |

Feet 15 · · · 10 · · · 5 · · · 0

Pine

Spruce

Birch

Oak

Hickory
Hemlock
Beech

Elm
Hop Hornbeam
Blackgum

Sweetgum
Cypress

Alder

Holly

Ericads
Grass

Compositae

Sedges

Figure 7.5

Generalized pollen diagram for Bladen County, North Carolina. Radio-carbon dating shows that much of this diagram was contemporary with glacial time, and that the top few feet represent the beginning of the postglacial period. Even without being able to deduce the exact vegetation represented by the various pollen spectra, it is obvious that there have been marked changes in the local vegetation over the time span covered. This sort of evidence seems incompatible with the Braun hypothesis that eastern American vegetation in Ice Age times was much as we see it now. (From Frey, 1953.)

trial pollen of its own. If a few water-plant pollen grains are added, these are easily identified and ignored. Thanks to the invention of piston samplers (Livingstone, 1954), lakes are actually easier to core than bogs (unless someone has drained and sectioned your bog like so many of the European ones). Lakes are now preferred sites for seeking pollen records, but they can present difficulties of their own. Mud may slump from steep sides underwater, thus mixing your sample. Burrowing larvae, such as tendipedids, may mix pollen through the sediment to a depth of about 10 centimeters (R. Davis, 1967), and some big pollen grains may be blown about on the surface before they become waterlogged, and concentrated at the edges. But these problems are only minor compared with the advantage of not having the pollen record of a region obscured by heavy pollen production at the site.

Much more serious is the problem posed by the statistics of percentages. The usual pollen method is to take a pollen sum of between 150 and 300 grains, then to express the content of each type as a percent of that sum. This means that a change in the actual number of pollen grains of any one taxon can affect the apparent proportion of all the rest. There are as many variables as there are pollen types, and a change in any one affects the percentage composition of all the others in the pollen diagram. The dilemma is apparently insoluble by statistical tests unless very large pollen sums are counted, in the tens of thousands of grains per interval, for which task a pollen analyst's life is too short. We have to accept that there can be an error of 5 or 10 percent in the apparent percentage of any one taxon. A conventional pollen diagram, in short, leaves us with worrying doubts about the actual changes in numbers of the parent plant implied by quite wide changes in pollen spectra. There is a way round this though, if a time-consuming one that cannot always be applied. We can avoid percentages all together. We count the number of each kind of grain found in unit volume of sediment and then calculate the sedimentation rate by a lavish use of expensive radiocarbon dates. This enables us to express the history of each taxon in terms of the number of grains falling on unit area of sediment surface each year, an absolute statistic which lets you study the fate of each taxon through time independently of its neighbors (M. Davis, 1963). There are as yet very few such diagrams available, and they still leave you with formidable difficulties of interpretation, for they do not help with the next problem, that different kinds of plants produce very different amounts of pollen.

Temperate forest trees, what pollen analysts of Europe and North America mostly work with, produce much pollen, many of them probably in comparable amounts. Which is why conventional pollen diagrams seem to make sense. The diagram shows you a mixture of taxa given the names of forest trees, and the mixture of tree names seems plausible when compared to a forest of real trees. But the suggestion of a familiar forest in the pollen diagram may be misleading, because some trees may produce much more pollen than others. Pine trees, for instance, are notorious for their copious production of pollen. If you park your car near a pine wood in spring you may find it in the evening colored yellow by a layer of pine pollen. By contrast the pollen production of maples or willows may seem tiny. A thought which haunts the dreams of a pollen analyst is that the 10 percent of pine pollen in his diagram may mean just one pine tree growing very close to his site. But, in fact, the analysts of temperate lands are fortunate in having to cope with no worse than this. Pollen analysts are now seeking their first long records from the equator, and have there come up against the fact that many tropical trees are insect pollinated and so produce very little pollen. The pollen cloud from tropical woodlands and savannah may be mostly herbs, even grass (Livingstone, 1967, 1971). This reveals rather starkly the difficulty of reconstructing vegetation from the pollen record.

The obvious way round this problem of differential production is to try to allow for it with conversion factors. You find, for instance, how much more pollen is produced by a typical pine tree than is produced by a typical oak and use this figure to adjust your pollen counts. But the difficulties are very great. Not only is the task of comparing the pollen productions of all the species in your pollen spectra daunting, or even next to impossible, but you must also allow for the differential dispersion of the pollen. If some blows further than other kinds, you will get more of it in your lake no matter that the parent tree may not have produced so much of it. A start has been made to tackle some of these problems (M. Davis, 1963), but we have not progressed very far yet.

A better way round the difficulty of differential production and dispersion seems to be that of taking "surface" samples, of trying to estimate the pollen cloud from modern vegetation and then comparing the fossil pollen spectra in the diagram with the modern spectra. If the spectra are the same, then you say that the ancient vegetation was like that which gave rise to the modern pollen spectrum, even though there might be no obvious similarity between the makeup of the modern pollen cloud and the percentage composition of trees in the modern forest. This is the best practical tool we have, even though the imprecisions are obvious. A vegetation of hundreds of species may be represented by, perhaps, a dozen pollen types in the spectrum. Even if the spectra match beautifully, it cannot mean that all the hundreds of species were there in the modern proportions. And there is still that 5 percent error inherent in the percentage method which might mean that your beautiful match is a lie. On top of these difficulties is the fact that what should have been the "modern" vegetation with which you will compare your ancient sample has been ploughed under for a century or two. We do have some likely surface spectra from eastern North America, but not very many (Ogden, 1969).

In spite of these difficulties of interpretation, it seems that, although all the North American vegetation reacted profoundly to the climatic events of the Ice Age, there was no general orderly retreat in latitudinal zones (Whitehead, 1965). Comparison of pollen diagrams from glaciated New England (Figure 7.6) and from the Dismal Swamp of Georgia (Figure 7.7) shows the sorts of evidence on which this is based. The New England diagram shows a postglacial succession somewhat analogous to the pollen sequences of northern Europe: a boreal start to the diagram with conifers predominating in the early days (although some diagrams show a short interval of treeless times comparable to tundra immediately after the glacial recession); then a development of broad-leaved forest which, with minor changes, persists until the English arrive with their axes. The coniferous trees seem to represent as much a stage in a succession as the pass-

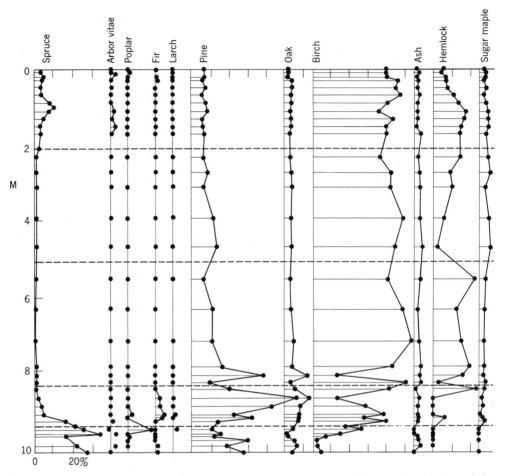

Figure 7.6   Pollen diagram from the mud of a pond in Vermont. This provides a history of the postglacial vegetation in eastern North America on a site that was once under glacial ice. A small bottom section, probably with the pollen of tundra plants as occurs in many New England diagrams, is missing. After this is a time of coniferous trees, of spruce and pine. This possibly represents the invasion of open ground by conifers as a stage in succession to hardwood forest, although the successional stage was prolonged. The divisions A, B, and C to the right of the diagram are the pollen zones for New England set up by Deevey (1939), letters being used to avoid confusion with the Roman numeral system of Europe. Percentages are calculated as percent total terrestrial pollen. (From M. Davis, 1965.)

ing through of a boreal forest latitudinal belt. In the early postglacial days at Dismal Swamp there are again clear signs of northern conifers like spruce (*Picea*) being present, but these occur with a different set of bedfellows. The record suggests, in short, that the colder times of the Ice Age required a resorting of the species of the American forests, so that combinations quite unlike those of the present day existed.

Pollen analysis thus seems to show that associations are not permanent entities. They represent temporary pacts between species, not permanent

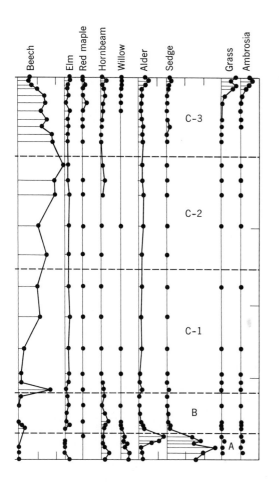

alliances. The views of Gleason must prevail over the views of Clements (Chapter 6). Although we go on to study ecosystems as functional units, we must always remember that these systems are fleeting entities built from mutual adjustments of individual species. The system can last only as long as the physical conditions of life are not altered.

Pollen analysts have often been called in by archaeologists to "look" at samples from their digs. The purpose has usually been dating, to use pollen analysis as a stratigraphic tool as the European sequence was for so long used. Sometimes pollen analysis is still the best answer for this, although radiometric methods are better if possible. Also stratigraphy seems to an ecologist to be a rather dull use of his tool. More exciting is the quest for man in the pollen diagrams themselves.

The development of forest clearing and shifting agriculture in Europe is beautifully shown in many diagrams, an event known as **landnam,** the Danish word for "land-taking" used for the phenomenon by Iversen (1949) who first spotted the significance of blips in the tree pollen curves,

POLLEN ANALYSIS AND THE RECORD OF MAN

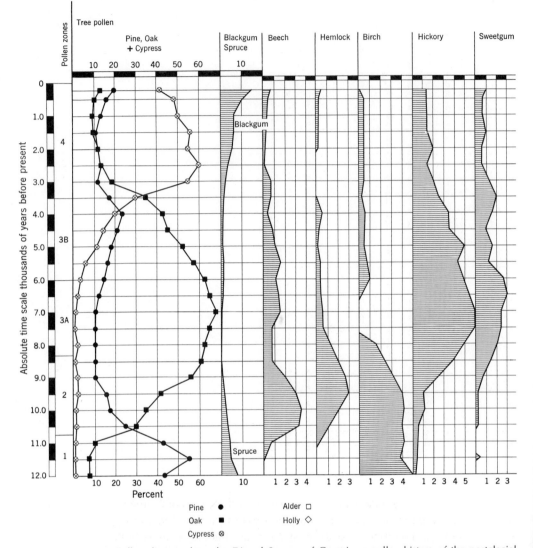

Figure 7.7   Pollen diagram from the Dismal Swamp of Georgia, a pollen history of the postglacial period at a site that was always south of the ice sheets. Although spruce appears at the bottom of the diagram the spectra of these early times are not really comparable with spectra further north (Figure 7.6). The comparison suggests that the vegetation of Georgia in late glacial times was different from any that can be seen in modern America. Pollen from some other plants such as grasses were found in the swamp but were not included in this count. (From Whitehead, 1965.)

coupled with upsurges of the pollen of agricultural weeds such as nettles and plantains. A landnam event has been found in many parts of Europe, everywhere first dated at about 5000 years ago and suggesting that men quickly learned the new ways from each other throughout the whole continent. Figure 7.8 shows the very clear example of a landnam phase in Ireland, taken from the work of Smith (1961).

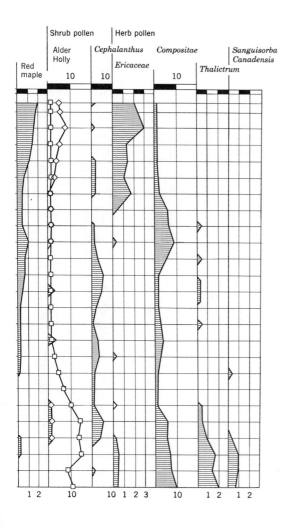

Elsewhere pollen seems to offer the first evidence for ancient agriculture. A record from Taiwan shows a disturbance 8000 or more years ago which might well have been the start of agriculture (Tsukada, 1966), although the interpretation still suffers from the difficulties to which pollen diagrams are always suspect. If correct, this is the oldest record of agriculture in Asia. There are also alluring possibilities for pollen analysts in trying to reconstruct the climates in which our ancestors lived. One of these concerns the colonization of America by the ancestors of the American Indians.

The first Americans were those whom Columbus mistakenly called "Indians." Anthropologists have long thought them to be people of Asiatic stock who had come from Asia long ago, certainly more than 14,000 years ago as undisputed radiocarbon dates show. All are agreed that there is a very high probability that they came by way of Siberia, the Bering

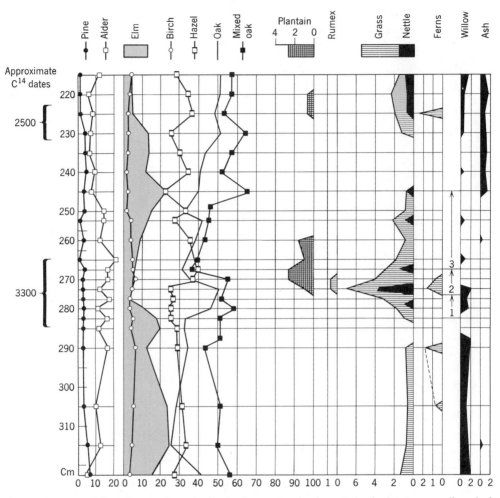

Figure 7.8 Pollen diagram from a mire in County Londonderry, Ireland, showing a well-marked *landnam* phase. The *landnam* event occurred between the arrows. There is a large increase in plantain and nettle pollen at a time when elm pollen fell sharply. Elm is not a heavy pollen producer, so it is likely that the forest before the *landnam* event had many elm trees in it. Men cut down the elms, grew some crops, and let the weeds grow; which doings have left a beautiful trace in the pollen record. (From Smith, 1961.)

Strait, and Alaska. We do not have definite proof of this, but there is much circumstantial evidence for this route and alternative routes involve migrations across the great oceans in Stone Age boats. A few adventurers might have made such journeys, but for the migration of whole peoples the Bering Strait route makes much better sense. And the Bering Strait more than 14,000 years ago was dry land (Hopkins, 1965).

More than 14,000 years ago takes us into the time of the last Ice Age, and at such times the level of the world oceans falls as water is continually taken from the sea to build glaciers. A considerable portion of the

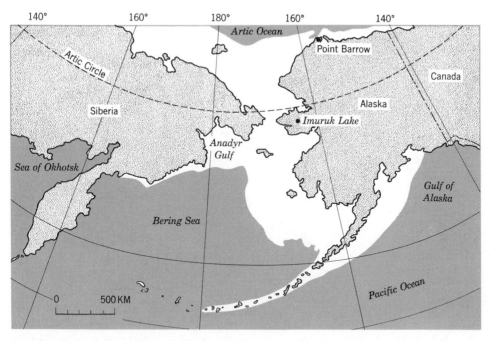

Map of Bering Sea region at the time of the last glacial maximum, showing the coasts of the Bering land bridge. This was the land exposed when sea level fell by 100 m. (Redrawn from Hopkins, 1965.)   **Figure 7.9**

oceans is thus held in cold storage on the land, enough during the last Ice Age to lower sea level by more than 100 meters. This drained all of the Bering and Chuchi seas on either side of the Bering Strait, causing Alaska to be fused to Siberia by a land bridge some 800 miles broad (Figure 7.9). This land was not under ice, presumably because there had been too little precipitation in the arctic for ice sheets to form even when they spread all over the continents to the south. And we know, from fossil finds and from the evidence of geographical distribution, that many animals lived on the Bering land bridge. Among them were the ancestors of the bison of the great plains, horses and the long extinct mammoths.

The ancient land of the Bering land bridge, now washed under the waves of the Bering Sea, is the probable original home of the first Americans. Their descendants included all the Indian Peoples of Canada and the Western Plains as well as those who built the civilizations of the Inca, Aztec, and Maya. But the first Americans occupied the Bering land bridge in glacial times. What was the environment of that ancient place like? Could primitive men, the half-naked savages sometimes conveyed by the words "stone age," have lived there? Or was the climate such that the ancestors of all the American Indian Peoples must have been culturally advanced? An answer to these questions could be had if a pollen sequence from a lake of the ancient land bridge were found. Such a lake, called

Figure 7.10   Imuruk Lake on Seward Peninsula, Alaska. The shrub tundra in the foreground consists of a turf of sedges, heaths, dwarf birches, and prostrate willows. The tallest plants in the photograph are willows which grow best along drainage channels and, as here, at the side of the lake. White dots in the left foreground in front of the reindeer are the fluffy seed heads of cotton grasses. Pollen records from the lake mud suggest that this tundra is the least arctic during the lake's history of tens of thousands of years.

Imuruk Lake, exists on Seward Peninsula, the part of Alaska that points like a finger to the Bering Strait (Figures 7.9 and 7.10). A geologist, David Hopkins (1959), surveyed the lake and realized its importance. It sat high on a lava flow, on top of a rise of land where the rate of infilling should be very slow, and was probably more than a 100,000 years old. I went and drilled its sediments from a rubber boat, and through the winter ice, obtaining the coveted record of land-bridge pollen (Colinvaux, 1963, 1964).

Figure 7.11 shows the top portion of the Imuruk Lake pollen diagram, a portion which spans more than 30,000 years. Imuruk Lake is surrounded only by tundra (Figure 7.10), the nearest spruce tree being 50 miles away. And yet there is spruce and alder pollen at the top of the diagram. This is explained because enough spruce pollen to make up 10 or 15 percent in a tundra pollen diagram can be blown many miles beyond the tree line. The pollen diagram ends with a time a few thousand years ago (the top sediments have been removed by wave action) when spruce and alder were near enough for their pollen to have been blown from the tree line to the lake. Earlier in late glacial time, zone K, spruce and alder pollen must have failed to reach Imuruk, suggesting that the tree line was still far

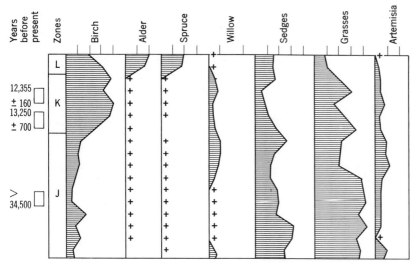

Figure 7.11

Pollen diagram from Imuruk Lake sediments, showing the time since the height of the last glaciation. The three-zone history shows first the invasion of the herb tundra by dwarf birches (present in the modern Imuruk Lake tundra) and then the arrival of spruce and alder trees near enough to cast their pollen to the lake. The Bering land bridge existed only during the earliest, most extremely arctic, period (zone J).

away. The birches which show as a maximum in this zone were tundra birches, dwarfs living among sedge tussocks, not trees. They are common around Imuruk Lake now and the apparent surge of their pollen in zone K time was because of the withdrawal of spruce and alder pollen from the pollen sum. This is an artifact of the pollen percentage method discussed in the preceding section. But back in the time of the land bridge, zone J, the dwarf birches had waned, and the pollen sum was largely made up of grasses, sedges, and some of the low sage plants of the arctic, *Artemisia*. Zone J pollen can be matched by surface pollen from Point Barrow, 400 miles to the north, a colder bleaker place by far than modern Seward Peninsula (Figure 7.9). Much of the Imuruk Lake pollen history has now been duplicated from other parts of Alaska and islands in the Bering Sea (Colinvaux, 1965, 1966), and the story is everywhere the same. The first Americans lived in a harsh arctic place comparable to the bleakest places in modern Alaska.

We know that a tundra such as is revealed by the pollen history of the Bering land bridge meant a cold arctic environment. But we also know that the land-bridge plains supported big game. We may thus reasonably infer that the ancestors of American Indians were big-game hunters of those ancient plains, possessed of the skills and social cohesion needed to live in such a place. Instead of being half-naked savages, they had boots and gloves and parkas of fur, as do the modern Eskimos who live on the arctic coast. More important to ecology, is the general implication of the way of life, that of nomad and hunter. There are many clues which suggest this way of life for Ice Age man the world over. As an ecologist ponders the adaptations, and limitations, of the big-game-hunting way of life for carnivorous animals, he can reflect that the limits he describes were once applied to his own ancestors. In the problems of our own day, we need to remember that we must still be preadapted to that hunting way of life, and perhaps not so completely suited to peasanthood or city dwelling.

The concept of dominance proved to be of little use to animal ecologists, but that of succession led to some interesting ideas. The animals of successive plant communities proved to be different, suggesting that there was a parallel succession of animal communities. The fish fauna of ponds apparently changed as the ponds aged, protozoan populations could be seen to change as their supporting cultures aged, and herds of different species of African game animals graze the same patches in succession. It was true that the animal communities accompanied changes in the plant communities, suggesting that the animals were merely passive followers of other events, yet the activities of the animals did alter the conditions of life in the habitat to some extent. Digging animals altered the soil. The excretions of fish and protozoa altered the chemical conditions for life. The grazing of game animals altered the composition of pasture. It could thus be claimed that animals prepared the habitat for the next community in succession just as the plants were supposed to do. But what came through more clearly from these studies was that animals moved about, settling only where the conditions for life were right for them. This led the pioneer animal ecologist, Shelford, to suggest that each animal species had its own special needs, which he called the *mores* of the animal, an idea which was later taken up by ecologists under the name of *niche*. Meanwhile, those who tried to describe plant communities realized that animals were an integral part of the units they sought to study. Attempts were made to replace the idea of the "plant association" with that of a more embracing concept, which should include both plants and animals, after which it was apparent that the new unit should also include the physical habitat in which

THE DAWNING
CONCEPT OF THE
ECOSYSTEM

115

the animals and plants lived. Ecologists groped for words to describe this new unit until the proper term was found for them with Tansley's *ecosystem.*

CHAPTER 8

THE DAWNING
CONCEPT OF THE
ECOSYSTEM

While European plant ecologists argued about the association, and Americans studied the climax, there was no comparable blooming of animal ecology. The habits and habitats of animals were, of course, studied as they had always been, but there was little sign of an exciting synthesis of the subject. As late as 1927 it was possible for Charles Elton to write in the preface to his famous textbook that, "The principles of animal ecology are seldom if ever mentioned in zoological courses in the universities" (Elton, 1927). In such circumstances it was natural for botanists and sympathetic zoologists to wonder how much the ideas of the plant sociologists were suited to the study of animals. Concepts such as **ecological dominance** and **ecological succession** were so interesting that it was natural to look for examples of them among animal communities.

It was, of course, evident that animals were important factors in the plant environment and that they modified the development of vegetation by eating plants, trampling plants, and dispersing the seeds of plants. An extreme example of these effects could be seen in any pasture or natural grassland, where the form and species composition of the vegetation was obviously determined by the grazing and trampling of herds of ungulates. The result of plant successions in such a place was actually controlled by the animals. But in most plant formations, and in the successions within them, the influence of animals was less obvious. Some animals, particularly insects, lived on forest trees without apparently determining the form of the vegetation; but might they not influence its development? It was necessary to ask if successions of plants were influenced by parallel successions of animal communities. The sand dunes of Lake Michigan again became the site of classical investigations into succession, this time into animal successions, and by a student of Cowles, Victor Shelford.

Shelford compared the animal communities in the seral stages of the plant successions in the sand dunes (Shelford, 1912). As expected, the varieties of animals were so numerous that it was not possible to master them all, but Shelford could show that some part of the animal community in each seral stage was unique to that stage. The poplars, the pines, and the beech-maple stages on the Lake Michigan sand dunes, for instance, seemed to be the homes of different species of tiger beetles (*Cicindela* sp). Other insects, reptiles, and even mammals seemed more or less restricted to one seral stage or another, and a few obviously played a role in modifying the environment in ways that could be construed as tending to encourage the next community of the plant succession and, hence, of the next animal community also. Digger wasps (*Sphex* sp) of the cottonwood stage buried their prey as food for their young, thus tending to enrich the sand with organic matter and to make of it a usable soil for pine seedlings. Other soil animals affected the development of the soils of later stages, principally earthworms and a variety of arthropods. But these effects of the animals seemed minor compared with the great changes brought about by the plants themselves. Shelford's study leaves the impression that although animals influenced plants, the composition of the animal communities was still more dictated by the physical presence of the plants.

It was possible to find a successional series in which the interactions between animals and plants need not be so one-sided, and Shelford found such a series in the rows of ponds left among the sand dunes by the retreating lake (Shelford, 1911). They were long narrow ponds, running parallel with the modern shore and separated from each other by ridges of sand (Figure 8.1). It was clear that they were arranged in order of age, with the youngest close to the lake and the oldest farther away, some thousands of years probably being represented. Until the builders of Chicago ran roads and railroads across them, there were water connections between the various ponds so that they could all be reached by fish and other organisms from the lake. Young ponds near the lake had little rooted vegetation and sandy bottoms; older ponds were progressively more choked with plant growth and with accumulated debris on the bottoms. Collections from a series of the ponds revealed a succession of fish communities, ranging from those that liked the bare bottom, clear-water, lakeside ponds which were still sometimes scoured by storm waves of the main lake sweeping over the dividing spit, to those living in the oldest, weed-choked and stagnant ponds farthest from the lake. Shelford reasoned that the progressive changes revealed by his series of ponds were brought about by the organisms living in them, as their decomposing parts and excrement both filled the basins and changed the chemistry of their waters. The activities of the fish could not be unimportant in this process. As the environment in an old pond changed, it was favored by different species of fish, all of which could come and go by the network of connecting channels.

Shelford had no doubt that this succession of fish communities came about because the fish had different physiologies, habits, behavior, and modes of life, a collection of parameters which he called the **mores** of the animals. Fish able to travel along ditches from one pond to another remained where conditions were best suited to their **mores.** Shelford was thus feeling toward an answer to the question, "Why do successions occur?" in the behavioral and physiological responses of the animals. In doing so he anticipated many of the trends of modern animal ecology. His word **mores,** which has an awkward and alien sound, disappeared, but the idea he was seeking to express by it later found an outlet in Elton's **"niche"** and became a central ecological concept of great intellectual reward. More than half a century after putting down in simple and com-

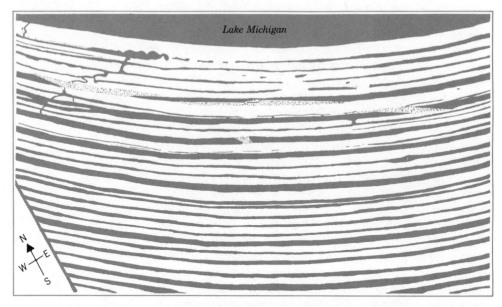

Lake Michigan

Figure 8.1   Plan of the sequence of parallel ponds near the Chicago shore which were studied by Shelford early in the century. The 25 ponds were all connected to each other and the lake by
ditches. They represent an age sequence, the southernmost being the oldest, which were left
dammed behind successive beach ridges as the lake retreated toward the north. Shelford
found successively different fish communities from the young bare ponds near Lake
Michigan to the old weed-choked ponds further away. (Redrawn from Shelford, 1911.)

pletely convincing prose his arguments showing that the fish succession
had occurred, Shelford was still writing notable works on ecology (Shelford, 1963), having lived to see the fruits of the new discipline of animal
ecology that his work in a sense foreshadowed. He died in 1967.

Shelford's work prompted other zoologists to consider the possibilities
of animal succession, but good examples were not common. In most
sites, the succession of plant communities was obviously the important
process and the related succession of animals minor, as Shelford had
himself found among the Lake Michigan plant communities. Other
common time-sequences were obviously related to the seasons, as
summer forms followed spring forms and so on, and thus did not change
the environment as was implied by the term "ecological succession." But
one curious apparent succession in animals had long been known; the
successive appearance of different species of protozoa in a hay infusion.

From the earliest days of microscopy, hay infusions had been used as
sources of protozoa for examination, providing the *animalcules* of the
first microscopists. Placing hay in a flask of pond water resulted, in a
week or two, in a cloud of swimming protozoa, but the cloud was made
of different animals from week to week and long experience had shown
that they appeared in roughly predictable order. In the flasks, there were
just hay, the bacteria which decomposed it, and the protozoa which fed

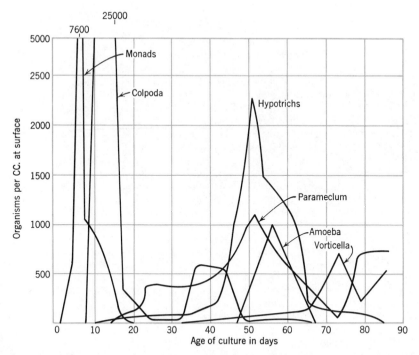

Figure 8.2

Succession of protozoa in hay infusions. Although all the different kinds of protozoa are present in the flasks at all times, they occupy the open water as massive blooms in turn. The sequence in which they become abundant is orderly and predictable, as apparently the stages of a secondary plant succession. (Redrawn from Woodruff, 1912.)

on the bacteria or on each other. There were no plants to interfere with the development of a succession of animals alone. A Yale biologist, L. L. Woodruff, having read Shelford's work with the fishes, set up a series of experiments to see if a definite ordered succession really occurred in hay infusions (Woodruff, 1912).

It was important that the infusions used for the experiments contained individuals or resting stages of a wide range of protozoa and, to make sure that this was so, Woodruff put in each of his flasks at the start hay, water, and a stock sample of protozoa mixed from his laboratory cultures. In numerous duplicated flasks the same series of events occurred; first one animal became immensely common so that the cloud in the water at the top of the flask was made up mostly of individuals of one kind, then these were replaced by another protozoan so that the cloud was made of individuals of another kind. In succession there appeared huge populations of monads, *Colpoda, Paramecium*, hypotrichs, amaebeae, and *Vorticella*, in an order that never varied. At any one time it was possible to find individuals of all these forms by diligent search, particularly among the debris at the bottom of the flasks, but they succeeded each other in massive abundance (Figure 8.2). There could be no doubt that the great abundance of one kind of animal resulted in the environments being al-

Figure 8.3

Three specialized grazers of the East African plains; zebra, wildebeest, and Thompson's gazelle. The animals have been seen to graze the same patch of grassland successively, in rotation, each taking a different portion of the total crop.

tered to its detriment but in a way that suited another animal, and that these changes of environment followed a regular sequence. Chemical excrement in the water, shortage or changing quality of food, and predation or parasitism could all be involved in promoting the decline of one form and the rise of another, but it was impossible to decide. Woodruff confessed that determining the exact changes involved in any of the successive replacements was too complex a task to be attempted, and he did not mar a classic paper by guessing. He had illustrated something closely comparable to Shelford's fish succession, but in miniature; the environment was continuously changed by the animals that lived there while there was always a reservoir of animals who could take advantage of the changed conditions. Although following orderly successions of stages, neither example led to a self-perpetuating climax as did plant successions, which should not be expected, since animals are always dependent on outside sources of food.

Recently there has come the story of a similar succession among big game animals of Africa, but one in which the environmental change brought about by each population in the succession is also known. In the Serengeti park herds of animals graze the same land in order; first zebra, then wildebeest, and then Thompson's gazelle (Figure 8.3). Shooting some of the animals in the act of grazing enabled Gwynne and Bell (1968) to obtain fresh stomach contents, and so find out what they had been eating. There was a variety of plant parts in all the stomachs, of course, but the zebras had eaten much more of the long flowering stalks of grass, the wildebeest had eaten the side leaves, and the gazelles had eaten a variety of low-lying leaves and fruits. Apparently the zebras were able, with their pinching equine incisors, to cut down the standing stalks of grass and did so, after which the wildebeest (with their bovine mouths untoothed in the upper jaw) could browse on the softer grass parts, finally clearing the way for the gazelles to find their more individual items of food. This still does not answer the question of why the zebras should stop at the dry stems and make way for the wildebeest. Why is the zebra so specialized in its feeding habits? This is the sort of question now being tackled by modern animal ecology.

The quest for ecological successions in animal communities thus led to some interesting lines of inquiry. The quest for an animal equivalent of the **ecological dominance** of plants was less fruitful, although it is possible, with a little ingenuity, to find uses for the word "dominant" when talking of animals. In a sense, predators dominate their prey. There is no violation of English meaning in talking about a large carnivore as a "dominant animal," but the presence of large carnivores does not overshadow all the animals of the place. Perhaps the carnivore dominates the things it actually eats, but it can scarcely be held to dominate the lives of the rest of the myriad animals living in the area. In fact, large herbivores are better candidates for the honor of dominant species. The bison grazing the plains must exert a powerful influence on the lives of all the small grass eaters: on the antelopes, jack rabbits, mice, insects, and mites which also feed on the prairie. Food and room to live for the smaller animals must, to some extent, depend on the actions of the **dominant** bison, but acknowledging this does not seem to get us very far. There has recently sprung up a habit of talking about animals with the most biomass as being the most "dominant"; thus, if you found a greater mass of caterpillars than of bugs on a cabbage, you say the caterpillars are dominant. This usage is an ill treatment of English which is likely to lead to wrong thinking. In general, the idea of ecological dominance has not been of much use to students of animal communities.

THE CONCEPT OF
THE ECOSYSTEM

While zoologists of Shelford's generation were beginning studies on communities of animals, botanists working with the more advanced discipline of plant ecology still faced the problem of how best to acknowledge the role of animals in the development of vegetation. Since much of the plant ecological work was devoted to the classification of communities, animals were assigned roles in the classifications. The formation had become the biome. Seral stages of plant successions were known to be the homes of animals, often small ones, which lived nowhere else, so these were called **index animals** of the communities in which they lived.

Associations and sociations, however defined, were also likely to afford living places to particular species of animals which had adapted to them and whose presence must have some impact on the vegetation, and various attempts were made to include the names of characteristic animal species in the association descriptions. This led to a standard ecologist's joke about the *spruce-moose association.*

Yet field men everywhere knew that the plots of land which they studied had characteristic living things, both animals and plants. Individual animals might be overshadowed by plants but the whole complex of animals of a place nevertheless must react very strongly on the plants. Some men began thinking not just of classifying vegetation, and then thinking of the animals, but of identifying functional units of all living things in which the plants were no more than equal partners. By 1914 we find them talking about units such as **biotic associations,** the boundaries of which were inclined to be drawn by habitat boundaries (Vestal, 1914).

Then it came to be realized that all the units chosen to be studied somehow came to be defined by habitat, by physical inanimate limits. Braun-Blanquet acknowledged that his associations were aggregations of plants which detected areas of uniform habitat; Clements had accepted that even his superorganism had bounds set to its domain by climate. Now those who, like Vestal, set out to study the animals and plants together were to find the same thing. Plants, animals, and the physical environment had their fortunes inextricably meshed together and, frighteningly complex though the meshes were, they must somehow be studied as wholes. But what to call the new units, what names for the wholes? Various names in German and Greek came out of central Europe for the new entity; the **holocene** of Friederichs, the **Naturcomplex** of Markus, the **Räume** or "landscape unit" of Passarge. We read of **biogeocenoses,** of **biotic districts,** of **biochores;** and the names come from Europe, Russia, and the Americas (Whittaker, 1962). An idea had found its time and wanted but a word that was, clear, expressive, and did not twist the tongue. It was found when Tansley gave us the **ecosystem** (Tansley, 1935; Figure 8.4).

Tansley occupied an influential position in the development of ecology. He was an informal leader of a British school of plant sociology which learned both from the continental disputes and from Clements' persuasive writing in English. British ecologists worked both with the man-scarred landscape of their island, when their problems were similar to those encountered by Braun-Blanquet in France and Switzerland, and with more virgin vegetations of British possessions in the Empire, where they often saw the things that the American disciples of Clements saw. They adopted some of the Clemencian ideas on succession, but without being sure that you must look for one areal climax; and they could use the descriptive techniques of Europe, perhaps, more those of Uppsala than of Zurich-Montpelier.

Tansley's **ecosystem** would be defined by habitat and vegetation, and it would include the animals of the space. You could choose any size area

THE ECOSYSTEM

I have already given my reasons for rejecting the terms "complex organism" and "biotic community." Clements' earlier term "biome" for the whole complex of organisms inhabiting a given region is unobjectionable, and for some purposes convenient. But the more fundamental conception is, as it seems to me, the whole *system* (in the sense of physics), including not only the organism-complex, but also the whole complex of physical factors forming what we call the environment of the biome—the habitat factors in the widest sense. Though the organisms may claim our primary interest, when we are trying to think fundamentally we cannot separate them from their special environment, with which they form one physical system.

It is the systems so formed which, from the point of view of the ecologist, are the basic units of nature on the face of the earth. Our natural human prejudices force us to consider the organisms (in the sense of the biologist) as the most important parts of these systems, but certainly the inorganic "factors" are also parts—there could be no systems without them, and there is constant interchange of the most various kinds within each system, not only between the organisms but between the organic and the inorganic. These *ecosystems*, as we may call them, are of the most various kinds and sizes. They form one category of the multitudinous physical systems of the universe, which range from the universe as a whole down to the atom. The whole method of science, as H. Levy ('32) has most convincingly pointed

[3] If this statement is applied to the individual organism, it of course involves the repudiation of belief in any form of vitalism. But I do not understand Professor Phillips to endow the "complex organism" with a "vital principle."

Figure 8.4 The first appearance of the word "ecosystem." The paper by Tansley in which it appears is in a special issue of *Ecology* dedicated to H. C. Cowles of the Michigan sand dunes, from whose work sprang many of the thoughts of Clements, Shelford, and others who brought ecology to the point of thinking in systems.

you liked to study, provided its boundaries were convenient; and then, freed from the problem of classifying imaginary units that graded one into another, you could set to finding how your chosen ecosystem worked. From this idea stems much of the progress of modern ecology, with its growing understanding of the functioning of the living world. The idea was born in the minds of many ecologists 50 years ago, lost as they then were to the politics of the "real" world, shut up in their ivory towers. But it fell to Tansley to come on the word which was to give substance to their thoughts. And they knighted Tansley for his services to British botany, so that he became Sir Alfred Tansley. This might have brought a chuckle to the lips of they who wore the golden spurs at Agincourt or Crécy, but it is not hard to see who left the greater mark on his posterity.

The earth is divided into a number of huge regions by the circulations in its atmosphere. Its surface is a patchwork quilt of air masses, each patch of continental proportions, each owing its position to the relative positions of continents and oceans. The limits of each patch wax and wane with the seasons and through secular time, but the mean positions of fronts change only gradually with the changing geography of the earth. Each major air mass provides a distinct environment for living things, and also a distinct regimen for the weathering of rocks.

The weather in the sway of an air mass determines what plants shall live. The plants then act as agents that modify the crust of the earth, driving their roots into the surface rock, mixing their dead parts with it, affecting its chemistry, making it into soil. But the weather also controls the soil directly, determining which minerals shall be leached away and which shall remain. These properties of the soil then influence the development of the vegetation in their turn.

Vegetation, soil, and weather together make up the complete environment in which animals must live. It follows that the patchwork of the earth's air masses is mirrored in a patchwork of forms of life. The enormous extent of each formation of plants, together with the soils and animal communities belonging to it, is set by the vast domain of an air mass. Often the borderlands between air masses are vaguely defined and then the borders of formations are vaguely defined also, so that we find gradients of vegetation across continents. But sometimes, as with the Arctic Front, the edge of an air mass is a clear-cut thing, for all that it wanders. Then formation boundaries may be made wonderfully distinct by the enduring qualities of forest trees, by the long span of their generations; and by their looming physical presence, resistant to short-term change and controlling the microenvironments in which their own offspring develop.

Within the bounds of each formation there are smaller patches of plants, but these are poorly defined. Attempts to map them as units of a small mosaic within the greater mosaic of the formations have always failed, for the reality is that they grade one into another. As the environment alters, both in time and space, so the vegetation alters, too, being made anew by the adjustments of the individual plants of which it is made. The labors of both plant sociologists and of pollen analysts lead to this same conclusion. The association is an ephemeral unit. And only when there is some discontinuity in the physical habitat can an apparent edge to an association be found.

But patches of land could be conceived as functional units, which, small or large, could be looked on as natural systems. These are the ecosystems. They are not items in some mystical natural mosaic, but convenient portions of a grand natural continuum chosen for study. Within each ecosystem, the physical environment determines what plants shall live, and the plants control the soil and the microclimate. The soil further determines the plants. And the plants, soil, and climate combine to determine what animals shall live there. And then the animals take a hand in determining what plants shall establish and persist. Everything that happens in an ecosystem affects everything else. Minerals are mined and circulated. Water and gases pass round and round between the habitat and living things. Much work is done in all this, and the energy needed to do this work comes from the unfailing radiation of the sun.

So geography led us to vegetation, and vegetation to the understanding that life was lived as individuals, each inevitably acting on the others, both directly and by the changes it brought in the physical conditions of life. If there was no future for geographers in the boundless gradients of land and vegetation within the greater formations, the concept of the ecosystem was a greater reward. For we can understand and analyze systems. If you move a plant, or an animal, or a rock, you alter the system. Man is now engaged in moving plants, animals, and rocks, in an almost Godlike, if sometimes evil, way. Often, as when he builds a city, a sub-division, a dam, or a dust bowl, he destroys ecosystems that were once there to replace them with something novel. Up to now he has neither understood how his new ecosystems would work, nor cared what the results of their working should be. The results we all see around us. A first step toward doing something about the mess we have made is under-standing the wonderfully viable natural ecosystems that we have in-herited, systems refined by the fine adjustments of long spans of geologic and evolutionary time.

PART 2

ANIMALS AND
PLANTS AS
PARTS OF
NATURAL
SYSTEMS

Because animals move about and cannot be seen in the mass zoologists, wishing to study how they lived in communities, were forced to begin by looking at the individual kinds of animal and asking themselves what the animal was doing; perhaps, more particularly, what it ate. Elton applied this approach to the rather simple communities of an arctic island where he found that foxes were tied both to the tundra and the sea by *food chains.* Animal communities thus existed, being firmly tied together by bonds of eating and being eaten, but they might overlap such distinct geographic boundaries as a sea coast. Once these communities were identified it was immediately apparent that the animals low on food chains were both much smaller and much more numerous than the animals further up the chains. Although exceptions could be found, this arrangement was seen to be true for all kinds of animal communities, presenting a general phenomenon that seemed to require a general explanation. Elton accounted for the fact that animals tended to be larger at each successive link up a food chain by the *principle of food size,* since an animal must always be large enough to both overpower and ingest its prey, but this explanation still left the relative scarcity of large animals to be explained. This rarity of large animals high on food chains was only satisfactorily understood later when the food requirements of animals were thought of in terms of calories and energy flow. Each time a portion of potential animal food is passed from one link in a food chain to another, some of the calories are used to do work, which is to say that energy is degraded according to the second law of thermodynamics. There is thus less energy available to each successive link in a food chain, meaning that less biomass can be made, with the final result

129

that large animals on the end of food chains are rare. The concepts of *food chains* and *energy flow* opened the way to analysis of natural systems based on measurements of such things as numbers, mass, and calories. From these studies also comes an understanding of some of the difficulties we have encountered with pesticides or trace pollutants, and of the development of human cultures since the Stone Age.

CHAPTER 9

ANIMAL
COMMUNITIES BY
FEEDING HABIT

FOOD CHAINS AND
THE PYRAMID OF
NUMBERS

Different kinds of animals generally live together in one place, but they are not all to be found in the neighborhood all the time. Many move about, feeding in different parts of their habitat or sometimes going away to a different place. As a result, the boundaries of animal communities are generally not as clearly delimited as those of plants, and attempts to map animal communities always end up by mapping vegetation boundaries instead. But if you cannot often map an animal community, you can yet be sure that it is there. Why do animals seem to exist in communities? Why are there big and little animals in a community instead of all being just one ideal size? Why are some members of the community always common whereas others are always rare? Answers to questions like these have been reached by looking at animal communities as groups of individuals connected by their common interest in food, and from seeing how they shared the available food. Biologists have long studied the feeding of animals, of course, but our understanding of how community structure results from feeding stems mainly from the discerning work of one man, Charles Elton (1927).

As a young man during the 1920s, Elton went with Oxford University expeditions to Bear Island, near Spitzbergen, then a favorite place for exploring parties; and there he studied the ecology of the tundra, a vegetation in which the larger animals were not hidden by trees (Summerhayes and Elton, 1923). The most conspicuous animals of such a place are the arctic foxes, which wander about in the open landscape and are so tame that they are easy to watch. On St. George Island of the Pribilofs, I have had one attempt to take sandwiches from my pocket as I sat on a rock. Elton studied the community of which the fox must be a part. He noticed that foxes caught the summer birds of the tundra, the ptarmigan, sandpipers, and buntings; then he observed what the birds ate. Ptarmigan ate berries and leaves of tundra plants so that the fox was connected to the tundra by a chain of feeding habit, a **food chain** that ran tundra-ptarmigan-fox. The sandpipers and buntings ate insects, which in turn ate tundra plants, so that the food chains connecting the fox to the tundra through these birds were longer, running tundra-insects-birds-fox; but the fox was still "tied" to the tundra by this chain of eating and being

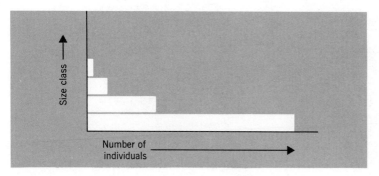

Figure 9.1

eaten. But foxes also ate seabirds, such as gulls, eider ducks, and auks. Through these birds the fox was chained not to the tundra but to the sea, the food chains running through the birds to sea animals and thence to sea plants. In winter the birds flew away, and Elton found that the foxes survived by eating polar bear dung and the remains of the seals that the polar bears killed. The food chain that supported the foxes then might run: sea plant-sea animal-seal-polar bear-fox. There was thus on Bear Island a community of animals, of which the foxes were a part, that was based on both the sea and the land, whose existence cut across as beautifully distinct a geographic boundary as one could find, a seacoast.

The foxes, which were the last link in the Bear Island food chains, were the largest land animals, for the seals and polar bears which hauled ashore, or wandered ashore in winter, were marine animals belonging to communities of the sea. The next lower link of the land food chains were the birds which the foxes caught, and these were at once much smaller than the foxes and also more numerous. Some of these birds were insect-eaters, so their food in turn comprised animals more numerous than themselves as well as being much smaller than themselves. As you worked up through a food chain it seemed that the animals became progressively larger and less abundant. A community could be separated into groups of animals of similar sizes. The animals in each of these groups would be feeding at about the same level in its food chain, and the smallest size classes would contain most animals. If you drew a graph of size class against number of individuals, your figure would look roughly like Figure 9.1 and each size class represented by a box in the figure would consist of animals feeding in the same level of a food chain, or at the same **trophic level** (from the Greek word meaning "nurse" or "suckle"). If you redraw the figure with the vertical axis in the middle instead of at the side you get the pyramidal diagram in Figure 9.2. Elton called this result, **"The pyramid of numbers."**

It seemed plain that there was a pyramid of numbers associated with all communities, since it is common experience that large animals are less abundant than small ones (Figure 9.3). Elton did not need to go out and make counts to prove this assertion. In the words of the Chinese prov-

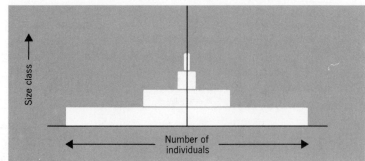

Figure 9.2

erb, "One hill cannot shelter two tigers," although it undoubtedly had many smaller animals; and in Elton's words, "If you are studying the fauna of an oak wood in summer, you will find vast numbers of small herbivorous insects like aphids, a large number of spiders and carnivorous ground beetles, a fair number of small warblers, and only one or two hawks" (Elton, 1927).

As animals became larger and rarer, they also had larger **home ranges;** the whole hill for the tiger, a patch of woodland floor for a spider, one shoot of a plant for a whole generation of aphids. So the distribution as well as the abundance of animals was reflected in the pyramid of numbers, the larger animals being more wide ranging as well as less numerous than the small animals.

FOOD SIZE AND THE
FOOD WEB

The large animals were those near the ends of food chains; why was this? Elton suggested that it must be because of the simple problems of catching and eating food. Generally an animal must be large enough for its mouth to engulf its prey; but, on the other hand, the prey must not be so small that it became impossible for the animal to catch enough of it. There is an advantage in an animal's being big, so that it can easily catch and eat its prey, but it must not be so big that it cannot catch enough small prey to keep itself alive; so there must be an optimum size for any animal, a size determined by the size and agility of its food. Animals of different trophic levels, therefore, are likely to be of distinctly different sizes. Elton summarized this conclusion as **the principle of food size.** It explained not only the increased size of animals in successively higher links of food chains but also some of the exceptions. Lions are no bigger, or are even smaller, than the ungulates on which they prey because they have evolved specialized techniques for catching large prey and reducing it to mouthfuls, but most animals have not done this and must have their own size adapted to the size of their food.

Any one animal usually eats a variety of food and thus, like the arctic foxes of the first example, may be a part of several food chains that intersect. On the other hand, food chains starting from a plant source must commonly radiate outward as the plant food is eaten by different herbivores, and as they are eaten by different carnivores, and so on. There is

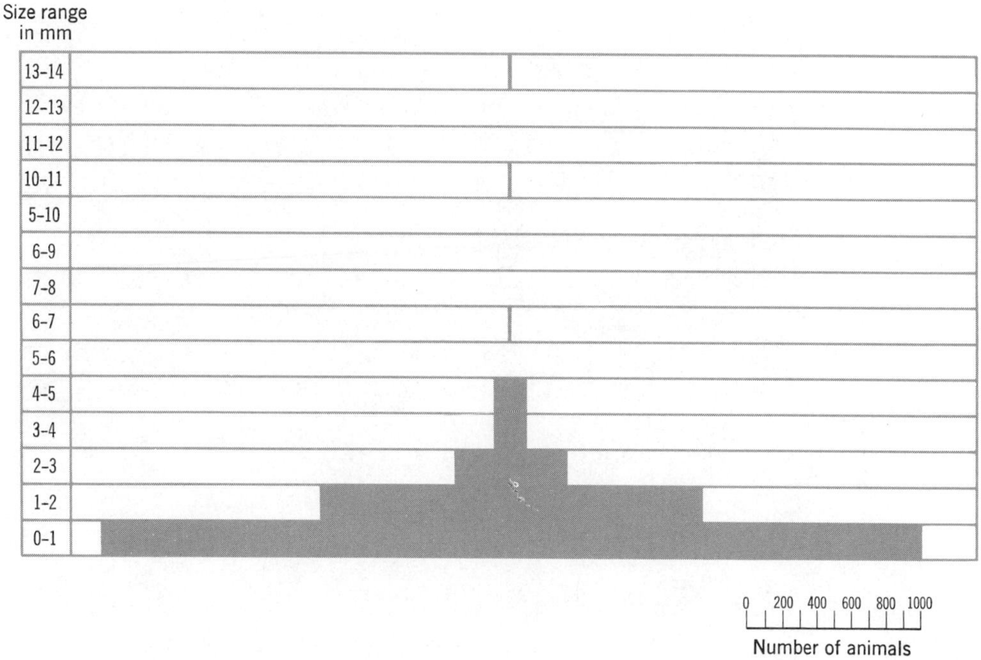

Size range
in mm

13–14	
12–13	
11–12	
10–11	
5–10	
6–9	
7–8	
6–7	
5–6	
4–5	
3–4	
2–3	
1–2	
0–1	

```
0    200  400  600  800  1000
```

Number of animals

Eltonian pyramid of numbers on the floor of a forest in Panama. Williams (1941) collected all the animals in small samples of the litter on the floor of the forest, counted the individuals of his catches, and sorted them into size fractions. The smallest and most numerous animals were Collembola (spring tails) and mites, both of which were herbivores or scavengers feeding in the litter. The larger rarer animals, such as ground beetles and spiders, were carnivores.

Figure 9.3

thus really a complex **food web** formed as the food chains first radiate outward from the plants and then come together toward the top of the pyramid (Figures 9.4 and 9.5).

SPECIES NUMBER
AND NICHE

The pyramid of numbers is a pyramid of species number, as well as of numbers of individuals. In the Bear Island example, the fox was the only species at its trophic level, but there were several kinds of birds on which it fed, and many more kinds of insects in the trophic level below that. Thinking about any community quickly convinces one that there are always fewer kinds of animals in the higher trophic levels. The hill sheltering its tiger shelters few, if any, other large carnivores, but there are many kinds of rodents and birds on a Chinese hill, and very many kinds of insects. That the food chains converge after initially radiating means that the animals of the higher trophic levels must have varied diets, that they must be less specialized in their feeding habits than the animals lower down. Position in a food chain and the pyramid thus places restraints on the feeding habits of an animal, with those at the top accepting many kinds of food. Feeding habits reflect all the adaptive qualities of an animal, its physiology, its mechanical abilities, its behavior; and all these qualities of an animal fit it for a role in life, a profession that enables it to

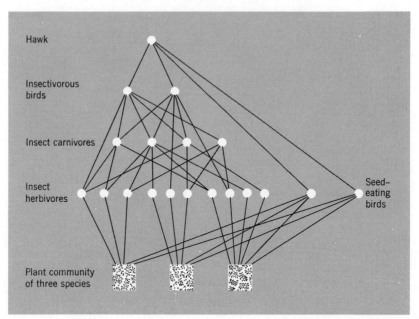

Hawk

Insectivorous
birds

Insect carnivores

Insect
herbivores

Seed–
eating
birds

Plant community
of three species

Figure 9.4

Hypothetical food web. It is assumed that there are three species of plants, ten species of insect herbivores, four insect carnivores, two bird herbivores, two bird insectivores, and one
hawk. In a real community, there would not only be many more species at each trophic
level but also many animals that feed at more than one level, or that change level as they
grow older. Some general conclusions emerge from even an oversimplified model like this
however. There is an initial diversity introduced by the numbers of plants. This diversity is
multiplied at the plant-eating level. At each subsequent level the diversity is reduced as the
food chains converge.

get its food. Elton called this profession of an animal its **niche,** and
defined this as **the animal's place in the biotic environment, its relation
to food and enemies.** Shelford had earlier been groping to the same end
with his **mores** (Chapter 8), but the ring of clarity was not in Shelford's
word, nor had he connected his description of the animal's profession
so closely with feeding habit. There is an immediate idea conveyed by
the simple English sentence "every animal has its niche" that defies
confusion.

THE FOOD CYCLE   The scattered pathways of the food web were the routes of food substances passing along the food chains and through the animal community. Food substances had their origin in the inorganic environment and
were finally returned to it. The animals acquired all the carbohydrates,
and proteins of which they were made from plants. All of them burned
sugars and breathed out the carbon dioxide that plants would someday
use again. They broke down proteins and excreted nitrogenous compounds into the environment, where these were further broken down by
bacteria. If animals were not eaten but died, their bodies were eaten by
scavenger animals, who might in turn die and be eaten by smaller scavenger animals, so that there were food chains of scavengers, later to be

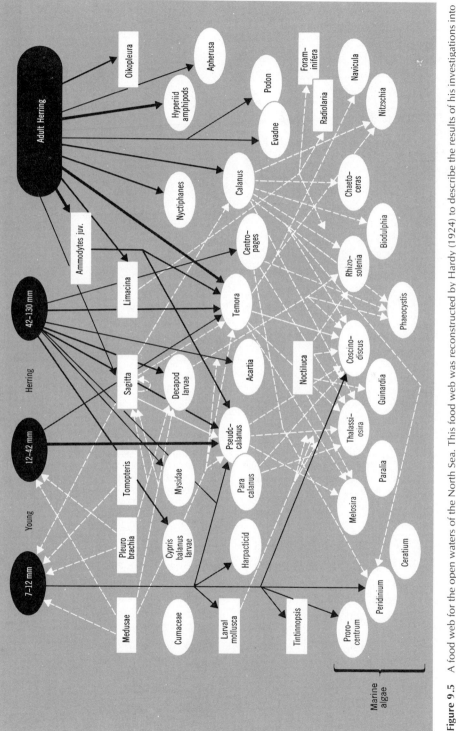

Figure 9.5 A food web for the open waters of the North Sea. This food web was reconstructed by Hardy (1924) to describe the results of his investigations into the herring fishery of the North Sea. The herring is a carnivore that feeds several links in its food chains away from the floating algae which are its ultimate source of food. The names in the bottom lozenges are of various genera of phytoplankters. Higher up in the food chains are planktonic animals, still larger animals, and then the herrings themselves. Notice that young herring feed on smaller food than the adults, as you would expect. As the herrings age, so they feed at higher and higher trophic levels.

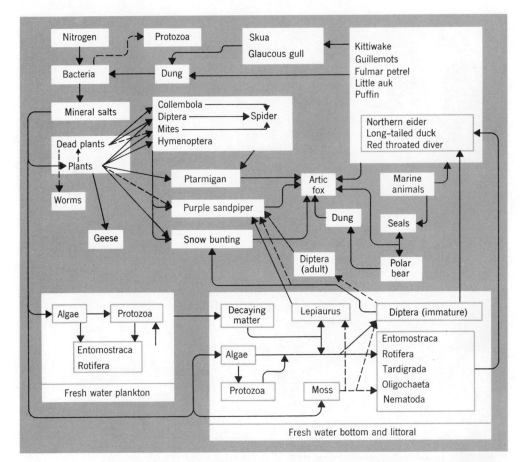

Figure 9.6   Food cycle among the animals on Bear Island. Elton worked out the details of this food cycle by seeing in the mind's eye the movement of nitrogen with the food through the community. Some of the pathways he was able to confirm by direct observation, others (shown as dotted lines) he could only infer. He notes that the best way to read the diagram is to start with marine mammals and to follow the arrows. (From Summerhayes and Elton, 1923.)

neatly called **saprophyte chains** by Odum. Particles of food were passed along these chains, too, until all had been broken down to simple inorganic constituents and cycled back to the physical environment. Other particles of food were deflected from traveling all the way up the food chains by being eaten by parasites. These particles would cross into another food chain if the parasite was eaten by a predator, as when a bird eats a tick off the back of a sheep, or they could pass into the parasite of a parasite, and thence through a **parasite chain.**

There can be parasite chains of several links in nature, as when a wasp parasitizes a caterpillar, has its own maggot parasitized by a second species of wasp as it lies within the first unfortunate caterpillar. The second wasp maggot might well be attacked by a protozoan parasite (the results appearing as a syndrome which we should call "disease"). And it is not too fanciful to expect that the protozoan might sometimes succumb to a bacterium, and the bacterium to a virus. The result is a food chain in which the animals of each successive link are smaller than those of the link above. Members of the successive links of parasite and saprophyte chains are always smaller and smaller as the last link is reached because the niches they occupy require them to be smaller than their food.

Elton talked of the passage of food through the animal community as the **food cycle.** It gave both an intellectual tool for understanding animal communities, and a practical tool for unraveling the complexity of a real one. Elton had been able to unravel the Bear Island food chains and food web partly by tracing the pathways followed by one ingredient in the food, nitrogen. It seemed likely that much of the nitrogen cycling through the island community as nitrate and ammonium salts came from the droppings of sea birds which abounded around the coast. This nitrogenous manure was used by plants. In the mind's eye, it was possible to see the nitrogen moving to herbivorous insects, thence to the land birds, thence to the fox, thence via its urine to the beaches, thence to the sea, and so on (Figure 9.6). The actual tracing of nutrients through a community to reveal all the manifold pathways of a food web was beyond the technology of the 1920s, but now it is possible to introduce radioactively labeled nutrients into a system and to monitor their appearance in successive links of the food chains and trophic levels of the pyramids.

The Eltonian pyramid and the concept of the food cycle should have enabled us to anticipate one of the disasters of our day, the slaughter of birds and animals by insecticides. Why, when we spray our crops with concentrations of DDT which will kill insects but which are much too low to hurt vertebrates, do many birds die? The DDT soaks into the insect cuticles or is eaten by them, so that they collect it in the fat of their bodies. Many die, but some are eaten by other insects, or by spiders, or by small birds, the predators of the next link in the food chain. These predators collect in the fat of their bodies the DDT from many insects, concentrating the insecticide there. They may excrete some of it, but this riddance is balanced by more coming with the next meal. The hoard in their bodies is maintained. When the predator of the next link eats them it receives a concentrated dose of the poison. If it is not killed by it, it goes on collecting more by eating other animals that have stored it in their fat, until its own fat reserves are riddled with it. A hawk or carnivorous mammal then eats these animals and receives a gradually fatal dose, or one that prevents it from breeding which has the same population effect. Peregrine falcons are vanishing all over the world. The skylarks that made a brave display over the English wheat fields in the years of Hitler's war may be in danger now. Bald eagles and pelicans are disappearing from America, and some of the woods are quieter in the spring. When DDT was invented we already had the information which would tell us that this must be, but no one had the wit to use it.

PESTICIDES AND THE
FOOD CYCLE

Elton's insight had shown how we should identify animal communities from the food chains and the pyramid of numbers; how we should study communities by means of the food cycle; and how these principles explained much of the distribution and abundance of different kinds of animals, why some were large and rare, and others small and common. But he did not provide a sufficient explanation for the pyramid of numbers itself. The animals high in food chains must be big, certainly, but why so few of them? Partly this might seem a result of simple geometry, that you can pack many small things in the space occupied by one big one, but thoughts about animals in the wild show that this is not a good enough explanation. In the open ocean, for example, there is apparently unlimited room for sharks but, in fact, they are spread very thinly through this space. Why *are* there so few top carnivores? Elton's answer was that small animals can reproduce more quickly than big ones, which means that they can keep up a vast supply of offspring even though many get eaten. The big animals breed more slowly, and so cannot support such large populations in the face of predation and accidental death. This explanation has a vaguely unsatisfactory ring about it. In time Elton's work prompted others to offer a more satisfying explanation, one that emerged from studies on lakes.

The animal communities that live in lakes have long been a favorite with ecologists because they, more than most communities, may be studied in some isolation from the other communities of the neighborhood, and because the accidents of weather are not often dramatically upsetting in a lake. Raymond Lindeman (1941a, 1941b) had worked for his doctorate during four years of the late 1930s on a small lake in Minnesota, trying to study the basin as a complete system, working out the past history of the local vegetation from borings around its edge, and watching the fortunes of the various trophic levels within the lake from year to year to give him an impression of the slow changes associated with the succession. He tried to understand the niche of each animal, how it fed and on what it fed. And he looked at food, not as particulate matter to be returned to the physical system as Elton had done, but as the energy supply of the animals. Food for him, as for many men before him who had been interested in **production** of living systems, was something to be measured in calories. When this **calorie food** was used up, it was not cycled back to the physical environment but was dissipated as heat.

When Lindeman had written his doctoral thesis on these matters, he went to work at Yale with the ecologist G. E. Hutchinson, who had independently been thinking on these lines. The result of their collaboration was published by Lindeman in what is one of the most famous papers in ecology (Lindeman, 1942). One part of this paper was to show how energy flowed through a community, how it was continuously dissipated as heat, and how this energy flow explained the Eltonian pyramid.

The energy that is the true food of the community all comes from the sun. In photosynthesis the plants fix energy, storing it as potential energy in their body substance, and this body substance is the food on which all

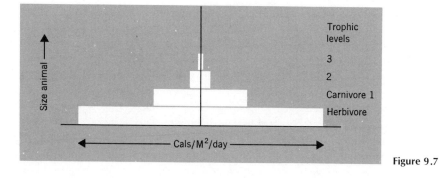

Figure 9.7

the animals of the community have to depend. When an animal eats a plant it uses energy to do so, since it performs work in the process. The end result of its doing this work is that some of the food, or potential energy, has been transformed into its own protoplasm, the potential energy of its own body. In the feeding process, energy had been transferred from one state (potential energy of plant protoplasm) into a second state (potential energy of animal protoplasm), and these energy transfers are, according to the second law of thermodynamics, never 100 percent efficient. They always involve degradation of some of the energy to a lower potential state, usually heat. The energy degraded by the herbivore was used to do the work of feeding. When a carnivore eats the herbivore there is a second loss of energy to the community, due to the degradation of energy in the second transfer, and so on up the food chains. Energy is constantly flowing into each trophic level from the trophic level below, and is constantly being dissipated as heat within each level. Since the energy is constantly flowing, the only way to measure or describe it is as the amount passing through in unit time, that is as a rate. It is possible to visualize the energy flowing through any trophic level as a rate in this way, say as the number of calories per square meter per day. It is obvious that the flow available to each successive trophic level is less than that available to the level below it. The whole process may be represented as a pyramid of energy as shown in Figure 9.7.

Since the energy flow of each higher trophic level is less than that of the level below, it is obvious that less protoplasm can be supported at each level, at least, if the protoplasm is about equally energetic. If the animals of each trophic level were of the same size, they would yet have to be rarer and rarer toward the top of the pyramid where energy is in shortest supply. Since the animals at the top are, in fact, the larger and most active animals, they are sharing a diminishing supply of energy among increasingly large and active bodies. The animals high in food chains, therefore, must be few, and the Eltonian pyramid is a necessary consequence of the second law of thermodynamics. This was the intellectually satisfying truth that Lindeman and Hutchinson saw from their studies of the life in lakes.

ON EXTREMELY
LARGE ANIMALS

It is the common experience of biology that predatory food chains never have more than four, five, or (rarely) six links, and that the last animals in these longest chains are large. On land, the largest true carnivores are the big cats. The biggest things in the sea that we think of as truly predatory, the largest sharks and killer whales, are even bigger than the biggest cats, but there seems a clear size limit to the predatory way of life in the sea too. Animals that preyed on big sharks and the biggest cats would have to be very large and active, and they would have to maintain this large active state on the restricted amount of energy that could be made to flow out of the shark-cat trophic level. Apparently the potential energy available to the niche of shark or lion-eater is too little to support a viable population of animals, and no such animals have evolved. The second law of thermodynamics thus apparently sets a practical limit to the length of food chains, and it also sets limits to the size of the largest viable carnivore. But there are animals that are much larger than lions and sharks, and there have been some big ones in the past too. Their niches can be examined in terms of the principle of energy flow also.

Elephants and big ungulates are larger than lions. In the past there have been bigger mammals still, such as giant ground sloths and *Titanotherium*, the largest mammal ever. There have also been the largest reptiles of the Mesozoic. These are all land animals and they have all been herbivores. Because land vegetation exists in large continuous chunks, big animals have been able to feed on it directly. All these large animals occupy the herbivore trophic level, the second link in their food chains, where the energy flow is large. Since there is much energy flowing into their trophic level, and since their way of life does not require the energy-consuming activity of hunting, it has been possible to support a large biomass of these animals, with the result that individuals could be large.

There is, perhaps, but one land animal that seems a little hard to explain in this way, because it was a carnivore: the huge *Tyranosaurus rex*, "the largest carnivore in history," and the horrific creature beloved of cartoonists and science fiction movies. This animal was certainly much larger than all the land carnivores of subsequent times; what enabled it to escape the constraints apparently placed on its successors by the second law of thermodynamics? In trying to answer this it is important to note first that the tyranosaur fed at the same level as its modern successors, the big cats. It fed on herbivores, relatively low in the food chain where the energy flow was still considerable. But we know that there were many kinds of very large herbivore about, animals that, in the absence of pack hunting predators like dogs, could only be overcome by very powerful attackers. So we might conclude that the necessity for mesozoic predators to be large and ferociously active is self-evident. Yet there is another possibility, one prompted by a recent revision of what a tyranosaur actually looked like. The classic picture of the hopping predacious tyrant is derived from nineteenth century reconstructions of the animal. A revised reconstruction of 1968 shows it to be a waddling, slow-moving beast, not at all the sort of thing one can imagine rushing with slavering fangs after a herd of galloping brontosauri. But it probably got them all the same, picking out the weak and the sick; and often getting them only as carrion. This is really not so very different from the methods of many modern predators, who rely extensively on weakened individuals from the herds of their prey (Chapter 28). The tyranosaur still supported a large mass by meat-eating, but it escaped the energy-consuming price of being active and having to overpower prime specimens of the giant prey it ate.

Apart from *Tyranosaurus* and some close relatives, then, the really large animals of the land are all relatively sluggish herbivores. The energy available in the carnivore trophic levels is not sufficient to permit both very large size and the activity required by predation. But the largest animals that have ever existed live in the sea, not on the land, and they are carnivores; the baleen whales. The plants of the open sea are minute. Nowhere in the open ocean are there lush pastures where a sluggish herbivore may feed, so that all herbivores of the open sea must be small enough to catch their tiny floating and swimming food. The way of life of a marine herbivore is, in fact, rather like that of a terrestrial carnivore in that it has to catch discrete moving food, with the fast dissipation of energy that this implies. But the small herbivores, the zooplankton, live suspended in a void with nowhere to hide, and baleen whales have invented a way of catching vast quantities of them with the minimum expenditure of energy; they net them out of the ocean with the sieves of whalebone which flank their mouths. Because they feed much lower in the Eltonian pyramid than do the sharks, they receive a larger energy flow; because their method of feeding is lethargic, they conserve energy; hence, there is enough energy available to their niche for the animals to be enormous and still be common.

Men were once hunters and gatherers of food. Throughout most of the Stone Ages and most of human history, small bands of men were spread across the earth, a sparse population of nomadic hunting peoples. Men were primarily carnivores in the Eltonian pyramids of the time, occupying a niche for which the energy supply was limited. And they used much of the energy they did win in pursuing active lives. Men were scarce. Then they learned to herd animals instead of hunting them, a trick that not only gave men a more certain supply of meat but that also let them squeeze out other top carnivores. The niche of herdsman yielded more energy than the niche of hunter because it ensured a larger supply of the game. Men could become more numerous. Then they tried another trick. They invented agriculture and started acting partly as sedentary herbivores. There was even more energy available to this niche, and men became more numerous still, founding cities and civilizations. So the history of our kind can be read in terms of the Eltonian principles. We started as hunters, active and rare. Then we climbed down the Eltonian pyramid, displacing first other carnivores by herding, and then reaching the bottom of the pyramid by becoming herbivores. We displaced very many of the original herbivores and appropriated their energy flows for ourselves. We used this energy to make more bodies with. And we conserved energy by

HUMAN HISTORY
AND THE ELTONIAN
PRINCIPLES

being sedentary, thus freeing still more energy to make bodies with. The end result of this process is the rice-eating peasantry of Asia, with the densest local populations known. The comparatively carnivorous Americans and Europeans are much thinner on the ground. But their numbers continue to rise. Soon they will give up meat-eating and adopt peasant habits eventually like those of Southeast Asia. Unless, of course, they prevent their populations from rising further.

ELTONIAN PRINCIPLES AND THE ECOSYSTEM

The four phenomena identified by Elton (who called them "principles") were:

Food chains and the food cycle
Size of food
Niches
The pyramid of numbers

and the fifth principle of Lindeman and Hutchinson is:

Energy flow

These five principles, put forward as guides to the understanding of the structure of animal communities, led to more than this. They provided an intuitive understanding of how ecosystems worked; linking the animals to the plants and the physical environment. Onto the surface area of any ecosystem there flowed a regular source of radiant energy. This energy source drove the climatic cycles that are manifested as weather, providing rainwater and other environmental necessities of life. Part of the energy supply was captured by green plants, enabling them to manufacture large molecules and start the food cycle. The food passed through various pathways, through decomposers, herbivores, and carnivores, and thence back to the physical environment. In the plant tissues and with every transfer of food, work was done, degrading energy to heat and radiating it out, first to the atmosphere and then to outer space. The food cycles were driven by the constant energy-flow of solar radiation. Like a man puffing at a toy windmill, the sun kept the cycles in motion; the systems only continue to go round and round because of the outside energy source. Animals successively further up food chains were geared more and more remotely to the same prime energy source, feeling only what was left of the puff after much of it had been dissipated by doing work for others.

At last, we had a conceptual model of the way the natural world worked, one that seemed to hold promise both for meaningful measurements of the doings of living things, and perhaps for an eventual predictive theory of ecology. Every ecosystem started with a finite and measurable energy source. What it did with this energy determined how many plants and animals there could be. We should measure the energy requirements of the ecosystem, piece by piece. We should try, through measuring (if we could) the parameters of the niches of animals to see what limited the efforts of any one kind to get energy. We should try to understand the forces that translate these restrictions on a species' energy-getting power into a population of a certain size: to understand the regulation of populations within an ecosystem. And we should try to measure the influence of one ecosystem on another, so that one day we might have a working idea of the globe as one ecosystem. Most of modern ecology can be seen as pursuits of these goals.

The realization that food energy is the final arbiter of the activities of animals and plants, and hence of their numbers and sizes, makes it important for an ecologist to know how efficient plants are at converting solar energy into organic molecules. Much has been learned from the work of plant physiologists. A first finding is that only half the available solar energy, essentially that comprising visible light, is used by plants in photosynthesis, apparently because only light of these wavelengths is sufficiently energetic to cause the required electron transitions. The efficiency at which the remaining wavelengths are used can be measured by monitoring the rate at which sugar is produced and comparing the calorific value of this sugar with the calories supplied to the plant as light. Rates of sugar production are estimated from gas exchange measurements. Laboratory studies suggest that all plants in conditions of dim light can convert solar energy into sugar with an efficiency of about 20 percent. In bright light, however, this efficiency falls progressively leveling off at about 8 percent, and this is true for all plants whether small algae or large land plants. This falling off of efficiency in bright light is apparently due to the scarcity of the essential raw material of photosynthesis, carbon dioxide. Physiologists thus have shown how well plants do in laboratory culture, but ecologists are more interested in knowing the efficiency of plants living under natural conditions and, indeed, of complete pieces of vegetation. Various field measures of growth and gas exchange provide data of this sort and suggest that average efficiencies on good land in the field are only about 2 percent. Actively growing plants of wild vegetation may achieve the 8 percent of laboratory studies, but the average for the whole vegetation throughout the growing season is only 2 percent. This is probably because there are times during the year when plant cover is not complete, thus allowing sunlight to fall on bare ground, or when shortage of water or other physical adversity prevents plants from working at maximum efficiency. Algae are basically no more efficient than land plants, although it is theoretically possible to maintain algal cultures at an efficiency of 8 percent whereas field crops attain efficiencies of no more than 2 percent. The energy cost of such algal culture, and the difficulty of harvest and maintenance, makes this a dubious advantage however. Misplaced belief that algae are more efficient than land plants has been partly carried over into an equally misplaced belief that the oceans are potentially highly productive places for farming. In truth the seas generally are comparable to terrestrial deserts, producing very little; a condition that results from a shortage of such essential nutrients as phosphorus. Improving terrestrial agriculture is likely to be much more fruitful than attempting to farm the oceans. It should yet be noted that the best agriculture does no more than approach the productivity of the vegetation that it replaces, and that our efforts have not been able to increase the efficiency of any plant to even the slightest extent. A high-yielding crop is merely one which lays down much of its production in parts that men like to eat.

CHAPTER 10

THE TRANSFORMATION OF ENERGY BY PLANTS

Of the solar energy that reaches the earth, some is absorbed in the upper atmosphere, but most passes through to reach the solid or liquid surface. Radiations that pass through the atmosphere are either reflected from bright surfaces or are absorbed in the earth's crusts and oceans. The radiant energy absorbed in rocks, water, and living things is degraded to heat and eventually radiated outward. These far-red radiations warm the atmosphere and do work driving the circulation of the air, the effects of which we know as weather. The warmed atmosphere itself radiates heat outward to space. The average daily input of energy from the sun to the earth is exactly balanced by the average daily energy lost from the earth by radiations to outer space, providing a condition known to physicists as a "steady state."

Some large part of the energy being degraded to heat at the surface of the earth is absorbed by inanimate objects, by rocks and water, serving only to raise their temperature. But part of the total solar energy is absorbed by green plants. The plants do work in the process of photosynthesis to convert some of the energy they absorb into the potential energy of glucose, and it is this portion of the total solar energy which is the ultimate energy source for all living things. The rate at which solar radiations are converted to the potential energy of organic compounds sets a limit to the activities of life and thus is the final arbiter of the distribution and abundance of animals and plants. An ecologist must attempt to gauge the limits set by this process of energy storage. How efficiently do the most efficient plants convert solar energy to glucose? Are some plants, or plant communities, more efficient than others? What limits this **ecological efficiency** of plants?

The results of photosynthesis can be summarized in the well-known photosynthesis equation as follows:

$$6CO_2 + 6H_2O \xrightarrow{\text{Energy}} C_6H_{12}O_6 + 6O_2$$

This equation is poor chemistry because it shows only the beginning and ending of a complex series of reactions in which many other molecules take part, but it is useful because it shows the proportions of gases given off and absorbed during the synthesis of glucose. The rate at which glucose is synthesized in a plant is hard to measure directly, but it is a comparatively simple matter to measure the rate at which oxygen is evolved by a plant or the rate at which carbon dioxide is absorbed, rates that are proportional to the synthesis of glucose and enable this to be calculated. In practice, allowance must always be made for the fact that plants are continually respiring, a process in which they absorb oxygen, release energy from glucose, and give off carbon dioxide, the exact opposite of photosynthesis, in fact, as follows:

$$C_6H_{12}O_6 + 6O_2 \xrightarrow{\text{Energy}} 6CO_2 + H_2O$$

The gas exchanges of respiration may be measured in the dark, and then allowance made for the respired gases when measurements are made on a photosynthesizing plant in the light. Simple chemical titrations, spectroscopic methods, rate of evolution of oxygen bubbles in water, and measurements of pressure changes produced by bubbles of oxygen over water have all been commonly used. The essence of the method may be summarized as follows:

1 Measure oxygen added to container in unit time in light.
2 Measure oxygen taken from container in unit time in dark.
3 Add volume O_2 *added* in (1) to volume O_2 *lost* in (2) to derive total O_2 produced by photosynthesis in unit time

OR

4 Add volume CO_2 *lost* in unit time in light to volume CO_2 *added* in unit time in dark to derive total CO_2 used by photosynthesis in unit time.

An obvious objection to all these methods is the assumption that respiration by day and night are the same. We calculate the volume of CO_2 respired in daylight from estimates of what the plant does in the dark. This is surely not fair. When the plant is working away at synthesis in the daylight hours it should be respiring more than at night when it is doing less obvious work. There is no easy way round this difficulty (Westlake, 1963). It means that our measures are probably marginally too low, but this is not important for the main argument.

Men have measured the rates of photosynthesis of algae and detached parts of large plants in laboratories by various methods based on gas exchange for about three-quarters of a century; under more and more varieties of experimental conditions and for an ever-increasing list of plants. There is thus in the literature of plant physiology a great mass of data on which one can draw to answer ecological questions about the efficiency of photosynthesis.

A first finding is that only visible light is used, in spite of the fact that half of the solar energy reaching the surface of the earth consists of infrared radiation. These red radiations may warm the plant, thus affecting the rate of photosynthesis, but they are not used for the synthesis itself. They are in a sense wasted, and we must in some way account for this. One explanation sometimes offered calls on the early history of evolution, for we believe that life, including photosynthetic life, began in water where radiations of the far red do not penetrate. Water absorbs red light much more strongly than blue light so that a water-plant is virtually shielded from infrared radiations. A form of synthesis evolved in water would not be adapted to red light, making it possible to claim that failure to use red light by land plants is a heritage of an aquatic ancestry. But this explanation requires that photosynthesis could not adapt to use energy of the red even over the time-span of several hundreds of millions of years during which plants have lived on land. This is something biologists should always have found hard to believe. And the matter is anyway satisfactorily explained by matters of quantum energy.

Different wavelengths of electromagnetic radiations both have different energies and different chemical effects on the molecules that absorb them. There is seven or eight times as much energy in a quantum of light at the blue end of the visible spectrum as in a quantum of light in the near infrared wave band in which much of the total solar radiation is received. And the more intense blue wavelengths produce highly specific electron transitions. When they are absorbed at particular chemical loci, their energy is largely used to raise the energetic state of electrons. Photosynthesis works by directing the way in which these excited electrons fall back to lower energy states. But much of the lower intensity radiation in the red wavelengths causes only diffuse vibrations when absorbed. There are then few specifically excited electrons for the plant to direct, and most of the energy becomes dissipated as heat. This state of affairs is illustrated in Table 10.1 Photosynthesis is only possible with radiations that largely result in electronic transitions. Low wavelengths produce mostly thermal vibrations, and the highest wavelengths are ionizing. About half the total solar energy, essentially visible light, produces the required electron transitions, and it is this half that is used in photosynthesis.

Solar Energy Distribution and Photochemistry   TABLE 10.1

Region	Wavelength range (μ)	Energy range (kcal/Einstein)	Fraction of solar spectrum (%)	Molecular changes
Far ultraviolet	0.01– 0.2	152. 2–304.4	0.02	Ionization
Ultraviolet	0.2 – 0.38	75. 3–152.2	7.27	Electronic transitions and ionizations
Visible	0.38– 0.78	36. 7– 75.3	51.73	Electronic transitions
Near infrared	0.78– 3	9. 5– 36.7	38.90	Electronic and vibrational transitions
Middle infrared	3 –30	0.95– 9.5	2.10	Rotational and vibrational transitions

Only radiations that result largely in electronic transitions can be used by photosynthesis. Ionizing radiations cannot be used, nor can low intensity radiations which result largely in vibrational transitions and dissipation as heat. As a result, photosynthesis effectively uses only visible light. (From Morowitz, 1968.)

Laboratory measurements of the rate of photosynthesis are usually made on microscopic algae in small containers of water or nutrient medium, or on whole leaves or disks cut from leaves of higher plants, or on whole seedlings. Artificial light sources enable the energy supplied to be measured directly, such physical parameters of the environment as temperature can be carefully controlled, and the gas exchanges of the photosynthesizing plants can be assessed with great accuracy. The rate of photosynthesis is given by the rate at which oxygen is evolved. At weak light intensities, this rate of photosynthesis is found to be proportional to the light quanta supplied; more light results in more photosynthesis, which suggests that the process at these light intensities is limited only by the amount of usable light energy available. The efficiency of energy conversion is given by the ratio of energy supplied per unit time to the energy equivalent of carbon compounds synthesized per unit time thus:

Efficiency of energy conversion

$$= \frac{\text{energy supplied per unit area per unit time}}{\text{energy equivalent of carbon compounds synthesized per unit area per unit time.}} \times 100$$

For small laboratory cultures under the best contrived conditions, and in low light intensities, the efficiency of energy conversion turns out to range from 15.7 percent for the red end of the visible spectrum (400 angstroms) to 27.5 percent at the blue end (700 angstroms) (Gaastra, 1958). The average efficiency for photosynthesis in ideal conditions and in natural sunlight at low intensity may thus be taken to be roughly 20 percent. Such efficiency seems good. Much energy must clearly be used to do the complex work of photosynthesis, suggesting that 20 percent is probably the best that can be allowed on thermodynamic grounds. Arguments about why plants can apparently never do better than 20 percent when

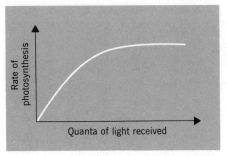

Figure 10.1

supplied with natural sunlight require a theoretical and quantum-dynamic approach. For an ecologist there is the more pressing question of why plants in nature in fact rarely achieve efficiencies comparable to their demonstrated capabilities of 20 percent; since they very seldom do attain such efficiencies.

If the light intensity of the experimental system is progressively raised there comes a time when the increase in the rate of photosynthesis begins to decline until finally the rate remains constant and cannot be raised, however much the quanta of light received are increased. Indeed, at very high light intensities the rate of photosynthesis may actually be depressed, an event that is thought to be due to photooxidation of some of the enzymes crucial to the process. That there should be an intensity of radiation which damages the plant system seems logical, but that there should be a limit to photosynthesis under quite moderate light intensities need not be anticipated on logical grounds. A curve of rate of photosynthesis plotted against quanta of light received within the normal range of radiation looks like Figure 10.1. It seems that, in light intensities above a critical level, light no longer sets limits to the rate of photosynthesis but that some other factor must do so instead. Obvious physical parameters such as temperature or pH do affect photosynthesis, but they affect the performance at low light intensities as well as at high, altering the values of points on the curve but not its overall shape. But changes in one factor, carbon dioxide, will move the leveling off point, raising the asymptote if supplied in excess and lowering it if in restricted supply. The effect of increasing the concentration of carbon dioxide is illustrated in the diagram in Figure 10.2. Thus no doubt exists that it is the supply of carbon dioxide which limits the rate of photosynthesis in bright light. Although in dim light the rate of photosynthesis is set by the energetics of the process, it is the supply of the raw material, carbon dioxide, that sets a limit when the energy flux is large. This need not be surprising, for carbon dioxide is a scarce component of the atmosphere, being present at about .03 percent by volume. There must obviously be a limit to the speed with which so rarefied a substance as that can be extracted. So we may give a partial answer to the question, "why do plants in nature

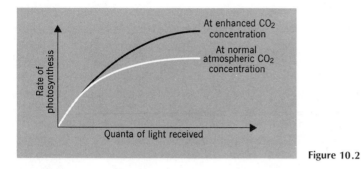

Figure 10.2

fail to fix energy at their theoretical efficiency of 20 percent?" by saying, "because the carbon dioxide needed for the process is in short supply."

Adding carbon dioxide to an experimental system cannot raise the rate of photosynthesis indefinitely, however, for there is always an asymptote beyond which the addition of more carbon dioxide fails to raise the rate whatever the light intensities used. A general explanation for this might be that the photosynthetic mechanism of plants is adapted to the range of carbon dioxide concentrations actually found in nature rather than to some theoretically ideal concentration. The carbon dioxide concentration near where plants live does vary a little, particularly, say, at night under the canopy of a forest where the respired gases of the living things of the forest collect, so that it is reasonable to expect that plants have adapted their mechanisms to take the best advantage of the carbon dioxide supply that may be present at any one time. This is why the efficiency of photosynthesis can be raised a little by pumping more of the gas into an experimental container. Such is a general explanation, but we can produce a more precise model, if a somewhat speculative one, following the analysis of Bonner (1962).

Green leaves, the homes of chloroplasts and the sites of photosynthesis, are stacked one above another in nature. A moment's thought about the inside of a forest shows how dramatically true this is; the layers of the canopy are piled above each other so effectively that only a dim green light may reach the floor, giving that stained-glass window effect noted by Stanley in his rain forest wanderings (Chapter 2). The same is true for crop plants, where light may have to penetrate between five and ten leaves before what is left of it reaches the soil, and the same is true even for microscopic algae floating above each other in a fertile pond. Under the bottom leaf or algal cell, the light is often so dim that a plant attempting photosynthesis there would not be able to fix energy as fast as it used energy for respiration, and so could not survive. Now this phenomenon of leaves shading each other means that many of them must operate in dim light, being highly efficient and making the most of both the light and the carbon dioxide around them, even when the upper leaves are inhibited by carbon dioxide shortage. But operating in dim light requires an elaborate array of absorption sites (the chlorophyll molecules) to make sure that quanta of light are absorbed fast enough to "feed" the sites of synthesis. Ten quanta of light seem to be needed for every mole of carbon dioxide reduced, and the 10 quanta must be supplied at once. All plants have been found to have what appears to be a surplus of chlorophyll molecules, for there may be hundreds or perhaps thousands of them for every synthesis site we can identify instead of just the 10 that seem to be needed to supply the site with its 10 quanta per mole of CO_2 reduced. Bonner suggests that this is because a large excess of absorption molecules is needed to maintain the necessary continuous supply of 10 synchronous quanta. The shaded leaves and chloroplasts thus work marvelously, but those exposed to bright light have more chlorophyll molecules than they need. These excess chlorophyll molecules absorb light from the bright supply but there are not enough synthesis sites to use it all, and the energy is wasted. Plants cannot, of course, afford to have too many molecules that are redundant, for the molecules themselves then represent a waste of energy. What natural selection must have done for plants is to arrange a compromise between the large number of absorption sites needed for dim-light work, and the lesser number needed for bright light. But the fundamental fact of plant life which has forced this compromise to favor the dim-light arrangement is the shortage of the essential raw material, carbon dioxide.

At high light intensities the efficiency of photosynthesis, as we have seen, may be comparatively low, (Figure 10.3). Cultures of the freshwater alga *Chlorella* (in Dutch experiments) which were 20 percent efficient in dim light in the laboratory were only about 8 percent efficient in full daylight, even when supplied with extra carbon dioxide (Wassink, 1959). It is important to remember, however, that the actual productivity of the plants in terms of dry matter produced in unit time was greater in full sunlight than in the shade; the efficiency of energy conversion might be lower but there was so much more energy available that this was more than compensated for. There has been some confusion in popular writings of late about the suitability of algae for food which bases hopes for the future on the high efficiencies of algae in dim-light cultures. In the bright lights of the real world these high efficiencies fall. Indeed, parts of higher plants can be made to show efficiencies equal to those of *Chlorella* cultures in dim light, although the whole plant exposed to the unimpeded sun seems relatively inefficient; just like the algae.

THE PRODUCTIVITY OF VEGETATION AND CROPS

Measuring the efficiencies and productivity of communities of plants in nature presents more difficulties than measuring the rate of photosynthesis of laboratory-sized fragments or collections, but reliable estimates can be made by roundabout means. The first and classic estimate of the kind was made in the American Midwest in 1926 by Nelson Transeau of The Ohio State University (Transeau, 1926). The worry of Transeau's day was not only food but fuel also. Fossil fuels on which modern civilizations are based obviously would not last forever, and the exhaustion of the world supply then seemed more imminent than it does now

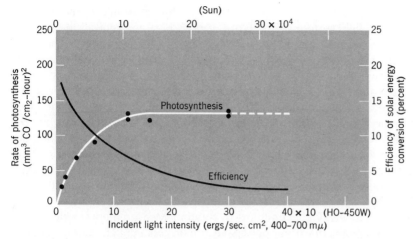

Figure 10.3

Rate and efficiency of photosynthesis in a sugar-beet leaf as a function of light intensity. The rate of photosynthesis could not be raised for light intensities much above 10 ergs per second per square centimeter, and the efficiency of the process accordingly progressively falls. It is important to note, however, that the yield of energy to the plant was highest at high light itensities even though the efficiency was low. Note also that this sugar-beet leaf in dim light was quite as efficient as a green alga culture in equally dim light. It is untrue to claim that algae are more efficient than other plants. (From Gaastra, 1958.)

because the total reserves were much underestimated. There was then no promise of atomic power to supply the needs of the future, and the only resource seemed to be to grow fuel; to use the energy stored by contemporary plants rather than by those long dead. It became important to know how efficiently crops fixed solar energy in order to calculate the possibilities of producing fuel for power stations and motor cars by agriculture. Transeau chose corn (*Zea mays*) as his crop plant because corn was apparently a highly productive plant, because it was an annual that grew progressively through the season making it easy to assess one season's growth, and because there was good data on corn plants to be culled from the literature. An acre of good corn land in Illinois (a site chosen because there was good local data on solar radiation) should contain about 10,000 corn plants which should grow from seed to maturity in 100 days. Accepting these generalized figures, and armed with a copy of an agricultural handbook, Transeau could proceed as follows:

Total dry weight of 10,000 corn plants (roots, stems, leaves, and fruits)	6000 kg
Total ash content of 10,000 corn plants (minerals from soil left after burning)	322 kg
Therefore, total organic content per acre	5678 kg
Average organic matter contains 44.58 percent of carbon, therefore carbon per acre equals	2675 kg

and 2675 kg of carbon is enough for 6678 kg glucose.
Therefore, the corn plants produced on one acre in 100 days represents 6678 kg of glucose fixed in photosynthesis.

This figure of 6687 kilogram represents the **net primary production** of the corn field in terms of glucose. It does not represent all the glucose fixed by photosynthesis because the plants were respiring all the time, burning some of the glucose that they had fixed to provide energy for their own metabolic needs. This amount of glucose respired must be assessed and added to that represented by the **standing crop** of plants at the end of the 100-day period to arrive at all the glucose fixed, or the **gross primary production** of the corn field. From this figure it would be possible to calculate all the energy stored by photosynthesis, and to compare this to the solar radiation received to arrive at the **ecological efficiency.** Transeau measured the respiration himself by keeping plants in dark chambers through which he passed continuous streams of air. He collected the carbon dioxide at the outlet in an alkali solution and estimated the quantity evolved in unit time by titration. Measurements made on typical corn plants of various ages gave him an average figure for respiration of 1 percent of the mass of each plant per day, which enabled him to complete his calculation as follows:

Since the crop at the end of the season weighed
 6000 kg, the average dry weight for the season was   3000 kg
Average respiration was 1 percent of this which   =   30 kg
Therefore, the total CO_2 released in 100 days is
 30 times 100   = 3000 kg
Carbon equivalent of 3000 kg CO_2 = 818 kg
Glucose equivalent of 818 kg carbon   = 2045 kg
Gross primary production of glucose equals net
 primary production plus respiration equals 6687 kg
 plus 2045 kg   = 8732 kg
But the energy required to produce 1 kilogram
 of glucose is 3760 Cals (a figure found by
 bomb calorimetry).
Therefore: Total energy consumed in photosynthesis of
 one acre of corn in 100 days equals 8732 times
 3760 equals approximately   33,000,000 Cals
Energy received by one acre of Illinois in
 100 days equals   = 2,043,000,000 Cals
Therefore, efficiency of photosynthesis equals

$$\frac{33 \times 10^6}{2043 \times 10^6} \times 100 = 1.6 \text{ percent}$$

This figure of 1.6 percent has since been shown to be in quite the normal range for agriculture. There have been many field trial experiments on a variety of crops following Transeau's general procedures, but with painstakingly accurate measurements on actual crops and of actual sunlight received by the crops, and the answers always come to something like 2 percent. This compares poorly, of course, with the results of short-term experiments with algal cultures or seedlings in bright sunlight, let alone with dim-light laboratory cultures, revealing efficiencies of 20 percent. The limit imposed by the supply of carbon dioxide should allow higher efficiencies than 2 percent; indeed, the efficiencies revealed by field trials are rather shocking. An ecologist must ask if such low efficiency is typical of natural plant communities as well as of agricultural crops and, if so, what is the reason for it.

For an agricultural crop of an annual plant it is easy to measure the **net primary production** during a season by measuring the **standing crop** at the end of the growing period as Transeau did, and knowledge of the average respiration rate of the crop plants then makes it easy to calculate the **gross primary production.** But in complex natural communities of plants the measurement of standing crop at no time gives the net production, because the plants are continually being eaten by animals and because part of the mass of some of them may have been produced in other seasons. A moment's reflection about the standing crop of grass in a well-grazed pasture is enough to show the truth of this, for much of the production of the pasture is eaten off daily by the cattle leaving only a little of the production of the field standing at any one time, and some of the grass roots were made the year before. Only where plants of the same age grow to maturity in a short time without being much eaten can the standing crop method be used in productivity measurements. The growth of annual plants in a desert after rain, and the development of some early successional communities may sometimes meet these conditions but, for most natural communities, the measurement of standing crop does not provide interesting information, and other methods must be used to assess productivity.

Estimating the productivity of a forest is a particularly daunting undertaking, although various ways have been tried. Howard Odum (1971) has tackled the problem by applying the gas exchange method to fairly large pieces of natural vegetation. He erects giant plastic tents or sleeves to enclose portions of the forest, then passes air through the sleeves while monitoring the carbon dioxide content of the inlet and outlet air. The carbon dioxide evolved in the respiration of both plants and animals by night is added to the net loss of carbon dioxide to the community in a similar period of daylight to give the total carbon dioxide absorbed, or the rate of **gross primary production.** Odum has even applied this method to the tropical rain forest (Figure 10.4), and others have applied it to various small pieces of vegetation. The carbon dioxide content of a stream of air can best be measured with an infrared gas analyzer, a spectrographic instrument, and for those who derive pleasure from operating electronics under adverse conditions, there is adventure to be found in carrying infrared gas analyzers to remote vegetations, like those of mountaintops, to determine productivity.

The plastic sleeve technique has the disadvantage that plants are in very unnatural surroundings, which must commonly make the results suspect. A perhaps more reliable alternative, if even more time-consuming, is that of **incremental harvest.** All the plants of the forest are sampled at frequent intervals to plot the actual rates of growth of their parts; of their stems, leaves, roots, and the like. Woodwell and Whittaker (1968) have made enough measurements of this kind on a forest at Brookhaven in New York State to be able to program the results for a computer and to

Figure 10.4

Odum's "sleeve" in a Puerto Rican rain forest. A portion of the forest is enclosed in this open sleeve. Air flows through it from top to bottom, and the content of carbon dioxide in the inlet and outlet streams is continuously monitored. Carbon dioxide lost by day added to carbon dioxide gained by night gives a measure of the gross primary production of this portion of the forest. The disadvantage of the method is that the sleeve may produce conditions so unnatural that the measure may not be applicable to undisturbed forest.

relate incremental growth rates to the stem sizes of trees. They have identified key measurements which are all that are required to supply their program with sufficient information to calculate the productivity of the whole forest. This method, of course, measures only the rate of **net production,** since it does not provide a measure of the respiration of the plants. A nightime figure of respiration of whole communities like that at Brookhaven can sometimes be made when the forest air is trapped under a low-lying inversion. In this event all the carbon dioxide respired is kept in the vicinity of the forest, so that respiration may be calculated from the local increment of carbon dioxide.

The habit has grown up of expressing the results of field determinations as grams of dry matter per square meter per day, instead of as calories or as efficiencies. The rate of production in gms/M²/day is referred to as the **productivity** of the plants, and the productivity of Transeau's 1.6 percent efficient corn in these units works out as about 21gms/M²/day.

Howard and Eugene Odum (1959) have collected together a number of determinations of productivity made by gas exchange and growth-increment studies and have summarized them as follows. Moist forests, moist grasslands, and communities of the secondary successions produce between 3 and 10 gms/M²/day, or commonly less than a field of corn. Natural grasslands and mountain forests produce only between 0.5 and 3 gms/M²/day, and deserts produce less that 0.5 gms/M²/day. It is clear that water limits the productivity of communities such as deserts, and that corn must be compared with plants growing where corn grows. But there is one immediate difference between the calculations for the cornfield and for the natural communities, which must be allowed for; the corn only grew for 100 days and the field was left bare for the rest of the time. If you divide the figure for gross production of corn by more days than 100 you can bring its apparent productivity down to something comparable to the performance of the moist forests and grasslands at up to only 10 gms/M²/day, suggesting that the efficiencies of these diverse communities of plants, agricultural and natural, are comparable. It is apparent that terrestrial plants growing naturally under the sun in apparently favorable habitats achieve only low efficiencies of energy conversion, some 2 percent of the usable energy received during the growing season, in spite of the fact that individual leaves or seedlings can do much better in the laboratory. Why is this?

An agricultural crop is obviously not as efficient at converting the solar energy falling on its field at all times in the growing season, since at the beginning most of the field is bare earth and at the close growth has nearly stopped. There must be some period of active growth when the productivity of the plants is much higher than the average. Work by Gaastra (1958) at the Dutch university at Wageningen, where much original work on photosynthesis has been performed, shows how different the average productivity can be from the best. The average efficiency of a sugar beet crop at Wageningen was about 2.2 percent, but in the middle of the season when the plants were of a fair size, almost covering the field and growing at their fastest, the efficiency was between 7 and 9 percent. This is equal to the best that can be obtained on a culture scale under full sunlight, even with algae or isolated leaves, and the rate of photosynthesis of the fast-growing beets during this important period of their lives must have been limited only by the supply of carbon dioxide. The apparently low efficiency of a field crop is thus a simple result of the energy wasted falling on bare ground at the start of the season, and of energy wasted falling on senescent plants toward the close of the season. During growth, the plants produce as fast as the environment allows.

Most plants of natural vegetation are not annuals that start the season on bare ground, so the explanation for the low annual efficiency of agricultural crops might be thought not to apply. In fact, the annual cycle of vegetation in temperate regions is closely analogous to that of a field crop, for in spring there are no leaves, and thus effectively there is bare ground, and with the approach of autumn the plants become progressively more dormant. It is likely that there is a period in early summer when a temperate woodland or pasture might well produce at something close to the rate set by the supply of carbon dioxide. That it does not continue to do so indefinitely is due to the changing environment for which the plant must prepare. For vegetations with the lowest productivities, deserts and arid grasslands, it is evident that productivity is set by shortage of water. It would be interesting to know the productivity achieved by desert plants during a short spell of rains; during the few days when the annual desert flowers spring up. It would not be surprising to find efficiencies of 7 or 8 percent, indicating rates of photosynthesis set by the available carbon dioxide.

It is thus possible to explain the low average ecological efficiencies of plants by a general hypothesis of environmental limitations; by the shortage of water, by the time taken to cover the ground in spring, by the dormancy that precedes the onset of winter. But this cannot be quite the whole story. An established woodland does not really take as long to cover the ground as does a field crop and, once the leaves are out, the ground cover is more complete so that little light is wasted. Trees even have lower "shade" leaves adapted to operate in the dim lights below the canopy, leaves that are highly efficient at these intensities. And the leaves of the canopy may function after the time of harvest of a crop. In spite of these apparent advantages, the average productivity of a deciduous woodland is, in fact, not so much higher than that of a field crop. On the other hand, rapidly developing natural communities like those of the secondary successions in warm sites well supplied with nutrients and water may be the most productive of all. These findings can be explained if plants are adapted to produce at the maximum possible rate only during the time needed for them to occupy space and achieve dominance; the period of rapid growth.

The pattern of efficiencies throughout a plant's life must be the result of natural selection, and selection operates by favoring individuals that leave the most offspring. To be able to reproduce at all, a terrestrial plant must win and hold space, so that selection must favor individuals which win space. This can only be done by rapid growth; and a limit to growth must be set by the supply of energy, which means by the rate of photosynthesis. It is not surprising, then, to find that plants generally achieve photosynthetic efficiencies during their period of vitally active growth at about the level which is set by the supply of carbon dioxide or other resources. Once space is won and held, as by a dominant tree that has set its leaves, the necessity for maximum energy supply may not be so crucial.

THE PRODUCTIVITY OF ALGAE, AND ALGAL CULTURE

Rapidly dividing unicellular algae, which live in a fluid medium, face different ecological problems to those of rooted terrestrial plants. They do not have to occupy space and achieve local dominance, but they do have to continually replace themselves. They live, in a sense, lives of perpetual youth in which no secure old age of dominance is achieved. As a result, they must always grow as fast as possible, for those who are slow will leave least offspring and will be selected against. Natural selection ensures that the algal population is made up of individuals that produce as fast as the limiting factors of the environment allow, which means that their productivity in full sunlight is always set by the carbon dioxide and other nutrient concentrations normally experienced. Throughout its short life, an individual microscopic algal plant maintains a rate of photosynthesis comparable to that of a terrestrial plant during the period of fastest growth.

There has been much interest in culturing algae to increase human food supplies, such that popular writings often talk of algae as providing the common food of the future. This interest has, in part, derived from the misconception that the high efficiencies, around 20 percent, of dim-light cultures could be maintained in field cultures. But it remains true that a properly managed algal culture should be able to maintain an efficiency of 8 percent for as long as sun and warmth allow, so that high productivity can be maintained all through the growing season, instead of just through part of it as with field crops. Eight percent is more than the 2 percent average of a field of sugar beet, giving four times as much food. Microscopic algae such as *Chlorella* can in theory be grown in large concrete tanks. The practical side of the matter is not so encouraging. To obtain high yields the tanks of algae have to be continually stirred, so that all the inmates get their share of nutrients and light, and such stirring requires energy, energy that has not been included in the published efficiencies. This point may be brought home by noting that men at the Wageningen station found that the most effective culturing technique was to grow the plants in the revolving chambers of washing machines, (Wassink, 1959) surely an energy-consuming process of some seriousness. Then again energy must be used in the frequent harvestings necessary to maintain the rapidly reproducing plants, and this energy must be included in realistic energy budgets. As the tanks get larger, it becomes impossible to keep out herbivorous plankters which are very willing to share the yield of the cultures with the men who grew them, and who are remarkably good at the task of cropping algae. Really large-scale operations have never been attempted. There may be a future for algal culture in places where conventional agriculture is not possible, say on rock slopes, but the general contribution of microscopic algae to human food supplies is likely to be marginal.

The plants of the open sea are microscopic algae similar to those used for culture purposes in freshwater, and may be expected to be as efficient. Culture experiments, summarized by Ryther (1959), show that marine planktonic algae are, indeed, as efficient as their freshwater counterparts. There should never be a water shortage in the sea, and the surface of the transparent ocean is well-lighted. The plants are insulated against violent changes in temperature, from day to night and from season to season.

PRIMARY PRODUCTION
IN THE SEA

Rivers have carried dissolved minerals to the sea throughout geologic time to such extent that the oceans are salty. Surely, then, the oceans should be highly productive.

One standard method of measuring the productivity of the sea evolved from the work of two Scandinavians, Gaarder and Gran (1927). They put seawater containing planktonic algae into bottles, one of clear glass and the other covered with black paint; then they lowered the bottles over the side of their ship and left them in the sea for a measured time. The light and dark bottles were equivalent to the light and dark tents used on land, but Gaarder and Gran estimated the oxygen content of each bottle rather than the carbon dioxide content, because dissolved oxygen is easily measured by simple titration. The light bottle had *gained* oxygen, due to photosynthesis, and the dark bottle had *lost* oxygen, due to respiration. The oxygen gained in the light bottle added to the oxygen lost in the dark bottle provided an estimate of the gross primary production of the time interval. It was thus possible to measure the productivity of the wild "vegetation" of the open ocean with much greater ease than the productivity of land vegetation. And then a simpler method still was evolved. A bottle of seawater could be supplied with a carbonate solution labeled with C^{14}, an isotope whose long half-life of more than 5600 years means that it effectively endures without loss over the span of normal experiment, and this C^{14} labeled carbonate is then used by the planktonic algae in the bottle as the raw material for photosynthesis. The amount of labeled carbonate that the plants have taken up can later be found by filtering out the plants and estimating their C^{14} content with a radiation counter. This method was worked out by Steeman-Nielsen (1952), and was used on the famous Galathea deep-sea expedition. He sailed equipped with vials of measured amounts of labeled carbonates, took flasks of seawater, inserted the contents of a vial with a hypodermic syringe, left the bottles in the sun for a period of one or more hours, then filtered out the plants for radiation determinations. Steeman-Nielsen's results were always lower than those obtained by the dark-bottle light-bottle method of Gaarder and Gran, and this is thought to be because plants respire some labeled $C^{14}O_2$ back to the water. The method gives a figure that is closer to net production than to gross production, although neither exactly one nor the other (Strickland, 1966). But the method is so simple and convenient that it has become standard all the same, so that oceanographic cruises the world over measure the productivity of the oceans in this way. The results give approximate figures, and have the great virtue of being comparable.

The results of many years of measurements by both methods (and some others) reveal most clearly that the open oceans are mostly deserts, the productivity of which is less than 1 gm/M²/day (Table 10.2). It is important to grasp this fact. Transeau's corn did more than twenty-one times as well as does the average ocean. In spite of the salty sea being flooded with light, and in spite of the plants being microscopic algae, the open oceans are deserts. This is contrary to popular belief that the oceans are marvelously fertile resources waiting to be tapped. Partly this idea originated with some early measurements which were grossly in error, and

Rates of Primary Production from Various Parts of the World Oceans   TABLE 10.2

Location	Season	grams carbon m²/day	Method
English Channel	June	0.50	^{14}C
North Sea	Annual Range	0.1–1.5	^{14}C
North Sea near English Coast	May	0.22	^{14}C
North Sea near English Coast	October	0.11	^{14}C
Kattegat (Baltic)	October	0.25	L + DB
Danish coastal waters	August	0.70	^{14}C
Danish coastal waters	March	0.30	^{14}C
Danish coastal waters	December	0.01	^{14}C
Western Barents Sea (Artic water)	May	1.30	^{14}C
Western Barents Sea (Atlantic water)	May	0.275	^{14}C
Mediterranean	Midsummer	0.03–0.04	^{14}C
Eastern Atlantic (15 miles offshore)	September	1.0	^{14}C
Eastern Atlantic (200 miles offshore)	September	0.15	^{14}C
Georges Bank (off New England)	April	0.95	L + DB
Sargasso Sea	May	0.04–0.05	^{14}C
Pacific off Ecuador (fishery)	Autumn	0.5–1.0	^{14}C
Equatorial Pacific	Autumn	0.01	^{14}C
Equatorial Pacific	March	0.10–0.25	^{14}C
Sea of Japan	May	2.0	^{14}C
Off Southwest Africa (inshore)	December	0.5–4.0	^{14}C
Arctic Ocean (ice island)	July	0.024	^{14}C
Transeau's corn field		21.00	–
Reef at Eniwelok		24.00	–

All these determinations were made by ^{14}C or the light and dark bottle method. Note that the productivity of the oceans varies from place to place as well as from season to season, but that everywhere the productivity is much less than that of fertile sites on land or of shallow water communties fixed in currents of nutrient-rich water. Data compiled by Strickland (1960) from various sources.

which may be found summarized in Steeman-Nielsen's (1952) paper. And partly it no doubt derives from the common misconception that algae are more productive than other plants. But even when the low productivity of the oceans per unit area was grasped, the idea of great resources in the sea has been maintained by reflection on the great size of the ocean basins. Seventy percent of the surface of the globe is sea so, even though each bit is poorly productive, the grand total must be large. So went the argument, but even this was wrong. Over the last 15 years progressively better and better measurements of productivity of all regions of the globe have steadily reduced the estimates of the contribution of the oceans, until the latest put the oceans as contributing only some 30 percent of the whole (Ryther, 1970). This figure seems generally in keeping with the discovery that the productivity of most of the ocean is low.

So the oceans are deserts and we must ask ourselves why. A first thought must obviously be that much of the light entering the sea is absorbed by the water or reflected and scattered back to space so that it is unavailable to the plants but, in fact, this is not so. Careful measurements from ramps over the sides of oceanographic ships, and with submersed instruments, have shown that direct backscattering and reflectance account for no more than 5 percent of the incident light, no matter what the height of the waves or the state of the ocean (Ryther, 1959). Light is only slowly absorbed by clear water so that a dense crop of planktonic plants in the surface waters would absorb almost all the incident light except for the 5 percent that is backscattered. It is true that in oceanic waters of typical low productivity much of the incident light is not absorbed by plants, but penetrates to great depths, being steadily absorbed by clear water the while, but this is a consequence of there being a low crop of algae in those waters and not a cause.

Carbon dioxide is usually present in solution in seawater in excess of plant demand, although local shortages may become limiting. Light, temperature, and water are apparently not limiting to photosynthesis in the oceans either, and the explanation of the low productivity is apparently that nutrients are limiting. The total mass of such essential dissolved nutrients as phosphorus in the oceans is immense, but this mass is spread through the volumes of the great ocean basins; is held in the millions of cubic kilometers of water that are not reached by light. The actual mass of solutes in the thin, lighted surface layer is low and is presumably limiting. The low solubility of phosphorus makes this vital nutrient particularly likely to be in short supply. Direct evidence that this is true comes from comparing the phosphate contents of unproductive open water with the sites of the few rich fisheries that are known. Phosphates are scarce in the open sea but abundant at the sites of rich fisheries. And the rich fisheries always turn out to be where there is an upwelling of deep water as some submerged current comes to the surface. This deep water is much colder than the normal sea surface, with its layer of sun-warmed water floating on the top. It is a peculiarity of phosphates that they are more soluble in cold water than in warmer water. So the cold waters of the upwelling are phosphorus-rich. And, moreover, they are continually replaced by the upwelling. The plants of such a place are provided with a continuously renewed supply of nutrient-rich water, and they may well produce at a rate set by the concentration of carbon dioxide like plants in favored places on land. Under the rich life of an upwelling there is a drifting cloud of corpses and other debris with, adsorbed to them, the phosphates which have made that rich life possible. The seafloor in such places becomes enriched with phosphate, providing, in the fullness of geological time, the deposits that we mine for fertilizer.

Another apparent exception to the general low productivity of the sea is suggested by the rich communities of coral reefs, where numerous polyps build the huge reef structures and where exotic marine animals live in great abundance. Howard and Eugene Odum (1955) went to study the reef at Eniwetok as part of the United States Atomic Energy Commission work on the effects of bomb tests in the Pacific, and they investigated this problem of the productivity of the reefs. The reef they chose, like others, had water continually flowing over it, and it had been suggested that reef animals lived by catching animals and plants of the plankton that drifted over the reef in the water from the outside. If this was so, the high productivity of the reef was apparent rather than real, representing energy brought in from outside by the flow of water. But the Odums sampled the plankton on both sides of the reef, upstream and downstream, and found that there was as much organic matter leaving the reef as there was being brought to it. The conspicuous animals of the reef must, therefore, rely on the energy fixed by the plants of the reef itself. This was a somewhat surprising conclusion because plants are not very evident in the structure of a reef. There were certainly some there, living as encrusting calcareous forms or buried as filaments in the surface of the reef matrix, but there did not seem to be many of them. The Odums set out to see if they were really present in significant numbers. They took samples of representative bits of reef, extracted chlorophyl from the bits with organic solvents, measured the concentrations of chlorophyl with a spectrophotometer, and calculated from this the mass of plant tissue in their samples. They found that, in fact, there was more living plant tissue in the reef than there was living animal tissue; the standing crop of plants was larger than the standing crop of animals. Many of these plants were the endosymbiotic algae that lived associated with the corals, the so-called *zoozanthellae*. Although standing crop was not a measure of production, it was now clearly reasonable to account for the reef production by the photosynthesis of the reef plants themselves. The next step was to measure the gross primary productivity of the reef plants, and the Odums did this by measuring the oxygen content of the inlet and outlet waters by day and by night. In daytime, the water crossing the reef gained oxygen during its passage by an amount proportional to the net photosynthesis of the reef plants. At night, there was a loss of oxygen during the passage over the reef due to the respiration of the plants and animals of the reef community. The oxygen gained in the daytime added to the oxygen lost during a similar span of night was a measure of the gross primary production of the reef during the interval. This was an ingenious adaptation of the light bottle/dark bottle technique, one in which the whole community was "in the bottle." The result of the Odums' calculations showed that the reef plants produced about 24 gms/M²/day, which is the sort of result suggested by laboratory experiments for plants working at maximum efficiency and limited only by the concentration of carbon dioxide. This productivity may be explained as due to the plants being

TABLE 10.3   **Efficiencies of Primary Production**

	Kcal/m²/day	Efficiency (in percent)
Transeau's cornfield	33	1.6
Gaastra's sugar-beet field	—	2.2
Sugar cane	74	1.8
Water hyacinths	20 to 40	1.5
Tropical forest plantation	28	0.7
Microscopic alga culture on pilot scale	72	3.0
Sewage pond on seven-day turnover	144	2.8
Tropical rain forest	131	3.5
Coral reefs	39 to 151	2.4
Tropical marine meadows	20 to 144	2.0
Gavelston Bay, Texas (fertilized by wastes)	80 to 232	2.5
Silver Springs, Florida (vegetated bottom)	70	2.7
Subtropical blue water (open sea)	2.9	0.09
Hot deserts	0.4	0.05
Arctic tundra	1.8	0.08

Transeau's and Gaastra's data as discussed in text. Remainder rearranged from Odum (1971).

supplied with a continual stream of nutrients by the water flowing over the reef, and is another vindication of our belief that the sea would be highly productive if it was not for the shortage of nutrients.

The trick to being highly productive in the sea seems to be to anchor yourself against a moving current of water that will bring you an unlimited supply of nutrients. Large benthic algae which live fixed to the bottom in regions of moving water meet this requirement, and might well be expected to produce as well as well-watered pastures on land. The longest vegetables on earth are, in fact, seaweeds that live anchored in cold currents, the giant *Nereocystis* and *Macrocystis* of California. Their flat fronds present large waving surfaces to the flow of water, so they are likely to be able to acquire all the nutrients necessary for photosynthesis. The low temperature of the water in which they live might be expected to depress their performance somewhat, but their condition of life should allow their productivity to be high all the same. It is reassuring to note that some first studies on California kelp beds (McFarland and Prescott, 1959) do suggest that their productivity is comparable to that of the Eniwetok reef.

If a way can be found to make satisfactory food out of kelp and similar large marine algae, it might be possible to find ways of anchoring attachments for some in places in the sea and so farm the oceans on a small scale. Otherwise the prospects for sea farming are not bright. The lighted layers of surface water are so deficient in nutrients that their productivity is comparable to that of terrestrial deserts. This is a hard fact that those who would extensively farm the sea must face. You cannot put enough fertilizer into the huge oceans to raise significantly their concentration of

TABLE 10.4
Annual Primary Production of Principal Regions of the Earth

	Area 10^6 km^2	Net primary productivity, per unit area dry g/m²/yr dry g/m²/yr		World net primary production 10^9 dry tons/yr
		normal range	mean	
Lake and stream	2	100–1,500	500	1.0
Swamp and marsh	2	800–4,000	2,000	4.0
Tropical forest	20	1,000–5,000	2,000	40.0
Temperate forest	18	600–2,500	1,300	23.4
Boreal forest	12	400–2,000	800	9.6
Woodland and shrubland	7	200–1,200	600	4.2
Savanna	15	200–2,000	700	10.5
Temperate grassland	9	150–1,500	500	4.5
Tundra and alpine	8	10– 400	140	1.1
Desert scrub	18	10– 250	70	1.3
Extreme desert, rock and ice	24	0– 10	3	0.07
Agricultural land	14	100–4,000	650	9.1
Total land	149		730	109.
Open ocean	332	2– 400	125	41.5
Continental shelf	27	200– 600	350	9.5
Attached algae and estuaries	2	500–4,000	2,000	4.0
Total ocean	361		155	55.
Total for earth	510		320	164.

Table after Whittaker and Likens, from Whittaker (1971).

essential solutes. Arranging to stir portions of oceans to bring up solutes from below is likely to require more energy than would be recovered in the additional production won.

The low productivity of the sea means that the energy supply per unit area in forms usable by animals is generally much lower than on land, a fact of importance when discussing the distribution and abundance of marine animals.

THE FOOD LIMITS OF THE EARTH

The absolute food limit for all animals, man among them, is set by the average efficiency of plants at the surface of the earth, on land and in the oceans, on fertile floodplains and on mountainsides or polar ice caps (Table 10.3). This average efficiency is hard to calculate, but it is certainly much lower than the 1.6 percent of Transeau's cornfield. Probably it is something like 0.25 or 0.5 percent. And the reason that the average productivity is so low is that photosynthesis is limited in most places not by carbon dioxide but by such things as water, nutrients, temperature, and the seasonality of climate. Sometimes farmers can make good some of these deficiencies, as by irrigation or the use of artificial fertilizer, but others like air temperatures and seasonal climates are beyond a farmer's control. They cannot warm the great ice caps (it might not be a very pleasant experiment if they could try), nor fertilize the even greater oceans. They cannot even water all the deserts, at least, without an expenditure in energy for distilling and transporting seawater which would be greater than the energy won in the food crop from the wetted sand. Much of a farmer's efforts to grow more food is still to make his crops as efficient as the vegetation they have displaced. And usually he is still far from succeeding, as the figures for mean productivity in Table 10.4 show. Crops can be made to be as efficient as the vegetation they replace, but only with heavy energy expenditures in the making of fertilizers and the driving of tractors.

Yet man has done nothing to increase the actual efficiency of plants themselves. The high yielding cereals and things that are often in the news, are not one jot more efficient than the wilder types they replace. All that the plant breeders have done is to make plants which put larger shares of their energy budgets into structures that we like to eat. A high-yielding corn plant produces more grain at the expense of leaves, stem, roots and, as we are beginning to find to our cost, at the expense of defense mechanisms that might protect the plant against diseases and pests. But we have not increased the energy-transforming powers of the corn plant at all. And, except perhaps with the watering of hot deserts, we can do very little to increase the real productivity of the earth as a whole. Indeed, agriculture has so far lowered the earth's productivity, not raised it.

The energy source of an animal community is in the form of high-energy food molecules and the *ecological efficiency* of animals reflects the proportion of this fuel supply that they can subvert for their own use. It is tempting to think that all animals live under conditions of restricted fuel supply so that the numbers and activities of animals must be set by the efficiency with which they can capture and digest food as fuel for their own activities. Then the conceptual device of the trophic level seems to offer a chance for describing, and even predicting, the size of animal communities on energetic grounds, for the size of the herbivore trophic level in a community, for instance, must reflect a percentage of the 2 percent of the solar energy that is transformed by the vegetation of the habitat. But practical and conceptual difficulties interfere with this approach. Real communities are not neatly separated into discrete trophic levels as the concept suggests, and measuring the efficiency with which animals capture and digest their food proves very difficult. After many years of work there is now a widespread opinion that efficiencies may vary widely from one kind of animal to another, even at the same trophic level, but that an average efficiency in the wild of about 10 percent is normal. This enhances our understanding of why animals at the ends of long food chains are rare, but is much too vague an estimate to serve as a basis for a predictive theory. Moreover, it is apparent that in many communities much of plant production is not eaten by animals at all, but goes in leaf litter to decomposers. There as yet seems no basis for attempting to predict the proportion of plant production that suffers this fate. The efficiency of animals as devices for attaining certain ends may also be investigated; for instance, their efficiency at producing young,

167

or their efficiency at attaining rapid growth, but measuring these efficiencies has not led to much enlightenment. Animal species have been preserved by natural selection to survive hazard and leave viable offspring, and we may expect them to be efficient at whatever specific tactic they have evolved for this end. A numerical estimate of their efficiency does not aid our understanding of this.

CHAPTER 11

THE TRANSFER OF ENERGY THROUGH THE ANIMAL COMMUNITY

The rate at which energy can be transformed by plants turned out to be set by the carbon dioxide concentration of the atmosphere, by the amount of space that could be flooded by light, and by various limiting factors of the environment such as the water of deserts or the nutrients of the sea. In dense natural vegetation nearly all the energy of sunlight available was actually absorbed, but the actual distribution and abundance of plants was thus set by the factors that determined how much of this total energy could be used. But animals may be thought to be in a different position. Their energy supply is already in the usable converted form of high-energy compounds. For them the energy supply is restricted to that portion of the available food that they can manage to eat. The efficiency of a whole animal trophic level must be a function of the food-getting and digesting powers of that trophic level, which is to say of its energy-absorbing powers. Vegetation may *absorb* most of the energy available to it, and have an efficiency determined by its converting powers. An animal trophic level can only *absorb* part of the available energy, but this energy is already converted to a usable form and is thus nearly all pure gain.

The energy available to the **primary consumers** (or herbivore trophic level) is a function of that portion of the vegetation which the animals can absorb, to whit *digest*. This is less than what they actually manage to eat because a substantial part of what they eat cannot be digested and must be defecated back to the environment. But a moment's thought about any part of the real world suggests that animals do not, in fact, eat all the vegetation, because much of this ends up as fallen leaves and rotting wood, food for decomposers not for herbivores. And then we remember that much of the energy converted by the vegetation was, in fact, used by the vegetation itself as respiration. The actual energy flow into herbivores is thus likely to be some small fraction of the total energy converted by plants.

Carnivores face similar but even more rigorous restrictions on their energy supplies, because their energy depends on how many of the herbivores they can manage to catch, eat, and digest. For secondary carnivores the restrictions must be more rigorous still, and so on through successive trophic levels. Is the energy supply of animals so critical that they have been adapted by evolution to make the most efficient possible use of it? If this were so, it might be that there is a characteristic number of links in food chains and that the standing crops of animals at adjacent trophic levels have characteristic ratios. Since energy is so important to animals, and is in such restricted supply, it is at least possible that the numbers and ways of life of all animals are principally set by the efficiency with which they handle energy, and that thermodynamic theory should enable us to predict the numbers, sizes, and trophic levels present in any community.

Before attempting to answer these rather exciting questions it is important to see that there are two general, and quite different, ways of looking at the efficiencies of animals. You may consider the efficiency of individual animals or kinds of animals, or you may consider the efficiency of whole trophic levels. The trophic level analysis is the one which might be expected to yield simplifying assumptions. Are there characteristic efficiencies for animals of whole trophic levels, so that a simple thermodynamic model might predict the numbers and activities of animals trophic level by trophic level? There is a glimpse here of the Eltonian pyramid on a computer, reconstructed from little more information than data on primary production and the efficiency of energy transfer level by level; a seductive thought.

But consider the practical and conceptual difficulties. We were able to measure the productivity of plants, and hence their efficiency, from yield of dry matter and respiration according to the following equation:

$$\text{Productivity} = (\text{yield of dry matter in calories}) + (\text{calories respired})$$

and to measure the productivity of an animal trophic level we must make similar measurements. We then find that animals are not so neatly separated into their "correct" trophic level as are plants, and also that many of them come and go, making fleeting appearances in their communities before going into some resting stage again. It is obviously a tough proposition to go out and measure the yield of bodies and the combined respiration of all the animals of an entity so ill-defined as a trophic level. We must in part be reduced to making measurements on individuals and then extrapolating to figures for an entire trophic level. At the start of our analysis, therefore, the two ways of looking at efficiency, at that of the individual and that of the entire community of a trophic level, get tied up with each other.

Then there is another conceptual difficulty. The energy assimilated by plants resulted in production of solid matter from gases, but animals are engaged in the reverse process, in the dissipation of energy and the breakdown of complex molecules. Saying that a plant is 8 percent efficient implies that 8 percent of the incoming energy is assimilated, most of which is fixed as visible production. But a herbivore which is 8 percent efficient must be one which absorbs from its gut 8 percent of all the energy potentially available to it as plant food. This energy absorbed is then used by the animal as fuel for its life processes. We seem to be talking of the productivity of plants in terms of something solid they have made, whereas the productivity of an animal is measured by its fuel supply. It is more natural to talk of an animal producing growth or young than it is to talk of its producing the energy which it absorbs. Nevertheless, natural productions of bodies and babies are all dependent on the energy-rich molecules that the animal absorbs, and a herbivore's efficiency as an energy converter is the efficiency with which it can absorb energy from the plants on which it feeds. The fuel input controls all its activities of building its body and constructing its young; all its constructions; and thus is the true measure of the productivity of the animal. But the awkwardness of describing productivity in terms of fuel supply remains, and is well illustrated by a remark of Howard Odum, who wrote, "This is no more pertinent than considering the rate of fuel consumption of a bonfire as the production of the bonfire" (Odum, 1956). Odum's sentence is fine and pithy but, alas, in the thermodynamic sense the bonfire *is* a device to dissipate energy, its production *is* the heat dissipated, and the fuel supply *is* a measure of its production. If only Hutchinson, who first called animal assimilation "production," had not chosen to twist English in the cause of thermodynamic logic, the difficulty would never have arisen (Hutchinson, 1942). It remains that the one energy flux of an animal community directly comparable to the gross productivity of plants is the energy assimilated; the food eaten less the food defecated; and that this is the only energy flux that can be used to compute a meaningful efficiency for a whole trophic level.

We may define the efficiency of herbivores as

$$\frac{\text{Rate of assimilation by herbivores}}{\text{Gross productivity of plants}} \times 100$$

and the efficiency of primary carnivores as

$$\frac{\text{Rate of assimilation of primary carnivores}}{\text{Rate of assimilation of herbivores}} \times 100$$

and so on for each successive trophic level (Table 11.1). Efficiencies calculated in this way are known as **Lindeman efficiencies** because he was the first to attempt the calculation of such efficiencies for real communities (Lindeman, 1942).

The practical problems of determining efficiencies of natural communities are formidable because it is necessary to monitor the food intake of varieties of animals. Sometimes it is possible to bring representative animals into the laboratory, to simulate natural conditions, and to estimate food intake, but this can require much ingenuity. Results from a famous example of success at this were used by Lindeman for his calculations; the measurement by the Russian fisheries scientist, Victor Ivlev, of the assimilation rate in a mud-living worm *Tubifex tubifex* (Ivlev, 1939). These worms feed by ingesting mud, by passing the mixture of silt and organic matter of lake bottoms through their alimentary canals and digesting what parts of it they can. Ivlev mixed some mud from the place where he found his worms with grains of platinum black so that the platinum black was evenly distributed throughout the mud. Then he put mud and worms in glass jars, the bottoms of which were made of gauze, and

TABLE 11.1
Common Ways of Calculating the Efficiency of Energy Transfer

Where I = energy intake and t = trophic level

$$\text{Ecological (Lindeman) efficiency} = \frac{I}{I_{t-1}} \times 100$$

$$\begin{array}{l}\text{Ecological efficiency of plants} \\ \text{(Lindeman and Transeau efficiencies)}\end{array} = \frac{\text{Rate of assimilation (photosynthesis) by plants}}{\text{Solar flux}} \times 100$$

$$\text{Lindeman efficiency of herbivores} = \frac{\text{Rate of assimilation by herbivores}}{\text{Gross productivity of plants}} \times 100$$

$$\text{Lindeman efficiency of primary carnivores} = \frac{\text{Rate of assimilation of primary carnivores}}{\text{Rate of assimilation of herbivores}} \times 100$$

$$\text{Growth efficiency} = \frac{\text{calorific value of organism}}{\text{calories expended in its development}}$$

$$\begin{array}{l}\text{Population efficiency} \\ \text{(of interest to predators)}\end{array} = \frac{\text{calorific value of population surplus}}{\text{calories expended to maintain population surplus}}$$

True ecological efficiencies (Lindeman efficiencies) are the ratio of energy flows in adjacent trophic levels. Except for plants and sunlight, these are almost impossible to measure for real communities, though a figure of 10 percent for animal trophic levels is commonly, though cautiously, used. Growth and population efficiencies can be measured with fair accuracy for animals in laboratory microcosms.

which were suspended in water. The worms positioned themselves so that their tails were at the bottom of the mud, projecting through the gauze into the water. All their fecal pellets fell down and could be collected in a watch glass underneath (Figure 11.1). Ivlev then only had to estimate the amount of platinum black in the droppings and in the original mud to be able to calculate the volume of sediment which had been ingested during the time that the worms were in the jars. By burning typical samples of the mud in a calorimeter he was able to estimate the energy present in unit volume of mud, and thus the rate of energy assimilation. He was also able to weigh representative worms before and after the feeding, to calculate the amount which they had grown, and to convert this amount to a number of calories. He now had both the number of calories ingested and the smaller number of calories represented by growth; the difference between these figures must represent the number of calories used in the animals' metabolism, and hence the respiration and excretion of the animals. Ivlev thus had rate of assimilation, rate of growth, and a combined rate of respiration and excretion for common bottom-living worms. Needless to say, for work which has stood the test of time, Ivlev ran many duplicates of his experiment and provided proper control experiments so that others could trust his conclusions.

Another example of a practical measurement is the estimation of the filtering rate of marine copepods by Gauld (1951). Copepods (Figure 11.2) are immensely important in aquatic systems, both fresh and salt, where they can be found in great numbers. There are many kinds, but there can

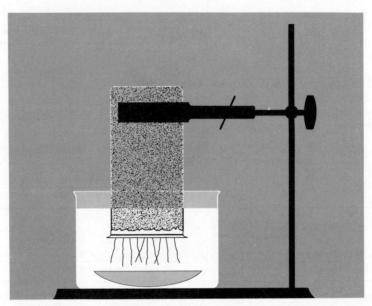

Figure 11.1

Tubificid feeding experiment of Ivlev. The inverted jar contains organic mud through which grains of platinum black have been evenly distributed. *Tubifex* worms placed in the jar orient themselves so that their tails project through the gauze that closes the bottom of the jar. Their fecal pellets can then all be collected in the watch glass below, and the amount of mud they had ingested can be computed from the concentration of grains of platinum black. By measuring the calorific value of the mud it was thus possible to determine the calories ingested by the worms in unit time. (After Ivlev, 1939.)

be little doubt that many at least feed on plants of the phytoplankton, for green pulp can be found in their alimentary tracts. But how do they feed? You cannot watch a copepod grabbing food for it is much too small, and its presumed food is microscopic. It has structures that look like filtering devices so that the animal can be readily imagined to swim through the invisible cloud of tiny plants, sweeping them from the water in a random way as it goes. Alternatively, it could be seeking out larger plants and food particles, then catching them in its sweeping devices; or it could be doing a mixture of both. How can you even be sure exactly how such a herbivore is feeding, let alone about how much food it takes? There is, in fact, a history of much work and much dispute about the feeding of such animals, but Gauld's experiments seem finally to have resolved the problem and to have calculated the food intake as well. He made up cultures of the tiny planktonic plant *Chlamydomonas,* which he found would swim at random throughout a bottle of water, thus making a nice even target for a filtering predator, and he counted the numbers of *Chlamydomonas* cells in subsamples so that he knew the concentration of cells in his experimental bottles. Then he put in a few chosen copepods, left them for 24 hours, and then again counted the *Chlamydomonas* cells to arrive at the new concentration. Control bottles

Copepod of the genus *Calanus*. This is a dorsal view. The large first antennae are used for swimming. The appendages projecting from the sides below the antennae create vortices which sweep water containing planktonic plants into filters on the ventral surface.

Figure 11.2

with *Chlamydomonas* but without copepods served as checks to see that the concentration of the plant cells did not change for some reason other than copepod grazing, which they did not. If the copepods fed by random filtering of seawater, then the concentration of *Chlamydomonas* cells should fall off exponentially throughout the duration of the experiment according to the equation:

$$C_t = C_o e^{-kt} \qquad (1)$$

where C_t is the concentration of *Chlamydomonas* at the end of the experiment
C_o is the concentration of *Chlamydomonas* at the beginning
k is constant denoting the ability of the animals to filter.

Gauld found that the decline in numbers of *Chlamydomonas* did show such an exponential decrease, letting him conclude that copepods were indeed random filter-feeders. But there was yet more to be gleaned from his experiments, for the filtering rate of each animal, giving the actual food eaten, could be calculated also. If v is the volume of water per animal, then vk is the volume of water swept free by one animal in unit time, so that the filtering rate F is given by

$$F = vk \qquad (2)$$

This expression can be evaluated from equation 1 as

$$F = \frac{\log_{10} C_o - \log_{10} C_t}{t \, \log_{10} e} \qquad (3)$$

And, since everything but F was measured in his experiments, the filtering rate could be calculated. Gauld's results are given in Table 11.2, which show that filtering rates, fairly enough, depend on the size of the animals. The way should then be clear to measure the amount of copepod-filterable food in unit volume of natural seawater and work out the food intake of wild copepods. Unfortunately, Gauld had to conclude that he did not have reliable enough data on copepod food in seawater to do this. Nor did he have any way of measuring copepod defecation, which would be necessary to compute the energy actually flowing into copepods.

We have very few elegant measurements like those of Ivlev or Gauld, and the truth is that we have little idea of how the majority of animals feed except in a very general way, that we do not know precisely what they eat, and that we have very little idea, indeed, about how much they eat. What we sometimes do have is a measure of standing crop, but this is a measure of limited usefulness. The standing crop of planktonic carnivores in the sea, for instance, is often much larger than the standing crop of planktonic plants which must be their main support (Figure 11.3). This may usually be explained as a result of the rapid turnover of the tiny plants, that they are short-lived and their corpses swiftly decomposed; but it may also be locally due to the zooplankters sampled having just eaten down a patch of phytoplankton in a patchy ocean (for oceans are

TABLE 11.2

Gauld's Results for the Filtering Abilities of Copepods

Series	No. of experiments in series	Species	No. of copepods per vessel	Volume of vessel (ml.)	Duration (hr.)	Temperature (°C.)	Mean volume swept clear per copepod (ml.)
A	21	*Pseudocalanus minutus*	1	10	24	10	4.28
B	45	*Temora longicornis*	10	150	24	10	8.38
C	8	*Centropages hamatus*	10	150	24	10	12.99
D	19	*Calanus finmarchicus* stage V	1	100	24	12.5	64.36
E	77	*C. finmarchicus* stage V	1	100	18	17	71.03
F	25	*C. finmarchicus* stage IV	2	100	18	17	36.65
G	13	*C. finmarchicus* stage III	3	100	18	17	22.24

The column "Mean volume swept clear per copepod" were calculated from equation 3. From Gauld, 1951.

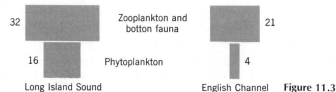

Long Island Sound   English Channel   **Figure 11.3**

Inverted biomass pyramids in the sea revealed by standing crop measurements. Values are in grams per square meter. A small standing crop of phytoplankton supports a larger standing crop of planktonic animals and bottom animals that presumably all derive their livelihoods from the small mass of plants. The plants turnover more quickly, so that there is more energy flowing through the phytoplankton trophic level than through the animal trophic levels that feed on them. This illustrates the low value of standing crop measurements for studies of energetics. (After Odum, 1959, from data of Riley, 1956, and Harvey, 1950.)

patchy). But, on either explanation, clearly the standing crop can be a misleading index of the energy flux in a community.

So we come to the realization that, glittering although the promise of a thermodynamic model of an energy pyramid might be, the obstacles are truly formidable. The pyramid itself involves the conceptual device of the trophic level when we know that trophic levels can seldom be distinct in real systems. Then we find that the measurements on real populations which we can make, those of standing crops, may be of little use to the task of estimating rates of energy flow. And even if we descend to making measurements on isolated animals, we yet encounter great technical difficulty. But the possibilities in such work are so alluring that attempts at measuring efficiencies of whole trophic levels have been made all the same.

The first, and classic, attempt was that of Lindeman (1942). He set out to calculate the ecological (Lindeman) efficiencies for all the trophic levels of a lake system that he had studied, and for comparable lakes from data in other people's papers. He had studied Cedar Bog Lake in Minnesota for 5 years, monitoring with painstaking care the fortunes of various communities throughout the seasons and year by year (Lindeman, 1941). He used nets and grabs to make standard samples, and then sorted each individual of his myriad catches to its proper link in a food chain or, at least, as closely as he could judge. At the end he had standing crop data, trophic level by trophic level, of an accuracy that has probably not been excelled. The problem was now to calculate the energy input for each trophic level over a typical year, and to do this from the standing crop measurements alone. Lindeman assumed approximate turnover rates on the basis of empirical studies of others; for instance, he assumed that the phytoplankton turned over every week in summer, that zooplankters turned over every two weeks, and that rooted pond weeds turned over only once a year. Multiplying standing crop measurements by turnover

THE EFFICIENCIES OF
WHOLE TROPHIC
LEVELS

times gave him figures that he called "productivities" but which were not the same as the total assimilations of the communities. The animals had done work in getting their food, work that had degraded energy and lost carbon to the population by respiration. The calories represented in the food-getting operation, one of the principal activities of the animals, were not accounted for when you multiplied standing crop by turnover time. Lindeman sought to correct this by adding figures for general respiration; figures that were based on laboratory studies like those of Ivlev; but in this he erred. These figures were for total respiration, and included the energy dissipated in the activity of turning over as well as that used for finding food. Lindeman's energy input was, therefore, too high by some large but unknown factor; and he then compounded the error by adding estimates for animals taken by predators and animals that died other deaths and went to the decomposers. The energy represented by these bodies had already been allowed for in the turnover calculation and were now appearing in the equation twice.

Lindeman had wanted to solve the following equation:

Productivity = (yield of dry matter in calories) + (calories respired)

but what he had actually done was to say:

Productivity = (standing crop in calories × turnover time)
           + (calories respired) + (calories lost in corpses)

which had both corpses and respiration appearing twice, once disguised as "turnover" and once in their proper places.

Lindeman's mistake was not noticed by the many students who read his classic paper over a span of 20 years, and was finally pointed out by Lawrence Slobodkin, who in many ways has been his successor in theoretical studies of energy transformations. Like Lindeman, Slobodkin developed some of his ideas at Yale where he was once a graduate student of Hutchinson's. In a review published in 1962 (Slobodkin, 1962), he noted that, if you could not measure the food input to all the animals of a community directly, you could arrive at it from the average respiration and yield of dead bodies, whether these dead bodies were eaten by predators or whether they were decomposed. There can be only two fates for a calorific molecule assimilated; either it is burned, in which event its passing is recorded by carbon dioxide respired, or it appears as a part of a dead body that is eaten by an animal of another trophic level. Yield of dead bodies had appeared twice in Lindeman's computation; once buried in the standing-crop turnover computation, and once as bodies eaten; and respiration had appeared one and a bit times; as total respiration and as part of the turnover figure. Without reliable estimates for the energy used for food-getting it is impossible to use standing-crop measurements to compute the energy assimilated by a population. In this event, Lindeman's calculations had no validity, and Slobodkin was forced to admit for our discipline that we have made virtually no progress in answering the pregnant questions about energy transfers between trophic levels which Lindeman asked more than 20 years before.

Lindeman's Calculated Efficiencies for the Animal   TABLE 11.3
Trophic Levels of Two American Lakes

	Cedar Bog Lake (in percent)	Lake Mendota (in percent)
Primary consumers (herbivores)	13.3	8.7
Secondary consumers (1° carnivores)	22.3	5.5
Tertiary consumers (2° carnivores)	No data	13.0

Lindeman had calculated these efficiencies from standing crop measurements, and estimates of turnover times and respiration rates. This resulted in his computing an energy flux in each trophic level which was too large by some unknown amount (see text). The calculated efficiencies are thus known to be in error, but the spread in his results are such as to suggest that the empirical error in measurements may be at least as great. The general suggestion conveyed by his work, that ecological efficiencies between trophic levels of about 10 percent are possible, is still accepted by ecologists. (From Lindeman, 1942.)

Lindeman's results (Table 11.3), although known to be in error, yet have a look of reasonableness about them. The range is wide, from 5 to 22 percent in the same trophic level of different lakes, but this rather suggests that inaccuracies in measurement (unavoidable in so pioneering a project) were, at least, as great as the errors introduced by the method of computation. It looks as if typical ecological efficiencies of an animal trophic level might be something like 10 percent, and ecologists have had this figure in the back of their minds ever since. This impression that whole animal trophic levels were 10 percent efficient has been reinforced by recent attempts to measure ecological efficiencies in simple laboratory systems.

Slobodkin (1962) and his students have taken advantage of the fact that in a laboratory microcosm it is possible to feed measured amounts of food of known caloric content, to measure the yield of dead bodies, and to monitor the size of the population. Respiration can be measured for typical animals under the temperature regimen of the laboratory container, and you have all that is needed to calculate ecological efficiencies. Then the **ecological (Lindeman) efficiency** equals

$$\frac{\text{(Yield of dead bodies and excretions as calories per unit area per unit time)} \times \text{(calories respired per unit area per unit time)}}{\text{Energy supplied as food per unit area per unit time}}$$

The animals that Slobodkin chose to use were water fleas (*Daphnia:* Cladocera) and *Hydra,* both animals that could be maintained in small aquaria and that would thrive on a diet of green flagellates. He developed micro-bomb calorimeters to measure accurately the calorific values of his small animals and their food, and he monitored many replicate systems. The food supply could be manipulated until a roughly constant yield of growth and fresh bodies was attained, when it was possible to measure the yield and respiration per unit time needed for a proper computation of ecological (Lindeman) efficiency. Much labor was required (Richman, 1958; Armstrong, 1960), but eventually reliable results were obtained. Maximum efficiencies (Slobodkin, 1962) were as follows:

Ecological (Lindeman) efficiency of *Hydra* population =   7 percent
Ecological (Lindeman) efficiency of *Daphnia* population = 13 percent

These results are close to the values found by Lindeman (Table 11.3), perhaps encouraging the view that his errors were not so serious in practice and that ecological efficiencies might generally be around 10 percent. It is a giant stride from measurements on single-species populations of small animals in aquaria to estimates of the efficiencies of whole trophic levels of many species in nature. But it is possible to argue, as Slobodkin has done, that all animals of comparable habit might well turn out to be roughly as efficient as each other. We can call on the experience of physiologists who found that all animals metabolize in about the same way, and at about the same rate, dependent only on the temperature, so perhaps all animals are about equally efficient. But there must also seem room for very wide variations in efficiency of different ways of life, even though of the same trophic level. A parasitic larva of a wasp, for instance, sitting comfortably inside the caterpillar of its butterfly host, is likely to be much more efficient at acquiring food energy than a cheetah running down gazelles. There has been hot debate among ecologists at their professional meetings over whether Slobodkin's suggested average ecological efficiency of 10 percent is valid, many of the protagonists indignantly pointing out such obvious anomalies as the parasites and the cheetahs. Most probably think efficiencies may range widely, perhaps from 5 to 30 percent. Probably in the great mixture, fluidity, and poor definition of populations making up the nearest thing to a real trophic level that nature is likely to provide, the anomalies cancel out and an average ecological efficiency of about 10 percent may well be common. It, at least, gives us a rule of thumb to demonstrate the successive drastic reductions in energy as you go up a food chain. Vegetation is only 2 percent efficient. The herbivores (such as beef cattle) can never do much better than 10 percent of that 2 percent. A primary carnivore can only have 10 percent of that 10 percent of 2 percent. And so on. The enforced rarity of feeding at the end of long food chains is very obvious.

But a rule of thumb is not good enough for a computer model, nor for a grand insight into the workings of nature. When Lindeman started his work, and when others followed in his footsteps, there was always the lure of producing a thermodynamic or mathematical model that should describe the distribution and abundance of all kinds of animals in nature and, latterly, perhaps to recreate the process of evolution on a computer. But this hope has faded. Not only has it been impossible to measure the ecological efficiencies of whole trophic levels but there has also come the increasing realization that energy may be partitioned in different ways by different animals. And over and above this is the fact that plants have evolved many clever ways of not being eaten by animals, by making themselves taste nasty or by being inedible wood. Much of the energy of terrestrial vegetation goes to decomposers, not into animal food chains at all. About all that can be said in general terms about the fate of the energy fixed by plants is that it virtually all is used by animals or decomposers in the end, with just a tiny trickle being stored as coal, oil, or peat. But the pathways followed by the energy depends on the kinds of living things available, and these depend, in turn, on the environment and its history.

The efficiency with which different kinds of animals handle their energy supplies can be looked at in a number of ways, depending on the results of the transfer. This can be understood by referring to the analogy of the bonfire used by Odum in his critique. The bonfire might be efficient at consuming fuel, and thus, as I have suggested, productive in the thermodynamic sense that Hutchinson used the word, but it also produces light and heat. The bonfire may thus be looked at from the point of view of someone using it to warm himself, in which sense the efficiency is a measure of the radiant heat output, or it may be looked at from the point of view of someone using it to light a cave, in which sense the efficiency is a measure of light output. The two efficiencies are quite different, although each must be related in some complex way to the basic fuel energy of the system, and each depends on a point of view. Slobodkin (1962) suggests that you can look at biological efficiency from different points of view also. From the viewpoint of an individual organism, the growth efficiency, which is the ratio of calories in the organism to the calories expended in its development (Table 11.1), might seem most important; but a predator might look on the population of the same animals as a source of bodies to eat, and must see its efficiency in terms of the rate at which it produces these bodies. Such efficiencies must be ultimately dependent on the common energy input to the population, which is measured by the Lindeman efficiency, but can be useful for special purposes, such as investigations of predation in experimental populations.

OTHER EFFICIENCIES IN SINGLE SPECIES POPULATIONS

Experimental populations that have been allowed to come to a steady state, with a constant input of food energy and a constant output of corpses, can be manipulated. With Slobodkin's water flea and hydra populations, it was possible to play at being predators, to see what happened when you removed animals of different ages, and to calculate appropriate efficiencies from the results. The **population efficiency,** the steady-state ratio of yield (in the sense of corpses removed) to the calories needed to maintain the population (Table 11.1), was highest when they took old animals. This immediately suggested that taking old animals might be the prudent thing to do for a predator, because then it would get more food on a sustained yield basis. Indeed, it is the general experience of game managers that large mammalian predators do usually take the old and the weak. Was there a hint here of a more-pervading influence of thermodynamics on the lives of animals? Has selection somehow produced species so that there can generally be wise management of resources, even down to such subtleties as eating old worn-out prey while foregoing the succulent young? In this example, at least, most ecologists do not think so. It seems much more likely that predators take the old and weak simply because they are easier to catch; nothing more subtle than this.

It is true to say that animals may profitably be viewed as devices for getting food energy. Those that exist today have been preserved and refined by natural selection; they have been chosen because they were better at getting food energy than other kinds that have become extinct. It must follow that they should use the food energy they do get efficiently. So far we are on safe ground, but then we come back to the question, "efficient at what?" The answer must always be, "efficient at surviving hazard and leaving babies to succeed you." It is of no value to predators to develop habits that preserve maximum energy for the future if they are chosen for their ability to get maximum energy *now,* to produce the most strong babies for the *next* succeeding generation. Being efficient at surviving hazard and breeding well in the good times may actually be best promoted by what looks like a squandering of energy, just as it is an advantage for a racing car to be inefficient in the use of fuel to make it go faster. There may thus be little intellectual fruit to be plucked by measuring things such as growth efficiencies and population efficiencies in the laboratory. Modern ecologists are making more progress by trying to put themselves in the place of animals in the field and asking themselves: How does this animal get its energy? Could it improve its supply by doing something else? What does it do with its energy when it has got it? What other choices are available to it? We can be sure that the animals are being "efficient" at whatever they are doing (just as the racing car was "efficient" at going fast), but it does not seem to help us very much to say so.

One of the grand phenomena of nature is that plants of the open sea are very small. They are so small that herbivores must be tiny also, meaning that all large animals of the sea are carnivores feeding many links of a food chain away from the plants and being necessarily rare. The smallness of marine plants is thus an important phenomenon for an ecologist to explain. Current hypotheses are that the large surface area earned by smallness helps the small plants to resist sinking and aids in the uptake of scarce nutrients, or that small size promotes rapid turnover and thus the efficient use of resources. These hypotheses do not meet the objection that large floating plants should attain such dominance, competing so effectively for light that the relative efficiency of small size would be unimportant, just as it is on land. It is probable that large floating plants do not exist, except in the Sargasso Sea, because drifting in winds and currents would impose unacceptable losses on their populations. The structure of marine animal communities which the smallness of plants imposes compounds the desert properties of the seas, which make the prospects for ocean farming poor. It is likely that a doubling of the present yields of world fisheries is the most that even enlightened ocean husbandry can attain.

ELTONIAN STRUCTURE IN THE OPEN SEA AND THE SMALLNESS OF MARINE PLANTS

181

CHAPTER 12

Casual inspection of life in the deep sea reveals a remarkable difference from life on land, and one that might not be expected; there are no large plants in the open sea. A nice warm piece of well-lighted ocean may be off shore from a tropical rain forest in a place where there is lots of water, lots of solar energy, and at least some nutrients; and yet there is not a big plant in sight. Indeed, if you swim in such a patch of ocean wearing a face mask, you will not be able to see any plants at all. You may see a big animal, a shark or marine mammal, but this is unlikely for such things are sparsely distributed over the ocean. Smaller fish will be fairly common, smaller ones still may be very common, and there will be myriads of tiny, darting specks, the zooplankton of the sea. This is an animal community of sparse big ones and common little ones, the whole neatly conforming to the concept of the Eltonian pyramid; but there will not be one plant big enough to be seen. Plants are present, and in great numbers, but they are all microscopic; the diatoms, dinoflagellates, and coccolithophorids of the open sea. The extremely small size of these plants has imposed on the animal community in a most complete way the effects of the principles of food size and energy flow. All herbivores must be small enough to eat microscopic food; all large animals must be carnivores, and very large animals must generally be many links of food chain removed from the tiny plants that are their ultimate energy source. Consequently large animals can obtain only a small energy flux and must be relatively rare. Even the filter-feeding whales and basking sharks cannot trap marine plants and must be carnivorous, feeding on planktonic animals, which in turn filter out the tiny plants. The curious absence of large plants from the open sea thus has far-reaching effects on the economy of the oceans, becoming a controlling influence in the lives of animals. Why are there no large plants in the open sea? How is it that a site offering water, warmth, and light good enough to support rain-forest trees on land has only plants of microscopic size?

Answers to these questions commonly offered in oceanographic texts depend on the fact that small bodies have large surface areas relative to their volume. The total surface of a population of microscopic plants is enormously greater than would be the surface of one plant of similar total mass, and this is so dramatic and obvious a result of small size that it is natural to seek in it the advantages of smallness.

A large surface area means that there will be relatively high frictional forces between the plant cell and the water, suggesting that it should sink more slowly than would a larger plant. Open-sea plants must stay up in the lighted water at the top of the ocean, and normal protoplasm is denser than seawater and will sink unless kept up by floats or swimming. It may be that reducing the tendency to sink by being very small gives a plant an advantage, but this explanation of smallness has an unsatisfactory sound to it. The advantage cannot be very great for even the smallest plants must still help support themselves in other ways; by containing bubbles of oil that act as floats, by lowering their ionic content (Lewin, 1962), or by swimming with flagellae. If small plants can use oil floats, why should large ones not use floats also? Indeed, we know that the large plants of coastal waters, the algae which are anchored to rocks, do float their fronds with air bladders, and that some of them may drift about the open sea when torn from their holds, when they may live and grow for a long time. Large ocean plants should thus have no difficulty in keeping themselves near the light, and any advantage a small plant has must be slight.

A second consequence of large surface area is that there is a large membrane surface in contact with the water which can be used for the uptake of nutrients. Essential nutrients, particularly phosphorus, are in short supply in the sea and set limits to the productivity of marine plants. Any superiority in winning the available nutrients should give a decisive advantage, and a large surface area may be thought to provide this advantage, thus explaining the smallness of oceanic plants. This is an attractive hypothesis, and one which has met with favor by marine biologists. It is the preferred explanation of the great English planktonologist, Alister Hardy (1956). And yet there are obvious objections to it. Many large plants do live in seawater, although not in the open sea. They live anchored to the bottom in the shallow waters off coasts, fixed but not rooted, deriving all their nutrients by absorption at the surfaces of their fronds. Among their number are the giant kelp of the North Pacific, the longest vegetables in the world. If these anchored giants can gather their necessary nutrients from the sea, why should not floating plants do the same, from similar seawater farther out from shore? It is true that anchored plants have the advantage of standing in a flow of water, but we also have the evidence of sargasso weed that nutrients can be sufficient for large plants even when not anchored. A small plant may have some advantage, but a flat leaf-shaped large plant should still expose a very large surface to the nutrient medium; surely enough to suffice.

Very small plants may be short-lived, and thus may turnover very quickly. That they do is well shown by measuring standing crops of zooplankton and phytoplankton from the same area of sea, when the standing crop of animals is usually found to be greater than the standing crop of plants on which they feed (Figure 11.3). The explanation for this is, of course, that the plants are turning over very quickly; that they reproduce quickly while being closely cropped by the animals. The energy flux through the plant trophic level is much greater than that through the herbivore trophic level, but the plants are rapidly degrading energy or passing it on to the animals as food. This high rate of turnover can be taken to mean that nutrients are not held for long in the bodies of the plants but are quickly returned to the system to be used again; which is to say that small plants use the sparse supply of nutrients more efficiently than would large plants. It is possible, therefore, to look at the smallness of oceanic plants as a device for the efficient use of nutrients. This is perhaps the most tempting hypothesis yet, for hypotheses of maximum efficiency always have a satisfying ring about them. But this hypothesis is also open to objections. Large plants which turn over slowly do live in the sea, when anchored. They can also go on living when torn up and floated out to sea, as a seaweed (*Sargassum* sp) does indefinitely when it is collected by currents into the Sargasso Sea. Then it may not be efficient to lock up nutrients in your own body, but once you have them, you have them; and you survive. Survival, not efficiency, is what counts in the natural world, something I have already suggested when discussing efficiencies of energy conversion. Any advantage given to small plants by the efficiency with which they used nutrients would probably be of little importance if there was real survival value in being large.

On land, the survival value of being large is apparently overriding, and the advantages to largeness are so apparent that we never need debate them. The large plant wins space, achieves dominance, takes for itself an energy supply which it denies to smaller plants beneath it. Such advantages should prove decisive in the sea also, if they could be won, and where plants can anchor themselves in shallow water, they apparently can be won. Large seaweeds do hold space, become dominant, and have canopies which, although waving, may be remarkably complete. In the sea, as on the land, the plant that can shade out the others has the real advantage. Anchored plants can win this advantage; floating plants cannot.

The floating habit does not in itself deny a plant the chance to win space and achieve dominance, since the water surface can be easily covered by floating plants. This actually happens in freshwater ponds covered with duckweeds (*Lemna* sp), which may cover the surface with a dense "canopy" as opaque as that of many a temperate forest so that little light penetrates to the water beneath. If such a layer of floating plants were established on the surface of the ocean, they should dominate the seas just as trees dominate the vegetation of the land. Tiny plants might still derive what advantage they could from their large surface areas and their high turnover, but it would avail them little in the darkness under the floating canopy. The reason that oceanic plants are small must be that floating canopies cannot be established; and it is not very hard to see why they cannot. The winds and currents of the sea would sweep them away to regions outside the tolerances of the plants. Duckweed does not cover large lakes, even though the plants may be plentiful round the edges, because wind piles them up against the shores. In the very much larger sea, such blowing away from preferred sites could always be expected. It seems that blowing away must be thought of as the selection pressure that denies the open sea to large plants, in spite of the advantages of dominance that large size would give them. Small plants must be continuously moved from their preferred habitat also, but their high turnover and low standing crop saves them, keeping the loss down to negligible proportions.

Support for this hypothesis comes from the phenomenon of the sargasso weed floating so densely in the Sargasso Sea. Accidents of wind and current concentrate floating objects in the Sargasso Sea, and here a floating canopy of plants collects; and does so despite the fact that the concentration of nutrients and fertility of the Sargasso Sea are known to be perhaps lower than in other parts of the ocean (Ryther, 1960). *Sargassum* grows and persists while afloat, as has long been known. In many tropical seas, not just the Sargasso, fronds of the weed can be seen floating as small clumps scattered over the ocean, but they are usually just pieces of shore-growing weeds that have been broken off by storm waves and carried out to sea. They do not normally reproduce away from their anchorage on shore rocks and so are not true pelagic plants. But in the Sargasso Sea, as is now known (Dawson, 1966) *Sargassum* completes its whole life cycle, living endless generations without recruitment from the coast. It is a true oceanic plant. Apparently, the peculiar current system that collects flotsam into the Sargasso Sea is an ancient one, having persisted long enough for the *Sargassum* there to have evolved reproductive systems suitable for the floating life.

There is a complication to the pattern of the smallness of marine plants, for they are not everywhere equally small. In 90 percent of the oceans, the deep, unproductive stretches of blue water, plants are very tiny indeed, the so-called **nannoplankton** (*nanno* meaning dwarf) whose cells are between 5 and 25 microns across. But in the more productive shallows over the continental shelves plant cells may be more than a 100 microns across (the **microplankton** and small-enough, for 100 microns means a tenth of a millimeter), and in rich productive upwellings there may be relatively huge filaments or colonial forms stretching for milli-meters or even centimeters (Ryther, 1969). What this probably suggests is that the selection for smallness to avoid loss by drifting is not quite so relentless in the more productive places so that various of the general ad-vantages of bigness come to be important. Sometimes, for instance, rela-tively large-size may have given some immunity from being eaten by the tiny zooplankters of the oceans. But such largeness is strictly relative, in no way modifying the general observation that oceanic plants are small. And this smallness if a direct consequence of the fluidity of the medium; a direct response to a physical parameter of the environment. It is responsi-ble, in turn, for the characteristic structure of the animal community of the sea, for the smallness of herbivores and the rareness of large animals.

SMALL PLANTS AND
OCEAN FARMING

The seas are mostly poorly productive deserts, intrinsically unsuited to farming (Chapter 10). To this fact must be added the consequences of the small sizes of ocean plants, which means that large animals, including men, must harvest them at the ends of long food chains. We cannot crop the tiny scattered phytoplankton, nor even the zooplankters which graze it, and must crop several food-chain links away where the original energy supply has been diminished several times over. This is a hard fact that those who would extensively farm the oceans have to face. We now have, in fact, enough data to put numbers on this reality and to compute the potential food resources of the sea, as has been done for us by Ryther (1969).

Ryther begins his analysis by noting that you may divide the oceans into roughly three kinds of place; the rich productive fishery areas of the upwellings, the less productive coastal and shallow regions, and the great blue desert of the bulk of the oceans.

The fishery sites of the upwellings produce with high efficiency, though

TABLE 12.1   Yield of Food From Parts of the World Oceans.

Province	Percentage of ocean	Area (km²)	Mean productivity (grams of carbon/m²/yr)	Total productivity (10⁹ tons of carbon/yr)
Open ocean	90	326×10^6	50	16.3
Coastal zone*	9.9	36×10^6	100	3.6
Upwelling areas	0.1	3.6×10^5	300	0.1
Total				20.0

* Includes offshore areas of high productivity.

they occupy only a small area, and the high primary production is in the form of relatively large small-plants, those whose filaments or colonies may be millimeters or centimeters long. It is possible for quite large fish to be herbivores in these places, so that fishermen can make their catch just one or two trophic levels from the plants. In coastal shallows, however, plants are reduced to the microplankton, and it takes several links of eating and being eaten before there are fish large enough for fishermen. And in the open sea, the nannoplankters are so tiny that five or more trophic levels may be heeded, perhaps going nannoplankters-carnivorous protozoan-carnivorous copepod-chaetognath-small fish-larger fish-larger fish still. Most of the open oceans are the homes of nannoplankters and long food chains, as well as being the places of least production.

Once the local primary productivity and the lengths of food chains are known, there is one thing more you must have if you are going to compute the yield of fish-food to men. This is the ecological (Lindeman) efficiency of energy transfer from one trophic level to the next (Chapter 11). Ryther accepted the general rule-of-thumb figure of 10 percent, but thought that efficiency might be less in the blue water where the dispersion of plants meant that much energy must go in hunting, and rather more in the productive upwellings where food was clumped and could be obtained with less effort. He chose 20 percent efficiency for upwellings, 15 percent for coastal regions, and 10 percent for the blue water. He could then compute the potential fish yield from each segment of the oceans, thus showing the total oceanic resources which men could hope to exploit. The results are in Table 12.1.

Ryther thus found that the production of fish by the world oceans is about 240 million tons annually. We could not, of course, harvest all this, because to take the lot would ruin the fisheries for subsequent years. Also we have competitors in other carnivores of the sea, such as guano birds, sea lions, whales, sharks, and the like, which we might not want to kill off

Primary production [tons (organic carbon)]	Trophic levels	Efficiency (%)	Fish production [tons (fresh wt.)]
16.3×10^9	5	10	16×10^5
3.6×10^9	3	15	12×10^7
0.1×10^9	1½	20	12×10^7
			24×10^7

The analysis suggests that the bulk of the oceans are not only deserts because they are short of nutrients (reflected in the productivity column) but that the small size of the plants of those places, and the low ecological efficiencies likely, compound this low productivity to ensure that the fish supplies which men can get from the oceans are small. Most fishing has to be done in the small productive places, and fish harvests already being taken cannot be more than doubled without ruin to the fisheries. Data from Ryther, 1969.

as thoroughly as we have killed off the carnivores of the land. So Ryther suggests that we should consider 100 million tons of that 240 million tons to be fair game. But the world fisheries of the moment average about 60 million tons a year, so that men can probably only hope to about double the yield of food from the sea. Even if this estimate is wrong by a factor of two or three, it still reveals the extremely dim prospects of ocean farming. We cannot significantly increase the food of a starving world by farming the oceans, because the oceans are nutrient-poor deserts, and because the plants are so small.

Lakes and the oceans always contain organic molecules in solution, as would be expected, because animals and plants are constantly excreting and dying there. The actual mass of such dissolved matter is very large being many times the solid particles, both living and dead, which are suspended in the water. This mass of dissolved matter represents a large potential energy resource, suggesting that solutes should be used for food by some sorts of animals. An early controversy, known as the Pütter-Krögh argument, developed over whether a large variety of aquatic animals relied on dissolved organic matter as their principal source of food. The argument was settled by the experiments of Krögh which showed that uptake of dissolved food by many aquatic animals was negligible. More recent work has shown that, first, much dissolved organic matter becomes synthesized into solid aggregates on the surfaces of bubbles after which it is used as food by filter-feeding animals and, second, that some bottom-living animals, notably the Phyllum Pogonophora, may well rely extensively on dissolved organic matter for food, as Pütter postulated. It should be interesting to know what part of the energy budget of marine life goes to maintain the reservoir of dissolved organic matter in the world oceans.

THE POTENTIAL
ENERGY OF
DISSOLVED FOOD:
A NICHE TO BE
FILLED

189

CHAPTER 13

THE POTENTIAL
ENERGY OF
DISSOLVED FOOD:
A NICHE TO BE
FILLED

Much organic matter is dissolved in the water of seas and lakes, as might be expected because animals and plants are always excreting or rotting in the water. What might not be expected is that the mass of dissolved substances is very great, certainly many times the mass of the living and dead solids suspended in the water.

Some of the pioneering studies in limnology were made by Birge and Juday (1934) on Lake Mendota, the large and beautiful lake on which the University of Wisconsin is built. They made standing-crop counts of all the animals and plants of the lake, the fish, the plankton, the rooted plants, and the various benthic forms from the bottom; and they screened out the dead organic particles that were floating or settling through the water. From these studies they could calculate the total solid organic matter present in the lake water, both living and dead. Then they filtered typical samples of water to remove all solids, and heated the filtrate with potassium permanganate. Any organic matter present in the filtrates would then be oxidized, reducing a proportional amount of the permanganate. Estimation of the remaining permanganate by titration then gave them an estimate of the total organic matter present in the sample, from which they could calculate the total mass of organic matter dissolved in the entire lake. They found that the figure for dissolved organic matter was about seven times that for the total mass of particulate organic matter, both living and dead.

Since Birge and Juday did their work, others have shown that such disproportionate amounts of dissolved organic matter are normal for lakes. In the seas the disproportion is even more striking, a figure for dissolved organic matter of 300 times the solids being recently suggested for the oceans as a whole. This striking discrepancy reflects the vast volume of bottom water in the oceans to which no light penetrates and in which plants cannot live. The reservoir of deep dark water presumably acts as a sump for dissolved matter, whereas most particles and living things are concentrated near the thin lighted zone at the top (the **photic zone** in Greek). But even in the photic zone of the sea dissolved matter certainly exceeds particles, just as in lakes, a fact which gives rise to the interesting ecological question: What happens to this dissolved organic matter? It represents energy. What niches have been developed to take advantage of this energy supply; or does it all go to waste?

These valid and significant ecological questions were asked by the German physiologist August Pütter as soon as the disproportionate amount of dissolved organic matter was known. A reasonable if tentative first answer should surely be that animals used it as food. Aquatic animals from copepods to fish could be thought of as living in a sort of dilute soup, so that it was not unreasonable to postulate that some of them simply absorbed the dissolved food supply through their skins. This was Pütter's hypothesis (Krögh, 1931). He was encouraged to believe in its soundness by the common knowledge that animals of all sizes were abundant in a relatively plantless sea. Plant food of marine animals was hard to find, but here was an alternative energy source, the dissolved organic matter, which could very well have come to the sea in rivers.

Pütter attempted to test his hypothesis by calculating the food requirements of aquatic animals. Animals most likely to absorb nutrients should be small ones of large surface area, many of which could be seen to direct currents of water through various bodily structures by means of cilia or appendages. It was commonly held that the structures and water currents of small animals were devices for catching solid food, but Pütter reasoned that the animals could just as well be extracting dissolved nutrients from the water instead. He could check on this by calculating the food requirements of the animals and then seeing if sufficient solid food was present in the water which an animal could fish clean or if the much greater mass of dissolved nutrients must be relied on. The animal must receive as many calories per hour from its food as it loses by respiration. Already the respiration rates of most types of small marine animals were known, and Pütter had himself been involved in some of the determinations; you put the animals in a closed jar of water for a set time and then determine by titration the oxygen loss. A small marine protozoan was an obvious choice for an experimental animal, since this type should have a large surface over which it directed water with its cilia, and Pütter chose *Collozoum inerme,* an animal of about 0.1 cc volume. Pütter concluded that such an animal would have to fish clean 9400 cc of water every hour just to find enough solid food to meet its respiration requirements, which is surely an impossibility. On the other hand, it should only have to pump 0.55 cc if it could rely on the dissolved organic matter. For Pütter this settled the matter, but others were not ready to accept his figures. He had made unfair assumptions about the sparseness of solids and the abundance of solutes in typical seawater to favor his arguments. When others did the same calculations on more generally accepted assumptions, the results were much less dramatic; being 760 cc per hour to be fished clean if feeding on solids, but only 6 cc if absorbing dissolved nutrients; but even these figures still favored the Pütter hypothesis, since it is unreasonable to expect so tiny an animal to fish clean even 760 cc per hour. But, strange to say, both Pütter and his critics had not noticed the most serious source of error in the calculations: no one had made proper allowance for the really tiny plants of the sea, the nannoplankton. The importance of these plants, the main primary producers of the sea, was not known until 1908 because the standard nets used to sample plankton up to that date were too coarse to catch them. Their presence had not been allowed for in Pütter's calculations. Once they were, the calculated fish-clean volumes became much more reasonable.

If Pütter had known about the nannoplankton from the start he would probably not have pursued the matter further, but by now he was committed, and he went on to seek experimental evidence for his hypothesis. He kept fish in aquaria of filtered water which was changed regularly, to see if fish could meet their energy requirements by absorbing dissolved food. He could calculate the respiration rate of his fishes at intervals by putting them in closed bottles for a time and measuring the oxygen loss, which gave him a figure for calories used in respiration during the experiment. If the respiration calories were more than could be accounted for by the loss of weight of the fish, then the excess must have come from the water. In Pütter's experiments there were apparently large calorific losses in excess of the weight loss of his fishes, and he considered his hypothesis proved. Unfortunately his experiments were not well-conducted. He used few fish and no controls. He "measured" the weight of his fish by measuring their lengths only and extrapolating; hardly a good estimate of fatness! And worst of all was his measurement of respiration. He fished out his fish with a net, presumably involving a chase round the tank, and then dumped them in the respiration chamber, where they swam quickly round in some sort of panic reaction. They were no doubt respiring very quickly during this time! and Pütter took that respiration rate as the average for the fish, whereas in the tank they were probably quietly starving to death and scarcely respiring at all.

Pütter made many other experiments of this kind, many of them having to modern ears an unsatisfactory sound, and published them in a book. Eventually the work attracted the attention of a more formidable experimenter who proceeded to perform more meticulous experiments and to criticize Pütter's. He was a Dane, August Krögh, one of the leading experimental biologists of the day, and winner of a Nobel prize for his experimental work. He made no mistake about his experimental methods. Animals were kept in tanks through which water flowed, and the organic matter in both the inlet and outlet streams was monitored. Krögh used fish, tadpoles, mussels, and cladocera; repeating each experiment many times. Always there was more dissolved organic matter in the outlet streams than the inlet, showing that the animals not only could not make a net gain of energy from the water but that they always lost to it by excretions. There might be some solute exchange across the animals' membranes, as seems natural, but always more was lost than gained. Krögh (1931) presented his findings in a classic review article, and pointed out with telling accuracy the inadequacies of Pütter's experiments. And Krögh had won the argument. His paper remains as a masterpiece of scholarly writing as well as an elegant example of the experimental method in biology. Pütter's methods were made to seem absurd, and he was routed from the fray. The animals of the sea lived on solid food. If some made use of solutes which happened to be in the water which they swallowed, the amount must be so small as not to contribute significantly to their energy requirements. It was possible that trace amounts of specially useful molecules such as vitamins could be acquired in this way, and this possibility has been actively investigated, particularly that many animals may need vitamin B_{12} from the water (Provasoli, 1958). But Krögh's paper seemed to make it plain that dissolved solutes could not be a significant source of energy for aquatic animals.

The ecological paradox remained: here apparently was a large source of potential energy with no animal using it. Had no animal developed a niche to take advantage of the dissolved organic food of the sea?—perhaps an answer in many minds was that bacteria, the decomposers, used this food, as they use so many other food molecules left over from the bodies of animals and plants. But then there came the results of work in oceanography which suggested a different answer.

There are often long shiny "slicks" on the surface of seas or lakes, which are apparently arranged along the lines where flows of circulating water meet. Oceanographers from Woods Hole and the Bermuda Biological Station studied such flows and slicks in the Atlantic. They suspected that the slicks were due to thin films of organic molecules collected in one place by the flow pattern, and looked into the physical processes that might produce suitable molecules for the thin films. Surface phenomena between air and water seemed logical; either at the sea-air interface, or on bubbles made by the turbulence of the waves. The three men, Sutcliffe, Baylor, and Menzel (1963), filtered samples of seawater to remove all living and dead particles, then bubbled air through the filtered water. A foam of organic particles then collected at the top, showing that organic solutes in the seawater had been adsorbed at the surface of the rising bubbles and built up into solid films that appeared as froth at the surface. The three men had, indeed, discovered the source of the "slicks," but they had done more than that. Folding and coalescing of the bubble-made films should produce aggregates similar to much of the detritus floating in the sea, detritus that had long been suspected as being an alternative source of food to the filter-feeding animals that normally ate phytoplankton. Much of this detritus had been thought to have come from the land; the debris of erosion at the earth's surface; but a more interesting suggestion was now possible. Perhaps the detritus in the sea was mainly formed from the dissolved organic matter, at the surface of the sea or on the bubbles formed by the churning waves, and perhaps the detritus so produced could be used as normal food by some of the filter-feeding "herbivores" of the sea. So they made themselves a supply of aggregates by bubbling air through columns of filtered seawater and collecting the froth, then they fed the material to cultures of brine shrimps (Baylor and Sutcliffe, 1963). Their experiments were not run like Pütter's; they had duplicates and controls. Young brine shrimps placed in filtered seawater without solids died, but young brine shrimps in similar vessels but supplied with the aggregates grew and developed almost as well as those supplied with phytoplankton as food. Clearly brine shrimps could use the energy supply of the bubble-made aggregates, and if brine shrimps could do so it seemed not unreasonable that other filter-feeding animals could do so also. Probably both the filter-feeding zooplankton and the larger filter feeders of the sea floor, such as oysters and clams, could eat the bubble-formed detritus, too.

Pütter's ecological reasoning had been sound. It was reasonable that the energy of dissolved molecules should be used by animals, but it was more than 50 years from his first concluding that they probably did until the discovery of the bubble mechanism that provided a niche for the use of this energy by many animals. And even with this discovery the fact still remained that there was a large reservoir of dissolved organic matter in the deep sea which offered a theoretical niche for an absorption-feeder.

Pütter may have overlooked some obvious animals that would have substantiated his hypothesis also, tiny ones that were not so evidently equipped as carnivores as was *Collozoum.* These are the colorless protists, near-relatives of some of the flagellated algae of the sea, and even perhaps some of their pigmented relatives themselves. That such organisms should feed, or supplement their diets, by absorbtion of organic solutes seems likely, although perhaps local concentrations of solutes are critical. Actively photosynthesizing planktonic plants may actually lose organic molecules to solution rather than the other way around (Ryther, 1969), although perhaps the process could be reversed when photosynthesis slowed. But we now have evidence that some large animals supplement their diets by absorbtion, too, in spite of Krögh, and just as Pütter postulated that they should. Not the kinds of animals around which the Pütter-Krögh argument raged though. Long after the argument had faded into an historical curiosity there was found in the deep sea a new phylum of animals, the Pogonophora. They apparently came to the knowledge of scientific man first as an annoying slime that clogged the deep-sea dredge of a British research ship, the scientists of which dubbed them "the gubbins" and threw them over the side. Later others pointed out that "the gubbins" were squashed animals of a new phylum, and they managed to retrieve some more or less intact animals. They were worm-shaped things which apparently lived in tubes buried in the surface mud of the deep ocean basin with one end sticking out into the water, and they had one great peculiarity. They did not have a gut, or not one that could be identified. A gutless animal is almost a contradiction in terms. The gubbins are not parasites in a broth of digesting food like tapeworms. Then how did they feed? Problems of such interest do not go long untackled among the ever-increasing numbers of modern scientists. Off Miami was found a relatively shallow water form which could be got up alive and kept in an aquarium for observation. A British research ship hauled others up from the deep sea and kept them (not called gubbins this time) for a short while in the ship's aquaria. Both groups thought of putting nutrients labeled with isotopes into the aquarium water to see if they were absorbed by the Pogonophora, and both used amino acids. Their papers were published side by side as "notes" in *Nature,* on June 1, 1968 (Little and Gupta, 1968: Southward and Southward, 1968) and both showed that the labeled dissolved amino acids appeared later in the bodies of the pogonophorans. They could, indeed, feed on dissolved nutrients.

Studies with labeled nutrients have also shown that various other marine animals can absorb some nutrients from the sea, notably starfishes and their stalked relatives, the crinoids (Fontaine and Chia, 1968), but these animals have other obvious methods of feeding which clearly supply most of their food; their absorption of nutrients may well not supply them with as much energy as they lose by excretion during the same time, as was shown to be normal in Krögh's experiments. What is particularly suggestive about the discovery of the Pogonophora is that

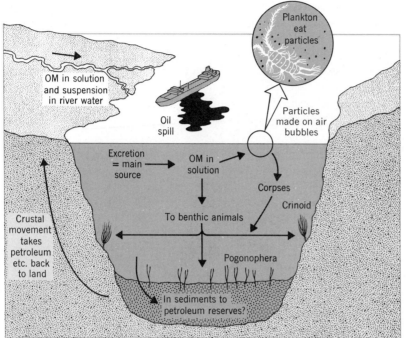

Figure 13.1

Fate of organic matter dissolved in the sea. Most organic production of the oceans is used as solid food by marine animals or saprobes, and is oxidized by them. But some is held in solution, being leaked from the cell walls of living algae, as soluble excretory products, or in the decomposition process. Extra dissolved organic matter is contributed by rivers and perhaps as a result of oil spills. Some is returned to living systems as solids formed at the surface of bubbles, some is absorbed by animals such as pogonophorans and echinoderms or by protists or even plants, and some perhaps is adsorbed by sediments and incorporated into fossil-fuel reserves. This figure is a descriptive model of a system that presumably results in a steady-state concentration of dissolved organic matter in the world oceans. We as yet have no measures of the rates of the various processes in the system.

they have no gut as evidence that they also feed on solid food. They do have tentacles, though, which might allow surface digestion of trapped solids. We have not yet shown that Pogonophora feed entirely, or even extensively, on dissolved nutrients, but the evidence does suggest that they might. It is ecologically right that the potential niche of solute-feeding in the deep sea should be filled. Pütter's original postulate may well be proved valid by the Pogonophora, a phylum of which he died in ignorance.

There is yet some very interesting ecology to be done on the solutes of the oceans. Perhaps the most exciting task would be to try to construct a model of the system by which the concentration of solutes is maintained. It looks as if a steady state might be established, with organic solutes being contributed by leakage from algae, by solution during decomposition, by excretion, and by runoff from the land; and that this input is balanced by losses to animals such as Pogonophora and protists, by the production of aggregates in surface turbulence, and perhaps by adsorbtion to sediments. It should be possible to make an energetic model of this steady state, which should show the rates of inputs and losses to that reservoir of dissolved organic matter which is 300 times the living things and all other organic solids of the ocean system. What part of the total energy flow of world oceans goes to maintain this steady state? (See Figure 13.1.)

Study of nutrients in ecosystems reveals how all elements on the crust of the earth are being moved about, cycled within ecosystems, and carried long distances by physical processes operating over geologic time. An inquiry into the apparent unsuitability of many tropical soils for Western agriculture illustrates this, for it is found that the nutrient reservoir of such a community as the tropical rain forest is held in the trees themselves. When, at the death of the trees, the nutrients find their way to the soil water they are quickly recovered by an efficient root network, thus completing a very effective nutrient cycle. In a temperate forest, however, much of the nutrient reservoir is maintained in the soil, both in slowly rotted organic matter and adsorbed to mineral and organic particles, but again the nutrients are retrieved when they reach the soil water by the root system. From both tropical and temperate forests there is a loss of nutrients in drainage waters which has to be made good by long-distance transport in the wind, or by mining the crust of the earth through weathering. Nutrients are constantly leaking from all terrestrial ecosystems in the drainage water, however efficient the short-term nutrient cycles may be, and these nutrient leaks all serve to enrich the sea, but it turns out that there is an opposite flow of nutrients from the sea back to the land which offsets the nutrient loss in drainage waters. Such a return was suggested by a study that compared the saltiness of the world oceans with the annual discharge of salt in the world's rivers. If the oceans are as old as radiometric dating methods suggest, and rivers have always discharged as much salt as they do at the moment, the oceans should be five times as salty as they now are. It is concluded that one-third of the salt brought down each year by rivers to   197 the sea is returned in wind and rain to the land and most of the remaining two-thirds is raised out of the sea annually in new sedimentary rocks by crustal movements. Wind, weathering, and crustal writhings, therefore, are constantly moving the raw materials of ecosystems about the surface of the globe, maintaining local concentrations at roughly constant levels over geologic time. This *unity of the biosphere* is further illustrated by the study of phosphorus cycles. Only a small portion of the annual phosphorus discharge in rivers, enhanced as it is by the phosphatic fertilizers of agriculture, is returned to the land by the wind, seabird droppings, and fish residues. The rest is deposited on the ocean floors and must await geologic processes for its return, or we must mine the seabed. There is a phosphorus cycle in lakes between the surface mud of the bottom and in the water, but heavily fertile polluted lakes deposit mud so quickly that the excess phosphorus is buried for the life of the lake and removed from the system. Polluted lakes are thus selfcleaning. The discovery that the biosphere must be regarded as a single ecosystem is less comforting, however, because any change or contaminant we introduce into one part of the earth may have repercussions everywhere else.

CHAPTER 14

THE CYCLING OF DISSOLVED NUTRIENTS

A well-known axiom of Western agriculture is that tropical soils are infertile. If you clear a typical piece of tropical land, plough it up, and sow your seed, your efforts are poorly rewarded. Within two or three years your crops fall off to uneconomic levels, and you may be forced to abandon the land. This pattern was so well-known to indigenous peoples of tropical places that they commonly resorted to abandoning the land after a few years and moving on to clear some more, a proceeding known as "shifting agriculture." It seems that trying to farm the same tropical place for more than a few years is not worth the effort. There are, of course, exceptional places in the tropics where you can make farming work, on the flood-plains of great rivers, for instance, but over large areas efforts to farm in the ways that sustain European civilization come to very little. So you say these tropical soils are infertile. There have been many postmortems on why this should be so, and they leave no doubts about the reason the crops would not grow. Many tropical soils are deficient in the essential nutrients of agriculture. When you send a sample of suspect soil to the lab, the report comes back that the sample was deficient in NPK, in nitrates, phosphates, and potassium. The mystery seems solved; but there is an intellectual catch here. Before you ploughed the land, you had to clear the native vegetation, and this might well have been rain forest. The soil is demonstrably infertile. It will not grow crops, and the lab says it is quite deficient in essential nutrients. And yet a short while ago that infertile soil was supporting the lushest form of vegetation known on earth. How could this be?

It is easy to see how a tropical soil might be deficient in essential nutrients, for these are all more or less soluble and the soil is being continually drenched and washed through by the warm tropical rains. This washing out of the tropical soil is so thorough that even the silicates in the soil are removed in time, and there is left just that red granular mixture of iron and aluminum oxides which makes the tropical ground red (Chapter 3). If even silicates are dissolved away, then the fate of phosphates and potassium salts must be certain. But if the nutrients are removed, how does rain forest manage to grow?

Examination of a living rain forest will show that the site during the life of the forest does, in fact, possess plenty of nutrients. But the supply is held in the trees themselves. In this ecosystem, the living things are their own reservoir of nutrients. It is, on the face of it, an uncertain reservoir, because it must be continuously raided by herbivores and by death. For when a tree dies, it is taken down by termites and fungi in an astonishingly short time, its substance rots, and its nutrients are washed out by the rain to vanish in the soil. And endlessly animals eat the trees, to hold nutrients in their bodies during the fleeting spans of their lives before donating them to the soil; or else they excrete the nutrients away within hours of snatching them from the trees. These raids on the reservoir must be a serious leak, one that should rapidly waste the entire hoard, unless the loss was made good in some way. But the loss seems to be made good by an efficient network of tree roots. Deep underground is spread a net of rootlets and associated fungal filaments so complete that the nutrients escaping in the soil water are nearly all recaptured. It is as if there was a dialyzing membrane spread beneath the forest to catch the precious nutrients before they are washed away to the sea; a dialyzing membrane made of the mycorrhiza and root hairs held by the roots of the forest. When a farmer clears the land, he destroys this nutrient retrieval system, and the whole hoard on which the forest was based can be sent to the sea in weeks or months. The farmer finds the soil infertile; no wonder, he made it so (Nye and Greenland, 1960).

This story reveals that nutrients are cycled in a rain forest. From the reservoir in living things, the nutrients steadily escape to the soil as the living things die or are eaten. The sojourn in the soil is very brief, for the roots catch the nutrients once more and send them back to the community above. The cycle seems to be remarkably efficient, but nothing is perfect and there must be some leaks. The roots and mycorrhiza must always miss a molecule or two, so that a trickle of nutrients runs away. This trickle must be made good, or the forest would slowly perish. There are two ways in which the loss can be made good, by the weathering of fresh rocks, and by washout from the atmosphere brought in with rain or as dust. Weathering of fresh rocks is essentially a mining operation and as such its success depends on the richness of the local "ore." But past

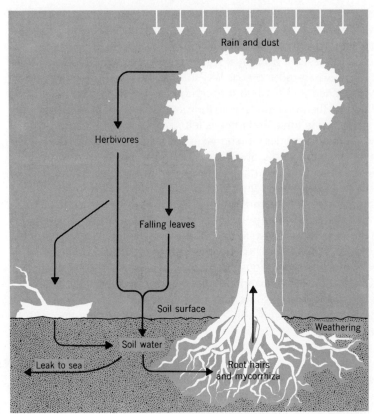

Figure 14.1

Nutrient cycle in a tropical forest. The nutrient reservoir in many tropical forests is held in the plants themselves. Nutrients entering the soil water by decomposition of plant or animal remains are quickly retrieved by a fine network of root hairs and mycorrhiza. This retrieval system may be so efficient that the loss of nutrients in the drainage waters is small enough to be offset by what fresh supplies are provided in rainwater or from ancient weathered rocks. Farming in such places destroys the living nutrient retrieval system with a consequent loss of the nutrient reserves in the drainage water.

weathering of rocks under a rain forest may have been so severe that there are little nutrients to be won by further mining, so that it may be that many a tropical ecosystem is dependent on the small supply of nutrients brought by wind and rain from distant places. But wind, rain, and weathering together must be just sufficient to offset the leak to the sea. There is a balance between input and output, with an efficient cycling in between, the condition called by physicists a **steady state.** Such cycling and steady-state flow is essential to the survival of the forest (Figure 14.1).

This explanation of the infertility of tropical soils raises further questions, the first of which is: Why, if tropical soils are infertile, is agriculture of sorts so successful in some parts of the lowland tropics, tropical rice culture, for instance? The answer to that one is probably that the sites on

which tropical agriculture is successful are generally deltas and flood-
plains of major rivers; the Mekong, the Ganges, the Nile. These are the
places where sediment in suspension is dumped by the rivers, sediments
that carry adsorbed to their surfaces the leaked nutrients from all the
ecosystems of the hinterlands. It is a very mixed blessing damming one of
the major tropical rivers, as has been done across the Nile at Aswan,
because you deny your floodplain its annual increment of nutrient-rich
sediment.

A second question is: How, if the soils without the forest are infertile,
did the forest get there in the first place? The answer to this must lie in a
long process of succession. There are plants that can live on soils considered by a farmer to be remarkably infertile, as is witnessed by the condition of abandoned land in the tropics; it is covered with plants. Pioneer
plants must be adapted to making do with lean nutrient supplies. They
cover the ground. The ecosystem of which they are a part cycles nutrients, inefficiently to be sure, but enough for the stores in their bodies to
build up. Other plants, of later stages of succession, are able to enter the
system. The cycles become more efficient, and so on. Such successions in
the tropics may sometimes be very long, perhaps requiring centuries or
millennia; we do not know. It is also possible that they might sometimes
require climatic conditions that have passed away. I suspect this may be
true in the rain forests I once studied in Northern Nigeria, forests that are
now receding before the doings of man to give way to savannahs in
which rain-forest trees no longer regenerate (Killick, 1957). Perhaps 5000
years ago, during the last wet period (pluvial) in Africa, succession to rain
forest took place. Ever since the forest maintained itself, controlling its
own microclimate, hoarding and cycling its nutrients, a leading force in
its ecosystem. Other ecosystems replace that forest now. Perhaps we
must wait for climatic change before the forest can be made to grow
again. This is speculation, but somewhere it may well be true.

A third question is of particular theoretical interest: Why, if nutrients
can be so speedily leached from soils, does agriculture work at all, even
in temperate regions? Temperate soils may be well-watered, too, and
their nutrients, being the same, are no less soluble. Why don't the nutrients of temperate lands vanish to the sea as soon as you start ploughing?
The answer seems to be in two parts. One is that organic decomposition
proceeds comparatively slowly in temperate climates. Low average temperatures keep the rate of decomposition down; the seasonality of
climate restricts the kinds and activities of soil organisms; and these facts
combined mean that there is always much organic matter in temperate
soils. This organic matter is a major reservoir of plant nutrients, one that
slowly releases them, providing a useful trickle for the plants. The temperate climate means that part of the ecosystem's nutrient reservoir can be
maintained in the corpses and slowly rotting excrement of the system, instead of having to be maintained in living tissues.

The second part of the answer lies in a physical property of temperate
soils. They commonly are rich in the minerals we collectively call "clay,"

minerals constructed on sheets of oxygen atoms to form lattices of very large surface area (Russel, 1950). The surface of a clay lattice is negatively charged, which means that cations are attracted to clay particles and held in the soil. Nutrients such as potassium, calcium, and many trace requirements such as cobalt and vanadium, are thus retained in these soils against loss in soil water. At the same time there are many sites on soil organic matter that are positively charged, and which, therefore, are able to retain anions such as nitrate and phosphate. The chemical properties of temperate soils thus also make them a suitable reservoir for the nutrients of the ecosystem of which they are a part.

There is a cycling of nutrients in temperate ecosystems no less than in tropical ones, but a large part of the total reservoir is held in the soil (Figure 14.2). Clearing the temperate forest and ploughing the land does increase the rate at which nutrients are leaked, but the new ecosystem of agriculture is yet able to maintain its cycles tolerably well. There is mining of the local rock by the weathering process of the ecosystem, and this rock provides a more plentiful source of fresh supplies than do the more thoroughly worked-over rocks of the tropics. The input of fresh nutrients from this source and precipitation may, perhaps, equal the leak from the agricultural ecosystem to rivers, maintaining a steady state and enabling temperate agriculture to continue indefinitely. The extra nutrient-leak represented by crops removed, however, must also be made good by manure and fertilizers, or agriculture proceeds by mining the soil.

There is a point of special interest to ecological theory which develops from the different nutrient cycles of tropical and temperate ecosystems. Temperate trees can rely on an accessible soil reservoir of nutrients, but tropical trees may not. There must be a struggle for nutrients among tropical trees which is far more intense than any similar struggle in temperate places. This is a fact that may have been of enormous importance in the evolution of the rich flora of a rain forest.

In summary, all the soluble nutrients needed by plants and animals, except for nitrates and carbon dioxide for which the ultimate source is the atmosphere, are replenished from the great reservoir that is the crust of the earth. They are mined by weathering and conserved by cycling. The nutrient cycles of temperate ecosystems allow for considerable sojourns in the soil, but tropical soils do not retain nutrients for long and they must be speedily retrieved by plants. Both natural systems are adjusted so that there is a steady-state input of nutrients from weathering balancing leaks of nutrients from the systems to the sea. All terrestrial ecosystems lose nutrients to the great sump of the sea. Although the reservoir of crustal rocks is available to ecosystems, a process of continual denudation of the land for the enrichment of the sea should have potent evolutionary consequences if carried on for long enough. It is, therefore, important to examine the fates of crustal nutrients over geologic time. What happens to the solutes that are washed to the sea? Does the sea always get richer at the expense of the land?

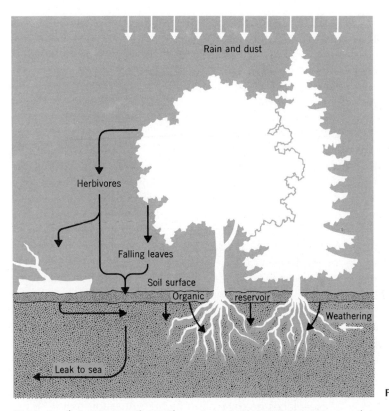

Figure 14.2

Nutrient cycle in temperate forest. The main nutrient reservoir in temperate soils is
in the soil itself. This is not subjected to such intense leaching as is a tropical soil,
and it holds much organic matter that, in turn, holds nutrients. Nutrient cycles are
slower, but there is still a steady-state supply. Wise agriculture may maintain this
steady state in the agricultural ecosystem.

The long-term results of the drain of nutrients from land to sea can be understood by following attempts to estimate the age of the oceans from
their saltiness. The oceans may be presumed to have started out as freshwater seas, and to have grown saltier over geologic time as the endless
washing of rain and rivers brought down salt from the eroding rocks cɨ the
land. If the process has gone on at a roughly constant rate since the seas
were first formed, as seems reasonable, then you can calculate an age for
the oceans by working out the annual tonnage of salt dumped into the sea
by all the rivers of the world and by dividing this into the total tonnage
known to be in the sea. The age long ago arrived at by this method, and
generally accepted by geologists, was 100 million years. There seemed
no general reason to question this figure until modern methods of
radiometric dating for ancient rocks were developed, when rocks of the
Cambrian formation, that is, rocks bearing ancient fossils that were yet
undoubtedly marine, were found to be not 100, but 500 million years

THE GEOCHEMICAL
CYCLE

old. The old calculation was out by a factor of five. If the leak rate of salt from land to sea had really been constant throughout the span of the fossil record, the sea should be five times as salty as it is. Had the leak rate changed recently? If not, where is the missing salt?

This problem recently attracted the attention of an ecologist who also had a geological and chemical turn of mind, Daniel Livingstone of Duke University. Livingstone had accepted a commission from the United States Geological Survey to bring up to date the standard work on the chemistry of lakes and rivers, in the course of which undertaking he acquired the most complete data yet assembled of the annual discharge of the rivers of the world (Livingstone, 1963a). He set himself to recalculate the age of the oceans on the basis of the latest figures for annual sodium input to the sea. And he came up with the same answer as his predecessors; 100 million years. Livingstone was sure that his figures for the annual input were not far wrong and that they could not be far from the average for the last few hundred million years. If such an input of sodium had gone on for 500 million years, as the radiometric age for the oceans said it must, then there should be five times as much sodium in the sea as there was. Where were the other four-fifths of it? Livingstone considered the theoretically possible storage places: saline liquids in the pore spaces of deep-sea sediments, the mass of sodium held adsorbed to the sediments themselves, the mass of sodium held by old sedimentary rocks that had once been under the sea but which had been lifted out by earth movements: but even the most generous estimates for sodium stored in all these places brought him nowhere near the required tonnage. Livingstone's (1963b) calculations so far were as follows:

Sodium carried by rivers to the sea during 500 million years of post-Cambrian time 119.4×10^{15} tons	
Sodium dissolved in world oceans	14.1×10^{15} tons
Sodium in deep-sea sediments	5.1×10^{15} tons
Sodium in other suboceanic sediments	5.4×10^{15} tons
Sodium in old sedimentary rocks now raised out of the oceans	2.6×10^{15} tons
Known reserves of rock salt	$\underline{0.4 \times 10^{15} \text{ tons}}$
Total sodium accounted for	27.6×10^{15} tons

Which still leaves nearly three quarters of that 119.4×10^{15} tons missing. Where was it? Or did it never exist?

Livingstone argued that the answer to the riddle must lie in some process of cycling. The sodium accounted for by his calculations must represent nearly the entire pool of sodium available to the oceans, and that this pool must be cycled between the oceans and the land. Most of that 119.4×10^{15} tons never existed. What reaches the sea each year in rivers must be roughly balanced by a comparable amount of sodium reaching the land from the sea. But how can soluble sodium get out of the sea and back to the headwaters of the world river systems? One obvious way is by travel in clouds and descent with rain. Salt spray is continually picked up by winds from breaking waves, and salt is carried with the winds to the clouds. Rainwater, particularly near coasts always has some

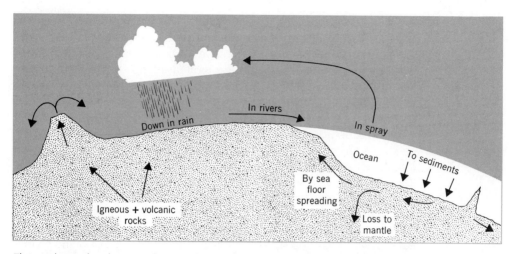

The geochemical cycle. Minerals are cycled between the oceans, the land, and the crust of the earth. Rivers carry soluble minerals to the sea but wind and rain carry some back and most of the rest are probably returned by crustal movements. Minerals in sediments are moved across the ocean bottoms by sea-floor spreading from the mid-ocean ridges (right of figure) to the continents. Loss to the mantle in the deep ocean trenches may be made good by the extrusion of fresh rocks.

Figure 14.3

salt in it, and the total mass of salt transported to the land in this way may be very large. Livingstone calculated the largest annual mass of sodium that could reasonably be expected to be returned to the land in this way, and then extrapolated to compute the return over the last 500 million years. About half the total sodium delivered to the sea by rivers in this time may, in fact, have been returned to the land by rain as follows:

Sodium carried to the sea in rivers	119.4×10^{15} tons
Sodium returned by the atmosphere to the land	55.2×10^{15} tons
Net transfer to the sea	64.2×10^{15} tons

Which means that atmospheric cycling, although important, cannot be the complete explanation. Livingstone suggested that there might be an even larger annual return due to movements of the earth's crust.

Sodium enters the sea in rivers, it spends a time there in solution, some of it is picked up by wind and returned to the land, but the rest eventually becomes concentrated in the interstitial spaces of the sea's sediments and adsorbed to its mud. As coastlines emerge, they raise sodium-rich sediments out of the sea, and these sediments become in time the consolidated rocks of the land from which the sodium is again weathered, perhaps to be cycled awhile in an ecosystem, and then sent once more on its way to the sea. This cycle may have been completed many times in the 500 million years that have elapsed since Cambrian time (Figure 14.3).

There have been two geologic discoveries since Livingstone published his conclusions which modify the story somewhat. The first is the discovery of sea-floor spreading (Longwell, Flint, and Sanders, 1969). The floors of all the major oceans move outward from the mid-ocean ridges and toward the continents like giant conveyor belts, finally plunging downward under the continents into the molten interior of the Earth (Figure 14.3). Ocean floor cannot be more than about a 100 million years old before it suffers this fate; so what of the sodium carried in its sediments? Is it lost to the system? And, if so, is its loss made good from new continental crust thrust up from below? It is perhaps not yet possible to be sure of the answers to these questions, but it seems likely that sodium return to the land is not seriously affected by the movements of the sea floor. As the floor begins its dive under a continent, its surface is likely to be scraped as with a giant bulldozer, so that sediments bearing sodium are piled up at the continental edges. Mountain ranges are thrust up along the coasts of collision, and these are likely to contain the sodium and other nutrients of the old seas. Sea-floor spreading has not altered the thrust of the Livingstone hypothesis, merely the method by which old-sea sediments become incorporated in the rocks of the dry land.

The second discovery was made by the deep-sea drilling operations from the *Glomar Challenger* of the JOIDES project (Ewing et al., 1968). There seem to be under some ocean floors huge dome-shaped deposits of sea salt, deposits containing cubic miles of the stuff. They seem to have been collected there in the youth of oceans when their waters might have been hot and particularly saline, like those of the modern Red Sea, and then to have been sealed in by later sediments so they remained isolated from the water of the later larger ocean. These deposits could account for some missing sodium after all. We do, however, know that much sodium is cycled back to the land, because we can find it in sedimentary rocks. The likelihood is still that the return form this source and the atmosphere is on the order of the loss in rivers. Perhaps the salt domes may be roughly equated with the extra supply provided to the land by the weathering of igneous rocks.

What is true for sodium must be true for all other soluble nutrients as well. Communities on land do not live only as a result of the mining of a finite resource of the earth's crust, but make use of nutrients that are cycled between the earth and the sea with the atmospheric cycle, or every 100 million years or so by crustal movements. There are two ecological and evolutionary implications of this discovery which may have far-reaching importance; that the nutrient supplies of modern land areas are typical of those throughout the long history of life on land, and that the concentration of solutes in the sea has been essentially constant since at least Cambrian times. The lives of plants and animals throughout the last 500 million years have thus been lived in places possessing nutrient supplies or salinities essentially the same as those of modern times.

THE PHOSPHORUS
CYCLE: PERHAPS THE
MOST CRITICAL FOR
AGRICULTURE

Phosphorus is perhaps the most ecologically interesting of the nutrients, because it is so vital to organisms and yet so scarce. With its peculiar capability of forming high energy bonds, phosphorus seems to be important to the energy transformations of all living things. The ratio of phosphorus to other elements in organisms is generally higher than the ratio in rocks, soils, and water, from which organisms must draw their supplies. Any leak of this element to the sea must be particularly serious to land plants, for whom the replacement rate achieved by crustal folding is uninterestingly slow. Man's hundred years of experience with phosphatic fertilizers is eloquent testimony that shortage of the element may commonly limit plant growth more than shortage of any other element. The low phosphorus content of the sea, particularly in the warmed upper waters where plants live, seems to be the principal reason that the fertility of the oceans is so low (Chapter 10). There is nothing we can do about this; not even dumping some mega-tonnage of phosphate in the sea (if we had it) would help because the problem is partly one of low solubility. The oceans deposit phosphate minerals in their sediments even at present concentrations. Which is one of the reasons why farming the oceans seems more a boondoggle than a reasonable undertaking to ecologists. It is, however, possible to test the general idea of what might happen if we could fertilize the sea by experimenting with putting phosphate in a low-phosphate lake. We have, of course, done this many times without meaning to. We call the results "pollution." But it has also been tried several times in a deliberate and controlled way, and the results have provided interesting general insights into the phosphorus cycle.

Crécy Lake is a small deep lake in southern New Brunswick, Canada, occupying a "kettle" hole left when a block of glacial ice melted in the ground after the retreat of the last continental glacier some 14,000 years ago. Like all such lakes it is rather unproductive, because it is deficient in nutrients such as phosphorus. M. W. Smith of the Fisheries Research Board of Canada used it for a fertilizing experiment in 1941, driving up with a truck load of superphosphate and dumping one and a quarter tons of it into the lake (Smith, 1948). This was sufficient fertilizer to raise the concentration of dissolved phosphorus by 2500 percent. The superphosphate all dissolved, but a few days later the concentration of phosphate in the water was only up 67 percent. The water had turned pea green for a while, evidence of a massive phytoplankton bloom, but then it returned roughly to normal. Three things could have happened to all that phosphate; it could have been held by rooted water plants, it could have been absorbed by the mud, or it could have mostly been used by phytoplankters in the tremendous bloom, and then been sent to the bottom in their dead bodies to become part of the lake mud. These possibilities could be examined by careful measurements of phosphorus in mud and plants over a period of time. Quantitative assays of this kind are not easy, but the way to their solution was soon available as the development of nuclear weapons in the 1940s provided us the by-product of radioactive elements. Radiophosphorus has a half-life of 14 days, which means that it lasts long enough for us to use a dose of it to trace marked elements through the phosphorus cycle of a season.

One of the first studies with labeled elements was an investigation into the phosphorus cycle of a small lake by G. E. Hutchinson (the same who first formulated the ideas of energy flow with Lindeman) (Hutchinson and

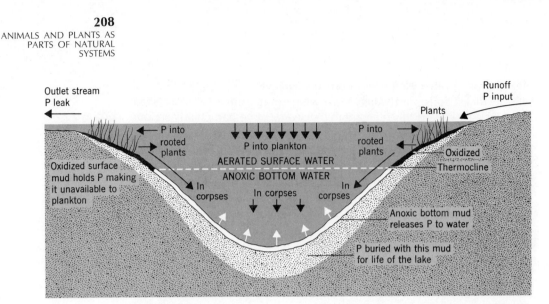

Figure 14.4   The phosphorus cycle in a lake. Phosphorus enters all lakes continuously in runoff water, the inlet streams, and from the air. Phosphorus is also continuously lost to the lake in the outlet streams and by incorporation in lake mud. When a lake has anoxic bottom water, as when a lake stratifies in summer (see Chapter 18), the top few millimeters of mud are chemically reduced, a condition that allows the mud to release phosphorus back to the water. The bottom water thus becomes phosphorus rich. Stirring of the lake by winter storms brings the phosphorus-rich water to the surface, completing an annual cycle, and fertilizing the lake for a spring plant bloom.

Bowen, 1950). Hutchinson and a companion rowed out onto Linsley Pond in Connecticut with some bottles of radioactive phosphate solution and poured them into the lake. In the subsequent days and weeks they sampled water, mud, and plants and were able to check the arrival of their labeled phosphorus by placing the samples under a radiation counter. It quickly went from the water, as did the massive dose of superphosphate in Crécy Lake, and as quickly appeared in the plants of the phytoplankton. Later it got to the rooted plants of the shallows, and then it went to the bottom mud in the falling bodies of the dead plank-tonic plants, or adsorbed to fine particles which settled through the lake after being brought there in the inlet stream. The deep muds of the lake, which in summer were in contact with water that had lost its oxygen because of biological activity, were chemically reduced, and they gave off ionic phosphate back into solution. The water circulation of the lake eventually brought this phosphate back to where plants could use it, thus completing the cycle (Figure 14.4).

Only a very thin layer of mud at the mud/water interface appears to be able to release phosphate ions back to the water, even under reducing conditions. If sediment-bearing phosphate is deposited very quickly, it is obvious that phosphorus could be lost to the system by being perma-nently buried. This may have happened to the tonnage dumped into Crécy Lake. Similar studies at Schleinsee in Germany by Einsele at about the same time (Hutchinson, 1957), but with more modest doses of fertil-izer and repeated a second year, suggested this still more convincingly. By September of the year following the second dose of fertilizer, the concentration of dissolved phosphorus in the lake was back to the level of before the start of the experiment. Hutchinson concluded from these studies that a lake was a self-regulatory system which should return to its original steady state when disturbed as by an increment of phosphate supply, and he considered this discovery to be "one of the most important in limnology." It holds cheer for those who wish to restore some of our polluted lakes to their original condition, suggesting that if we leave them alone for a while they will deal with the excess phosphate (a principal pollutant) by themselves; by burying it in their mud (Chapter 19).

Plant communities of the land do not deposit sediments as do those of lakes, but they do deposit organic matter in the litter layers of the "A" horizons, which can act as regulators of the phosphorus supply. In addition, iron and aluminum minerals of soils strongly adsorb phosphate ions so that an equilibrium is established between the concentration of phosphate ion in the soil water and that in the soil minerals. Such a system can buffer the phosphate supply to the plants, helping to promote a steady-state flow of the mineral into and out of the plant community. The ease with which phosphates may be taken and held by soil minerals is an important check on the possibilities of using phosphatic fertilizers, since soil minerals may remove any of the fertilizer not quickly used by plants, reducing the availability of phosphorus to that prevailing before the fertilizer was applied.

In the years following the first general realization of the importance of phosphorus in the middle of the nineteenth century came the discoveries of the guano deposits on some Pacific islands. These were huge mounds, island wide and tens of feet thick, of the droppings of seabirds, and they were valuable both as sources of nitrates for making explosives and as phosphatic fertilizers. The birds ate fish, whose bodies were phosphorus-rich, and much phosphate was voided with the birds' droppings. It was obvious that an enormous tonnage of phosphorus had been returned from the sea to the land in this way, and that similar, if less spectacular, returns of phosphorus were being made on all coastal areas of the world. In addition to this return, was the work of human fisheries in carrying large tonnages of marine fishes far into the continental masses before consuming them. It seemed possible that some considerable part of the leak of phosphorus to the sea might be offset in this way, a possibility that G. E. Hutchinson (1950) checked by calculating the annual tonnage of world fisheries, the annual droppings of seabirds of the world, and the annual phosphorus discharge of the world's rivers. He found that, whereas rivers discharged nearly 14 million tons of phosphorus annually, the fisheries and seabirds of the world combined could only return 70,000 tons. Clearly the leak goes on in spite of the fishermen. Indeed, man's activities have almost certainly made the leak worse because the ploughing of once forested lands removes mechanisms of phosphorus retention in the litter and the trees, and at the same time allows erosion of mineral particles, which are washed to the sea in turgid rivers, each particle with its little load of adsorbed phosphorus.

Not only have we increased the leaks of phosphates from ecosystems, so that the ecosystems of agriculture are losing phosphates faster than they can mine fresh supplies from rocks, but we are also hastening the mining process. We make phosphatic fertilizer by mining the richest deposits we can find, and we dump this on our land. Most of it ends up in the sea, whence it goes, via an algal bloom or two, to the mud of the sea floor. We are working out the easily mined phosphate deposits, and must soon use lower grade ores (or at least the underdeveloped countries will have to if they want some). These, if we can find the energy to mine them, should be enough for the forseeable future, after which we must start on the sea floor ourselves. The deposits we mine were once marine sediments, those that collected under upwellings where the phosphate-rich corpses from small patches of productive ocean rained down. They would have been used and cycled in the fullness of time by natural ecosystems, being used sparingly, and at a rate that would have kept pace with their replenishment from the sea by crustal processes. They should have lasted a few geologic epochs or forever. We are making them last just a few centuries or so; merely so that we can feed an uncomfortable crowd of people and go on having large families at the same time.

THE UNITY OF THE
BIOSPHERE

Study of the way in which the principal plant nutrients are moved about the earth reveals ways in which all elements are moved. The great geochemical cycle between oceans and evolving continents not only keeps the sea at a roughly constant saltiness but it also moves all the elements of the earth's crust between sea and land. When winds from the sea blow back to the land something like a third of the sodium brought down by rivers each year, they also blow back other elements brought to the sea. There are no dead spaces at the surface of the earth where elements may vanish or "be got rid of." Things are only moved around; sooner or later they will be back. This applies to fertilizers and pollutants and radioactive isotopes alike. Everything is cycled. If the element or radical is one that has a biological effect, it may be singled out for special cycling. Concentrations of biologically important compounds which may be achieved in the living parts of ecosystems may be between a thousand and a millionfold. When radioactive isotopes are chemically similar to stable isotopes, they too may be concentrated and to a like extent. Long-lived poisons like DDT also may be concentrated as they are cycled within systems and moved from place to place. The surface of the earth, the oceans, and the air above them form one grand ecosystem for living things, the **biosphere.** Anything done to one part of it may eventually affect any other part.

An element cycle of special interest is that of nitrogen because nitrogen is both a principal constituent of living things and of the atmosphere. The immediate nitrogen reservoir of an ecosystem is held in organic molecules or as nitrate and ammonium ions. In these combined forms nitrogen is cycled as are other soluble nutrients. Combined nitrogen in solution is likewise leaked from the ecosystem in drainage waters and this leak may again be massively increased by agriculture, but the nitrogen system is unique in that the leak must be made good from the atmosphere. Nitrogen compounds are synthesized from inert nitrogen gas by a variety of microorganisms whose synthetic activities are responsible for the greater part of the nitrogen made available to life. The old agricultural practice of leaving fields without crops for a season every few years allowed nitrogen-fixing organisms to replenish the supply of combined nitrogen, but the immediate effect of the unceasing synthesis of nitrogen compounds by microorganisms should be the accumulation of combined nitrogen deposits on the crust of the earth and the depletion of the atmosphere. This does not happen because other organisms living in anoxic environments reduce nitrates and release nitrogen gas. The atmosphere of the entire planet is, therefore, maintained by organisms some of which live in rather special places.

211

CHAPTER 15

THE NITROGEN
CYCLE

In the agriculture of the middle ages it was the custom to grow crops on the fields for two or three years, and then to leave the land alone for a year, a practice called "fallowing." If, instead, you just went on cropping year after year without a break, you soon found that your yields were low. The land only gave you a good crop for two or three years, then it seemed tired of your extortion, and it struck. But all the encouragement it needed to start producing for you again was a year's rest. If you left the land fallow every three or four years, you might expect good crops from it indefinitely. Medieval man said you had to "give it a rest sometime," and this simple analogy to the course of human labors let him practice good land management. But why should "resting" the land let it get "better?" The answer turned out to be that you allowed time for a store of nitrates to be built up in the soil.

Nitrogen in quantity is as necessary to organisms as is phosphorus in quantity, although nitrogen, because of its abundance in the atmosphere, is not rare at the surface of the earth like phosphorus. Living things need it principally for the properties of the amino radical, the NH_2 on which the structures of proteins, and hence of their bodies, are based. As with other nutrients, an established community may enjoy a steady-state flow of usable nitrogen with an input and output to the terrestrial hoard. But the hoard is in the atmosphere, not in the crustal rocks. All ecosystems are bathed in a fluid, the air, which is about 80 percent nitrogen. Not in our wildest excesses of mismanagement have we found a way of physically separating an ecosystem from its nitrogen reservoir. We can and do bull-doze rain forests and let their phosphorus reservoirs flow away to the sea, but we have not yet found a way of transporting their nitrogen. There is a catch, though. Nitrogen is an inert gas, so inert that it cannot be used in the metabolism of most organisms. They must have it in combined form, as nitrate or ammonium ions. These ions are cycled in ecosystems like the ions of other water-soluble nutrients, like phosphate and potassium. They go from the trees, to the litter, to the soil, to the trees. Farming denudes the supply with particular speed, because farmers are hungry for protein-rich crops. They cart the nitrates away. They savor them briefly in their palates, then dump them in the privy or, as now, by a hygenic flush, into rivers, lakes, and the sea. The ecosystem can make good this loss, because there are pathways through which nitrates are synthesized from nitrogen gas, but it takes time. Hence the success of leaving the land fallow for a year.

The nitrogen in living bodies is present in large, energy-rich molecules. When the body becomes a corpse, decomposers break down proteins and other nitrogenous molecules to release their potential energy for their own lives, a process called **deamination.** When plants are eaten by animals, similar energy-yielding chemistry is undertaken, and the end products consigned to the local soil as such nitrogenous molecules as urea. More energy still is derived by bacteria who break down these molecules to the final reduced stage of ammonia. In an oxidizing environment, such as is usual in soil, even ammonia represents a potential source of energy, and a group of bacteria, the **nitrifying** bacteria have evolved to take advantage of it. They oxidize ammonium ions to nitrite and nitrate ions.

Nitrate and nitrite ions are chemically stable. There is no energy to be had by tampering with them further. In normal terrestrial ecosystems, oxidizing conditions prevail, and ionic oxides of nitrogen may persist. They are a form of nitrogen acceptable to plants, are selected out of the soil water by roots, and are cycled within the ecosystem much as are potassium or phosphate ions. Like phosphates, they are anions, and thus are not held by clay minerals. They must be retained by the positive charges on soil organic matter, but failing such organic matter or efficient root systems must quickly migrate through the soil to the water table. The normal fate of these ions is, of course, to enter the sea, but if the groundwater flows toward a desert place, nitrates may accumulate there. This is apparently what has happened in Chile (Delwiche, 1965). Nitrates have traveled in percolating waters from the moist vegetated slopes of the High Andes and have collected in the dry, oxidizing, desert environment below. The supply was big enough for the Western Powers to make all the explosives they needed for the First World War, stern testimony to the nitrate-making power of the Andes ecosystem. This explosive resource, like the nitrate or ammonia which made good the leaks from other ecosystems, had to be created by making nitrogen of the inert atmospheric pool combine to form ions usable by plants.

The direct oxidation reactions of the series

$$N_2O \longrightarrow N_2O_3 \longrightarrow N_2O_5$$

can be made to occur only with difficulty, and they require energy. Some oxides are produced by the tremendous energy released by lightning flashes, and these reach the ground in rain. Other oxides are emitted from volcanic vents, but probably only in tiny quantities. Neither of these sources are likely to produce amounts of nitrate significant to plants except in large, barren, stormy places like Antarctica where the supply from lightning may be important. Instead, organisms have to perform the oxidations for themselves, supplying the energy and a catalyst for the difficult series of reactions or finding an alternative chemical route. The series of reactions needed to make nitric acid (HNO_3) out of water, gaseous oxygen, and gaseous nitrogen are more promising, since there is a net release of energy involved, and this may be the chosen route of some **nitrogen-fixing** organisms. But whether nitrogen is oxidized to nitrate, with energy being supplied by the oxidation of carbohydrates, or whether nitric acid is produced with a liberation of free energy, the production of nitrogenous compounds from molecular nitrogen by organisms is the only significant return of nitrogen nutrients to living systems from the atmospheric sink. And the curious thing is that of all the array of plants and animals in desperate need of nitrogenous nutrients, only a variety of tiny ones can perform the necessary nitrogen fixation. The large ones are absolutely dependent on the activities of such things as nitrogen-

Figure 15.1   Root nodules, the sites of nitrogen fixation. Above on haricot bean, right on alder. Symbiotic bacteria such as *Rhizobium* live in these nodules, drawing their carbohydrate supplies from the associated plant and leaking nitrates through their cell walls. The nodules

develop from infected root hairs (Nutman, 1956). In addition to apparently all species of the
family Leguminosea, nodules with symbiotic bacteria-fixing nitrogen have been found in
the genera *Alnus, Aurcasurina, Coriaria, Ceonothus, Dryas,* and *Myrica.*

fixing bacteria or blue-green algae for their input of fresh nitrogenous nutrients. Why the necessary enzymes have not been developed, or have been lost, by all taxa of higher plants is a major mystery of biology for which there must someday come some subtle answer from the laboratories of molecular biologists. The best that some plants have been able to do is to provide in their roots a good living place for nitrogen-fixing bacteria, thus concentrating a supply of nitrates in the immediate neighborhood. The famous root nodules of legumes are the best-known example of this, but nodules of this kind are also possessed by other plants, such as alders and the tiny herbaceous rose, *Dryas,* of the arctic tundra (Figure 15.1). The nitrogen-fixing bacteria live in the tissues of these nodules using some of the plant's store of carbohydrates for their own metabolism but giving a selective advantage to the plant which supports them in the nitrates they produce. Such *living together* is called a **symbiosis,** which means the same thing in Greek.

When an established plant community, enjoying its steady-state flow of nitrogenous nutrients, is cleared and crops grown in its place, there is an immediate loss of nutrient to the drainage channels and in the bodies of plants taken away. But the reservoir of nitrogen is near at hand in the air rather than in rocks that must be slowly eroded. Nitrogenous nutrients can be replaced as quickly as the nitrogen-fixing bacteria work, and these work whenever there are suitable living places in the cleared land. The plant communities of early succession stages apparently serve well enough the bacteria's purpose so that nitrogenous nutrients will begin to accumulate in the early successional communities. This is the reason why leaving fields fallow lets them get back their fertility.

Modern Western populations, with their ever-growing demands for food, are not so patient as to wait for ecosystems to replenish their own nitrates from the air. They do it for them by the Haber process, using much fossil-fuel energy and a platinum catalyst to make nitric acid from atmospheric nitrogen, then compounding it into fertilizer, and spreading it on the land. Fritz Haber invented the process just in time to provide the explosives which made the First World War possible for Germany. That epic struggle was partly a battle between the Andes ecosystem working through the mines of Chile and the fruits of Haber's genius.

The immediate needs of an ecosystem are met when their nitrogen-fixing bacteria, or the disciples of Fritz Haber, make good their nitrate losses. But how shall the atmosphere's loss be made good? The ecosystem robs the air of nitrogen to make nitrates, but it does not leak nitrogen back to the air, it just leaks nitrates. These are stable ions, as the mines of Chile and elsewhere demonstrate. Why, then, has all the nitrogen not been taken from the air long ago and dumped somewhere else on the earth as a huge deposit of nitrate rock? The answer is that nitrogen is released from nitrates by the activities of yet another group of microorganisms in the profession of doing something with nitrogen, the **denitrifying bacteria** and other small things which work in the same way (Figure 15.2).

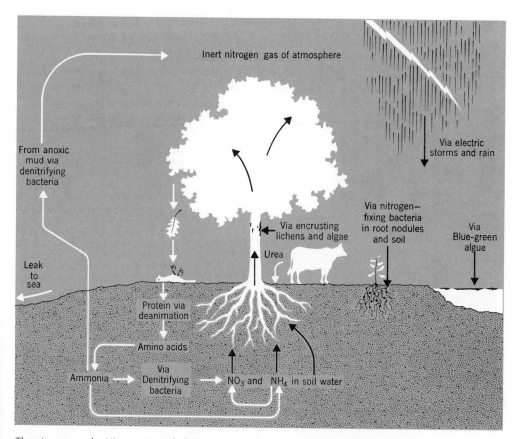

The nitrogen cycle. Nitrogen is cycled through the ecosystem in combined form either as oxides or reduced as in amommonia, amino acids and proteins, and the like. There is a continuous leak from the system in flowing waters that must be made good by fresh supplies from the atmospheric reservoir. Many kinds of microorganisms fix nitrogen, principally by reducing it to ammonium compounds, and there may be additional supplies (as oxides of nitrogen) in rainwater which have been synthesized in electric storms. Nitrogen is returned to the atmospheric reservoir by denitrifying bacteria living in reduced muds of fertile waters and bogs. Black arrows are paths of nitrogen synthesis and white arrows are paths of nitrogen release from more complex molecules.

Figure 15.2

The denitrifiers live in places without oxygen, in estuaries, the bottoms of fertile lakes, swamps, and parts of the sea floor, or in wet patches in the soil. Their sources of energy are various, but they all face the common problem that they must oxidize without oxygen. In the language of chemistry, they need a hydrogen acceptor, and they use nitrate for this purpose, reducing it first to ammonia and then to the atmospheric gas, which eventually bubbles back to the atmosphere. An organism typical of this group is the sulfur bacterium, *Thiobacillus thioparus,* which oxidizes the sulfides of reduced mud into sulfates. This reaction releases energy which the bacteria use for their life purposes. It is for them what oxidizing a carbohydrate is for an animal. But they need a hydrogen acceptor to make it possible, and they take advantage of the nitrates collected in their muddy home from the leakage of the world's ecosystems. And they complete the nitrogen cycle. This is no doubt a matter of indifference to all foreseeable generations of sulfur bacteria, and other microorganisms who work the same trick. But without them the earth's atmosphere would run out both of nitrogen and of the oxygen with which it combines.

The ways in which life processes are necessary for the maintenance of
the air has given rise to speculation that human interference with these
critical life processes might bring about catastrophe. Study of the
sulfur cycle reveals that oxygen is being lost from the atmosphere
incorporated in stable sulfate ions and that this oxygen is only returned
to the atmosphere by microorganisms living in anaerobic muds of
swamps and marshes. Our activities in draining swamps may reduce the
number of sites in which these microorganisms can live. More important still is the fact that much oxygen is apparently returned to the
atmosphere each year by the photosynthesis of green plants both on
land and in the seas. The thesis has been put forward that if we kill the
plants of the oceans with pollutants this will result in a catastrophic
decline in the oxygen of our air. This thesis mistakenly supposes that a
major portion of the total photosynthesis of the earth occurs in the
oceans, whereas, in fact, only a third or a quarter does so, but even if all
the plants of the earth were killed the effect on the air would be very
slight because only a fraction of one percent of the total oxygen of the
atmosphere is turned over annually by living systems. The loss of oxygen
hypothesis is, therefore, nonsense. More realistic are fears that the continued burning of fossil fuels will raise the carbon dioxide of the atmosphere to dangerous levels. In spite of the carbon dioxide buffer
provided by the world oceans, about half the carbon dioxide produced
by the combustion of fossil fuels is now known to stay in the atmosphere, where it will probably remain for the one or two hundred years
required for the oceans to be sufficiently mixed to absorb it. The likelihood is that the carbon dioxide content of our air will be more than

doubled by the time we use up our fossil fuel reserves. Various consequences have been suggested for this alteration of our air. For a biologist the main worry must be that we have very little idea of the long-term consequences for living things of an atmosphere enriched with carbon dioxide.

CHAPTER 16

THE MAINTENANCE
OF THE AIR

Study of the nitrogen cycle reveals how dependent the atmosphere is on the activities of microorganisms. Nitrogen is taken from it on a grand scale by nitrogen-fixing bacteria, and then is used by ecosystems to make nitrates. If the nitrogen in the nitrates was not eventually released by the denitrifiers, the nitrogen of the atmosphere should eventually disappear. We need nitrogen in the air we breathe. We need oxygen even more. And the release of nitrogen from nitrates left the oxygen tied up in other ions, principally perhaps sulfate ions. This oxygen must be returned to the air somehow, or we should have not oxygen to breathe, but a giant deposit of sulfate tucked away somewhere on earth instead. To see the release of oxygen from sulfates, it is convenient to follow the sulfur cycle.

Sulfur is another of the elements apparently needed by all living things in fair supply, mainly for its structural role in a number of amino acids. Plants acquire it from the soil in the form of sulfate ions, and it is then cycled within the ecosystem in the familiar way. There is a leak into the drainage waters which, since sulfate is an anion like nitrate, may be particularly severe in soils with a low content of organic molecules to provide the positive charges that might retain it. And this leak has to be made good to maintain a steady state. Making good the leak appears to be mainly a mining operation, like that which provides phosphate and potassium, although nowadays the skies rain sulfuric acid instead of pure water from the effluent of our fossil fuel economy, so few ecosystems are likely to be short of sulfur. The sulfur which is mined from rocks by weathering, as well as that which is mined from coal mines by miners, is probably mostly in the reduced form, sulfide, but this is spontaneously oxidized by the free oxygen of the atmosphere. The sulfate leak from any ecosystem is thus made good by the oxidized product of weathered rocks.

But the sulfates that leach away in drainage waters represent part of the air's oxygen supply. The steady leak of nitrates had raised the possibility of the earth's atmosphere ending up as a mound of nitrate rock; the leak of sulfates raises the possibility that the oxygen of the air should wind up in a pile of sulfate rock instead. This may seem doubly likely, since the solution to the nitrate problem was partly provided by microorganisms who lived by oxidizing sulfides, which is to say by making more sulfate,

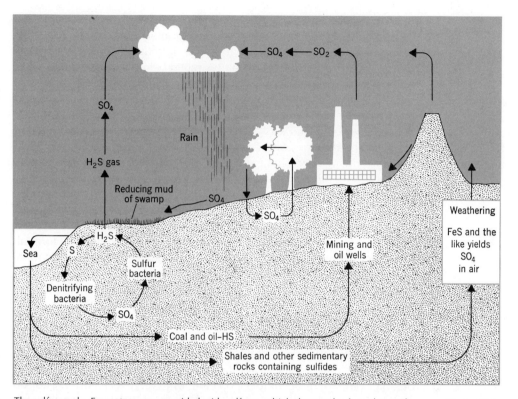

The sulfur cycle. Ecosystems are provided with sulfates, which they cycle, from the oxidation of the products of weathering. In reduced muds, sulfur is cycled between sulfur bacteria which reduce sulfates and others, like some denitrifying bacteria, which oxidize sulfides. H_2S returned to the atmosphere is spontaneously oxidized and delivered by rain. Sulfides incorporated in fossil fuels and sedimentary rocks are eventually oxidized following human combustion or crustal movements and weathering.

Figure 16.1

and subverting the energy released thereby. But the global ecosystem is saved from this dilemma by another group of anaerobic organisms, the **sulfate-reducing bacteria.** These live in all the same sorts of places in which the denitrifiers live; the mud of fertile lakes, the mud of estuaries, the mud of swamps. They, too, are killed by free oxygen. But they are photosynthetic bacteria, deriving their energy from the sun like a green plant. They live buried under the algal scum of the mud surface where light but not oxygen reaches, or under anoxic waters. There is no free oxygen for them to use as an hydrogen acceptor, as do green plants, so they use sulfates instead. The final outcomes of their labors are muds that smell of sulfurated hydrogen, H_2S gas released to the atmosphere, more sulphide for denitrifying bacteria to use, and oxygen made mobile for living things again (Figure 16.1).

Without denitrifying and sulfate-reducing organisms, the atmosphere would lose its oxygen, for there are no other processes, physical or biological, which reduce nitrates and sulfates at the surface of the earth. Air-breathing life is dependent on these two groups of organisms. And these organisms are, in turn, dependent on having reducing muds in which to live. E. S. Deevey (1970) has recently argued before a congressional committee that the conservation of mud is the really important side of the conservation of water. Modern societies drain marshes, fill in estuaries, and straighten rivers. Whenever they do so, they reduce the available supply of anoxic, smelly, vital, oxygen-giving mud. Deevey suggests that half the marshes and estuaries of the primeval United States have already gone, and that, in the very long run, we may be endangering our oxygen supplies by continuing that process of drainage. At least, the argument gives you a reply when you are faced by the boor who can look into the misty distance of some lovely sea marsh and say to you, "What use is it?" You can tell him that it provides him with oxygen to breathe.

We look on reducer organisms of swamps, who so obligingly forestall those mountains of nitrate and sulfate, as producers of oxygen. But more impressive producers of oxygen, of course, are green plants. We all know that animal life is "only possible" because green plants provide the oxygen which animals breathe to burn their carbohydrates. There is an oxygen cycle within every ecosystem, with energy being degraded to reduce carbon dioxide, and the oxygen released then being used to oxidize the resulting carbohydrates for the benefit of the animals. This is apparently a nicely balanced business because carbohydrates and other organic compounds accumulate on earth only very, very slowly. They do accumulate a little, otherwise we should have no petroleum and coal to squander, but the entire hoard of fossil fuels represents storage of only a tiny annual amount of organic carbon. The annual storage of the moment appears to be about four parts in 10,000 of what is being produced by photosynthesis (Broecker, 1970). So we conclude that production of oxygen by photosynthesis is nicely balanced by the oxygen removal of respiration. The oxygen of our atmosphere remains impressively and comfortingly constant at 20.946 percent by volume in dry air (Machta and Hughes, 1970), surely reasonable enough when the use and production of oxygen are so nicely balanced. But what if something happened to the plants, to the producers of oxygen?

There is a possibility that something might happen to many of the plants, just as something is beginning to happen to the homes of those other producers of oxygen, the sulfate and nitrate bacteria. We deliberately spray 2,4-D and 2,4,5-T, wonderfully deadly plant killers, into the biosphere with such success that it was recently argued in an Iowa court of law that 2,4-D should be considered a normal part of the Iowa atmosphere in summer. 2,4-D is already being succeeded by even more deadly plant killers such as picloram. Some of these are being carried about the world in ships. Supposing one of the ships sinks? Or several of these ships sink? Would the result be the destruction of all the plants in a large part of the sea? Some large proportion of the total photosynthesis of the earth has long been thought to occur in the sea, such that introductory

The Comparitive Production on Land and Sea   TABLE 16.1

	Net production (10^9 tons/year)	Estimated respiration (% gross production)	Gross production (10^9 tons/year)	Percent of total global gross production
Land	109	60	272	75
Sea	55	40	92	25
Total	164	100	364	100

From data compiled by Whittaker and Likens in Whittaker 1970.

textbooks still talk of the oceans being responsible for 70 percent or even more of the annual oxygen return. What if a picloram spill, or other poisoning, should block the sea's input of oxygen to the atmosphere? And if we can imagine blocking the photosynthesis of the oceans in this way, might we not also pollute the land that photosynthesis stops there also? We are already loading the atmosphere with some 3000 novel compounds at the moment, many of them of unknown long-term biological effect, and there are many more to come (Cole, 1971). Perhaps we shall finally be clever enough to block the photosynthesis of a whole generation of plants. What then? We should, of course, starve to death, but what of the atmosphere while we were starving? Might we not anticipate starving by succumbing from oxygen deficiency?

This possibility of catastrophic oxygen loss following catastrophic plant death has been much discussed in these years of environmental concern. In its simplest form, the hypothesis imagines disaster being brought on merely by the destruction of marine plants. Seventy percent of the world's photosynthesis goes on in the oceans, say the textbooks, so all that is required for disaster is destruction of oceanic life. But the oceans are not responsible for 70 percent of the earth's production. The claim for 70 percent, and the even larger figures still being quoted in some textbooks rests on early miscalculations that are being uncritically quoted long after they were shown to be in error. The oceans are now known to be unproductive deserts (Chapters 10 and 12) which, in spite of their occupying 70 percent of the surface of the globe, probably account for no more than 25 or 30 percent of the earth's production (Table 16.1). It is apparent that killing off marine plants cannot block most of the oxygen return, vile and calamitous though such an act would be.

But would the killing of all plants, on land as well as in the sea, produce catastrophic changes in the atmosphere as this general thesis requires? The hypothesis has always failed to take into account the truly immense size of the oxygen reserve in the atmosphere. If only a tiny part of a huge reserve is actually cycled each year, then the total volume will seem constant despite small fluctuations in the cycles. Furthermore, tampering with the cycles, as we are undoubtedly doing, will not make very serious differences in the great mass of the reserve. This, in fact, seems to be the truth of the matter, as recent calculations by the geochemist Wallace Broeker (1970) show. There are 60,000 moles of oxygen over every square meter of the earth's surface. But photosynthesis of each square meter produces only 8 moles in a year; 8 out of a total reserve of 60,000! Of the 8 moles produced annually, 7 and 9,996/10,000 thousandths moles are used in respiration, and the remaining 4 ten-thousandths are probably used to oxidize things like sulfides in rocks. They give a measure of the actual contributions of the organisms of muddy bottoms; important over a geologic epoch or two but unlikely to matter in the short term of human generations.

If all photosynthesis were to stop, oxygen would be taken out of the atmosphere by respiration and not replaced. But the most oxygen that could be removed in this way would be what was required to oxidize all the organic carbon available to living things at the surface of the earth. Once this was done, the drain on the oxygen hoard must stop. But the available organic supply is small, little more than that 8 moles-worth, and Broeker calculates that this supply would be gone when only a tiny fraction of one percent of the atmospheric oxygen was used up. We can starve ourselves by killing plants. We can increase the sum total of human misery by further voiding plant poisons into the biosphere, and I expect we shall. But we are not going to suffocate ourselves by doing so. Nor by draining marshes either. There are so many good reasons for not spraying 2,4-D and picloram, for not draining marshes, for not pumping lead out of automobile exhausts, that there is no need to grasp at dramatic but erroneous possibilities for catastrophe. I wish those who have been responsible for the loss-of-oxygen and other end-of-life-on-earth scares would heed the fable of the little boy who cried, "Wolf."

THE CARBON CYCLE   The atmospheric reservoirs of both oxygen and nitrogen are so immense that it is beyond the powers of our folly to inferfere with them. But what of carbon dioxide? This vital gas is present in the atmosphere as a minor element, being on the average about 0.03 percent by volume. Here there is no great reservoir to smother the effects of our interference. It is regularly cycled by photosynthesis and respiration, and we are introducing extra carbon dioxide into the contemporary atmosphere by burning fossil fuels. What effects are our tamperings having on the atmospheric carbon dioxide supply, and what might any such effects mean for our future? To answer these questions properly it is necessary to understand carbon cycles on three different scales.

The cycling of carbon within an ecosystem is, over the course of a year, remarkably complete. Some tiny quantities are stored as organic matter for a while in reducing environments such as bogs and lake bottoms, but essentially nearly all the carbon taken from carbon dioxide by plants of the ecosystem is respired back to carbon dioxide within the year, generally within the same ecosystem. So this living part of the carbon cycle should normally itself work to keep the carbon dioxide content of the atmosphere roughly constant. But there is a second carbon cycle that should work even more effectively to the same end, the cycle between air and sea. Carbon dioxide dissolves readily in water, so that there is an equilibrium established between the carbon dioxide of the atmosphere and the carbonate-bicarbonate system of the oceans. There are more than 50 atmospheres-worth of carbon dioxide in solution in the modern oceans, a huge reservoir which acts as a giant shock absorber for the carbon dioxide content of the atmosphere (Hutchinson, 1948). Enrich the atmosphere, and much of the enrichment should be taken up by the oceans; remove carbon dioxide from the atmosphere, and the oceans should replace most of what you remove. Apparently there is a great reservoir of carbon dioxide, just as there is of oxygen and nitrogen, but this is in the sea rather than the air. Nor is this all; the shock absorber is itself shock absorbed, by the carbonates of carbonate rock.

There are some 40,000 atmospheres-worth of carbon dioxide held in the limestones and dolomites of the crust of the earth, and these have all been deposited by the sea at some time or other, as the foraminiferal-globigerina ooze of the deep sea, as chemical deposits, and in the skeletons of corals and other calcareous structures such as mollusk shells. The limestone hoard of the earth is itself cycled over geologic time, as weathering erodes carbonates and washes them back to the sea, after which they are redeposited as fresh limestone, raised by crustal warping, and so on. But those 40,000 atmospheres-worth must have come from somewhere, and the most likely source is volcanoes. Carbon dioxide is being continuously emitted from volcanic vents, representing a fresh supply from deep in the interior of the earth. We assume that this venting, continued ever since the formation of the earth, has now amounted to 40,000 atmospheres of carbon dioxide. But the continual venting has not affected contemporary atmospheres much, since the atmosphere has always been buffered by the ocean. The ocean, in turn, has buffered itself by depositing fresh limestones at a rate proportional to the volcanic emissions, and the concentrations of carbon dioxide in the ocean-atmosphere system have been maintained constant over geologic time (Figure 16.2).

But there is yet another cycle within the carbon story, that which slowly loses some corpses to peats, coal, and oil deposits, and which is only now being completed as man burns the ancient bodies. In most conditions of life, corpses are quickly and completely decomposed, but where they lie in reducing environments, as in bogs or in some ocean sites, fragments may escape the decomposers, become further reduced, and persist indefinitely as the inert molecules of fossil fuels. Carbon has been lost to the free system in this way throughout the duration of life on earth, but the rate of loss must have been small and was no doubt always more than made good by the emissions of volcanoes. The total accumulation of the fossil-fuel store, however, became large in time, finally representing several times the total tonnage of carbon in the atmosphere. There seems little doubt that we intend to use as much of this fuel as we can get and that we will probably succeed in using all that is accessible in

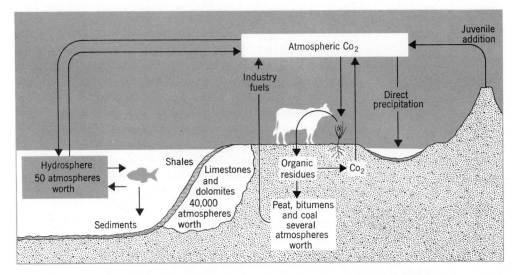

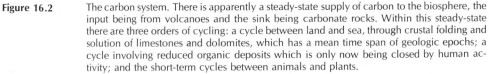

Figure 16.2 The carbon system. There is apparently a steady-state supply of carbon to the biosphere, the input being from volcanoes and the sink being carbonate rocks. Within this steady-state there are three orders of cycling: a cycle between land and sea, through crustal folding and solution of limestones and dolomites, which has a mean time span of geologic epochs; a cycle involving reduced organic deposits which is only now being closed by human activity; and the short-term cycles between animals and plants.

one or two centuries at most. We are thus in the process of adding several atmospheres-worth of carbon dioxide to our air.

Until recently there seemed no reason for believing that the ocean shock-absorber and carbonate-sediment systems would not be able to cope with a mere few atmospheres-worth of extra carbon dioxide spread over a century or two, and this belief was encouraged by our not being able to find evidence that pumping fossil carbon dioxide into the atmosphere made any detectable difference. We did know that the carbon dioxide content fluctuated slightly from 0.03 percent (in the third decimal place) from place to place and from season to season, but this should be expected as the coming of winter or the pouring of concrete where once there was forest altered the local rates of photosynthesis. But since 1958 very sensitive measurements of carbon dioxide have been made with infrared gas analysers kept on top of mountains in Hawaii and in Antarctica, both places with good atmospheric mixing and far from the seats of industry. The analyses (Revelle, 1965) show that the carbon dioxide content of the atmosphere is rising fast, and by an amount equal to half the discharge from fossil fuels. The ocean shock absorber is thus coping with only half the extra. Confirmation for this comes from the work of one of the men who has studied the radiocarbon of the atmosphere, Hans Suess (Revelle, 1965). Fossil-fuel carbon has been out of circulation for so long that it is radioactively dead, but carbon of the normal atmosphere should contain a proportion of radioactive carbon 14 due to the cosmic bombardment of the earth. If much of the fossil-fuel carbon stays in the atmosphere the relative amount of radioactive carbon should be reduced. It is; and Suess has accurately measured this "Suess effect" so that he can calculate the volume of fossil-fuel carbon remaining in the atmosphere. This work agrees with the direct measurements in showing that half remains, leading us to ask why the oceanic shock absorber is not working as expected. The answer is probably that the ocean itself does not circulate fast enough. Only the top few hundred meters of the oceans are effectively in contact with the air. In a few centuries, the slow mixing of the waters of the sea will work to absorb the exodus of fossil-fuel carbon, but by then carbon dioxide will be present in two or three times its normal amount, with consequences for life on earth that we find it hard to foresee.

One possible consequence which has been much discussed is that global air temperatures should rise, because carbon dioxide in the atmosphere absorbs infrared radiations from the ground (the **greenhouse effect**). Increasing the amount of carbon dioxide must thus increase the heat absorbed by the atmosphere, raising the temperature of the air. It has been realized all along that any actual global increase in temperature from this cause could only be a few degrees, but a few degrees might well be sufficient to trigger major climatic change. The most intriguing suggestion has been that the extra warmth might melt the Antarctic ice sheets, thus raising the world sea level and flooding places like Florida. This has seemed an alarming possibility to people living in these places. A biologist must take a more balanced view, noting that such events would not be catastrophic to the generality of life on earth; he might regret the loss of the Everglades, but he would probably offset the loss against the drowning of Miami Beach. But tragedy or no, recent calculations suggest that there will be no melting of Antarctica because the warming of an increased greenhouse effect will be more than offset by cooling resulting from our adding of aerosols to the atmosphere.

The burning of fossil fuels does more than put gases into the atmosphere; it also emits finely dispersed droplets of liquids which remain in the atmosphere for very long periods of time. These act together to form a reflecting mirror. They increase the earth's albedo, reflecting solar energy which once passed through the atmosphere to heat the earth. And at the same time they are transparent to the infrared radiations emitted from the earth, so that they do not join carbon dioxide in increasing the greenhouse effect. The effect of aerosols, therefore, must be to let the earth cool until a new steady-state temperature is achieved, one in balance with the reduced solar insolation that penetrates an atmosphere laden with aerosols. Rasool and Schneider (1971) of the Goddard Space Flight Center have made careful calculations of what the net effect of carbon dioxide warming and aerosol cooling will be, if our increased use of fossil fuels proceeds as expected. They find that the warming through the greenhouse effect is small compared with the cooling due to aerosols, and that global temperatures must fall, not rise. It seems likely that a fall in mean global temperature of some 3°C may result unless we are able to abandon the use of fossil fuels in the foreseeable future. Cooling the earth by 3°C is generally considered to be sufficient to trigger the next ice age. Rasool and Schneider suggest that an ice age might well be triggered within the next 50 years.

Although the affect of increased carbon dioxide on temperature will be unimportant compared with that of the aerosols, it may have serious biological consequences. The concentration of an important constituent of the atmosphere is to be doubled, and this must have effects. They will probably not be catastrophic effects, but they will be there. One we know about; we shall increase for a while the productivity of the earth. The rate of photosynthesis is commonly limited by the supply of carbon dioxide (Chapter 10), so doubling the supply will increase the rate. Possibly the school of agricultural thought which defends the use of DDT can therefore urge the further pollution of the atmosphere with fossil fuels in the same cause of increased production. Such a course would hold out an intriguing prospect of famine for the people living in the years when the supply of fossil fuels is exhausted, because then the yields of agriculture should fall remorselessly, year after year. But it is also likely that raising the concentration of carbon dioxide will have other effects which will actively interfere with production, offsetting any gain in the rate of photosynthesis. The biota and communities of the earth must now be adapted to the carbon dioxide concentrations experienced by their ancestors, and cannot be so well adapted to a different concentration. What other effects on the working of the biosphere might there be? The worry for any intelligent man must be that we do not know what all the effects might be. We are embarked on the most colossal ecological experiment of all time; doubling the concentration in the atmosphere of an entire planet of one of its most important gases; and we really have little idea of what might happen.

The concept of the ecosystem illustrates our general idea of how nature works, but we must ask if the concept is also useful as an aid to further theoretical understanding or in practical work. Conventional ecosystem diagrams show work being done by energy which cascades like falls of water, or which is directed down a network of pipes like water through plumbing. These hydraulic analogies tend to give misleading impressions of real ecosystems because the flow of energy in a community is not directed as by a network of pipes, but follows diverse pathways that are set by accidents of environment and history. Perhaps the one interesting general prediction which has developed from the ecosystem concept is that complex communities, those with large numbers of species in them, should be more stable than simple communities. This prediction is developed from the assumption that ecosystems are directly comparable to physical systems which can be subjected to mathematical analysis. The prediction is both plausible and attractive because it seems to account for the frequent outbreaks of pests of agricultural crops, claiming that these outbreaks represent the instability of the artificially simple agricultural ecosystems. It remains to be proved, however, that ecosystems do in fact operate like the physical systems with which they are compared, and it is also not yet certain that populations in very complex tropical ecosystems are, in fact, always more stable than those of impoverished northern climes. But if the theoretical fruits of the ecosystem concept are meager, its practical value may be very great. Natural communities may be assumed to operate as systems and then described by the techniques of systems analysis. Such techniques guide the collection of useful data and enable the actual functioning of natural communities to be simulated by computer models. Such simulations can be used to investigate the effects of changes on the local community and so can become tools for wise management.

CHAPTER 17

THE ECOSYSTEM AS
A PRACTICAL MODEL

Energy cascades, food and nutrients cycle; such, in short, is the ecosystem concept. It lets us draw satisfying diagrams to describe the workings of the real world, diagrams that show the blazing sun as the ultimate driving force being intercepted by plants, and the plants doing work with this energy to collect nutrients from the mineral soil and gases from the air to start the food cycles. But herbivores intercept some of the energy from plants, taking their places in the food cycles as they do, and carnivores intercept part of the energies of herbivores, and so on. Figure 17.1 shows one such diagram, taken from a recent review by the British ecologist MacFadyen (1962). When you draw a figure of this kind, you have at once organized your thoughts and undertaken an analysis of the complex system which you must study. Probably all working ecologists have sketched their own version of such a diagram on the back of an envelope at some time or another, or have at least carried the image of it in their brains. Now you can decide which of those little boxes (labeled "herbivores" or "decomposers" or something) you want to study, noting that you must keep in mind the arrows which tie the chosen one to cycles and energy flow.

So much is now routine in the thinking of any ecologist, the animals and plants he studies are parts of natural systems whose workings can be roughly sketched. But the ecosystem concept must surely take us further than this. We know for instance, something of the scale of activities in our various boxes, or at least we do if we express their inmates and their doings in units of energy. Energy cascades through our system from trophic level to trophic level, and must lose something of its ability to do work at each bounce in that vital cascade, just as the second law of thermodynamics tells us it does. As the energy flow proceeds, so the flow gets less. This concept of the ecosystem degrading energy according to the second law of thermodynamics, was included in the celebrated hydraulic analogy diagrams of Howard Odum (Figure 17.2). Instead of simple arrows joining the boxes, we have pipes of descending thickness to indicate the lesser flow of energy at each successive trophic level. All the pipes converge to a common outlet labeled heat. If the ecosystem was perfectly closed, if it had absolute boundaries separating it from all other ecosystems, the diameter of the "heat" outlet pipe should be the same as that of the "sunlight" inlet pipe, conforming to the other (first) law of thermodynamics, which says that energy can be transformed but not created or

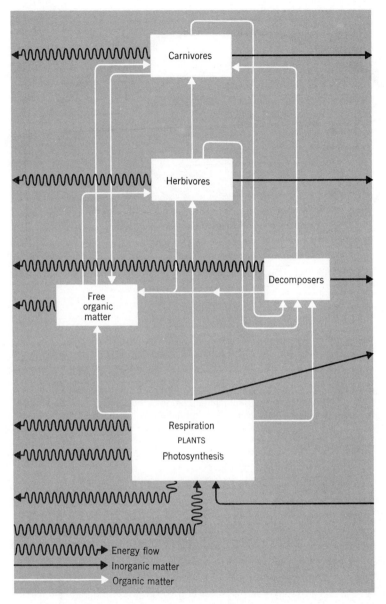

Figure 17.1

Theoretical energy flow scheme in an isolated ecosystem. In this view of an ecosystem the various trophic levels, together with corpses and decomposers, are seen as discrete entities which you can draw as boxes. Energy enters by way of the plants, then cascades from box to box, being radiated from the system the while. Matter circulates as energy cascades. Such a diagram forms a useful conceptual device, and one that suggests what a practical man should measure, but it is important to remember that nature does not come packaged in such discrete boxes. The animals, plants, and decomposers of real systems often have many and changing roles to play. (Redrawn from MacFadyan, 1962.)

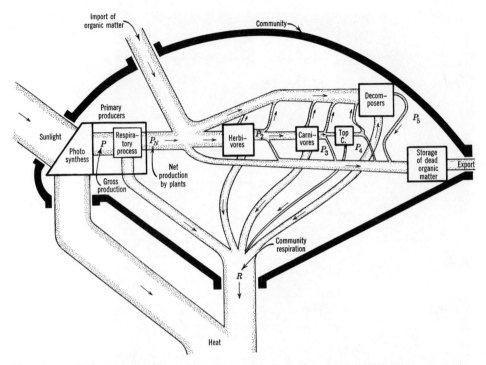

Figure 17.2 The hydraulic analogy of Howard Odum. In this analogy the energy cascade is imagined as being channeled through pipes whose thickness is proportional to the rates of energy flow. A prism placed at the entrance (or some hydraulic equivalent of a prism) deflects most of the sunlight from the community to represent that proportion of incident light not used in photosynthesis. From then on the degradation of energy at each trophic level is shown by pipes running to the heat outlet. The diagram is an excellent illustration of the ecosystem concept, but is poor as a practical tool because energy in real systems is not channeled and regimented as if by pipes. (After Odum, 1956.)

destroyed. In any real biological system, which must be partly open, potential energy, as some form of organic molecule, will be physically transported in and out. The Odum diagram takes account of this by the provision of extra inlet and outlet pipes labeled "import" and "export."

The hydraulic analogy can lead to a tempting vista of theoretical possibilities. We see energy as water bouncing through the ecosystem in a cascade, losing energy at every bounce; or flowing down pipes whose widths are measures of its ability to do work. Can we not express this cascading flow in formal mathematical terms and thus derive a thermodynamic theory for an ecosystem? There is no doubt that ultimately the relationships between beings in an ecosystem should be describable by thermodynamic theory, just as the relationships between molecules in a chemical reaction can be described by thermodynamics. But there are theoretical, and even conceptual, difficulties in applying this approach to ecosystems that have yet to be overcome. The hydraulic analogy itself gives a misleading impression of what really happens in an ecosystem.

Energy is seen as being directed down pipes, but there are no pipes in an ecosystem; indeed, there are no channels at all. Work is done as food is passed from mouth to mouth, and as animals do work to maintain themselves as organized beings, but there is no physical direction to this doing of work, like the direction which a pipe gives to a flow of water. All the hydraulic analogy tells us is that some animal habits provide less energy to their owners than do other habits. An eater of the meat of herbivores, like a tiger hunting buffalos, accepts that much work has already been done to prepare its food for it, and that tiger food is limited as a consequence to some small calorific fraction of food eaten by buffalo. But what fraction of solar energy entering an ecosystem is actually going to be available to a tiger as buffalo meat is determined by an as yet unmeasurably large number of fateful meetings between other animals. Far from being directed along pipes through a formal series of Eltonian trophic levels, the ability to do work is bounced about a complicated food web in the form of food molecules. Organisms take meals at one trophic level and then another, like a fox which eats roots, and beetles, and berries, and eggs, and mice, and carrion, and even beetles that were eating the carrion. If the ability to do work can really be said to flow through an ecosystem, it is only as an exceedingly turbid flow, with no directing pipes, but with innumerable feedback loops instead. As Slobodkin (1962) has pointed out, even the second law of thermodynamics itself becomes little more than a crude analogy of this process of energy degradation. There seems to be as yet no formal theory of thermodynamics advanced enough to model the sorts of energetic relationships prevalent in a real ecosystem.

But there is a worse difficulty still, which was pointed out in the discussion of ecological efficiencies (Chapter 11). It is a mistake to believe that animals and plants have all evolved primarily as efficient converters of energy. The pressures of natural selection are pressures for survival, and survival may sometimes be more concerned with the efficient use of nutrients, ensuring that individuals mate, safe overwintering, or swift growth and dispersal than with the efficient use, or even collection, of energy. Many choices made by natural selection must compromise food collecting for the safety of the organism and its progeny, so that food is wasted by the herbivores and carnivores of a system, going straight to the decomposers by default. And unless some predictive measure of this food wastage can be built into a thermodynamic model, the model should have no predictive value and would be useless. But natural selection has made the choices that control this wastage through a long history of environmental change and in face of the accidents of invasion by other animals. There is no way in which we can allow for, or anticipate, these choices in our thermodynamic model. Such models, therefore, are never likely to be a realistic possibility. We must be content with attempting empirical models instead, models that deal in the calories represented by numbers of organisms, and in masses of nutrients, rather than in thermodynamic theory.

THE HYDRAULIC
ANALOGY FOR
COMPARATIVE
DESCRIPTIONS

Diagrams based on the hydraulic analogy can best be used as a descriptive device to compare the fates of plant production in different ecosystems. This is because plants as a trophic level really are distinct from all other trophic levels. Nothing feeds on sunlight except green plants and photosynthetic bacteria, and there are only three possible fates for the tissues of plants and bacteria; to be stored in sediments, to be eaten, or to be decomposed. Only negligible amounts are stored. We can measure the productivity of plants by gas exchange, by cropping, or by growth increments (Chapter 10), so we can draw a pipe of width proportionate to this rate of production. We can also measure the respiration of the plants and we can measure incident solar energy and draw appropriate pipes. This simple series of pipes (Figure 17.3) represents the things of most interest to a practical farmer or range manager; the efficiency and productivity of the vegetation. We have more difficulty relating these indexes of the working of vegetation to the productivity of the herbivores, because of the practical difficulties of converting standing-crop measurements into productivity measurements (Chapter 11), but we can work out the mass of vegetation eaten and decomposed each year and superimpose these onto our pipe diagram as flows of organic matter expressed as energy. In this way we can make charts that describe the main paths of energy in systems so that different ecosystems may be compared.

MacFayden (1962) has developed simple diagrams of this sort (Figure 17.3). A square block in the middle labeled "stock" represents the average annual standing crop of plants in the ecosystem. Pipes out of this at twelve and two o'clock represent the average annual flow of calories in organic matter to herbivores and decomposers. It is important to notice that the two scales of the diagram, representing rate of energy flow and proportions of average annual standing crops, cannot be derived one from another. The relationship between flowing energy in pipes and what looks like a reservoir (stock) tank in the drawing is entirely arbitrary. But, once this convention is established, the flow diagram can be used to show the different fates in the first trophic levels of the variety of ecosystems for which MacFayden could collect reliable data (Figure 17.4).

MacFayden's diagrams do bring home some of the essential differences between diverse ecosystems such as forest, grassland, and marine plankton. The extreme difference in standing crop between, say, forest and phytoplankton of the open sea, is shown very neatly by the relative sizes of "stock" tanks, all drawn to scale. Revelation of such disparity is hardly novel, but other revelations of the diagrams are not so obvious without them. The similar metabolisms of the various plant communities, as shown by the pipes for plant respiration, would perhaps not be so generally known. The rate for the algal communities, with their tiny standing crops, for instance, is comparable to those of land and sea-grass mats. This is taken by MacFadyen to reflect the fact that respiration is a function of the surface to volume ratio, and so is relatively high for each tiny planktonic plant.

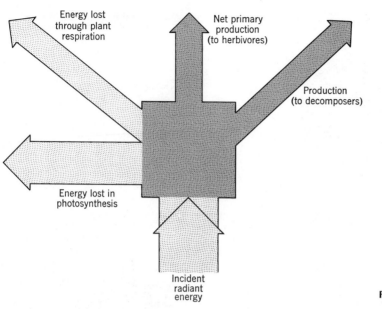

Figure 17.3

MacFadyan's version of the hydraulic analogy. Note that there are two scales on the diagram, that to which the pipes are drawn so that their widths are proportional to the rate of flow of energy and that to which the area of the stock tank is drawn. For more details see text. (Modified from MacFadyen, 1962.)

Such dramatic presentation of data can serve its purpose, but it must be admitted that use of the hydraulic analogy in such diagrams as these serves no better function than illustration. The analogy is useful as a tool for teaching and persuasion, as a means of convincing student and resource developer alike of the limits that thermodynamics and efficiencies place on what can be expected of an ecosystem. But the analogy itself does nothing to help our real understanding of ecosystems. As a practical tool, it is hampered not only by the inadequacy of thermodynamic theory but also by the fact that we cannot monitor the passage of energy through an ecosystem. We cannot mark it, and watch it flow, as we could if it was really like water. We have not even learned how to place volt meters of some kind at what we take to be strategic points, in order to monitor the passage of the ability to do work. When we set out to trace a real energy path we are always reduced to following food and nutrients through parts of the food cycle first, and then we try to calculate the work that must have been done to move those molecules of food and nutrients along. The raw data of any real ecosystem study are thus always measures of mass; of parts per million, of grams per square meter, of tons per acre; or of numbers; numbers of young, numbers of dead, numbers of shoots. Only when we have burned representative samples in calorimeters, and converted our mass measurements into rates of flow of calories, do we invoke our conceptual model of the energy cascade.

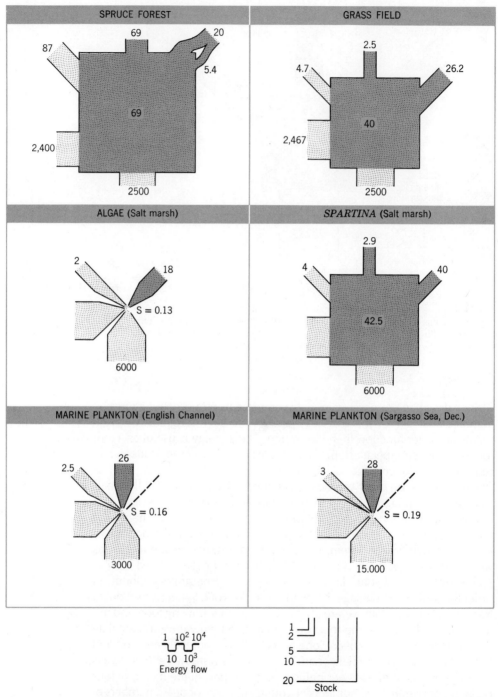

Figure 17.4 Hydraulic analogy diagrams to compare plant systems. Conventions are as in Figure 17.3. Units are Calories × 10⁶ per hectare. Mean stock is proportional to the area of the central square. Such diagrams afford a striking way of presenting data, of showing such things as how similar productivities may be achieved by quite different standing crops under different conditions. They also reveal some of the uncertainties in productivity data, since it seems doubtful to some ecologists that a spruce forest respires at 20 times the rate of a salt marsh, or that consumption exceeds decomposition in a spruce forest as suggested by the diagrams. (Redrawn from MacFadyen, 1962.)

A complex system is one with many parts and many interactions between those parts. Such a complex system tends to be maintained even when stressed. The impact of outside forces which might change the system is dissipated among its many interacting parts, and the system may be said to be stable. A simple system of few parts and few interactions must be relatively unstable; disturbing forces are not dissipated and the system may be easily changed from one state to another. That complex systems should be more stable than similar simple systems is susceptible to formal mathematical proof using information theory, and may be accepted as true. Yet it is worth pondering for it does not at first sight seem to correspond with everyday experience. We are used to thinking that complex gadgetry is inherently likely to go wrong and that for reliability we should choose the simple thing. Perhaps the clearest illustration of this is in the analogy of the sensitive watch that keeps marvelously accurate time but that easily goes wrong, compared with the simple rugged alarm clock that seems to resist abuse indefinitely. But with natural systems it is the other way around. Complexity in a watch means that all depends on the correct function of every part; a complex watch is really a very simple system. But complexity in a system means that the functions of every part depend on the functions of many parts so that change in any one of them has a minimum effect on the working of the rest.

But the ecosystem concept requires that the parts of natural communities function together as systems. If they really do so, then the workings of complex communities should be more stable than the workings of simple communities. Now stability in a community means that the numbers of individual living things be roughly constant from year to year without being given to sudden changes, outbreaks, or plagues. A complex community has many species of animals and plants living in it. A simple community has few, so we may think that a simple community of few species would be much more liable to things such as plagues and outbreaks than a complex community. A first test of the hypothesis that complex communities represent complex ecosystems with high stability is thus the experimental replacement of complex communities with simple ones. This experiment has been carried out for us many times by farmers who replace the complex communities of natural vegetation with crops. And farming is notoriously prone to the attacks of pests, to outbreaks or irruptions of herbivorous insects or weeds. This experience suggests rather strongly to ecologists that the hypothesis is sound, that complex communities with many species work as complex ecosystems, and that these ecosystems are more stable than simple ecosystems of few species.

The parts of ecosystems that do the interacting are the species of animals and plants and they interact by being eaten or eating each other. Suppose that a community has only one or two kinds of predator and very few kinds of herbivores on which they can feed, as commonly occurs in the Arctic. If one of the herbivores, a prey species, becomes rare for some reason, perhaps due to disease or some catastrophe of weather, its predator may be hard pressed to find sufficient alternative food to support its own population. Conversely, if one kind of prey becomes very numerous, the predators may be faced with a glut which they cannot handle. The instability of the simple ecosystem of which these animals were a part would show up as wild fluctuations in numbers of predators and prey. And it is the common experience of ecologists that numbers do fluctuate wildly in the Arctic. But if the community is complex, if there are many kinds of alternative prey with many kinds of alternative predator, the dilemmas of food shortage or glut may not arise. Ecologists note that they have fewer reports of fluctuating numbers in tropical regions where communities are much richer in species, and are inclined to attribute this in large part to the fact that there are more species in the tropics and that these more numerous species combine together to form more complex, and hence more stable, ecosystems. The matter is not so clear-cut as it might seem, however, because other things like stable climate or lack of changing seasons might be responsible for the apparent stability of tropical numbers and the recorded fluctuations in the Arctic. Also the observations might result merely from there being less ecologists in the tropics. These various possibilities are considered in Part 3 of this book.

The information-theory approach which led to a formal proof that true complex systems are stable was developed by Shannon and Weaver (1963). MacArthur (1955) has adapted this approach to see if ecosystems functioning in the way we imagine, essentially by eating and being eaten, should indeed conform to the general model. A stable ecosystem would be one in which some abnormal abundance in one kind of animal had little effect on the numbers of all the others. An unstable system would be one in which an abnormal abundance of one species produced repercussions and fluctuations throughout the system. Stating this hypothesis in more quantitative terms, we might say that a system in which one species of predator fed on just one species of prey would have zero stability, and that a system with many predators sharing among themselves many species of prey had great stability.

There appeared to MacArthur to be two ways in which increasing the number of species in a food web could increase stability; by increasing the number of links in the system, which is to say increasing the lengths of the food chains, or by increasing the number of choices at each level, which is to say increasing the number of animals at each trophic level. MacArthur was able to show that the two systems had equal stability (see

Figure 17.5

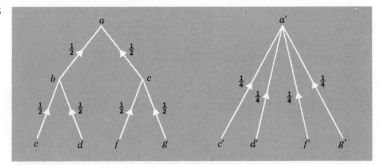

Figure 17.5) in which the halves and quarters denote the proportions of the available energy transfer along each pathway. The stability function is

$$S = -\Sigma \; Pi \; \log \; Pi$$

where

S is the stability of the system
Pi are the numbers in the diagrams.

For the diagram at the right

$$S = -4 \; (1/4 \; \log \; 1/4) = \log \; 4.$$

MacArthur's information theory brings us to the point where we can say that, if predators and prey interact in the ways in which we have every reason for believing that they do, then the ecosystem with more species will be more stable. But it remains only an hypothesis that real simple systems, existing in real environments which do fluctuate, owe their apparent instability to their paucity of numbers and not to some other cause. No one has yet found a way of testing in the field the hypothesis that observed instability is, in fact, due to ecosystem simplicity. And the state of our ecological art leaves us in a worse position even than this. We have not yet been able to show beyond reasonable doubt that complex natural systems are, in fact, more stable than simple ones, although we may believe very strongly that they are. We may show, as indeed we have, that the simple system of the Arctic is unstable. But we have nothing like such convincing evidence that the more complex ecosystems of the tropics do not suffer comparable population fluctuations because our knowledge of numbers in the tropics is so meager. What might mouse numbers be doing in the Amazonian forests? A recent suggestive review appearing in *Science* (Smith, 1970) by a man then living in Panama who noted that he had seen as many irruptions in his four years in Panama as he had seen in many past years spent in the Arctic. This, of course, might mean no more than that the parts of the ecosystems in which Panama rodents and other irruptive forms live are just as simple as Arctic ecosystems, even though other parts of the system, those involving insects and birds, for instance, are more complex and more stable. Recent study of forest insects in Malaysia (Chapter 29) do suggest great stability for this complex ecosystem.

It is worth again pointing out that we as yet have no formal proof that complex natural ecosystems are, indeed, more stable than simple ones because of that very complexity, as E. P. Odum (1966) has recently cautioned. But the circumstantial evidence that they are seems strong enough to convince many ecologists (including Odum). The mechanism which we postulate to bring about this increased stability, the increasing interactions between predator and prey, has been proved by MacArthur's work to be mathematically sound. The only reasonable working hypothesis is thus to expect complexity of an ecosystem to lead to stability. This conclusion is of profound significance as a guide to man's exploitation of nature.

The parts of the earth in which men practice most of their agriculture were formerly occupied by such complex ecosystems as deciduous forest, the tropical forests of deltas, and prairies, ecosystems even small patches of which had species lists running into hundreds or thousands. Man replaces these with monoculture. Instead of, say, a beech-maple forest of 130 plant species (Chapter 5) he plants with corn, with one species of plant, although in practice two or three others which he calls weeds manage to survive in spite of him. Any cornfield also supports a number of invertebrates, and a few birds and mammals, but this represents an ecosystem of a few dozen species at most where once there were hundreds. To an ecologist, this is replacing stability with instability. A rise of the number of one species by some accident is thus likely to cause repercussions throughout the system, with dramatic fluctuations in the numbers of others to be expected. A plague of any one of the few animals left may have such effects on the numbers of the others that the crop is destroyed. This is a process far from unknown to farmers, who try to meet it with poisons, thus further reducing the stability and making another outbreak more likely. But the planting of the crop itself represents an abnormal increase in the numbers of one of the species of the ecosystem, which must be expected to promote the damaging fluctuations in the numbers of the animals of the place. It is easy to see why pest outbreaks occur.

Ecologists hope for agriculture of more diversity, of more complex ecosystems, albeit artificial, which should achieve moderate stability. Such agriculture may be argued for in economic terms, in that pest outbreaks are much less likely, thus saving the cost of prevention or of their ravages. Monoculture on too grand a scale invites disaster, and we now react to these disasters when they occur with mass poisoning of the land. Such mass poisoning leads to other disasters, like the deadly accumulation of DDT and mercury through food chains. These disasters, costly both to the quality of life and to individual budgets, are the indirect consequences of farming such simple ecosystems that catastrophic instabilities are invited. An ecologist, in requesting a more diversified landscape, one of many crops and of many shelter-belt woods and hedgerows, feels that he is arguing not only for what is aesthetically desirable but also for something that makes long-term economic sense. He wants stable ecosystems in which pest outbreaks are naturally unlikely.

THE APPROACH OF SYSTEMS ANALYSIS

We should be able to study a system by systems-analysis, an ecosystem no less than a weapons system or an industrial plant. When we think of anything as part of a system, we think of its effects on other parts of that system, and the return effects of the other parts of the system on it; which is to say that we think in terms of relationships. And these relationships are dependent on the states of the things in the system, whether they are hot or cold, many or few, active or sleeping, hungry or satiated. Furthermore, the states, and hence the relationships, change with time. Finally,

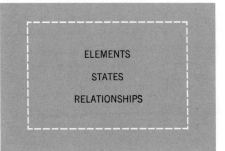

Figure 17.6

the system occupies space, which is to say that it has a boundary, although the boundary may be a permeable one so that the system is affected by the state of things that cross its boundary. Calling the things of a system which occupy part of its space for a period of time "elements," Schultz (1969) proposes the diagramatic model in Figure 17.6 for an ecosystem. The boundary is drawn with a dotted line to indicate that there is exchange of elements with neighboring ecosystems.

The **elements** of any real ecosystem are very numerous, being all the different kinds of animals and plants that live there as well as all the different kinds of inanimate things, such as water and nitrate ions. Each of them may have many **states,** providing for many **relationships.** The relationships may further change as the elements are moved from place to place within the ecosystem space. Every change of relationship and every change of state will affect the state of every element of the system. The whole presents a pattern of frightening complexity, a thing whose inordinate detail cannot be comprehended at once and together by a human mind. But two developments allow us the ambition of dissecting such complex systems, of understanding how they work, and of predicting what might happen in different circumstances (putting it bluntly, if we muck around with them). We may identify processes, or subsystems, that may be particularly important for our purpose, discarding from our model relationships that are obviously of lesser importance. And we may then call on the prodigious memories of modern digital computers to keep track of a vast number of changing states as we simulate in seconds the passing relationships of years.

To make a computer simulation of an ecosystem, you must begin by deciding what aspect of the system really interests you. Usually this is a simple task. Your work is going to be so very expensive that you will not get the money to follow a process that is not obviously interesting. Generally the object will also have the peculiarly fetching advantage of promising to save somebody money. It will concern yield of food, or timber, or minerals. With your objective settled, you must then make a map of what you already know about the important relationships in the ecosystem. Usually you know much already about what is important, about the principal predators, or mineral deficiencies, or disasters which regularly ef-

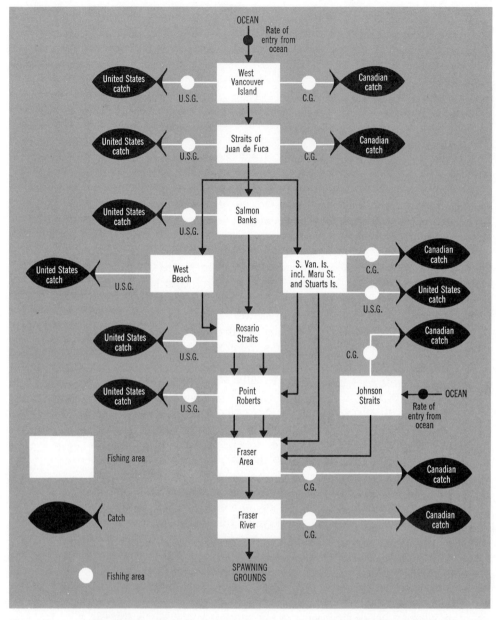

Figure 17.7   Flowchart for salmon simulation study. For details see text. (Data of Royce et al., 1963, from Watt, 1968.)

fect your yield, so that this first step is a simple one to take. What you do not know, and what you hope to simulate, is the detailed interactions of all these main elements and relationships, one upon another.

Figure 17.7 is the starting map, or flow chart, of one of the first successful and useful computer models of part of an ecosystem, that of the Pacific Salmon fisheries by Royce et al. (1963) of the University of Washington. They first divided their problem up into areas fished which, of course, also represented populations of fish. The numbers of fish in each area would depend on the numbers that had come in from the ocean, the numbers which had gone somewhere else, the numbers which had been taken out by fishermen and, of course, on the numbers which had perished in sundry minor ways. Fishing experience suggested that the only really important cause of death on the fishing grounds was, in fact, fishing, so the minor causes of death could be overlooked for the first simplified model. But the kind of fishing gear used did make a big difference, and was thus a major **element** in the ecosystem. Then there was that final vital **state,** the rate at which young salmon returned to the ocean reservoirs from their spawning grounds. These main **relationships,** known from the start to be of key importance, are shown in the flow chart. Along each pathway so shown is a pool of fish, a fishing device of various efficiency, and an annual catch of fish. The investigators wanted to know what would happen to each catch if fishing was continued at its present level, if the methods were changed, or if the effort was increased. Designing a computer model let them do so without possibly disastrous large-scale experiment. They wrote computer programs for each one of the sets of relationships shown by each of the arrows in their flow chart. In each program the computer could be set to repetitively add effects, of more fish coming in, of more fish being caught, of more fish swimming on, digesting great tables of fish-landing statistics in the process. Then all programs could be integrated to show the effects of every one set of relationships on every other. The work showed that only a 50 percent increase in fishing effort would destroy the fishery entirely, so the cost of running the computers was decidedly less than would have been the cost of launching more gear and taking away livelihoods from fishermen.

There are two specially interesting things about computer simulation studies. The first is that the pseudoalgebraic languages used, like FORTRAN, quite fortuitously, allow you to mimic biological processes very closely (Watt 1968). They allow you to add effects, over and over again. This is just how the running of a real ecosystem goes. There is an effect, say A eats B; there is now less of B than before, but an A again eats a B; but meanwhile a B has been adding to a B population by eating something else, which will add to some other element of the system. The computer languages allow you to describe very large numbers of steps and parallel relationships of this sort. You run a continuous inventory of your ecosystem. The languages are good at running inventories; it is what they were designed to do.

The second noteworthy thing is that computers are hungry for data. To write programs you must have representative numbers of everything, taken everywhere with the same methods and with the same reliability. The "mindless-machine-modeling-the-ecosystem" sounds as if it might drive the real earth and water out of a biologist's life. But the reverse is true. Any biologist who knows the country in which he works could nowadays settle down to draw a flow chart for the communities he knows best, almost certainly being able to include all the relationships that can finally be proved to be decisive for the normal functioning of the system. But he would know as he drew the chart that he could not hazard even a reasonable guess of how many of each element there were marking the states of key relationships. How many beetles are there over there in February? You would have to go and find out things like this before you could write your computer program. This means designing appropriate sampling techniques, then knuckling down to doing the work. The old literature of the plant sociologists, particularly those of the Uppsala School, was replete with papers on sampling technique, of how to make random samples and when you can safely "stratify" your random samples, and so on. The same old terms and the same old problems now bother the systems analysts, only they need much more data. A recent paper on the organization of an ecosystem study, quite a simple one of grasslands, suggests that you need to set out with an organization of 100 men, 30 of them with the Ph.D. degree (Coupland, 1969).

The biological community is just now (1971) beginning to organize the teams needed to collect the data for describing whole ecosystems. This is the task that is being organized under the International Biological Program. And it is surely high time that we did. We do not yet have enough data on any ecosystem on earth as large as say a small pond to make a realistic computer simulation. The best we have yet done is to study parts of a few ecosystems which have especial resource value, like that of the Pacific salmon which involved only four species of fish. And yet this is the epoch in which the most complicated and interesting systems are being removed from the earth by our own deliberate act. We consciously are setting out to destroy the last wild and complex natural systems to make the land yield to agriculture, an endeavor driven by the belief that only in this way can we solve the problem of hungry people. But this solution can be no more than a fleeting thing, a meeting of the appetites of perhaps two or three generations of expanding populations at the most. Then we must "solve" the problem of hungry people in the way that it should have been "solved" all along, by checking the growth of population. Meanwhile, and for such a fleeting reward, the last stable, complex, beautiful ecosystems of the world will have gone. Some of the things that will have gone with those ecosystems may be unknown and incalculable. And the more interesting of the systems will not even have been described before they were swept away.

I find men of my own university going to advise the Brazilian government on the soils of the Amazon basin, so that they can bulldoze the wonderful forest, then plough it up. We do not even have a species list for the trees of the forest. One day some surviving race of truly enlightened men might want to try and rebuild a rain forest for the delight of their descendants. But we shall not even have left them a description which intelligent men could use to reconstruct one in their mind's eye, or even in

practice if enough plant species survive somewhere to give them raw
materials. They will only be able to wonder at the glimpses which travelers' accounts give them of the glories which once were, as a modern his-
torian must long for the great library at Alexandria, destroyed when Julius
Caesar set fire to a fleet. We should be doing better to describe the Brazilian rain forest than in helping its contemporary custodians to cut it down.
It would be expensive to build a team of biologists to do so. Perhaps it
would cost as much as one of the moon shots.

Lakes are particularly useful subjects for ecological study because they house communities that are unusually well defined in space, and whose fortunes are strongly influenced by fairly easily understood physical events. Study of lakes is also useful because it leads to an understanding of one of the more obvious kinds of pollution. Since the sun shines on a lake from above, the surface water both becomes heated so that an upper warm layer, the *epilimnion*, comes to float over a bottom layer of cold water, the *hypolimnion*, and becomes the place where the plant life of the lake is concentrated. In lakes of low productivity (*oligotrophic lakes*) this floating of warm water over cold has few dramatic consequences, but if a lake is very fertile (a *eutrophic lake*) the concentration of plant life in the warm surface waters may be so great as to form an effective canopy leaving the bottom waters in darkness. The cold, dark bottom water is then both cut off from the oxygen of the air above and denied the oxygen resource that would have been provided by photosynthesizing plants. A rain of corpses comes down from the life zone above. They decompose, and this decomposition uses up the oxygen reserve. The bottom waters of fertile lakes in summer thus become anoxic. Another special quality of lakes as units for study is that they age; the basin progressively fills with sediment until the lake is eventually extinguished. There must, therefore, always be progressive changes in the conditions for life throughout the existence of a water-filled basin. Early students of this physical ageing process in lakes were mistakenly led to believe that the process was partly controlled by the life of the water, that a lake ecosystem evolved from youth to old age by some inner dynamic of its own. This view, although now known to be in

247

error, has been partly responsible for a widespread misapprehension of the present day which likens polluting a lake with fertilizer to the natural ageing of a lake. The false suggestion is often made that polluted lakes are dying lakes. In fact, the unpleasant syndrome of excess fertility caused by dumping sewage or phosphates into lakes is easily reversed. Lakes bury excess fertilizer in their mud and will spontaneously come clean when the source of pollutant is diverted. The way in which the conditions for life in a lake are dominated by physical processes acting from outside makes lakes particularly suitable to ecosystem studies by systems analysis.

CHAPTER 18

LAKES AND THEIR
DEVELOPMENT AS
ECOSYSTEMS

Students of communities and ecosystems alike face the difficulty of indeterminate boundaries. No community lives in isolation from neighboring communities, and no conceived ecosystem smaller than the biosphere can be modeled as if closed. But these difficulties of measurement and conception may be least to the student of lakes, because a lake is bounded by a rather definite edge. It is true that the fortunes and history of a lake are much effected, and even controlled, by events in its drainage basin, by erosion and the life of the banks, and by the air above, but for some purposes lakes can be more easily studied in isolation than can other areas on the earth's surface of similar size. **Limnology** (Greek *Limnos,* a lake), therefore, has sometimes had special contributions to make to ecological thought. Lakes also have the advantage to the student of having lives of definite spans. They start as hollows filled with water which will gradually be displaced with sediment that fills the basin. All lakes must be young once, must grow older, and must finally be extinguished. Furthermore, this history of infilling is recorded in the mud of ancient basins and may be reconstructed from drill cores. And on top of this long history of infilling is a seasonal history, perhaps more definite and clear-cut than the seasonal history of other places because of the dampening effect of a large mass of water. This seasonal history affords special opportunities for study and measurement.

During the warm part of the year, a temperate lake is warmed from above by the sun. The surface layers of water are, of course, warmed more quickly than the deep water; but warm water is less dense than cold water and thus stays on top in the path of the sun's rays to become warmer still. A temperate lake thus acquires in summer a layer of warm water, the **epilimnion** (Greek for "over lake") over a layer of cold water, the **hypolimnion** (Greek for "under lake"). Separating the two is a fairly thin layer across which there is a sharp temperature gradient, the **thermocline** (Figure 18.1). Such a system of warm water over cold water is physically stable and will persist unless considerable force is applied to

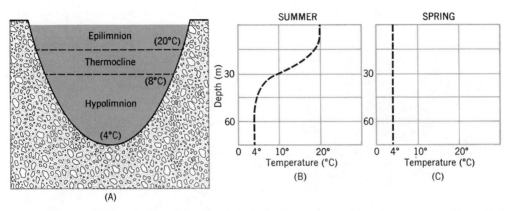

Temperature stratification of a lake of temperate latitudes in summer. Figure A is a diagrammatic cross section of a stratified lake. A warm epilimnion floats over the cold water of the hypolimnion below, and the two are separated by a layer, the thermocline, of rapid temperature change. Figure B is temperature profile of the very deep Senneca Lake in Wisconsin as it appears in August, drawn to the same scale as Figure A. The practical effect of such stratification is to isolate completely the bottom water from the air until cooling and the winds of the fall effect a mixing of the lake once more. In the following spring there will be a time when the lake water has a temperature of 4°C at all depths, as in Figure C. (Data for Senneca Lake from Birge and Juday in Ruttner, 1963.)

Figure 18.1

overturn it. The necessary force usually only comes from the winds of autumn gales, when the surface of the lake is already cooling so that a good blow can overturn the lake, completely mixing the waters of epilimnion and hypolimnion. There is thus an annual cycle in temperate lakes, in which the top and bottom waters are completely separated from each other for part of the year.

If a lake is highly productive, thermal stratification has some interesting consequences for the bottom water. Photosynthesis thrives in the top water near the light, thus keeping the epilimnion well supplied with oxygen. But in the darker waters of the deep hypolimnion there may be almost no photosynthesis, and thus no oxygen produced. The animals for whom the deep water is home must still respire, however, and the oxygen store of the hypolimnion must begin to be used up. The resulting **oxygen deficit** will be made worse by the continual action of decomposers, for the life of the lighted surface will be continually dropping debris and corpses into the hypolimnion, and these will be attacked by bacteria as they rain down. Bacterial respiration can quickly reduce the oxygen dissolved in the deep water to virtually zero, and there is no way of replacing the lost oxygen until the autumn overturn mixes the lake once more. Reducing conditions prevail in the hypolimnion of a fertile lake all summer (Figure 18.2). A thin layer of bottom mud is also reduced, and is then able to give off adsorbed nutrients to solution in the water, particularly phosphorus which is less soluble when oxidized. At the end of the summer, therefore, the hypolimnion tends to be nutrient rich, although nearly lifeless because devoid of oxygen. At the fall overturn the stored

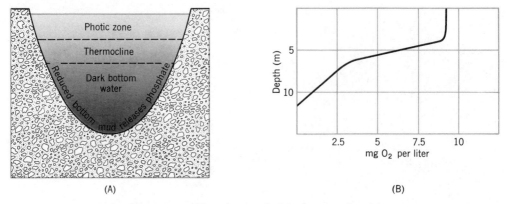

(A)　　　　　　　　　　　　　　　　　(B)

Figure 18.2　Light and oxygen profiles of a stratified fertile (eutrophic) lake in summer. (A) Section through the lake showing that light has been almost completely absorbed by the plankton of the top few meters so that too little light penetrates to the thermocline and beyond to support photosynthesis. But there is a rain of corpses into the deep water, whose decomposition requires oxygen. Since the deep water is cut off from the air until fall overturn, there develops an oxygen deficit in the deep water, and the bottom mud is reduced. An oxygen profile typical for such a lake is given in (B).

nutrients are distributed throughout the lake to be available for the spring plankton bloom.

If a lake is poorly supplied with nutrients, and so is not fertile, there may not be a thick bloom of phytoplankton near the surface. Light may continue to penetrate to deep water so that photosynthesis can occur below the thermocline, providing the hypolimnion with a continual input of oxygen. The low fertility of the upper layers also means that there is a relatively slight rain of debris into the deep water so that bacterial respiration is slow. In such a lake there may never be a deficit of oxygen in the hypolimnion, and the bottom mud remains oxidized all summer long (Figure 18.3). There is much less return of nutrient, particularly phosphorus, from oxidized mud and there is thus a tendency for the lake to remain infertile.

A high-fertility lake or, in the Greek terminology invented by the German limnologist August Thienemann, a **eutrophic** (good nursing) lake, is thus more prone to aquire an oxygen deficit in the deep water during summer than a low-fertility or **oligotrophic** (few nursing) lake. It is easier to measure the oxygen content of lake water than it is to measure its fertility. In practice, therefore, oxygen measurements are often used as indicators of the lake's fertility, and **eutrophy** comes to be inferred from the fact that there is an oxygen deficit in the deep water. But fertility is not the only parameter that can effect the oxygen tension in the depths. There is a lake-size effect that can be superimposed on the fertility effect to control the concentration of oxygen in the hypolimnion. If a lake is shallow, the bottom water of the hypolimnion is of small volume and thus cannot contain much oxygen. A shallow lake, therefore, does not have to be so fertile as a deep lake to acquire reduced bottom water and anoxic mud. But

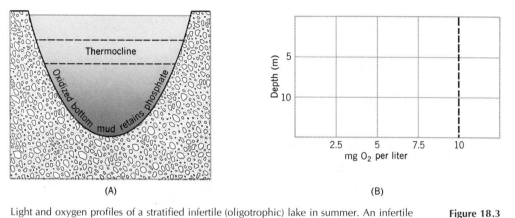

(A)   (B)

Light and oxygen profiles of a stratified infertile (oligotrophic) lake in summer. An infertile
lake may stratify in summer, with consequent isolation of the bottom water from the air, but
there is so little plankton floating in the top water that light penetrates deep into the lake,
permitting photosynthesis and oxygen generation even in the hypolimnion. Also there is less
oxygen demand in the deep water, since there is less detritus coming from above to be
decomposed. (A) is a cross section of such a lake; (B) is an hypothetical oxygen profile.

Figure 18.3

all lakes get shallower as they age, because they fill in. Their epilimnions
remain at about the same volume throughout their lives, but their
hypolimnions steadily shrink as their bottoms are raised steadily upward
toward the thermocline. It is possible, therefore, for an oligotrophic lake
to acquire a deep-water oxygen deficit simply by growing old (Figure
18.4). This does not mean that the lake has the other qualities of eutrophy, although it may also be true that fertility is aquired with age as nutrients accumulate. Weathering continually introduces nutrients to a lake,
and cycling by the lake's biota may then conserve them, allowing some
increase in the total available. So there are some indications that all lakes
take on more of the qualities of eutrophic, fertile lakes as they grow old.
This has led to two different lines of thought in the studies of lakes. In
recent years the mass pollution of lakes with fertilizers has come to be
likened to their growing fertile through old age. We talk of artificial aging.
And it has led a school of theoretical ecologists to ponder the possibilities
of an ecosystem aging as the result of some dynamic of its own.

Linsley Pond (Figure 18.5) near Yale University in Connecticut is a small
deep kettle lake, one of the many left behind by the continental glaciers.
It now has all the characteristics of a fertile eutrophic lake, possessing in
summer a hypolimnion from which nearly all the oxygen is removed and
a bottom of organic mud which is chemically reduced. In the late 1930s,
E. S. Deevey of Yale drilled through the mud of Linsley Pond from an
anchored rowboat, with a simple peat-borer fitted with extensible rods.
He got through 43 feet of mud, a thickness which represents the infill of
the entire history of the lake since the glacier left it as a melting block of
ice. His pollen analysis of this mud (Deevey, 1939) from top to bottom
revealed a history of plant formations ranging from something close to
tundra at the bottom through various intermediate vegetations to the

THE HERESY OF LAKE
AS A SELF-DEVELOPING
MICROCOSM

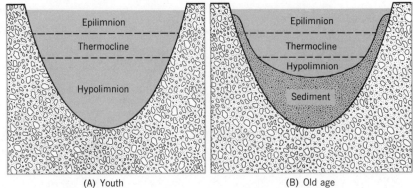

Figure 18.4

(A) Youth   (B) Old age

The effect of aging on an infertile lake. The young lake at the left (A) is infertile (oligotrophic), retaining oxygen in its deep water all summer. The drawing at right (B) is the same lake when infilling of the basin is far advanced. The thermocline is in the same place, and the epilimnion is scarcely altered. But the hypolimnion is a vestige of its former size. The oxygen reserve of the reduced hypolimnion may be too small to support as much decomposition as before, and the bottom becomes anoxic. As defined by the oxygen content of the bottom water, the oligotrophic lake has become eutrophic by aging.

Figure 18.5   Linsley Pond. Linsley Pond is a kettle lake, occupying the hole left when a block of glacial ice melted in a till deposit. It started life as an oligotrophic lake. Now, in its middle age, it is eutrophic.

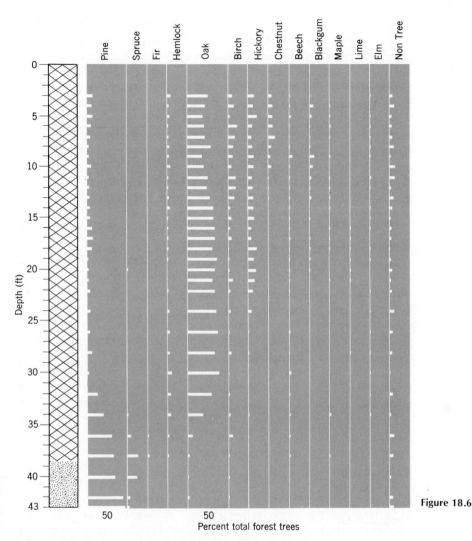

Figure 18.6

Pollen diagram from Linsley Pond, Connecticut. Pollen is shown as percent forest tree pollen. The last column on the right is of pollen other than trees as percent total tree pollen. The history spans from an early date when the young basin was still collecting largely inorganic sediment (dotted on the monolith at the left) until the date of the boring. In early days spruce trees grew near the site. Scattered pine trees, which produce vast amounts of pollen, accounted for the large pine pollen input. Later studies have shown that samples from the very bottom of New England lakes contain pollen of vegetation almost devoid of trees. The coniferous forest that surrounded Linsley Pond in its early days later gave way to deciduous forest that has been subjected to minor changes right up to the present day.

deciduous forest which the pioneer New Englanders knew. This pollen diagram (Figure 18.6) confirmed that the mud core recorded most of the history of the lake since tundra times closely followed the retreating ice. The lake ecosystem had endured all the climatic changes associated with this dramatic history.

Hutchinson and Wollak (1940) performed a number of chemical analyses on the mud from Deevey's boring, in particular analyses that recorded the proportions of various fractions of organic matter present in the mud. The earliest lake mud had little organic matter, suggesting that the lake had not been very fertile then. It seemed that eutrophic Linsley Pond had started life as an oligotrophic lake; that it had evolved from a state of low fertility to one of high fertility. If this was a true finding, the early Linsley Pond must have had free oxygen in the hypolimnion all summer, which should have affected the kinds of mud-burrowing animals found there. Deevey (1962) found the evidence that confirmed this change. He extracted from the mud tiny fossil fragments of midge larvae. In the early lake the fragments belonged to species which are known to require a rich oxygen environment, but they were later replaced by the species now living in the lake, which can stand the oxygen deficit of the summer months.

That a lake should change from infertile to fertile while the surrounding vegetation changed from near-tundra to hardwood forest need not be surprising, since it could be no more than a temperature effect. But, in fact, the fertility changes were completed early in the lake's history, at a rate that suggested no correlation with climatic change. Nor had fertility changes anything to do with shrinking of the hypolimnion in the classic manner, because they had apparently been completed before the midge larval population changed from one species to the other. The lake had become fertile early on but had so large a hypolimnion that there was no oxygen deficit until much later when the lake floor had been raised closer to the thermocline. Since the environment and physical shape of the lake had apparently nothing to do with the animal production of organic matter, the way was clear for the several investigators to entertain a tantalizing possibility; that the increase in productivity of the early lake was a result of the activities of the plants and animals alone; that the ecosystem had developed from one of low productivity to one of high productivity by the action of living things. This view was encouraged by the fact that the curve showing the increase of organic matter in the sediment from nothing to the high value prevailing throughout most of the lake's history was "S"-shaped (Figure 18.7). The productivity of the lake had apparently grown as a population of organisms grows; slowly at first, but then faster and faster, until it began to tail off toward the steady state of a population in balance. The communities of the lake were acting together to grow together like a superorganism. It was Clements' idea all over again. This time it was not just the formation which grew, "As an organism," but all the biota, and thus the whole ecosystem. And, in its new guise, the superorganism grew not just to limits set by some vaguely defined entity like climate, but to the limits of productivity; that is, to limits set by energy conversion. From this dawning idea of energy controlling the growth of the living parts of an ecosystem, developed at Yale, by Hutchinson, Deevey, and their students, there grew the many thoughts about the restrictions of energy on life which have characterized much of ecological thought for the last 30 years. Lindemann (1942) was soon to

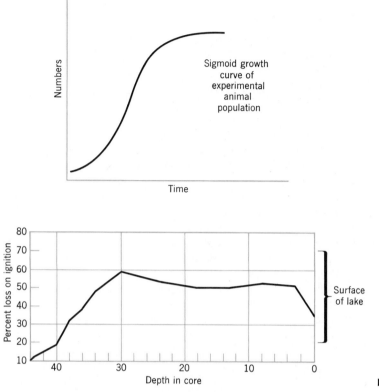

Figure 18.7

Sigmoid growth in Linsley Pond. The accumulation of organic matter in Linsley Pond was once compared to a sigmoid growth curve of an animal population. A good estimate of the organic content of sediments is given by weighing a dried sample, burning it, and then reweighing. The percent *loss on ignition* that can then be calculated is a useful measure of the organic content. The increase of organic matter in the early sediments of Linsley Pond was sigmoid, imitating the shape of a population growth curve. This allowed speculations that the whole lake ecosystem somehow grew to productive maturity like a population. Later work showed that the sigmoid pattern merely reflected decreased erosion in the drainage basin which lessened the input of inorganic sediment to the lake.

work with Hutchinson and to reexplain Hutchinson's ideas on the control of ecosystems and communities by energy flow as part of his famous paper on trophic dynamics (Chapter 11, for Clements' ideas see chapter 6).

It was another Hutchinson and Deevey student, Daniel Livingstone (1957) (he of the sodium cycle), who finally pointed out the errors of interpretation of the Linsley Pond record which invalidated the conclusions of superorganismic growth. Livingstone made new borings of Linsley Pond, using the piston sampler which he designed himself and which now carries his name. Unlike Deevey's peat-borer, the Livingstone sampler raised an undisturbed core with all the fine structure of the sediment preserved. Fate decreed that there should be fine banding all through the thickness of deposit in which the S-shaped "growth" of organic matter occurred, and that the bands could be used in a crude way to record the passing of years. It was possible to calculate the organic content of the mud as mass deposited per year instead of just as mass per unit sediment, and when this was done the S-shaped curve disappeared. It was clear that there was, in fact, very little increase in the productivity of the lake during the period remarked on by Hutchinson and Deevey. All that had happened was that early on the organic matter had been diluted by silt and clay washed into the lake from outside and that this had decreased with time. This was reasonable enough when you reflect that the barren tundra country round about was being gradually covered with richer vegatation which should control erosion in the lake's drainage basin. The changing input of silt and clay could also be used to explain other chemical changes noted by Hutchinson in the early lake, through such mechanisms as the swift burial of nutrients. There was no doubt that no superorganismic quality of the biota had been involved in changing the early lake deposits. All was due to changes in the physical condition of the ecosystem. No reparceling of energy was involved; merely an un-muddying of the water.

Livingstone's conclusions were accepted at Yale, as elsewhere, and it is now generally conceded that lake ecosystems do not grow like orga-nisms. Clements' idea had had a second birth, but also a second death, and most ecologists have stopped trying to see a deterministic growth in the systems that they study. But the matter is reviewed more fully in Chapter 40 of this book.

LAKES AND WATER
POLLUTION

Modern industrial societies let much refuse run into lakes, an action which produces a syndrome of effects that we call "pollution." These ef-fects are considered to be unwelcome. The word "pollution" itself is a simple statement of our dislike of the process, because it comes from the latin verb *pollutere,* meaning to make dirty, or perhaps even more liter-ally "to make unwashed." And yet, if we set aside actual poisons like DDT or mercury, what we have done to a polluted lake is essentially what a worthy farmer does to his fields. We run into lakes a mixture of sewage and chemical fertilizer, either from our cities or from croplands.

The immediate effect of this fertilizer in the water is, as its name implies, to promote life. Putting either sewage or chemical fertilizer on farmland makes the crops grow, and putting them into water makes things grow there, too. Popular literature talks of making lakes die by put-ting sewage into them. The reverse is really true. A polluted lake is com-monly crawling with life. That is what is wrong with it. Massive blooms of algae are generated by the fertilizer, but such massive blooms are ap-parently so rare in nature that herbivore communities able to handle them have not been evolved. The algae, not being eaten, must die; their corpses are cast up on shore, decompose where they float, or sink to the bottom to rot. The fertilizer had produced life; the life had produced corpses; and the corpses rotted. Which is why polluted waters stink.

But the idea of a polluted lake dying has a firm hold on the public

Figure 18.8

Dead alewives in Lake Michigan. The probable cause of this mass death was oxygen deficiency due to excessive respiration in the lake.

imagination. Popular magazines carry articles with titles like "Who Killed Lake Erie?" Lake Erie is not dead, nor is it polluted beyond reclaim. But it is now very productive where it was once not so productive. Parts of the lake, particularly the large shallow embayments of the west end, become thermally stratified in summer, the rain of corpses from the algal bloom above causes an excess of decomposition in the hypolimnion, and the bottom waters become anoxic. Trout and other bottom living game fish then suffocate. Even the fish of the upper waters of a polluted lake may die from lack of oxygen also, in spite of the fact that the water teams with plant life. The great algal mats produce much oxygen by day in the course of photosynthesis, but at night they can only respire. If bacteria are also working over the algal corpses floating in the surface waters, and perhaps sewage and garbage as well, then all this respiration combined can make the epilimnion anoxic as well as the hypolimnion. If this happens for only a short while in warm weather, it is enough to kill nearly all the fish in the lake, surface fish as well as deep-water fish. Piles of alewives on the shores of Lake Michigan are one of the fruits of this process (Figure 18.8). There is then calamitous and visible death in the lake. But this is not the same thing as saying that the lake itself has died.

It is the change from oligotrophy to eutrophy, simulating as it does the natural aging of a lake, which gives rise to the idea that polluted lakes are dying or actually dead. A lake naturally becomes eutrophic in its old age for the reasons cited earlier in this chapter; because nutrients accumulate from runoff, because plants of the shallows hoard nutrients, and because the volume of the lake shrinks until the hypolimnion becomes so small that an oxygen deficit in summer is inevitable. And old age cannot be reversed. A lake almost extinguished by having its basin filled with sediment is dying, for all that it becomes more fertile at the last. But it is not reasonable to argue from this that a lake made fertile with an influx of fertilizer is dying also.

If we had dumped enough solid waste into Erie as to nearly fill up the hole it occupies, then the comparison would be valid. But we haven't. We have merely made its waters fertile, creating an oxygen deficit in summer. The resulting productive state of its water, including the associated deaths of fishes we value, will continue as long as the water remains fertile. But the water will not remain fertile unless we keep on adding more pollutants. For this form of eutrophy, called in ugly jargon **"cultural eutrophy,"** is reversible, unlike true aging.

A sizable portion of the corpses and detritus which rain through the water of a fertile (eutrophic) lake is not decomposed at all but is incorporated into the sediment as organic mud (the **gyttja** of the limnologist). Included in this mud is much of the fertilizer which promoted the life which preceded the corpses. Only the mud of the top few millimeters can yield its nutrients back to the water, which means that adsorbed nutrient which is buried more deeply is lost to the lake system forever. Lakes are thus self-cleaning. If you stop dumping sewage, fertilizers, and garbage into any polluted lake, its waters will become clean again, by themselves, without any benefits of technology, in a very few years.

The cheerful conclusion that lakes can clean themselves of excess fertilizers does not need to rest only on theoretical considerations and pilot experiments, because there have already been large-scale demonstrations that it is true. The most striking of them has been the successful campaign to clean up Lake Washington in Seattle. Lake Washington is a large lake, but in 1963 the surrounding suburban communities were pouring 75,600 cubic meters (about 19 million gallons) of sewage effluent into it everyday. The lake showed the typical symptoms of being polluted: there were algal blooms, the bottom waters were anoxic in summer, analysis showed the water to be massively enriched with phosphates and nitrates, and it stank. Local citizens responded by seeing that sewage was diverted elsewhere (a task easier for Seattle than some places because of the local geography), and 99 percent of the daily input had been diverted in the six years prior to 1969 (Edmondson, 1970). The

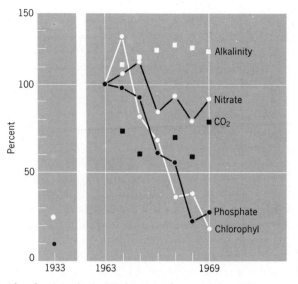

Figure 18.9

The cleaning of Lake Washington. These were crucial measurements on the waters of Lake Washington over the period of its recovery from cultural eutrophy. At the left are measurements of phosphate and chlorophyll in the water before the massive pollution, fortunately recorded in 1933. Between 1965 and 1969 most of the sewage discharge into the lake was diverted. The phosphate content of the water plunged, presumably as the phosphate originally in the polluted lake was buried in the mud. The decline of chlorophyll records the decline of the algae. The record clearly indites phosphate as the primary pollutant, for the lake lost the symptoms of pollution while nitrates and other potential fertilizers remained in large amount. (From Edmondson, 1970.)

effect was all that had been hoped. The smell went away; the algal nuisance came no more; the water became more transparent; and, most conclusive of all, the phosphorous concentration fell drastically to near levels that had been recorded in 1933 before the days of Seattle's growth (Figure 18.9). What can be done for one lake can be done for others. Lake Erie, perhaps the most notorious of the polluted lakes, has the extra benefit of having a river flowing through it, which will help to flush the pollutants out to sea. If we stopped dumping sewage and garbage into Erie, we could expect it to come clean in a few years.

On the outline so far presented limnologists are in wide agreement. Sewage and garbage enrich lakes with nutrients, which then promote excess plant life with the familiar unpleasant consequences. But the nutrients that do the mischief are quickly buried in the lake mud so that the disagreeable symptoms will only remain as long as you keep topping up with sewage and garbage. But there is not such complete general agreement about which particular nutrients are the culprits. Probably all agree that phosphorus can often be the villain in chief, and for many lakes you can demonstrate its importance. Edmonson's study of Lake Washington (Figure 18.9) provides very striking evidence that phosphorus was the controlling nutrient in that lake, for instance. The correlation between phosphate in the water during winter and chlorophyll present during summer blooms over the years in which the sewage effluents were diverted is dramatic. There was no correlation between chlorophyll and the other nutrients measured. It is also noteworthy that phosphate disappeared from the water much faster than other nutrients, which seems reasonable because of its liability to be adsorbed to various minerals which might be deposited in the sediments. So, for Lake Washington we can safely suggest that the history of pollution and subsequent cleaning recorded the concentration of phosphorus in the lake waters. But the chemistry of sewage and garbage is complex, and aquatic life needs more than phosphorus. Sometimes some other mineral nutrient may be in most critical supply; potassium, perhaps, or nitrates, or even gallium. At other times it might be dissolved carbon, or vitamins. This allows a very real possibility that it may sometimes be a pollutant other than phosphate which is the prime cause of trouble with the local lake, a possibility which is being actively canvassed by the detergent industry in its defense. It is true that there might be far more dangerous things than benign phosphorus, sought by all organisms and so easily disposed of in lake mud. We should be careful about phosphate substitutes in detergents. But the case that other pollutants may commonly be the main cause of excess fertility is weak. The possibility may be put in perspective by comparing the fertilizers of lakes with the fertilizers used by agriculture. There are fields in nature which may be in critical need of some other element in preference to phosphate, but we have not heard cries from the agricultural industry that phosphatic fertilizers do not work as a consequence. Phosphatic fertilizers generally work on lakes, too, and must usually be the prime cause of excess fertility.

LAKES AS IDEAL SUBJECTS FOR SYSTEMS ANALYSIS

The state of a lake changes with the seasons in easily measurable ways, for temperature and oxygen profiles tell you much about the state. And the range of states possible are also set by easily measured things, by the shapes of basins and the history of infilling. In the regulation of productivity in a lake there turns out to be a large inanimate control, the rate of input of dissolved nutrients and the rate of burial in the lake mud. Far from evolving along a predestined path, as the early studies on Linsley Pond seemed to suggest, lakes as systems can be seen to react promptly and directly to physical changes in the drainage basin. More, perhaps, than other patches of the earth's surface they can be examined as systems; discretely bounded, with biota that respond quickly to changes in the physical state, susceptible to manipulation and measurement. Computer simulations of lake systems may thus come to show the way to more ambitious analyses of parts of the land surface.

Ecologists are able to reach a special understanding of some of the causes of pollution, but it is also true that there are aspects of pollution about which an ecologist has no more expertise than anyone else. A study of air pollution shows this. Although we pollute the air with many novel chemical compounds, and these compounds then undergo further chemical changes through the influence of sunlight, the effect on the atmosphere as a whole is negligible. Nor can interference with biological processes have more than a negligible effect on the air. Air pollution only becomes serious when pollutants are concentrated under inversions. Undoing the damage of air pollution is more a subject for engineers and meteorologists than it is for ecologists. About the pollution of water an ecologist has more to say because the changes that annoy us are changes in living systems resulting from the pollution. The immediate counsel of ecologists on this matter is cheerful, being that natural waters will cleanse themselves as soon as we stop adding more pollutants. This, at once, raises the central practical problem of pollution control. What are we to do with the pollutants if we are to stop discharging them at random into our environment? Ecologists give what must be the only realistic long-term solution when they say "recycle," but recycling may involve immense economic costs, which communities may not be able to bear in the immediate future. Such relatively trivial nuisances as stinking lakes might be preferable to seeing part of one's community living in poverty. What concerns an ecologist most are not temporary troubles, like the pollution of lakes and the air which we now witness, but the permanent forms of damage that can never be made good. These are the great extinctions of plants and animals that we are

261

witnessing, and will witness even more in the decades to come. Long-lived pesticides, and, perhaps, weed killers, may well eliminate animals and plants far from the site of application; perhaps particularly in the deep oceans. Reserves and game refuges cannot remain as populations expand, for first the dangerous animals are killed and then the land is taken over for irrigation and agriculture. Conservation movements based on charity are powerless against the demands of needy people for land. The only thing that can halt this process of permanent elimination of the earth's riches is the early imposition of a limit to the size of the human population.

CHAPTER 19

AN ECOLOGIST'S
VIEW OF POLLUTION

There are aspects of the pollution and conservation debates on which ecologists have professional knowledge and may claim a special right to be heard. The pollution of lakes with fertilizers is one of them, for this is a process that disrupts the working of a natural system. Ecology is very much concerned with such a process, can report on it, and can advise on courses of action from the bastion of professional experience. But there are other forms of pollution on which an ecologist has no more expertise than anyone else. The no-deposit-no-return bottles which litter our road-sides are fairly called a form of pollution, but they have little effect on the ecosystems into which they are discharged. An ecologist's dislike of them is no more important than the equal dislike of any other civilized man. There are also many extreme decisions of development which modern societies take which, although they disrupt natural communities, may still be such that an ecologist's view is no better than another's. The decision to put houses on the last open space in a great city is one of them; the natural community will be destroyed, but the decision over whether it shall be destroyed or no is an aesthetic one and not based on fine points of natural balance. It is important at the outset of any discussion of pollution and the use to which we put the earth to separate problems which an ecologist can tackle with professional skills from those for which he is just another concerned citizen. Bottles by the roadside are nothing to do with ecology, unless you think of them as traps for water in which mosquitos may breed.

AN ECOLOGIST'S VIEW
OF AIR POLLUTION

The volume of the atmosphere is essentially fixed, the concentrations of its gases buffered on a geological time scale by geochemical and biological processes (Chapters 15 and 16). Only the carbon dioxide and water vapor concentrations seem to have fluctuated to any measurable degree before the advent of modern technology, carbon dioxide with the seasons or the emissions of volcanoes, although only slightly, and water vapor with wide amplitudes. But the development of a society based on the energy of fossil fuels has resulted in the sudden addition to the atmosphere of many novel compounds, the combustion products of those fuels. They are carbon dioxide in sufficient amount to override the ocean buffering system, the real significance of which we find hard to estimate (Chapter 16); carbon monoxide (a very toxic gas); nitrogen oxides; sulfur dioxide (which becomes spontaneously oxidized to sulfuric acid); and a great range of hydrocarbons, of varying inertness or malignancy. The whole constitutes an atmospheric soup of remarkable and incompletely probed variety. And this dilute chemistry is suspended in a transparent fluid medium flooded with the energy of sunlight. Complex chemical reactions proceed, particularly based on nitric oxide and various of the hydrocarbons, to produce fresh ingredients for the atmospheric soup, some of them very unwelcome. This process is called **photochemical air pollution.** Notable among its products are nitrogen dioxide, ozone, formaldehyde, and peroxyacyl nitrates (familiar enough to have been given the cognomen PAN, a term recurrent in the jargon of pollution engineers), all of which injure plants. Formaldehyde and PAN make eyes smart, being the most noxious ingredients of the infamous Los Angeles smog.

The output of the raw materials of this photochemistry by a modern technical society can be prodigious. Leighton (1966) estimates that a standard American car cruising at 60 miles an hour emits 3 liters of nitrogen oxides each minute. Oxides of nitrogen become a nuisance when they are in concentrations of about 0.05 ppm (0.05×10^{-6}), which means that 60 million liters (60×10^6) of air are required to dilute the pollutants produced by the car every minute. This volume of air would meet the breathing requirements of between 5 and 10 million people for that same minute. The annual output of cars by American industry is said to be about 8 million. Fortunately they are not all driven at 60 miles an hour all the time, but they are driven quite enough to matter. Power stations, factories, and home furnaces all add their contributions, too.

But the atmosphere, like a lake, is a self-cleaning system. Some of its contaminants, like the sulfuric acid that derives from the sulfurous fumes of coal-burning power stations, are soluble in water and last only until the next rainstorm. Others are eventually formed into solid particles which may serve as nuclei for water droplets, and hence may be carried to the ground. Eventually all must be returned to the earth's surface, probably into its waters (where they may not be particularly welcome), and probably over time spans measured in weeks rather than in years. And the volume of the atmosphere is so large that the mean concentration of the worst contaminants at any one time is likely to be uninterestingly low. So why should this pollution of the atmosphere produce adverse effects on living things? It does so because atmospheric mixing is often prevented so that pollutants are concentrated in small volumes of air that cling close to the ground.

Typically there is a temperature lapse with altitude, but it is sometimes possible to have cold air flowing under a layer of warm air. This may happen, for instance, in a hollow between mountains when cold air,

Figure 19.1   Results of air pollution in the Copper Basin of Tennessee. An inversion commonly overlies the valley, the product of cold air running down the mountainsides into the basin. A smelter discharged fumes containing sulfur and arsenic into the valley for 50 years. All plants died and erosion completed the devastation.

being dense, flows down the mountainsides and collects in the hollow, or it may be the result of more widespread movements of air masses. When it happens a stable system is established with warm air floating over cold air, something analogous to the condition of a stratified lake in which the warm epilimnion floats on the cold hypolimnion. The boundary layer, which in a lake is called the thermocline, is called by weathermen an **inversion layer.** Much energy is needed to overturn an inversion, energy such as the arrival of a front or other source of strong winds, so that the inversion may persist for days or even weeks. If an inversion settles over a city, the people of that city are in a plight analogous to that of a trout in a polluted lake. The people are trapped in their bottom air, with no way of replacing it. The filth from their chimneys and exhausts then collects around them, the energy flux of sunlight promotes chemical reactions among the concentrated pollutants, and you have the results for which Los Angeles is so well-known.

The most spectacularly unpleasant results of voiding pollutants into inversions come from the records of the mining industry, and the most spectacular of these is, perhaps, the story of the copper basin, near Ducktown in Tennessee (Leighton, 1954). The basin collects cold air from the mountain slopes with which it is ringed, and a copper smelter got rid of all the sulfur and arsenic in ore by voiding it into the air under the inversion as gas. Fifty years or so of this killed all the vegetation in the valley, which is still a dead place (Figure 19.1).

All these examples of chronic air pollution are purely local phenomena, due to voiding noxious gases in places of restricted circulation of the air. Now that the cause is known, the cure is generally straightforward, if expensive. Technology can devise stacks and combustion processes which do not void chemicals (other than carbon dioxide) into the atmosphere, and it is possible to consult meteorologists about the likelihood of local inversions before building plants or expanding cities. On all these matters an ecologist has no special competence. The skills at systems analysis which some of them possess are no better than similar skills possessed by engineers and meteorologists. And the purely ecological thesis that replenishment of the atmosphere with oxygen is a delicate biological process which can be upset to produce a sudden disruption of the air has been demonstrated to be false (Chapter 16).

The useful observations that an ecologist can make about this comparitively minor, local, and "fixable" problem of air pollution are those concerning the impact of human numbers and energy costs which are discussed in a later section.

Common water pollution is caused by fertilizing lakes with sewage and agricultural runoff. The cure is simply to divert the flow of pollutants, when the lakes will slowly clean themselves (Chapter 18). But there is a second form of water pollution, pollution by toxic chemicals, and these must seem more menacing than the simple nutrients of sewage and garbage: DDT, 2,4,D, mercury, arsenic, the sulfurous effluents of paper mills, acid mine wastes; many such things pollute by poisoning. They may do so when present only in trace amounts, either directly, as when DDT kills a lake's insects or even phytoplankton, or because they are concentrated in food chains (Chapter 9), either killing the top carnivores or making their flesh unfit for human consumption. There seems something particularly forbidding about a trace constituent doing such damage, because how are you to remove trace poisons from a lake? In fact, there is good reason to believe that natural waters will rid themselves of trace poisons in the same way that they rid themselves of excess nutrients; by burying them in the mud.

The poisons which kill in small amounts owe much of their persistence to their very deadliness. Organisms find them hard to handle with safety, for they may damage excretory systems no less than other systems. A common device of all organisms is thus to store some of the dangerous material in more-or-less inactive parts of their systems, as when DDT is stored in fat. This phenomenon is what makes possible the dangerous concentrations through food chains. But it also provides a way in which lakes may rid themselves of the contaminating poisons, for the stores which animals and plants have built up in their tissues must often fall to the mud with their corpses. Mineral poisons such as mercury and arsenic must often be adsorbed to falling mineral particles as well, and this may

AN ECOLOGIST'S VIEW
OF WATER POLLUTION

also happen to some of the organic poisons, further consigning them to the mud. And it now appears that bacterial flocs may likewise adsorb such things as insecticides and carry them to the bottom.

Some bacteria of fertile waters form flocs, granular particles made of polymers that the bacteria synthesize outside their bodies. We do not know what is the critical selective advantage of this activity to the bacteria, although it is possible to suggest that the flocs give them physical protection, are excretory products, or serve to concentrate nutrients in their neighborhood. But the flocs are electrically charged and may adsorb a variety of nutrients or other molecules. Recently, microbiologists at The Ohio State University have shown that the pesticide aldrin attaches itself to these flocs, and is carried with them to the bottom mud (Leshniowsky et al. 1970). Probably many other dissolved substances find their way to the mud on bacterial flocs.

Once held at the surface of the bottom mud, the contaminants are not yet safely disposed of, however. They may be reintroduced into the lake system, either by resolution or when bottom mud is eaten by detritus feeders. This latter process is likely to be particularly effective, because many organisms make their livings by working over the corpses and detritus which forms the mud; larvae of insects such as midges, gastrotrichs, worms, crustaceans, and so on. These animals then form the basis of new food chains which carry the contaminants abroad in the cosystem once again. We have practical evidence that this can be so in studies with labeled DDT in a marsh near Lake Erie (Peterle, 1969). But both solution and scavenging do not appear to penetrate the mud very deeply, so that it is only the top few millimeters of mud which can return pollutants to the ecosystem. If mud collects quickly, poisons can be buried out of reach, just as excess fertilizer is buried. Polluting a lake with fertilizer may thus help clean it of poisons, since the fertilizer both promotes the growth of bacteria which form flocs and causes a rapid rain of corpses which swiftly become deeply buried.

The message of the limnologists about water pollution is thus that all you need to do is to stop polluting and the lakes and streams will come clean by themselves. This is a counsel of cheer and hope to set against some of the counsels of ecological despair. But, unfortunately, it is not the end of what an ecologist can say about these pollutants. For, if you do not put sewage, toxic wastes, and garbage into lakes and streams, you must do something else with them. What should this be? For most garbage an ecologist has a ready answer, "Put it back where it came from or use it again." This simple answer is surely the only right one in the long term. Put sewage back on the land to fertilize what you want fertilized. Recycle the minerals. This leaves only the long-lived deadly products of organic chemistry, like DDT, of which an ecologist can only say, "Refrain from making them in the first place." But these seemingly simple answers are not so simple to practice. They exact costs; both familiar economic costs and more serious physical costs which must be paid in the currency of energy.

The economic costs of restraining polluters are not things that an ecologist is particularly competent to assess, but he should be aware that they can be great. It would be very expensive, for instance, to divert all sewage from Lake Erie, and this expense is to be measured in terms of what other goods we deny ourselves for the sake of a good in the form of a clean Lake Erie. The goods foregone could easily be better schools in the South or more decent homes for slum dwellers in cities near the lake. The improvidence of past generations, which has left us with both slums and polluted lakes, may force us to make some unpleasant choices. Even if we can face the economic costs of what we desire, we shall still be faced with the fundamental energy cost, and this may in the long run be harder to meet.

If, instead of letting our sewage and garbage fall quietly into rivers, there to be transported for us by that portion of solar energy driving the hydrologic cycle, we resolve to take it back where it came from, we must do extra work. This requires a source of high-grade energy. For a few decades we can provide this energy cheaply from fossil fuels, after which we must increasingly resort to some other source. If energy from uranium fission plants, breeder reactors, or thermonuclear fusion does not come to our aid quickly, we may well decide to live with polluted waters and use our dwindling fossil-fuel hoard for other things.

Nuclear power stations themselves have a forbidding sound to a student of lakes and streams, because all designs so far discussed require the waters of lakes and streams to cool them. These stations are all devices for releasing nuclear energy as heat, and this heat has to be harnessed by making it do work such as boiling water to run turbines, which in turn run generators. The second law of thermodynamics requires that heat be released from the system at all these stages, and this heat must be carried away from the power station. The only effective way we have found to do this so far is to carry it away in flowing water. This has an effect on the aquatic systems so used that we have come to know as **thermal pollution.** Heating water obviously alters the conditions of life within it, and some of the first casualties are cold-water fish like trout, which happen to be those favored by anglers and gastronomes alike. So thermal pollution is at once unpopular with powerful lobbies. But there is more to it than this. Warm water floats. If you discharge warm water into a lake, you create a floating layer of warm water on top, an artificial epilimnion. This artificial epilimnion may increase the length of time during which the bottom waters are isolated from the oxygen of the air, increasing the chances of their becoming anoxic, and bringing on many of the unpleasant symptoms of a lake polluted with fertilizer. And even this is not the full extent of the woe. Extra-warm surface waters promote the growth of blue-green algae over other forms of algae; and blue-green algae are nitrogen fixers. Promoting their growth through thermal pollution thus results in fertilizing the water with nitrates, bringing on the well-known polluted condition of a stinking lake.

Polluted lakes, rivers, and air are all essentially local conditions and are all reversible by natural processes. Pollution of ancient lakes with rare or

THE PROBLEM OF
GLOBAL POLLUTION

endemic faunas can involve great losses to the human heritage; it would be tragic if Lake Baikal in the Soviet Union or Lake Tanganyika in Africa, with their swarms of endemic species, became so polluted that their wonderful faunas were lost; but otherwise the brief centuries of man's excesses will appear in the sedimentary history of the earth as a thin phosphate-rich band separating essentially similar microfossil assemblages. An ecologist, treasuring the diversity of terrestrial life and wishing to pass on to his descendents this rich treasure intact, must be more concerned with what does irreparable harm to the world ecosystem. Any action of ours which exterminates species does damage that those who come after us cannot repair, however enlightened and skillful they may become, and it is by the extinctions we cause that we do most lasting harm. Extinction by overhunting is now well understood. It is theoretically possible to prevent it by setting aside reserves and regulating hunting pressures, or by inventing ways of hunting that make the chase more difficult. The English system of hunting foxes with a pack of hounds is a highly efficient conservation measure, letting a large number of men get yearnings derived from their Pleistocene hunting niche out of their systems with the minimum risk to the fox population. Reserving grizzly bear skins for those willing to hunt their grizzly armed only with spear would be another illustration of the principle. But measures for preserving game can only work as long as it is possible to set aside land and production for the game to use. Pressures of rising population require that reserves soon vanish. Tigers have virtually disappeared from India since the Second World War, not because of hunters or fur traders but because peasants required tiger food and tiger land for themselves (Schaller, 1967). This story will be repeated everywhere on earth if population continues to rise. First to go are the big predators, not only because they need the largest productive base but also because they are dangerous to men and farm animals moving in on their land. But soon the wild herbivores and the forest must yield also. It is curious that rich societies of the West imagine that they can preserve the game refuges of Africa and other places by charity collections, when the wealth that provides their charity came by subduing their own wild and their own wild beasts. When the populations of Africa are large enough, they will press in on the refuges, demand protection from the dangerous animals, insist on grazing their cows, and finally demand irrigation so that they may plough. This process is gathering momentum all over the earth. It results in extinction of all the animals and plants of complex communities, allowing to survive only those generalist species that can endure troubled times: the weeds, the rodents, and the sparrows.

The gross environmental pollutants have this same effect of encouraging generalist species, those of the early successional stages. This is true even for such exotic pollutants as gamma radiations, as was beautifully shown in studies in the Brookhaven National Forest on Long Island by Woodwell (1970). The demonstration was very simple, merely the hanging of a 9500 curie cesium 137 source from a tower among the trees (Figure 19.2). The trees near the source soon died, as perhaps should be

Figure 19.2

The effects of chronic gamma radiation on a Long Island forest. A 9500-curie ^{137}Cs source is suspended from a central tower. After 8 years there has developed a concentric series of communities running from a simple association of mosses and lichens near the tower to the undisturbed pine-oak forest beyond the reach of the radiation. The series of communities duplicates a successional series resulting from other disturbances in the area.

expected, and there was a small bare patch in the forest. But as time passed a concentric pattern appeared, with plants of varing degree of sensitivity persisting in rings around the source of radiation. Some mosses and lichens were able to endure massive doses of radiation, and colonized the bare ground near the tower. Then there were various herbs, followed by various shrubs; then some trees, and finally the original forest, pressed back to a safe distance from the tower. The series of communities from tower to forest was a familiar one to students of succession. The pioneer communites endured the stress the best, and the effect of radiations had been to favor the generalist pioneer species of the successions at the expense of the climax forms of complex vegetation. Radiation stress effected plants as if it were a natural stress like fire, drought, or flooding.

Pollutants of air or water have similar effects. The algal blooms of a polluted lake are simple communities, apparently of pioneers invading a habitat which is no longer tolerable to the species of more complex communities which once lived there. The plants which come in after strip

mining, or on land contaminated with acid mine waste are likewise pioneers. Woodlands fade away in the polluted air coming from cities to be
replaced by scrub. It seems that always our pollutants are treated by natural communities as stresses to which there is a common response. Pioneer
communities adapted to recurring natural stresses survive, and the climax
or more complex communities are removed. The spread of pollution over
the earth, even pollution at levels not obviously fatal to men, steadily
removes complex communities. It thus becomes an extra agent of extinction wielded by expanding populations, one allied to the immediate
needs of the expanding populations for the land itself.

But to the twin direct pressures of expanding population and gross
pollution is added an agent of extinction that may be the most pressing of
all, the broadcasting of persistent poisons that are effective in trace
amount. These kill selectively, at the tops of food chains, again acting
against the species of complex communities. Pioneer plants and animals
again seem able to survive stresses of heavy metals and insecticides the
best, probably because there is more genetic variability in their populations on which selection can work. And these trace poisons, released into
the biosphere, can work their extinctions far away from the edges of expanding populations, in the places where ploughs and smog have not yet
reached. These agents of extinction worry ecologists second only to the
expansion of population itself.

The alarm has sounded on the elemental poisons so well that their dis-
bursement is likely to stop. We can ban lead from gasolines, and regulate
emissions of mercury and arsenic. Those who pollute with such things
have few friends, so that pollution of this kind will probably be stopped.
But the massive release of the more insidiously damaging long-lived insecticides is backed by economic interests so powerful that they do no
more than twist under the pressure of logic.

DDT is typical of the family of chlorinated hydrocarbon insecticides. It
has so stable a chemical structure, that we can find little evidence that
much of the DDT that we have released over the last couple of decades
has been degraded to innocuous substances. It is true that vertebrates do
excrete a large part of their intake, in addition to that which they store in
their fat, and that they do metabolize it slightly in the process. But the
most common metabolite, called DDE, is nearly as toxic as DDT itself,
and is not very different from it. DDE is largely to blame for the vanishing
of predatory birds due to their inability to make shells for their eggs (Keith,
1966: Hickey, 1969). Only minor metabolites are thought to be relatively
harmless, and there is no field data to suggest that DDT is broken down
by physical process. The disappearance of DDT from fields after a
number of years has been noted, but there is no reason to believe that
anything more has happened to this than movement to other ecosystems
in air and water; for DDT both dissolves and evaporates. It travels the
world and has even been found in Antarctic snow (Peterle, 1970).

DDT works its deadly way but slowly up food chains, so that a recent
computer simulation of Wisconsin ecosystems suggests that we have nine

years to wait before what is already in the state appears in its most massive concentrations in the top links of the native food chains (Harrison et al., 1970). After that, it, together with the DDT from the rest of the world, and all its ugly relatives, are headed for the world oceans. It is possible that some of the wonderfully diverse life of the deep ocean floors (Chapter 40), to say nothing of creatures of the more familiar sea, are doomed by poisons already spread. Compared with this, to worry about phosphates in detergents seems trivial.

THE ENERGY BASE OF POLLUTION

We do work to pollute. Lakes and streams lose their pollutants to mud and discharge waters so quickly that we have to work really hard to keep them topped up with our garbage. The air pollutants trapped over a city by an inversion are always swept away in the wind before long, and all the work to concentrate them has to be done over again. We do this work with fossil fuels. If we wish to move pollutants more gently to places of our own choosing, we must do yet more work, for which we must again draw on the energy of fossil fuels. We do even more work to concentrate men in giant cities, to make new motor cars so that the old ones can be thrown away, and to make no-deposit-no-return bottles.

It also takes much work to replace complex ecosystems with crops, and to defend these crops against herbivores and competitors. We make our work task even harder by developing high-yielding strains of crop plants, since they put the maximum of their trapped solar energy into food stores which we eat instead of using their energy to compete and defeat herbivores as the wild plants do. We must then defend our tender plants and their energy stores with an alternate energy source which, of course, we derive from fossil fuels. We plough and hoe to defend plants no longer able to defend themselves. In a real sense, therefore, we are eating fossil fuels, since fossil-fuel energy as well as solar energy goes into the support and production of our crops. But we have found that we can defend our crops from competitors and herbivores more easily, which is to say with less expenditure of energy, by spraying insecticides and herbicides. This is the sort of action that reveals the fundamental and ecological basis of the environmental crisis. It takes less work if you pollute.

If we chose to live in a less-polluted world, we must first accept the fact that it will take much work to do it. The total work we can do for the immediate future is measured by the flow of fossil fuels, and their supply is both strictly limited and the property of the world community. There may not be enough for what we plan. To this is sometimes made the cheerful answer that soon we shall have breeder-reactors or fusion reactors, which will give us unlimited power. But this is not true. Suppose, which I tend to expect, that fusion reactors are made, that they are essentially clean, and that we resign ourselves to living with thermal pollution of water by measures such as banning the power plants to the Arctic Ocean where there are not many people; this still does not provide us with unlimited power. These power plants will heat the earth. In effect, we will be adding a small sun to the solar system. The temperature of the earth is set by a steady state established between the rate of solar insolation and the rate at which the earth radiates energy back to space. As power plants are built, so will fresh steady states be established at ever higher temperatures. A practical limit to power output by fusion plants will be reached when the effects on things such as weather of the raised temperature of the earth become intolerable. No doubt we can generate many times our present output of power before this point is reached, but this is not the same thing as claiming that fusion plants will give us unlimited power. Perhaps fusion plants would let us provide for every citizen of the modern earth the power consumption that a North American now enjoys without raising the earth's temperature more than would be tolerable. But population goes on doubling, and so do the per capita desires for power. Control of pollution must require still greater per capita consumption of power, because we must do more work in the controlling of it. An ecologist must remain unimpressed by all technical solutions to pollution, whether it be the minor affairs of algae-covered lakes and smog or the more serious things like DDT and heavy metals, as long as population continues to rise. The limits to the quality of life which any man can enjoy are set by the numbers of other men, or the energy we release, or both, and various of the syndromes that we call pollution are inevitable. In anything but the very short term, the real cost of controlling pollution is setting a limit to the numbers of people.

One way of looking at plants and animals is as sets of tolerances. Each kind has a preferred place in which to live and cannot tolerate the conditions for life in different places. A refinement of this idea is that there may be places that are nearly suitable for a particular animal or plant and that might need to be changed in but a single factor to make the place suitable. This refinement is essentially a simplifying assumption, for it must be obvious that very many among even physical factors must interact to form the environment of a plant or animal. Careful measurements of things such as temperature and nutrient supplies in field crops, and in growth chambers, may yield a set of apparently optimum conditions which should be provided for good husbandry, but this is an approach that leads to very little ecological understanding. Wild plants and animals usually have their distributions set by the activities of other plants and animals more than by the physical conditions for life. But more serious still is the fact that this approach tends to regard species as created entities doomed to find what living they may in such physical habitats as they can tolerate. A proper ecological viewpoint is to think that species have evolved to get a living, to take advantage of particular resources, and that they have evolved such physical tolerances as are required for this chosen way of life. There is much that is interesting and useful in finding out how this trick of physical adaptation is achieved, a sort of study that is often called "physiological ecology," but again such work does not help the more important ecological question of *why* these adaptations were evolved. The same criticism may be applied to recent attempts to define the tolerances of animals and plants by their heat budgets. Any particular design of animal may well be shown to be   273 limited to a certain range of environments because it cannot balance its heat budget anywhere else, but once more the more interesting ecological question is to ask why that particular design of animal was adopted by evolution, and the answer to this must be sought through an understanding of the resources made available to the animal by that particular physical design. Thinking of animals and plants as sets of tolerances is useful for the practical purpose of systems analysis, but it is more interesting to try to understand why such animals and plants came to evolve their ability to live where they do. The habitat is best seen as a stage on which an evolutionary play is performed, a play which we see at one moment in its development as numbers of plants and animals adapted to various ways of life.

CHAPTER 20

ON LIMITS AND
TOLERANCE

It is a commonplace of natural history that each kind of plant or animal can tolerate only a certain set of conditions, and that the occurrence or commonness of any kind must be set by the frequency with which its favored conditions are met. Animals requiring brackish water must only be looked for where there is a brackish water; plants that require frost to break the dormancy of their seeds or bulbs can only be found in places with frosty winters; frogs that spend part of their life cycle in water cannot be found in deserts, and so on. The distribution of many animals and plants must be sharply affected by such simple considerations, and may often be satisfactorily explained by them.

Sometimes just one overriding critical factor may appear to be limiting, as was first pointed out by the great German chemist Liebig, in the middle of the last century. Liebig discovered the importance of phosphorus to agriculture, noting that if a field was poorly supplied with phosphorus, crops would not do well on it however splendid a place it was in other respects. He had, of course, stumbled on the most generally limiting of the nutrients needed by all life, the one really vital thing that was in general short supply because of its comparitive rarity in the earth's crust, but the principle should hold good for any other nutrient that might be in critically short supply. Liebig wrote in a day when you must express your major discoveries as "laws" for them to penetrate the corporate consciousness of science, and so he expressed his finding as his famous "Law of the Minimum" which states, "Growth of a plant depends on the amount of food-stuff which is presented to it in limiting quantity." The impressiveness of this statement is such that biologists have long been mesmerized by it. In fact it says "plants are limited by what limits them," which is not so impressive.

In practice, the growth of most crops probably depends on their having "presented" to them, as Liebig put it, enough of just three nutrients that are commonly in short supply nitrate, phosphorus, and potassium; the familiar N:P:K of the farmer, although it is even better to give them water. It is possible for soils to be deficient in less likely nutrients than N:P:K, such elements as boron and iodine, for instance. There is no a priori reason for doubting that the whole periodic table may not be needed in some tiny proportion, but it is doubtful if the fortunes of agriculture are much affected by shortages of most of the elements.

Wild plants must be limited by restrictions in the supply of essential nutrients no less than crop plants, but wild plants must also be severely limited by the activities of other plants and of animals which eat them, and these pressures may be much more "limiting" than simple nutrient shortage. Critical provision of minimal amounts of nitrates, phosphates, and potassium, principally by mineral cycling, is, however, important to wild plants, and may restrict their growth and distribution. Some plants may not live on bare ground at all because nutrients may be critically limiting; and this may be, as we have seen, an important cause of plant succession. Others may grow poorly in a place, reducing the productivity of the habitat and imposing restrictions on the numbers and kinds of animals that can live there. The life of the world's oceans appears to be critically limited by phosphorus or iron in this way.

On land, water is perhaps the most important of nutrients that may be present in critically limiting amounts, there being, for instance, no doubt that the paucity of plants and animals in hot deserts is due to scarcity of water. The growth of any desert plant depends, as Liebig would have said, on the amount of water presented to it; and we may further reason that the growth of desert plants so limited determines the energy flux available to desert animals; which must, in turn, determine the numbers and sizes of those animals. But desert plants grow best if you take them out of deserts and give them lots of water, an experiment that the world's thousands of cactus enthusiasts have demonstrated many times. Apparently what is really important to the distribution and abundance of desert plants is that they can tolerate life in a place where water is scarce whereas most plants cannot. The opposite must also be true; that the desert plants cannot tolerate life in the well-watered parts of the earth. Since desert plants like water as well as other plants, their exclusion from watered places must be due to some other factor of the environment which they cannot tolerate. They may not be able to survive the different seasonal cycles of watered places, or such physical stresses as rain, frost, or developed soils; or (and this is the most likely explanation) they may be excluded by some doings of the other plants and animals of the place. But it is clear that a set of tolerances of some sort places limits to the distribution of desert plants. Their tolerance of low water supply enables them to live in deserts, and their failure to tolerate some other aspect of life in watered places restricts them to the deserts. Like all plants and animals, desert plants are limited in one direction by their minimum requirement and, in the other, by the adversity that they can tolerate.

V. E. Shelford is credited with having first pointed out the importance of tolerances in explaining the distribution of individual species of animals. He concluded that the distribution of tiger beetles within the Lake Michigan sand dunes was determined by soil moisture, strangely enough, for the beetles run and fly under the sun and over the hottest sand in the course of their predatory lives. But their eggs are laid in holes in the ground, and the ground had to be just right; damp enough, not too hot, not too cold, and just the right texture. The distribution of the beetles was thus set by a very narrow range of environmental limits, and these applied for one short phase of the animal's life cycle (Shelford, 1908).

Shelford (1913), probably in conscious imitation of Liebig, described his views on tolerance as the "law of tolerance," pointing out that animals were limited to environments which they could endure, and that Liebig's Law of the Minimum defined only one of the parameters of what they could endure. Liebig had said that plants were limited by what limited them, now Shelford said that animals tolerated what they could tolerate.

It is sound ecological thinking to ask, "What are the things of the environment which limit this organism?" It is perhaps not so interesting a question as, "Why has this organism been adapted to this set of limits?" but it is a valid ecological question all the same. Shelford was trying to answer the first of these questions with his work on the tolerances of tiger beetles. He thought of animals as being defined by sets of tolerances. This view is forced to turn away from the really interesting side of an animal's life, of how it gets its food and how it adapts to the presence of other animals, of the most vital parameters of the Eltonian niche. But concentrating on physical tolerances does have the practical advantage that you have something to measure. Physical tolerance must be to things like temperature, or water, or salinity, or Liebig's minimal nutrients. Thinking of animals and plants as being narrowly limited by the physical environment gives you a license to go out and collect data and, when you are tired of field work, you can bring your organism into the laboratory, make it live in growth chambers which formidably mimic all possible states of its native environment; then measure its tolerances with beautiful precision.

Measuring the simple physical stresses which an organism can endure in this way may have practical value. You really can define a nutrient mixture in which some plants will do well in a certain climate, for instance, and this knowledge is useful if you want to grow the plants as a farmer. But this practical usefulness has been marred by the way in which ideas of tolerances and limits have been put forward. "Plants are limited by what limits them" and, "Animals can tolerate what they can tolerate," are not useful statements. But they have been venerated as laws, and they have conveyed the idea that all you had to do in your field studies was to go out and measure things until you found something which seemed to be limiting, and that you had then explained something significant. The chances were that you had done nothing of the sort. You may have come

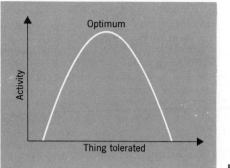

Figure 20.1

across some temporary condition in which an animal had bumped against some real limit or even that the correlation with an apparent restraint was entirely fortuitous. It is easy, for instance to believe that phosphorus is limiting a plant crop, and even to show that the crop grows better with more phosphorus, when in fact it was much more limited by shortage of water. Thinking of the environment as sets of limits has practical use, but the impression given by the meaningless "laws" of the minimum and tolerance, that single factors were often limiting, is false. Many practical field workers have been led astray by this.

From the original ideas of tolerance have developed schools of ecology devoted to defining the activities, and hence the distributions of animals and, particularly, plants from environmental measurement. Many growth chamber and field measurements about many individuals yield what may be called a range of tolerance for a population, and the findings can be plotted as a bell-shaped curve as shown in Figure 20.1. The optimum tolerance for the population seems to be the point at the top of the curve. Measure lots of parameters and you can draw lots of bell-shaped curves, then you can sum these by sophisticated mathematics to define in one equation all the tolerances of an animal or plant that seem most important. Such an approach is usually implied when work is described as **autecology,** the study of the habits and habitats of individuals or individual species. Treated in this way, the study of tolerances is, at least, more useful than that quest for critical limiting factors of the environment which is so dangerously suggested by the statements of Liebig. But it ignores the fact that probably most of natural distributions are more critically determined by the relationships of living thing to living thing, of eating and being eaten. Tolerances are just things that animals and plants have evolved in order to get their food.

The student of tolerance quickly finds that some animals and plants have broad tolerances whereas others have narrow tolerances. When he finds tolerance to be broad he describes it with the Greek prefix **eury-;** when the tolerance is narrow, with the Greek prefix **steno-.** Thus, when temperature or salinity may fluctuate widely without seriously affecting

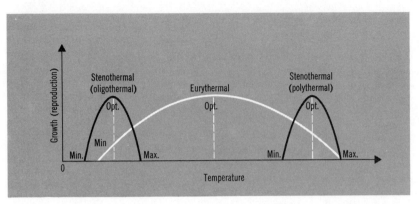

Figure 20.2

Illustration of ranges of temperature tolerance in stenothermal and eurythermal animals. All animals are thought to be most active at some optimum condition, in this example, temperature, but some have narrow tolerances and others broad tolerances. (Redrawn from Ruttner, 1963.)

individuals, the species are called **eurythermal** or **euryhaline;** when slight changes of temperature or salinity are fatal to animals or plants, they are called **stenothermal** or **stenohaline.** These four terms are used so widely in the ecological literature as to become necessary jargon of the trade. It is, of course, possible to call on bastard Greek for more elaborate classifications of kinds of tolerance (see Figure 20.2 and Table 20.1), but these do not have wide currency.

THE EVOLUTIONARY
VIEW OF TOLERANCE

The idea of critically limiting physical factors may serve only to obstruct a theoretical ecologist in his quest for a true understanding of nature. His task is to explain the distribution and abundance of animals, the reasons species exist, and why there are so many of them. To say that animals live where their tolerances let them live has an uninteresting sound to it. It implies that animals have been designed by some arbitrary engineer according to some preconceived sets of tolerances, and that they then have to make do with whatever places on the face of the earth will provide enough of the required factors. But these animals have evolved their sets of tolerances to meet particular environmental conditions. The real question for ecology is: Why have animals and plants evolved the particular

TABLE 20.1   **Terms Sometimes Used To Describe Various Kinds of Tolerance**

stenothermal — eurythermal	refers to temperature
stenohydric — euryhydric	refers to water
stenohaline euryhaline	refers to salinity
stenophagic — euryphagic	refers to food
stenoecious — euryecious	refers to habitat selection

The suffixes *-thermal* and *-haline* are used quite commonly. The others (fortunately) less commonly.

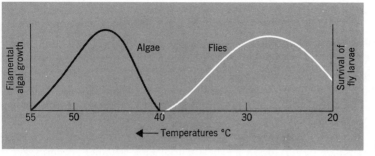

Figure 20.3

sets of tolerances to which they do own? Measuring tolerances takes ecology no way at all on the road to answering this question.

The intellectual fallacy that may result from too much faith in a limiting factor approach is best illustrated by an example. The bottoms and sides of streams running from hot springs in Yellowstone Park are lined by algal mats. There is a temperature gradient down the stream channels, as you would expect, and there are different taxa of algae to be found in various temperature ranges. Between about 55°C and 40°C the algal mats are largely made up of filamentous blue-green algae, but these plants are rather stenothermal and will not actively grow at temperatures below 40°C, although the mats can persist at lower temperatures as long as they are not eaten or otherwise disrupted. The mats are eaten, however, by a species of fly which raises its maggots on them. The fly is very skillful at seeking out algal mats to lay large numbers of eggs on them, but the fly larvae cannot tolerate the temperatures at which the algal mats will actively grow, and hence must make do with moribund colonies in cooling water. This state of affairs can be understood from the graph shown in Figure 20.3, based on a sketch drawn for me by R. D. Mitchell, who has been investigating the phenomenon (Wiegert and Mitchell, 1972). The tolerances of the flies are such that they are apparently disbarred from eating what one might think should be their main source of food. Why should they have evolved such tolerances? Or is it that the fly engineer somehow could not make a fly which would tolerate warmer water?

The answer proposed by Wiegert and Mitchell rests on the discovery that most of the standing crop of the algae at any one time is present as moribund mat at temperatures which are near the optimum for the fly larvae. This comes about because of some peculiar properties of the spring system; the algal mats acts as insulators so that the middle of an old mat becomes too cool for active growth. Furthermore, algal mats serve to divert the flow of the shallow streams, deflecting warm water, and cooling the mats still more. It thus comes about that the main food resource available to the alga-eating flies is at a temperature too cold for algal growth. The ecological strategy of the flies has been to

adopt temperature tolerances that will enable them to feed on this resource. It has apparently not been worthwhile in an evolutionary sense
for flies to have developed tolerance to warmer water, because the resources in that warmer water were too small to support a viable population. Knowing the temperature tolerances of those flies would be valuable
to a man setting out to kill them, but is not very illuminating to someone
wanting to know the flies' role in the hot spring ecosystem, and is almost
useless in aiding an understanding of why that species of fly evolved.

The ecological moral in this story is that we must prefer asking ourselves what an animal or plant is doing, to asking what it can stand. For it
will have evolved to stand whatever is necessary in order to do what gives
it a livelihood. Ask first not, "what is its temperature?" but rather, "what
is its resource?".

The most noted of the descriptive plant geographers of the last century, A.
F. W. Schimper, based much of his work on the relationship of plant form
to water supplies. One of his most discerning studies was on the true nature of the epiphytes of trees in the tropical rain forest (Schimper, 1903).
These plants lived high on the branches of the great trees, exposed with
them to the endless damp and the sousing equatorial rains, and yet they
were often shaped like plants Schimper had seen in deserts. They had
thick fleshy leaves, with dark green shiny cuticles and sunken stomata,
and they often had spongy mats of suspended root or curious cup-shaped
leaf bases (Figure 20.4). In the desert, sunken stomata, heavy cuticles, and
fleshy leaves are clearly devices to reduce water losses from evaporation.
Very well, they must serve the same function in the tropical rain forest.
The water-catching leaf bases and spongy roots clinched the argument;
the rain forest epiphytes were living in a drought. And of course they
were; they had no contact with the ground and must conserve what water
they could of that which went flowing by. Schimper was able to go on to
draw some intriguing conclusions about the relationships of epiphytes
and desert plants, suggesting that suitability for life in one place should
make a plant suitable for life in the other and that there might be found
common evolutionary histories in some of the epiphyte and dry-site
groups.

Once the ability to tolerate is revealed, there comes the task of discovering how the trick is done. Desert animals have been studied inten-
sively from this point of view. Some save water by hiding in the ground all
day, only coming out in the cool of night, and some even sleep away long
periods of the worst hot weather in some shady hole, a physiological phenomenon called **aestivation** and an analog of the **hibernation** by which
some animals survive the northern winters. And all desert animals have
efficient kidneys that are good at concentrating urine and so reducing
water loss during necessary excretion.

One of the most beautiful of studies on desert animals is probably that
of the Schmidt-Nielsens, Bodil and Knut, on the question: How do
camels manage to cross the desert? (Schmidt-Nielsen, 1964). From the
earliest times, it has apparently been the general belief that camels stored

Similarity of epiphytes to desert plants. Both groups are adapted to scant water supplies. Notice the similarity of the epiphytic bromeliad (A) to the Yucca plant (B) (page 282) of the
deserts of the S.E. United States. Spiny bromeliads may live on trunks of trees in the rain-
soused tropical rain forest but they have no roots to the ground and so are on short water
commons all the same. This one has spines, which presumably diminish the attentions of
herbivores, thick cuticle, small leaf area, and cupped leaf bases which, together with the
spongy root mass, trap water. The plant in C (page 284) is another bromeliad, this one with
particularly well developed cupped leaf bases, and with defenses against herbivores again
reminiscent of desert plants.

Figure 20.4

B

Yucca in the desert

C

Bromeliad epiphyte in rain forest

water, presumably in their hump. But do they? The only certain ecological fact was that they could travel long distances without it; that they
could tolerate drought. Knut Schmidt-Nielsen was a student of August
Krögh (he of the Pütter-Krögh argument); Bodil was Krögh's daughter; the
problem of the camels was quite interesting enough to have attracted
Krögh himself. The Schmidt-Nielsens journeyed to Africa, bought some
camels of their own, and looked at the bodies of those butchered for meat
by the Arabs. The humps of camels, as had been long known, contained
fat, not water, but it had often been said that the fat itself was a water reservoir because oxidation of fat produces water as a by-product, and there
is certainly no shortage of oxygen in the Sahara. A 40-kilogram hump
should yield 40 litres of water (2 jerry cans full); surely a nice reserve with

which to cross the desert. But taking in the oxygen necessary for fat hydrolysis meant ventilating the animal's lungs, and gas exchange across the lung surface must involve much evaporation. The Schmidt-Nielsens estimated this water loss, finding that more than 40 litres of water would evaporate while enough oxygen to hydrolize 40 kilograms of fat was acquired. So a camel could not get useful water from the fat in its hump. Neither did it store water as water anywhere else, which the Schmidt-Nielsens ascertained by dissecting the corpses of thirteen butchered animals. But their camels showed how the trick was done. They weighed a camel, put it in a pen under the Sahara sun, gave it just dates and hay to eat but no water, and sat back to watch. Day after day it seemed content, losing weight but not overheating, and looking a perfectly healthy camel. After 17 days of it the Schmidt-Nielsens' natural worry for their animal caused them alarm or, as Knut puts it in his book, "We felt rather uncertain about where we stood." So they decided to give the camel a drink. It drank 40 litres (9 gallons) of water in 10 minutes, after which it weighed about what it had at the beginning of the 17 days. The camel had survived under the sun all that time not by storing water, but simply by going without. This was confirmed from blood samples of thirsty camels, which showed their blood to have unusually high electrolyte contents. The camels' adaptation to desert life was a tolerance of physiological drought much higher than that of other animals, coupled with a good thick insulating fur to keep the sun's heat out and an ability to drink quickly when opportunity occurred. Neither camels nor any other animals apparently store water to cross deserts.

Physiologic studies like those on desert animals are necessary to answer almost all the "how" questions of ecology. The approach is to accept a species as a fact, a collection of individuals having a common way of life, and then to discover the mechanisms which let them live as and where they do. Such studies are often exciting and emotionally satisfying to biologists; the camels that cross deserts; the whales that plunge deep into the sea without getting the bends; the tiny hummingbirds who can muster in their scraps of bodies the energy for transoceanic flight; the bears who sleep their prodigious sleeps hidden in a winter's cave. Studies of such performances clearly concern the habits and habitats of animals, and hence are truly part of ecology. They are often considered as a seperate subdiscipline called **physiological ecology**. But it is important to remember that they do not help to answer the grand ecological questions. The Schmidt-Nielsens showed splendidly *how* camels crossed deserts, trenchantly slaying the sacred cows of folklore as they did so. But they had not shown *why* camels crossed deserts. The feats of engineering by which camels overcome the hazards of crossing the desert are of less interest to ecology than noting that camels exist because there is some desert resource which they are able to exploit.

Animals do work, which degrades energy to heat, which warms them up. Plants live under the sun, absorbing its rays, which warms them up. Neither warming process can go on indefinitely or animals and plants must

LIMITS AND THE BALANCED HEAT BUDGET

become too hot to live. They must balance their heat intake with heat losses, must balance their **heat budgets.** Whether they can maintain a balanced budget depends on the energy flowing in their environment and the environmental conditions which determine their rates of heat loss. An ultimate limiting factor for any plant or animal is thus some function of the energy resources of a place and the energy losses occasioned by living there. This idea is familiar to zoologists, who have always intuitively understood the importance of cold weather, wind-chill factors, and regulation of body temperature. But the idea is also one that allows formal analysis.

An animal standing under the sun (Figure 20.5) receives energy from many sources; directly from the sun, considered as a point source; by the radiation of skylight, considered as a hemispherical bowl above the animal; by reflection from its surroundings; and by infrared radiations from the ground, surrounding objects, and the clouds. If the air is hotter than the animal, it may also receive heat by convection, though convection is usually a source of heat loss in real environments. And an animal receives much energy from its food, which we call metabolic energy. On the other side of the heat budget, the animal loses energy by black body radiation, by convection, and by evaporation of water at the respiratory or skin surfaces. We may draw up a heat budget for an animal as follows:

$$
\begin{pmatrix} \text{Metabolic} \\ \text{energy} \end{pmatrix} + \begin{pmatrix} \text{energy absorbed} \\ \text{from direct} \\ \text{sunlight} \end{pmatrix} + \begin{pmatrix} \text{energy absorbed} \\ \text{from} \\ \text{sky light} \end{pmatrix}
$$

$$
+ \begin{pmatrix} \text{energy absorbed} \\ \text{from radiations} \\ \text{coming from ground} \end{pmatrix} + \begin{pmatrix} \text{energy absorbed} \\ \text{from radiations} \\ \text{coming from clouds} \end{pmatrix}
$$

$$
= \begin{pmatrix} \text{energy lost} \\ \text{by black body} \\ \text{radiation} \end{pmatrix} + \begin{pmatrix} \text{energy lost} \\ \text{by} \\ \text{convection} \end{pmatrix} + \begin{pmatrix} \text{energy lost} \\ \text{by} \\ \text{evaporation} \end{pmatrix}
$$

The metabolic energy available to an animal is essentially measured by the resource which the animal can utilize. As such, it is a property of the environment. The various radiant heat sources are also properties of the environment, so it is apparent that, apart from being skillful at harvesting its resource and choosing to lie in sun or shade, the animal can do nothing to control its energy input. The energy losses, however, may be partially within the control of the animal. The loss by black body radiation is proportional to the fourth power of the absolute temperature; so the animal can modify this loss by insulating itself with fat or fur, a stratagem which keeps the radiant surface at a lower temperature than the body core. The energy lost by convection also depends on the temperature of the surface, and may likewise be limited by insulation. The loss by evaporation cannot be reduced below a threshold set by the need to ventilate

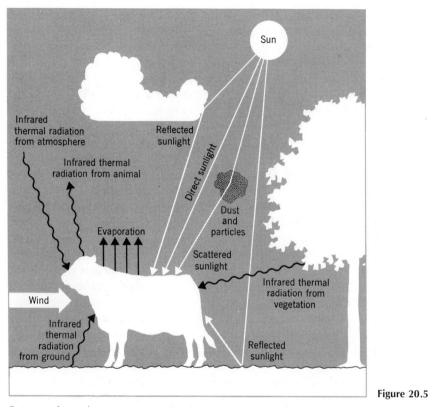

Figure 20.5

Energy exchange between an animal and the environment. (After Gates, 1968a.)

the lungs, but it can be increased by panting, by sweating, or by rolling in wet mud as pigs do.

The energy lost by evaporation can be very large, so that an animal provided with unlimited water can virtually always balance its heat budget in a warm climate, thus avoiding the risk of getting hotter and hotter until it dies. Camels, as the Schmidt-Nielsens have shown, use up their water much more completely than do other animals in their efforts to balance their heat budgets. Camels also use insulation to avoid absorbing energy by conduction both from the high surface temperature produced by radiation from desert sand and desert sky, and from convection in hot desert air. But when the camel's stretched water supply does give out, its temperature must rise and the beast must die.

There are more cold-to-life places on earth than there are hot-to-life places. In these, an animal's problem is unbalanced energy losses that accumulate until it dies; or, more exactly from the ecological point of view, until it no longer can do the work of reproduction. An animal with fur or feathers can control its energy losses much better than one without such insulation. Being fatty helps, so does being large, since this reduces your radiating surface in relation to the volume of your body core. Having behavior well enough developed to get you out of the wind is also useful. We know that arctic faunas include many birds and mammals, which have insulation, and not lizards, which do not. Then it has long been known that animals generally get bigger as you go toward the poles, the so-called **Bergman's rule.** And animals like lemmings, which must over-winter in the Arctic, live under the snow instead of on top of it. These adaptations of arctic animals can be expected from simple considerations of heat budgets.

But it is important to appreciate the real ecological significance of these adaptations to cold-living. The subject is commonly discussed in biology in relation to warm-bloodedness and cold-bloodedness; to **homoithermy** and **poikilothermy** (two terms Greek enough to convey a proper sense of wisdom). But whether an animal's tactic in life is to hold its body temperature steady or to change it as the occasion demands is largely irrelevant to its ability to survive in any particular place. If the niche of large top carnivore of tropical rivers could best be filled by an animal that controlled its temperature within narrow limits, I suspect that crocodiles would long since have been replaced by mammalian fishers. Poikilothermous lizards are not excluded from the Arctic because they are poikilothermous, but because, on the average, they cannot balance their heat budgets if the ambient temperature is low. Mammals and birds survive in the Arctic because, on the average, they can.

Thinking of an animal as a balancer of a heat budget lets us extend the concept of limiting factors to climate, not just in the general ways in which climate has long been used by biogeographers to explain distributions, but in concrete and potentially measurable ways. Any animal is limited to the places in which it can balance its heat budget. Its loss of heat will depend on various measures of radiation, on wind speeds, and on relative humidity; all directly measurable parameters of the environment. But to apply these data to our animals we must measure to what degree these parameters represent unmanagable drains on the budget. Measuring the heat budgets of real animals is not so easy, since it must involve things such as measuring the rate of metabolism, radiations from the several parts of the body, convective losses, and the like. It is not impossible to do this sort of thing, but it can be immensely difficult. Fortunately, as David Gates, of the University of Michigan Biological Station, has shown, the necessary data can be drawn from temperature measurements of various parts of the animal alone.

Gates (1968 a and b) starts his analysis with the simplifying assumption that an animal is cylindrical, and with negligible appendages; not really too desperate an oversimplification if you think of the shape of a deer. The cylindrical animal then consists of a body core, an insulating sleeve of fat bounded by skin, and there may be a sleeve of fur or feathers separating this from the radiating surface of the animal as shown in Figure 20.6.

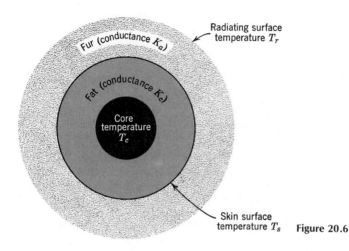

Radiating surface temperature T_r

Fur (conductance K_a)

Fat (conductance K_c)

Core temperature T_c

Skin surface temperature T_s **Figure 20.6**

where:

M = metabolic energy as cal. cm.$^{-2}$ min.$^{-1}$ generated at skin surface

E_1 = energy lost by sweating cal. cm.$^{-2}$ min.$^{-1}$

E_2 = energy lost by evaporation at respiratory surface cal. cm.$^{-2}$ min.$^{-1}$

K_a = conductance of insulation (fur or feathers)

K_c = conductance of body fat

T_c = temperature of body core

T_s = temperature of body surface

Q = radiation absorbed by body surface

hc is the convection coefficient

T_a = air temperature

ϵ = emmissivity of the skin surface

δ is Stefan- Boltzmann constant for radiation

T_r = temperature of effective radiating surface at outside of insulator.

Then the energy conducted from the body core to the skin surface is

$$\begin{pmatrix} \text{Metabolic energy} \\ \text{generated at} \\ \text{skin surface} \end{pmatrix} = \begin{pmatrix} \text{energy passed by} \\ \text{simple conduction} \end{pmatrix} + \begin{pmatrix} \text{energy lost at} \\ \text{respiratory surface} \end{pmatrix}$$

$$M \qquad = \qquad K_c(T_c - T_s) \quad + \qquad E_2 \qquad (1)$$

if the animal has no fur or feathers the energy budget of the external surface is

$$\binom{\text{Energy passed}}{\text{by simple conduction}} + \binom{\text{radiations absorbed from environment}}$$

$$K_c(T_c - T_s) \quad + \quad Q$$

$$= \binom{\text{radiant energy lost as black body}} + \binom{\text{convective heat loss}} + \binom{\text{energy lost from skin evaporation}}$$

$$= \quad \epsilon\delta T_s{}^4 \quad + \quad h_c(T_s - T_a) \quad + \quad E_1 \tag{2}$$

eliminating $K_c(T_c - T_s)$ from eq. 2 we get

$$M + Q = \epsilon\delta T_s{}^4 + hc(T_s - T_a) + E_1 + E_2 \tag{3}$$

if the animal has fur or feathers the energy budget at the radiating surface is given by

$$\binom{\text{Energy reaching skin surface by conduction}} = \binom{\text{energy lost from skin evaporation}} + \binom{\text{energy conducted across insulating layer}}$$

$$K_c(T_c - T_s) \quad = \quad E_1 \quad + \quad K_a(T_s - T_r) \tag{4}$$

and the energy budget for the radiating surface is given by

$$\binom{\text{Energy conducted across insulating layer}} + \binom{\text{radiations absorbed from environment}}$$

$$K_a(T_s - T_r) \quad + \quad Q$$

$$= \binom{\text{radiant energy lost as black body}} + \binom{\text{convective heat loss}}$$

$$= \quad \epsilon\delta T_r{}^4 \quad + \quad h_c(T_r - T_a) \tag{5}$$

This (eq. 5) describes how the environmental energy flow may affect the heat budget of the animal, but it is not easy to put numbers against various terms in the equation. Consider the difficulty of measuring K_a or ϵ for a real animal, for instance. But we may avoid this difficulty as follows:
By eliminating $K_c(T_c - T_s)$ from eqs. 1 and 4 we get

$$M - E_2 - E_1 = K_a(T_s - T_r) \tag{6}$$

solving for T_s, and substituting back into eq. 1 we get

$$T_c - T_r = \frac{M - E_2 - E_1}{K_a} + \frac{M - E_2}{K_c} \tag{7}$$

now

$$\frac{M - E_2}{K_c}$$

is the temperature difference between the core and the skin surface and

$$\frac{M - E_2 - E_1}{K_a}$$

is the temperature difference between the skin and the radiant surface so that

$$\frac{M - E_2}{K_c} + \frac{M - E_2 - E_1}{K_a}$$

is the temperature difference between core and radiant surface, as eq. 7 states.

Measuring the temperature at the radiant surface of the animal and its core temperature under various environmental regimens thus enables you to measure its ability to balance its heat budget under those regimens. Gates and his students (op. cit. and 1962) have collected these data for a few animals of different taxa, and can show the environmental extremes tolerated by them. Not surprisingly, they conclude that birds and beef cattle can stand greater variations in the energy flowing through their environment that can locusts or lizards.

The heat budget of a plant is simpler to understand because, for much of its life, at least, the metabolic heat derives directly from the solar radiation absorbed, and is in fact only a small portion of the incident solar radiation (Chapter 10). A plant receives all its energy, except when metabolizing the reserves of previous years (as when a potato sprouts from a tuber), from radiations in the contemporary environment, or by convection (Figure 20.7). To avoid being cooked, it must balance this incoming radiation by losing an equal amount. The energy flux which it arrests by storage in plant tissue is usually a very small amount of the total budget. The rest it must lose by black body radiation, by convection, and by evaporation promoted by transpiring through the stomates. Most plant leaves are such thin fragile structures that regulation of the budget through insulation is negligible. A plant cannot alter its size or shape from minute to minute as the sun plays hide-and-seek with the clouds, leaving only one possible expedient for balancing the short-term energy budget, the transpiration stream. This is under the plant's control by opening and closing the stomates, and temperature regulation achieved in this way may be remarkable. Gates found that the monkey flowers (*Mimulus lewisii*) growing at 10,600 feet in the Nevada Range on one day of still air at 19°C had leaf temperatures of between 25 and 28°C. He drove hurriedly down the mountain to 1300 feet where the air temperature was 37°C and found that the leaf temperature of local monkey flowers (*Mimulus cardinalis*) was between 30 and 35°C, or very close to the temperature of those in the cold air near the mountain top [Gates (1965); Figure 20.8].

The temperature-regulating feats of the monkey flowers may be likened to the dessication tolerance of camels, they are adaptations which let the plants live in places which they inhabit for some good reason. An ecologist is again more interested in the reason for living there, than in the physiological adaptation that makes it possible. Monkey flowers must live where they do because they are able to divert a resource to their own ends, they must be able to win space against the pressures of other plants or by avoiding the pressure of animals.

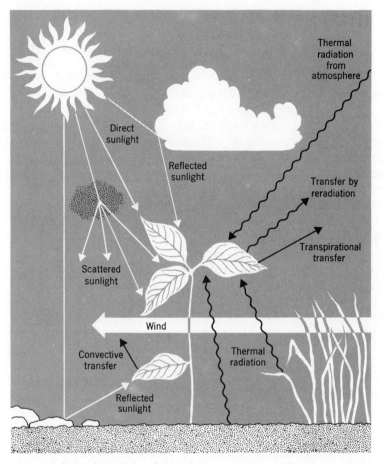

Figure 20.7

Energy exchange between a plant and its environment. (After Gates 1968b.)

The aspects of a plant's heat budget which are more interesting to ecology are those which are more or less fixed for the species, those concerning the shape and positions of leaves. A flat leaf, for instance, presents a large absorptive surface to the sun, which might be expected to overload the input side of the heat budget in some circumstances. This may be true in deserts (think of a cactus). But a broad flat leaf must also be a potent radiator to the black body of the night sky, and we find that the trees of the northern boreal forest commonly have needle-shaped leaves which have the minimum surface directed to that cold black body of space. The shape of a leaf also influences the rate of heat exchange by convection, something which Gates has investigated by making silver casts of leaves and studying airflows round them in a wind tunnel (1965). The rate per unit area of energy exchanged by a pine tree needle was many times greater than that of a poplar leaf.

The shape of its leaves thus sets limits to the environmental regi-

Figure 20.8

Monkey Flower, Mimulus sp. Plants can regulate their temperatures to some extent by opening and closing the stomates. Monkey flowers at over 10,000 feet in the Nevada Range were found to be at nearly the same temperature as those 9000 feet lower down.

mens in which plants can live, and the limits are expressed in the ability of a plant to balance its heat budget. For plants no less than animals, therefore, the various parameters of the environment which influence heat budgets; wind speed, day length, relative humidity, and the like; can be thought of as setting limits which may theoretically be measured for any kind of plant in an objective and physical way. But, when we have measured these limits, we are still faced with the prime ecological question; Why was it worthwhile for the plant to evolve adaptations to this set of limits? The true ecological answer must again be given in terms of resources which could be acquired by that particular set of tactics. Gates ends one of his papers (1965) by saying:

"A question that remains unanswered is why it is that different plants show opimum efficiency at different temperatures. The eventual answer to this question will probably involve the catalytic activity of plant enzymes; when it is found, it should open the door to the understanding of a fundamental mechanism in the ecological system of our planet."

That is a physiologist speaking. An ecologist should say that the eventual answer will show that plants evolved enzymes of different catalytic activities so that they could enjoy the use of adequate and private resources.

But study of the energy costs of having leaves of various shapes allows an elegant answer to be made to some of the outstanding problems of plant geography (Chapter 2). The most striking problems of plant geography were ones of form, that the shapes of plants changed with latitude, apparently with climate. Gates, and others thinking as he does, have shown how the limits of climate do express themselves in form, and through the most fundamental agency of energy budgets. Consider specifically the question: why are there no trees in the Arctic? It was never enough to say, "Oh, it is just too harsh for trees there," because succulent meadows of fragile flowers can be found in the Arctic. Nor could you say it was too windy, because there are places in that treeless expanse where it is no windier than other areas further south. Nor could you claim extremely cold winters, for there are treeless arctic places like the Pribilof Islands with relatively mild winters, and there are places with forests but winters of thirty below. Permanently frozen ground does not inhibit trees completely either, since there are forests growing over permafrost in the valley of the Yukon, although only in places where the ground thaws deeply. You can go some way toward an answer by pointing out (Wilson, 1968) that maintaining large woody structures requires work, and that this may be prohibitively expensive of the resources of a plant living in a poorly productive place. But a more thorough understanding is given when you consider that the air more than a foot or two above the arctic ground is both cold and moving, threatening to unbalance the heat budget of any plant that tries to raise itself into it. Close to the ground the air moves slowly and is warmed by red radiations from the soil and the plants themselves. It is mild if you do not leave the soil by more than a few inches. You can balance your heat budget down there, carrying on photosynthesis and transpiration without being fatally cooled. But you cannot balance your budget if you stand tall. That is why there are no trees in the Arctic (also see Figure 20.9).

LIMITING FACTORS
AND THE ECOSYSTEM:
CONCLUSION AND
CRITIQUE

When a systems analyst views a collection of living things, dwelling as they usually seem to be in some sort of harmony, he sees them as functional units going about their appointed tasks. They have their places in the physical world, and they also have their places in that other dynamic world of life. Their places in the physical world seem ordained by sets of

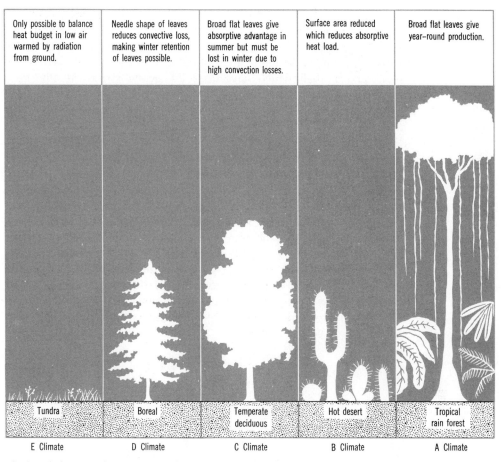

Only possible to balance heat budget in low air warmed by radiation from ground.	Needle shape of leaves reduces convective loss, making winter retention of leaves possible.	Broad flat leaves give absorptive advantage in summer but must be lost in winter due to high convection losses.	Surface area reduced which reduces absorptive heat load.	Broad flat leaves give year–round production.
Tundra	Boreal	Temperate deciduous	Hot desert	Tropical rain forest
E Climate	D Climate	C Climate	B Climate	A Climate

The heat-balance explanation for the life-forms of major vegetation types.

Figure 20.9

tolerances, and these can be understood, investigated, and described. Their responses to the presence of other living things can also be investigated in an analogous manner, an approach reflected in the term "biotic factors" which appeared so prominently in many textbooks of the past. The effects of biota could be measured just as could the nutrient supply or the wind speed. Programmed entities, the species, were then ready made and waiting, first for the concept of the ecosystem, and then for the analysts with the big computers who should mimic them for the practical benefit of mankind. As a pragmatic approach for understanding *how* nature works, this examining living things as sets of factors acting on each other to form a system has been immensely fruitful. Indeed, much of the fruit remains to be plucked. But the approach diverges from the greater intellectual task of explaining *why* species with these sets of tolerances should have evolved. Why do all these species exist, living together as parts of ecosystems?

Thinking in terms of physical limits, even in the refined energetic manner of Gates, does not aid us in the answering of this question. Nor does the concept of the ecosystem itself help us. Ecosystems do not evolve. They must be thought of simply as the results of the actions of plants and animals which have done the evolving, individually, by a selection that always favored those who obtained most resources.

Ecosystems are in the eye of the beholder, concepts, products of a way of looking at nature. When we make practical ecosystem studies we define our study plot by area, in fact, as a habitat. And in this habitat there may be many different kinds of plants and animals. The physical limits of the habitat do not explain the numbers of kinds that are there. To look for an answer in "biotic factors" merely begs the question. When embarking on the great task of explaining the numbers and kinds of plants and animals, the habitat is best thought of not as a set of physical limits, but as a stage on which plants and animals interact to produce the observed distributions. The numbers and kinds we see are the result of the actions which have taken place on that stage. G. E. Hutchinson (1965) has neatly phrased this concept with a title for one of his essays, "The Ecological Theatre and the Evolutionary Play." An ecologist must examine living things working on one another within the confines of the world habitat to explain the diversity of nature.

PART 3

THE NATURAL
CHECKS ON
NUMBERS

The idea of competition is central to the theory of evolution. We think of a struggle for existence, of survival for the fittest, of natural selection, all of which convey the idea of animals and plants competing with one another for the limited raw materials of life. An ecologist wants to explain why species exist, and to show how the acting of living things on the stage of the world habitat leads to so many of them. It seems that an ecologist must study competition, but there are dangers. What a field biologist sees in times of change are replacements. A new arrival replaces a familiar animal, surely what we should see if one animal was successfully outcompeting the other? But we must ask ourselves if such replacements, so familiar to contemporary biologists, are truly the result of competition. We now study a world undergoing massive changes in vegetation and landscape due to human activities, and these changes should lead to replacements of animals and plants in the normal course of succession. Even such apparently simple histories as the replacement of a native British squirrel by an introduced American species cannot definitely be shown to have resulted from competition. Experiences of this kind show that the concept of competition is of no practical use for field biologists and resource managers, that, indeed, they are likely to be misled by it. But for theoretical ecology the concept is vital. On the assumption that competition occurs in nature, both between species and between individuals of one species, it has been possible to make simple models of the natural world which have led to our most perceptive understandings of how species are separated, and kept distinct; and of how the populations of many animals and plants are restricted in nature. Part 3 of this book follows the fruitful intellectual journey that was started for ecologists when Darwin put forward his theory of evolution by natural selection, with its inherent idea of competition for the necessities for life.

CHAPTER 21

COMPETITION AND
THE REAL WORLD

In the Galapagos Archipelago giant tortoises of different islands are distinct enough to be told apart by their shells. Darwin noted this during his celebrated visit and allowed it to direct his thought toward the theory of evolution. One of the oddest and most strikingly distinct of the tortoises was that of the small remote island of Abingdon. In 1962 a party was sent out from the Charles Darwin Station to report on the Abingdon tortoise herd, so important to the history of biological thought, but two weeks of searching the small island convinced the field party that the tortoises were extinct. But they had not been extinct long because trapped in fissures of the highest parts of the island were remains of tortoises that could not have been dead more than a year or two. And the reason for this calamity was also evident; all the browse that lumbering tortoises could reach had been utterly consumed by goats. The last food had been in the highest parts of the interior of the island, and here the tortoises had journeyed only to die of starvation and mishap in the fissures. The agony to the human heritage of this loss of the harmless Abingdon tortoises can be gauged from the fact that it is known that goats were first introduced to the island by a party of fishermen in 1957. But the interest of the story to biology lies in the knowledge that there can be no doubt that the tortoises were eliminated by competition with goats for food.

"Competition" is a word with a clear meaning, valid and hallowed in English usage. There is competition whenever two or more individuals or groups "strive together" (the literal meaning of the Latin roots) for something in short supply. Men compete for prizes, and only one man, or one group of those competing, can win a prize. That animals and plants should also compete seems reasonable, and it also seems likely that the final thing for which they compete should be food energy. We know that the range or success of many species must be controlled by "biotic factors," by interaction with other animals and plants, and it seems logical to conclude that such interaction is competition for a resource, presumably energy or space in which to live. But it is regrettably easy to apply the concept of competition too loosely and to talk of animals competing when we have no real evidence that they do. Since the concept, properly applied, has led to some of the most exciting advances in ecological understanding, it is important to realize how difficult it sometimes is to establish that competition really occurs in nature. The goats and tortoises of Abingdon are one of the rare examples when there can be no doubt of the matter. The following history of British squirrels is an example of the type of thing we see far more often.

There is only one squirrel native to the British Isles, a small red squirrel *Sciurus vulgaris*, which until the end of the last century lived all over Britain, both in the oak and beechwoods of the south and the pinewoods of the north. But at about the turn of the century there were several introductions of the common American gray squirrel, *Sciurus carolinensis*. This animal becomes easily accustomed to people in gardens and parks, sitting up in the engaging squirrel manner to be fed. The native Britisher was more standoffish, and was accordingly not so amenable to park life, and the gray squirrel was brought over to supply friendliness to the estates of the 13 or so people who thought it would be nice to have squirrels on their lawns. *Sciurus carolinensis* thrived exceedingly in England, reproducing and spreading as if under some biblical injunction until within about 30 years it had occupied nearly every wood in the kingdom. The native red squirrel vanished from its old haunts, maintaining, like other British populations of the past, viable populations only in Scotland and Wales. To the ordinary countrymen there was no doubt that the invading gray squirrel had driven the now lamented red one out, that there had been competition by tooth and claw. The gray squirrel was declared a public enemy, a bounty was placed on its head (or rather tail), and persecuting musketry rang out the length of England's green and pleasant land, in season and out, year after year. But it made scarcely a bit of difference to the numbers of gray squirrels. The game animal of the American Midwest, whose numbers have to be maintained by stringently short open seasons, seemed to shrug off the vast toll of British resentment. It is there still, no doubt a permanent part of the British fauna.

It is seductively tempting to think of this tale of one squirrel replacing another in terms of competition, but we have no evidence that there was competition. There is, indeed, scarcely even any suggestion that the two populations interacted. All we know is that one replaced the other. It is possible that coincidence decreed that the red squirrels should be suffering a deadly epidemic (things do; even men) at the time when the grays were ready to enter the empty woods, or perhaps some human interference with the landscape made the island unsuitable for reds but tolerable for grays. Such things are possible, but there is no record of squirrel disease, and it is hard to see what sufficiently drastic changes had happened to timeless rural England in days before the great blight of the motor car. Perhaps it was direct competition for food after all, a sort of grand stealing of nuts, but if so the great robbery happened without our seeing it happen. There is not one shred of evidence that competition occurred between the native red squirrel and the invader. For 50 years the squirrel story has been known to every naturalist and biologist in England, and the doctoral dissertation that would clear up the mystery should at once have made the reputation of the person fortunate enough to submit it. And yet Elton could write in 1958, "The truth is that we simply do not know."

A reasonable man must find the squirrel case nonproven, and refrain from calling it "competition," for if he does he commits the sin of thinking the problem solved and ceases to think about it. Any apparent interaction of animals can tempt one into this error, and many have fallen. I recently attended a seminar given by the manager of a huge game refuge of a western American state, in which he described the main problem which had burdened many years: a species of ungulate which the state hunters liked to shoot was being replaced by one they did not like to shoot. The manager said that the wrong one was outcompeting the right one, and the word "competition" flew around the seminar room like a buzzing bee, interrupting the flow of his words and his thoughts. And yet it was clear from the evidence that great changes in vegetation, directly consequent on human interference, were happening, suggesting that one ungulate followed the other as the vegetation was changed by ecological succession. When the manager said "competition" he meant only "replacement," and until he got his language straight he was never likely to face the truth of his problem. There are many similar stories to relate, so many that Australian ecologists H. G. Andrewartha and L. C. Birch (1954), in a spirited introduction to their book, expressed the hope that one day the word would drop out of the ecological vocabulary. This is going too far; but it is vital that "competition" be used with care, only when it has certainly occurred, and in contexts of which Webster and the Oxford English Dictionary would approve.

Sometimes you can be suspicious that real competition has occurred without being able to demonstrate it; and then it is best to refrain from the word too. A fine story of this kind is told by Elton in his textbook. In the lower reaches of a tidal stream near Liverpool in England, there lives a species of copepod, *Eurytemora lacinulata;* and this animal is not found either at the mouth of the stream in really saltwater nor upstream in freshwater. Its range is the brackish tidal region only. A simple hypothesis to explain this distribution is that *Eurytemora* is adapted to a narrow range of salinities, being able to tolerate neither freshwater nor the sea. This is the simple limiting factor approach. For the sea end of the distribution, it is almost certainly correct because *Eurytemora* has never been found in the sea anywhere, but it will not work at the upstream end of the distribution because the animal is found in freshwater ponds and streams all over England. *Eurytemora* can live in freshwater, but it cannot live in that particular Liverpool stream except where the water is brackish; why? Elton found a hint of what might be happening from his studies on copepods in another part of England, those in ponds and filter beds of the Oxford municipal sewage works. *Eurytemora* abounded in the filter beds, and these were regularly drained and cleaned out. It was clear, therefore, that *Eurytemora* must be good at getting itself from pond to pond; it was an opportunist who could take advantage of any transient pond that might be available. In the more permanent ponds of the region, however, Elton never found *Eurytemora* but found another copepod, *Diaptomus gracilis.* *Diaptomus* was never in the ephemeral filter beds; *Eurytemora* was never in ponds of any age, although we may imagine that it was there once when they were very young. The observations can be explained if we suppose that whenever *Diaptomus* reaches a pond it takes over, and the preexisting population of *Eurytemora* dies out. *Eurytemora* survives by keeping one jump ahead of *Diaptomus*, when it can rely on the sole use of a pond until chance brings the first colonists of the *Diaptomus* population. This looks very like a story of true competition like that of the Abingdon goats and tortoises, but one in which the losing species can survive by moving away as the tortoises could not do. From these observations in Oxford, Elton was able to suggest an answer to the Liverpool distribution. *Eurytemora* used the brackish water as a refuge, just as it used the sewage filter beds for a refuge, and some other factor, very likely a rival animal, prevented it from living upstream. There were no *Diaptomus* in that Liverpool stream, and I do not think that Elton ever determined what was there to stop the *Eurytemora* spreading into the fresher water. But he thought it most likely that some biotic factor was at work, quite possibly direct competition with some animal whose identity we do not know. Without being sure it is best to watch our language and to say no more than that competition *might* have been involved.

In another class of organisms, the decomposers, we have good circumstantial evidence that competition is widespread. Antibiotics come from decomposers. By excreting these substances into their immediate neighborhood, fungi suppress bacteria; which is, of course, why we borrow the antibiotics from the fungi. Both fungi and bacteria around them derive their energy from the debris of plant and animal corpses, and they feed in similar ways by simply digesting the substrate on which they live. It is reasonable to expect that such decomposers should compete, because the food is both restricted and concentrated. The fungi are apparently waging chemical warfare on bacteria with their antibiotics so that they preserve a respectable proportion of the available energy for their own use. Excretions of substances like antibiotics are common in most classes of decomposers, and may be taken as evidence of competition of a rather dramatic kind.

Large flowering plants may wage chemical warfare too, as has long been suspected in agriculture (Börner, 1960). Barley is known as a "smother crop" because it keeps down weeds in the field, sparing the farmer the need to cultivate or use chemical weed killers. This it does with root secretions, weed killers of its own (Overland, 1966). Other plants well-known for inhibiting their neighbors are walnut trees, and it is common knowledge among growers of the American Midwest, where black walnuts (*Juglans nigra*) are common, that you cannot grow tomatoes near one. Even more striking are the rings of bare ground found round clumps of shrubs on the dry slopes of the California Coast Range (Figure 21.1). The rings are apparently kept bare by seed-inhibiting chemicals dripping down from the leaves of the shrubs and collecting in the soil (Muller et al., 1964). It now seems likely that chemical inhibitions like these are very widespread, being facts of plant life from the tundra to

Figure 21.1

Rings of bare ground caused by chemical inhibitors in the California coast ranges. Clumps of bushes in grasslands may be ringed with bare ground in which no plants grow, not even seedlings of the bushes (closeup of ring above, vertical view below). The region is one of intermittent rainfall. It has been shown that secretions from the leaves of the bushes collect in the soil and inhibited all plants from germinating there. Is this competition, or is it an accidental effect of a physiological mechanism which regulates the growth of the bushes themselves?

the tropical rain forest, a conclusion which seems a powerful reason for believing that plants commonly compete by chemical warfare. But there may be doubts that even chemical warfare in plants can always truly be called competition. Went (1970) has pointed out that the excretions of plants may actually inhibit their own seedlings more than those of other species, which seems an odd way of waging war. The bare rings round clumps of California shrubs are not kept free for their own use; they are denied to their own seedlings as well as to other plants. Went suggests that the true selective significance of the secretions is that they inhibit root growth of the parent plant in rare heavy rains, thus curbing exhuberant growth which may leave the plant overextended when the next drought arrives. The inhibition of other plants and their own seedlings may be no more than accidental effects. It remains true that some plant excretions may well be agents of competition, helping to win space and achieve dominance, but even chemical warfare must be examined very closely before it is dubbed "competition."

Plant succession is, perhaps, one of the most dramatic of phenomena which may be satisfactorily explained by competition, and one of the clearest examples we have of the working of the process. The well-watered parts of the earth are carpeted green, at least, in the spring, giving us clear assurance that most available space is occupied by plants and that any tiny patch left over will be the object of competition. Such a patch is provided by a farmer's abandoned field. Pioneer plants arrive. They are never very big, but they do not have to be because they go where there is lots of room for a little plant to thrive unmolested. The quality they do have is the power of rapid dispersal; they are opportunist species like the copepod *Eurytemora*. For a few months the pioneer plants bask in the glow of uncompeted-for energy. Perhaps they do not do too well for dissolved nutrients, but they are small and do not need to produce much nutrient-consuming biomass to go on basking in the sun. But then winter comes, the annual plants die, and the perennials withdraw to their buried rootstocks and rhizomes. Next summer the perennials cover the field because they can make rapid growth in the spring from their underground stores of energy and so occupy space before annual plants can find it. Then shrubs spread their branches over the perennial herbs, denying them light, competing successfully with them for energy. Trees shade out the shrubs, and taller trees the first trees, until finally only the dominant species of the climax are left. There can be no doubt that this is a story of true competition.

Competition in succession is competition between species, or **interspecific** competition. Competition between plants of the same species, or **intraspecific** competition, is also common, and is familiar to any gardener or farmer. Young plants growing thickly together must be thinned lest they "choke" each other. If you keep a crop thinned, however, the plants, relieved of competition, continue to grow well. The reality of intraspecific competition in plants is thus self-evident.

Intraspecific competition in animals is harder to demonstrate because animals are not easy to observe over long periods of time, and the immediate effects of competition must often be avoided or masked by their moving about. Also their resources of food energy are often various so that members of a crowded population may sometimes seem to avoid competition by changing food. But there are many known examples where animals of simple requirements have been crowded enough to suffer from the resulting competition. Too many fruit fly larvae reared in small bottles of banana mash grow slowly and die like crowded plant seedlings, and meat fly maggots in a carcass may be so numerous that the same thing happens (Nicholson, 1954). Perhaps the finest example from the wild is that of the barnacles studied on a Scottish seashore by Joseph Connell (1961). Connell watched a bare rock in the intertidal zone as barnacles colonized it. He used a glass plate to make a map of the population, holding it in a fixed position over the colony and making a mark to record the presence of every fresh arrival. As the population became crowded, some barnacles were crushed, prized off, or buried by their more vigorously growing relatives, until the rock was finally completely covered with only the proper number of full-grown barnacles. This history was much like that of a growing forest in which many trees die and only a few are finally left to expand over all the available space.

Connell (1961) also found that different species of barnacles competed by this same crude process of crushing or prizing one another off the rock. On his section of Scottish shoreline there were two species of barnacle, one, *Chthamalus stellatus*, living high on the rocks, and a second, *Balanus balanoides*, living lower down in a zone which was underwater for longer during each high tide. At dead lower water the two colonies appear as bands along the shore, one above the other. Connell saw that *Chthamalus* larvae often settled and started to develop on rocks down in the *Balanus* zone but that they were always squeezed out by aggressively growing individuals of the *Balanus* colony. The distribution of the *Chthamalus* population within the intertidal zone was thus apparently restricted at the top by a simple limiting factor of the environment; failure of the tide to push water any higher for long enough periods; and at the bottom, by competition with *Balanus*.

These limits to the *Chthamalus* colony may be compared with those proposed by Elton for his tidal stream copepod, *Eurytemora*, but Connell was able to observe directly the competition that kept his animal from the most favorable places, whereas Elton could do no more than surmise that the presence of another animal in the fresher water limited his.

For plants and fixed animals like barnacles, competition for space must be the normal condition of existence. But things that move can take advantage of the fact that further away there may be unoccupied space where an untapped source of energy may be won. The environment is an ever-changing thing, ebbing and flowing with the seasons and expanding and contracting with less predictable events like fires, plagues, and shifting climates. Animals, and many plants too, may well be so occupied in keeping pace with the vicissitudes of life in a home which is always changing that they rarely become crowded enough to compete with their fellows. There can be no doubt that all animals and plants may sometimes compete among themselves or with other kinds, but it can also be argued that those that are most agile; those that are prompt at finding new places on the changing earth; live much of their lives free from competition with their own kind.

In our own day, we see a world subjected not only to the normal vicissitudes of weather, natural castastrophes, and plagues, but also to the mighty changes that our own human doings bring about. In such times we should expect large fluctuations in the numbers of animals and plants.

The practical problems of range management commonly must be caused by these environmental changes, particularly as ranges slowly adjust to the first massive effects of man. Histories of one kind of animal replacing another within human memory or historical time must almost certainly be functions of these environmental changes. In the study of these changes the concept of "competition" has almost no place. "Competition" is thus a word that needs to be used very sparingly by game managers and naturalists seeking to explain the happenings of their own lifetimes.

But as a theoretical concept for studying the great biological questions of speciation and population control, competition has proved very useful. The central theory of biology, that of evolution by natural selection, portrays the origin of species as resulting from a "struggle for existence," which is to say from competition. Much of the insight of modern ecology has come from attempts to examine ways in which species might compete, and we have gone at least some way to explaining why there are so many species on the assumption that they occupy niches whose bounds are set by competition. Theoretical models of how species may compete with each other have, in turn, been based on models of how crowded animals of the same kind may so compete among themselves that their population growths are curbed and their crowds controlled. Much valuable understanding of the regulation of populations in nature has come from such studies.

The paths to these understandings have led from observations, to theoretical models, to tests with laboratory populations, and thence back to observations of animals and plants in the wild. The extent to which the distribution and abundance of animals and plants really can be explained by the body of theory built on the central concept of "competition" is still in some doubt, being indeed the subject of hot debate among ecologists. But that a very great deal has been learned from the efforts of those who have studied competition is not in dispute. A proper understanding of modern ecology must therefore involve consideration of the logistic model of population growth, equations of competition between species, the many experimental tests of these models on laboratory animals, and the long quest for parallels of the experimental results in wild populations.

But he who would truly understand the real world must also be ready to assess the pressures on populations maintained by inanimate nature, by the endless vagaries of weather. The relentless pressures of competition are accompanied by the no less powerful imminence of unexpected disaster. An essay into the natural checks on populations must delve into both the workings of competition and into the endless shifts of the physical conditions of life.

To a certain extent the numbers of animals and plants around us do not seem to change much, an observation that we refer to as the balance of nature. One of the jobs for ecology, therefore, is to explain how this natural balance is maintained, and we seem to be given a clue in the histories of populations of simple animals kept in the laboratory. If you place a few healthy individuals of both sexes in suitably comfortable laboratory quarters they will reproduce, and their numbers will increase until a time comes when they are very numerous but when their population seems to increase no more. Yet the population maintains itself, and we may say that it is in a state of balance. Such balanced populations in confined quarters can be arranged for many small organisms, including yeasts, bacteria, protozoa, and even insects. Ecologists have tried to understand how such balances in experimental populations can be maintained, in the hope that their studies should also reveal how that wider balance in nature is worked. Ascribing the balance to some property of the environment, which resisted the growth of the population, failed to yield useful ideas, but the suggestion that population growth was checked by the results of competition between individuals for limited resources was more fruitful. This latter hypothesis could be restated in mathematical terms by a simple descriptive equation, the logistic equation. Formal statement of the hypothesis of control through competition led many ecologists to devise experiments which should reveal ways in which experimental animals actually did compete. That competition for food did reduce the rate of reproduction has been convincingly shown in experiments with fruit flies, but the same experiments also suggested that more subtle forms of competition were at work. Many such subtle   309 mechanisms have since been demonstrated in experiments with flour beetles, but all the mechanisms revealed by such work can be thought of as arising from the conditions in which the animals have been made to live by the experimenter. That animals forced to crowd in the laboratory do, in fact, compete for food, or are otherwise checked by density-dependent pressures on their ways of life, is not evidence that wild animals are likewise checked. A real test of the logistic hypothesis had to be found either by direct observation of wild populations, or by manipulating the logistic equation to predict the outcome of contrived experiments in competition which might then lead to some useful generalization.

CHAPTER 22

AN INTRODUCTION TO CROWDING: THE LOGISTIC MODEL OF POPULATION INCREASE

Naturalists have long held the belief that the numbers of animals about them do not change very much from year to year. There are, it is true, good years and bad years, years in which a rarity is seen excitingly often or when familiar things are scarce, but there is still a comforting normality about the fauna of any place. It is predictable. We know what is common and what is rare, and we may tell our children about these things, confident that what we say will still be generally true when it is time for them to tell their children. A world in which this was not so, a world in which the numbers of animals were always violently changing, would be an uncomfortable place in which to live, one full of unpleasant surprises. The very idea savors of nightmare. And yet we know that all animals can reproduce much more quickly than is needed for their own replacement; that birds usually rear several fledglings a year, that mice produce large families every few weeks, that insects lay eggs by the thousands, and salmon by the million. Somehow the frightening implications of these large families must be offset to give us that comforting normalcy of animal populations which change little from year to year. We say that there is a balance in nature. How is this seeming balance attained?

A general idea of what may be involved in establishing a natural balance can come from observing what happens when a few pairs of simple animals are provided with a comfortable but limited home in a laboratory. Many such observations have been made, those of the Russian biologist G. F. Gause (you pronounce the final e making the word "Gauzer") with species of *Paramecium* being particularly well known (Gause, 1934). Paramecia are protozoans that feed on bacteria and other small particles. Gause found that he could make a standard brew of oatmeal which served as an excellent nutrient medium for paramecia, and one which he could add to their water in measured amounts. In a typical experiment, he placed 40 individuals of *Paramecium caudatum* in small tubes containing 10 cc of water and a few drops of his oatmeal medium. Each day thereafter he would take out a subsample with a pipette to count the paramecia present, which gave him an estimate of the total population. He would also spin the tubes in a centrifuge, which drove the animals to the bottom and allowed him to pour off the old water with its unused dissolved food, in order to replace it with fresh. There was thus a daily constant but limited input of food to the system. Under these conditions, the paramecia reproduced quickly at first, their numbers increasing exponentially until the water was cloudy with them. But at last, when there were about 4000 animals in each tube, the rate of growth leveled off, and the population did not seem to grow any more. Apparently the rate of replacement was finally about equal to the death rate, and the population was in balance.

This kind of population history can be duplicated with a variety of simple animals in simple systems; with yeasts, molds, bacteria, various protozoa, and even insects, such as fruit flies in banana mash and beetles in flour. The animals breed riotously when introduced into an empty habitat, but then seem to come under some sort of control, reaching a dynamic equilibrium or balance. From the first it seemed likely that the balance reached by the experimental populations was equivalent to the general balance apparently existing in nature, so that, if we could understand the one, we should also understand the other.

An apparently simple way of looking at population control and the establishment of balance was put forward in 1928 by R. N. Chapman of the Rothampstead Experiment Station in England. He thought it profitable to look at the sort of curbing of population growth which happened in Gause's *Paramecium* cultures in terms of the resistance of the environment to the population pressure. He was attracted by the analogy of electric current flowing in a wire, as described by Ohm's law, to that of a population of animals. Ohm had shown that the current flowing at any point in a wire was equal to the potential divided by the resistance of the wire (a constant which we now measure in ohms). The resistance was given by the following equation:

$$\text{Resistance (ohms)} = \frac{\text{potential (volts)}}{\text{current (amperes)}}$$

A population of animals might be said to have a potential in their innate ability to reproduce, which we call the **intrinsic rate of increase,** and the number of animals present when the population was in balance could be compared to the current flowing. If this number-of-animals current was set by an environmental resistance, then Chapman should be able to write:

$$\text{Environmental resistance} = \frac{\text{intrinsic rate of increase}}{\text{population at equilibrium}}$$

Since the intrinsic rate of increase and the population could be measured for real animals, it should be possible to solve for environmental resistance and to describe this as a number. There would then open the rather heady prospect of describing all the environments of animals in terms of simple numbers, the possibilities of which become clear when you think how it might affect everyday problems like that of a game manager who wishes to know the carrying capacity of a certain type of range for a certain type of game; the manager should only need to look up an index of environmental resistance in a book of tables. Unfortunately, as was soon realized, Chapman's analogy was unsound. An electric current flows through a wire, but there is no wire down which a population flows. To liken an environment, the total experience of an animal in which it finds its niche, to a wire is absurd. There is nothing fixed or permanent about an environment, and such key parts of it as the food supply are used up as the population "flows." Stating a numerical environmental resistance was in practice merely a cumbersome way of stating census data, like answering a question about the number of sheep in a flock by reporting the number of sheep's legs and asking your questioner to divide by four.

Chapman's ideas have left behind the term **environmental resistance** in the ecological literature which, for all its doubtful meaning, has so catchy a sound that it is still sometimes used. But a much more useful conception has been that populations are brought into balance as a result of competition. If animals breed riotously when there is lots of room and lots of food, but cease to do so when crowded, it is obviously a reasonable hypothesis to suggest that they do so as a result of increasing intraspecific competition for the resources of the habitat, or that they become suppressed by some effect of the environment which hinders more strongly as the population grows denser. This hypothesis enables us to make predictive models of population growth and control, which we can test by experiment and observation.

The growth of a simple population in a confined space with a limited input of energy is simply described by a graph that always looks sigmoid (Greek for "S"-shaped) (see Figure 22.1). The early rapid growth is easy to understand. The animals are blessed with plenty of space, and plenty of food; they grow and breed as fast as they can, attaining something close to their theoretical maximum, that same **intrinsic natural rate of increase** which Chapman used. As time passes, more and more animals are reproducing at this maximum rate, producing an exponential increase in the population size. We may write of this part of the curve:

$$R = rN$$

where

R = the observed rate of increase at any one time
r = the intrinsic rate of increase
N = the number of animals

r is, of course, constant, but N is rapidly increasing, so R increases also.

But in time the population slowly ceases to increase until an upper limit to the number of animals is reached. We may call this upper limit the saturation value, K. K is a constant, a property of the container in which

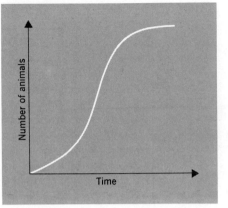

Figure 22.1

the animals are kept and of the food introduced as an energy source. It will differ for every system and can only be found by experiment.

All the properties assigned so far to the system (R, r, N, and K) are descriptive, the result of pure observation, but it is now necessary to account for the leveling off of the rate of growth of the population so that R eventually becomes zero and the population is stabilized at K. Something must be suppressing the growth rate R when the population is crowded, and our hypothesis states that this something is competition for crowded space and vanishing food. The dampening effect of competition presumably gets stronger as the animals get more numerous, until finally, at $N = K$, competition is so strong as to bring the rate of increase, R, down to zero. On this assumption, we may describe the complete history of the sigmoid growth of the population with the following equation:

$$R = rN \left(1 - \frac{N}{K}\right)$$

When N is small, as in the early days of population growth, the quotient N/K is negligible and the term $[1 - (N/K)]$ is, therefore, close to 1. The rate of population increase R is then essentially given by rN. But as N grows large, the term $[1 - (N/K)]$ becomes significantly less than 1, until at $N = K$, it becomes $[1 - (K/K)]$, which is zero. The population growth rate R then also becomes zero, and the system reaches equilibrium. This equation, which so neatly describes the growth of a population under our postulated conditions, is called the **logistic equation,** and it has been known since the work of the Belgian mathematician P. F. Verhulst in the middle of the last century.

The logistic equation describes changes in rates of growth of populations. Growth is slow at first, becomes faster and faster until a maximum is reached, and then falls away finally to zero. A graphical plot of the logistic equation therefore yields a parabola as shown in Figure 22.2.

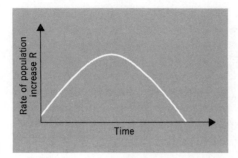

Figure 22.2

But a graph of the numbers living at any instant is, as we have seen, apparently sigmoid. Sigmoid curves are described by exponential equations, in form quite unlike the logistic equation. An exponential equation, defining the numbers of animals living at any instant of time, can be derived from the logistic equation, although the mathematics is not simple. One form of the derivation is as follows:

Where

R = rate of increase of whole population
r = intrinsic rate of increase
N = number of individuals
K = maximum supportable number of individuals
t = time

$$R = rN \left(1 - \frac{N}{K}\right) \tag{1}$$

or

$$\frac{dN}{dt} = rN \left(1 - \frac{N}{K}\right) = rN - \frac{rN^2}{K}$$

Separate variables

$$dt = \frac{dN}{rN - \frac{r}{K}N^2} = \frac{dN}{\frac{r}{K}N(K - N)}$$

Expand right-hand side by partial fractions

$$\frac{1}{N(K - N)} = \frac{A}{N} + \frac{B}{K - N}$$

and

$$1 = A(K - N) + BN$$

Let

$$N = 0 \implies A = K^{-1}$$

Let

$$N = K \implies B = K^{-1}$$

Thus

$$dt = \frac{K}{r} \left[\frac{1}{NK} + \frac{1}{K(K-N)} \right] dN$$

$$= \frac{1}{r} \left(\frac{1}{N} + \frac{1}{K-N} \right) dN$$

and

$$rdt = \left(\frac{1}{N} + \frac{1}{K-N} \right) dN$$

Boundary conditions

at

$$t = 0, \qquad N = N_0, \qquad N_0 \neq 0$$
$$t = t', \qquad N = N'$$

Integrating

$$\int_0^{t'} rdt = \int_{N_0}^{N'} \left(\frac{1}{N} + \frac{1}{K-N} \right) dN$$

$$rt \int_0^{t'} = \ln N - \ln(K-N) \int_{N_0}^{N'}$$

$$rt' = \ln \frac{N'}{N_0} - [\ln(K-N') - \ln(K-N_0)]$$

Take antilogarithm of both sides

$$e^{rt'} = \frac{N'(K-N_0)}{N_0(K-N')}$$

Solve for N'

$$(K-N')e^{rt'} = \frac{N'(K-N_0)}{N_0} = Ke^{rt'} - N'e^{rt'}$$

$$N' \left(e^{rt'} + \frac{K-N_0}{N_0} \right) = Ke^{rt'}$$

$$\therefore N' = \frac{Ke^{rt'}}{e^{rt'} + \dfrac{K-N_0}{N_0}} = \frac{K}{1 + \left(\dfrac{K-N_0}{N_0} \right) e^{-rt'}} \qquad (2)$$

Equation 2 defines a sigmoid curve, but there is no need to follow this formal derivation in order to understand the significance of the logistic hypothesis to ecology.

When constructing the logistic equation we assumed that population growth was directly influenced by the extent of crowding. It is our hypothesis that competition and other repressive effects will become proportionately stronger as the population becomes more crowded, which is to say as it becomes denser. Ecologists say that such effects are **density dependent.** Any direct evidence that populations of animals or plants are normally under density-dependent control, should support the belief that the logistic equation is a realistic model of natural events, even if a crude one. Conversely, any successful manipulation of the logistic equation to predict natural events should encourage the belief that populations in nature are, indeed, regulated by density-dependent factors, particularly perhaps by competition for food. Ecologists have worked hard at both these approaches; sometimes seeking evidence for competition and other forms of control by crowding, at other times erecting mathmatical models derived from the logistic equation so that these models may be tested by experiment or against the experience of naturalists. Both approaches have involved extensive work with populations of simple animals kept in the laboratory. It is convenient to consider first the approach of using experimental animals in attempts to discover how the control assumed by the logistic model might work in practice.

STUDIES WITH FRUIT FLIES

The geneticist's fruit fly, *Drosophila melanogaster*, thrives under simple care. The maggots grow well in a mash of banana or in a synthetic medium based on agar jelly. The adult flies need yeast plants, both as food and as a stimulus to lay eggs, but yeasts also grow well on banana mash or agar. An enclosure for healthy fruit-fly living can be made by pouring jellied maggot food into half-pint milk bottles, letting it set, and placing an inoculum of yeast on top. Fruit flies are so tiny that the air space in a half-pint milk bottle apparently provides comfortable room for hundreds of them. They come down to the surface of the medium to feed on the growing yeasts. They mate readily, and lay their first eggs the day after emerging from their pupae. The eggs hatch after one day, the larvae are big enough to pupate in four days and four days after that (4 days for larvae + 4 days for pupae) the next generation of flies emerges. If a culture is prepared by putting two pairs of fruit flies in such a bottle, closing the top with gauze, and then leaving the bottle quite alone, there will be a population history like that described by Bodenheimer (1938) as follows:

> "In the normal culture the number of adults begins to increase on the tenth day. The population maximum of 230 adult flies is reached on the 23rd day and is maintained with small fluctuations until the 31st day. The population then decreases until it is almost zero on the 50th day. The agar has dwindled away to a thin hard mass by this time and is entirely unfit as a physical environment for the larvae."

This population history includes rapid growth, leveling off, and final decline and death. The decline and death are obviously results of starvation when the food is all gone, but the leveling-off process that preceeded it should be due to the effects of crowding, on competition for limited resource if our logistic model is correct. The simplicity of the milk-bottle universe lets us test for ways in which the effects of crowding might work.

As the generations of flies in a milk bottle succeed each other they certainly get more crowded, but there must also be other changes due to aging of the cultures; changes in growth of the yeast, reduction of the medium, accumulation of corpses and excretory products, and other things like these. A first step in the analysis of crowding should be to try to separate the effects of the crowd itself from the accumulated effects of those who have gone before. An early experiment achieved this. In 1922 Pearl and Parker made a number of fresh *Drosophila* cultures some of which were crowded, and some not, right from the first day. They made up a series of bottles stocked with different numbers of flies ranging from one to 50 pairs in a bottle. The living conditions of the crowded flies were thus quite the same as those of the lonely flies; except that they were members of a crowd. After 10 days the flies of the next generation began to hatch from their pupae, and Pearl and Parker could keep count of the breeding success of the different cultures in terms of adult offspring produced. There were, not surprisingly, many more new flies in the bottles which had started with most parents, but their numbers were not proportionately more. When Pearl and Parker calculated the yield of flies per female, they found that the number was less in the crowded bottles. There was an inverse relationship between population density and yield of flies, showing that crowding in otherwise identical conditions was alone sufficient to depress reproductive success.

But *Drosophila melanogaster* is a holometabolous insect, which is to say that it has a complicated life history of egg, larva, pupa, and adult. Any of these stages could be crowded, so that changes in the yield of adult flies could reflect suppression of eggs, or larvae, or pupae, or the parent flies themselves. There seemed little reason to believe that crowded pupae should hatch any less successfully than lonely pupae, and observation of the cultures produced no evidence of untimely deaths of the crowded adult flies; which left suppressions of the eggs or larvae as being the most likely explanation of the effects of crowding.

Pearl (1932) took advantage of the fact that the eggs were laid in plain view at the surface of the medium to see if the effects of crowding were reflected in success at laying and hatching eggs. This time he arranged his series of crowded and uncrowded cultures in inverted bottles with his culture medium in watch glasses on the table. Each day he could tap the glass of his bottles, which made the flies fly up to the top, and then gently lift them off the watch glasses without losing any flies. It was then a simple matter to search for eggs on the watch glass with a dissecting microscope, to count the daily increment, and to keep a record of hatching. The eggs from the crowded flies hatched just as well as those from lonely flies, showing that there was no physiological response of crowded flies which made their eggs unfertile. But the crowded flies laid significantly fewer eggs per female. In some way, the fact of being crowded reduced egg production.

It seemed likely to Pearl that crowding might be felt most severely when the flies were actually egg laying on the little disk of medium, or when they were feeding there, but it was also possible that the effects were felt when they flew about and mated in the air volume above. He set up cultures in different-sized bottles so that the effects of air volume and medium surface could be separated. Changing the air volume did not affect the findings, and it was possible for Pearl to conclude that a prime effect of crowding was reduction in egg laying, which was, in turn, due to some sort of disturbance of the flies while they were feeding on the medium or during the egg-laying act itself.

There were still two ways of looking at Pearl's results: either the flies were cutting their egg production in response to subtle pressures of companionship on the agar, a social response that should have deep implications for population theory, or the flies simply found it hard to get enough to eat on that crowded dinner pail and, undernourished, failed to lay as many eggs as usual. These possibilities were investigated many years later in a beautifully designed series of experiments by Robertson and Sang (1944).

When Robertson and Sang began their studies in the 1940s, much more was known about making good food media for *Drosophila*, from the "cookbook" sorts of experience which had come from the many laboratories that used *Drosophila* cultures for experiments in genetics. Robertson and Sang compared the successes of series of similar cultures on different food media, being able to show particularly that the kind of yeast used made much difference to the nutrition of the flies. The flies preferred brewer's yeast to the baker's variety. And it was clear that the medium Pearl had used afforded a rather poor diet to the adult flies, so that any competition for food should quickly lead to malnutrition. When Robertson and Sang crowded flies on the best of their media, there was hardly any decrease in the number of eggs a female laid.

Pearl's results could only be duplicated by deliberately depriving the animals of food in such ways as covering part of it with cellophane. The reduced yield of eggs in Pearl's cultures was thus clearly due to malnutrition brought about by too many flies trying to feed at too small a trough. Here was apparently unequivocal evidence that control was brought about by competition for food acting through depression of the birth rate. Control was, indeed, density dependent, and a simple reflection of the energy available to the flies.

If adults suffered malnutrition when crowded, it seemed likely that the larvae should also. This was more difficult to check, because the maggots burrowed through the medium and were hard to watch. But Bodenheimer (1938) of Jerusalem University, with much labor, managed to follow the history of all stages in the life cycle of cultures of flies from the moment of introducing the two pairs of adults into fresh containers until the final death of the last flies in the exhausted bottles. Bodenheimer used a transparent agar medium for food. To make his counts, he would cut the agar into thin slices, then count the larvae and pupae in each slice. The counts destroyed the cultures, of course, so he established 90 similar cultures at the start of his experiment, and thereafter sacrificed two each day on the altar of quantitative ecology. The results of the pairs of daily counts were so similar that he felt confident that his cultures were developing at parallel rates, so that he could use his final daily estimates as if they had been made on a single growing culture without disturbing the inmates.

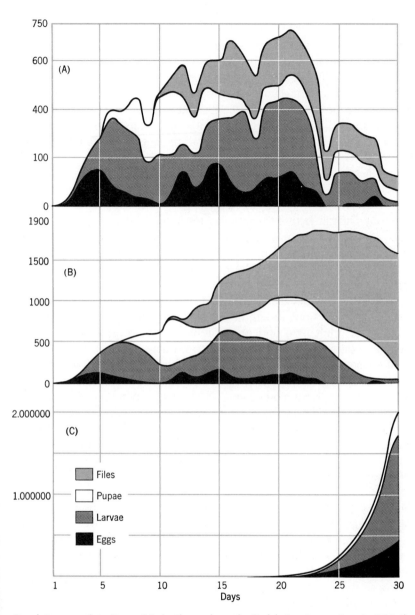

Figure 22.3

Population growth in *Drosophila* bottles as shown by Bodeheimer's experiments. (A) is the reconstructed actual history of a population of flies in a bottle containing culture medium from the initial innoculation until the starvation that ensues when all the food is gone. About five generations are spanned. Notice that the survival of all stages in the life history declines as the population becomes dense. (B) is a calculation of the population history if all the eggs laid were able to develop into adult flies. In this model the only check on numbers is reduced egg laying. (C) is the calculated population growth if all young survive and egg laying proceeds at the rate achieved by the flies of the initial innoculum. Notice that the scales on the vertical axis of each graph has had to be changed in order to draw all three on the same figure.

His results showed that larval death rates, and apparently even pupal death rates, increased as the cultures became crowded.

Bodenheimer expressed his results in the form of a graph (Figure 22.3a). Pulses of increased egg-laying record the passage of four generations of flies before the final catastrophe, and the decreased survival of maggots, pupae, and flies with the passage of time are evident. The data also enabled Bodenheimer to calculate how many of each later stage should have been present each day if all the eggs laid hatched and survived to become adult flies (Figure 22.3b), a proceeding that showed dramatically how severe must have been the mortality at later stages. Finally Bodenheimer calculated the population each day on the assumption that there had been no reduction of eggs laid per female, as well as no mortality of later stages before the female flies had laid their normal quota of eggs. The result (Figure 22.3c) is eloquent testimony to the effectiveness of the crowding mechanism, since Bodenheimer had to shrink his verticle axis 2000 times to draw his third graph on the same-sized paper as his first.

In the fruit-fly work there is strong evidence, amounting to proof, that the crucial reductions in egg laying by crowded flies was both due to shortage of adult food and proportional to the density of the crowd. Deaths of other stages are also seen to be density dependent, but they have not been shown to be due only to shortage of food, although there is a clear implication that this is the ultimate cause. Crowded larvae could be interfering with each other, and with pupae, in other ways, by banging into each other or because their numerous activities poisoned their medium with excretions. Such interference could conceivably control population growth even before malnutrition became general. The possibility of control in ways like these has been extensively studied with different experimental animals, flour beetles.

STUDIES WITH FLOUR BEETLES

Flour beetles, such as *Tribolium confusum,* are well-known pests of flour mills. They pass their whole lives in flour; burrowing through it, eating it, defecating in it, laying eggs in it, hatching, growing as maggots, pupating, emerging, and mating in it. A heavy infestation spoils the flour, making it grey and stinking from the accumulation of feces, excretions, and corpses. Like the fruit flies, flour beetles breed quickly; and they are even easier to keep for all they need is a box of flour at roughly room temperature. Females live and lay eggs for many months. In spite of a life spent hidden in flour, the beetles are easy to count for they can be sieved out daily, after which the eggs, larvae, pupae, and adults can be counted and either returned to the flour from whence they came or put in fresh flour. And this sieving and counting seems to have no effect on their daily lives; altogether most accommodating laboratory animals.

Flour beetles were introduced to ecologists when Chapman (1928) used them for experiments designed to test his ideas of environmental resistance. He grew his beetles in weighed amounts of flour, seeding all samples at the rate of two beetles per 4 grams of flour. Every 15 days he would sieve them out of their flour and count them. There was always lots of flour left after 15 days, so the animals could not be running short of food, but Chapman would put them back in fresh flour all the same. He continued the experiment for 156 days. His results (Table 22.1, Figure 22.4) showed that growth in all cultures could be described by the familiar sigmoid curve. In all cultures, population growth was at first rapid, but then tapered off, finally ceasing when there were about 4.4 animals per gram of flour. At this density, the populations were apparently in balance and were likely to stay so, as long as Chapman cared to change their flour every 15 days. But the beetles were not short of food, since they were still surrounded with masses of flour, with many times their own weight of it, in fact. If the populations were truly controlled by competition, by factors of their environment acting in density-dependent ways, the control must be something more subtle than simple squabbling for food. Chapman's use of these experiments to measure "environmental resistance" proved to be a blind alley of ecology, but his revelation of the conundrum of population control in flour beetles stimulated others to lifetimes of work on the subject. His paper, "The Quantitative Analysis of Environmental Factors," in which flour beetles made their ecological debut, ranks as a classic of ecology (Chapman, 1928).

Chapman's own answer to the question of control in his flour-beetle cultures was "cannibalism." The beetles were known to eat their own eggs if they came on them during their blundering passage through the flour. If every time a beetle came on an egg, it ate it, and if both the travels of beetles and the eggs were distributed randomly throughout the flour, it was obvious that the proportion of eggs eaten before they could hatch should go up as both beetles and eggs became commoner, until an equilibrium was reached in which the number of eggs eaten was equivalent to the number laid in the same time. This process should lead to just the sigmoid population history which Chapman observed. To test this hypothesis, Chapman kept monastic colonies of males only, adding daily to their flour the numbers of eggs which he thought should have been laid if half their number were females. He changed all the eggs often enough to make sure that none could hatch, and could thus assume that all missing eggs had been eaten by the beetles. His results showed that more and more eggs did, indeed, vanish as the populations of male beetles were increased, apparently enough of them to account for all the falling off in the apparent egg production of normal colonies of male and female beetles together. Here was a simple enough effect of crowding, and one quite in accord with the premises of the logistic model.

But the cannibalism hypothesis, for all the care of Chapman's work, was not the end of the story. Several people independently became curious about another possibility; that dirty flour might in some way inhibit the activities of the beetles. Thomas Park of the University of Chicago performed a famous series of experiments to test this possibility (Allee et al., 1949). He kept parallel colonies of beetles in fresh flour and in flour in which colonies of beetles had already lived for weeks, and he found that there were always fewer eggs in the old-flour cultures than in the fresh flour. This was so even at low beetle densities when cannibalism

TABLE 22.1
Chapman's results for population growth in flour beetles

Days	4 grams				8 grams				16 grams				32 grams				64 grams				128 grams			
	Eggs	Larvae	Pupae	Adults	Eggs	Larvae	Pupae	Adults	Eggs	Larvae	Pupae	Adults	Eggs	Larvae	Pupae	Adults	Eggs	Larvae	Pupae	Adults	Eggs	Larvae	Pupae	Adults
0				2				4				8				16				32				64
15	41	17	0	2	62	71	0	4	127	187	0	8	263	280	0	16	631	686	0	32	854	1543	0	64
30	44	74	0	2	30	168	0	3	103	314	0	8	188	509	0	16	369	1118	2	32	393	2371	99	61
50	42	45	21	31	47	75	51	90	89	178	79	167	383	310	114	332	402	792	220	639	1265	1204	503	1405
64	64	20	14	59	107	47	12	144	205	78	36	220	497	180	58	414	1145	400	157	842	2215	541	198	1832
78	60	10	6	65	114	11	20	144	330	16	39	158	636	14	91	428	1254	67	159	875	2705	230	256	1857
101	89	5	1	66	185	30	7	156	390	46	1	174	861	94	3	445	2086	146	1	928	2672	318	2	1906
114	125	2	0	66	180	20	0	156	368	21	13	174	846	56	20	449	1530	97	16	904	2943	218	48	1914
134	81	0	0	66	257	3	0	159	460	24	1	174	842	13	0	452	2143	32	4	908	3805	63	4	1905
156	89	0	0	65	236	2	0	157	544	8	6	173	837	6	4	445	1912	45	7	902	4097	63	10	1882

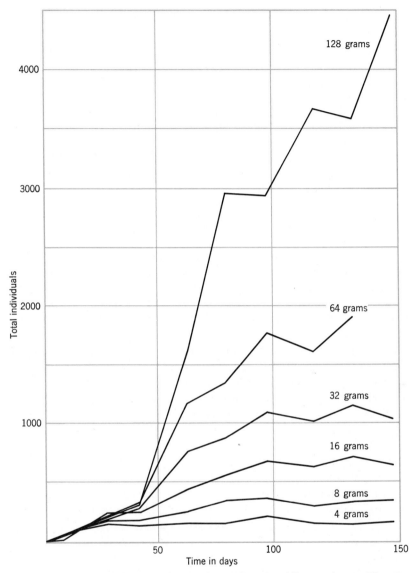

Figure 22.4

Chapman's results for the growth of flour beetle populations in different volumes of flour. In these experiments the flour was always changed at each count, so that in none of the cultures were the animals ever short of food. Yet population growth always stopped at about the same density, 4.4 beetles per gram of flour. Chapman concluded that egg cannibalism was the main density-dependent factor leading to population equilibrium. Later work showed that reduced fecundity was more important than cannibalism and that many other factors were also involved.

should have been unimportant, and Park concluded that something in the old flour so interfered with female beetle physiology that they laid few eggs. He did not at first know what this something was, so he called it a "conditioning" of the flour, to show that some particular factor in the dirtiness left behind by the beetles was probably responsible. Its effect was temporary, because when he moved beetles from conditioned flour to fresh flour, the yield of eggs quickly rose to be equivalent to that of similar cultures that had been in fresh flour all the time. Eventually studies in Park's laboratory showed that the conditioning factor was the chemical ethylquinone, and that this substance was secreted from special glands on male beetles. The males were particularly prone to secrete ethylquinone when disturbed in the mating act by the arrival of a third beetle. The substance was soaked up by the flour and accumulated in the habitat, there to serve as a population inhibitor which, like cannibalism, acted in a density-dependent way.

Both cannibalism and conditioning may be thought of as aberrations resulting from the peculiar habitat of a box of flour. It is quite certain that the ancestors of what we call "flour beetles" did not live in flour, because a world without human millers is a world without flour. Presumably, the ancestors of the modern *Tribolium* populations were content with the humbler role of eating seeds. Egg cannibalism might have been a rare event then; certainly a happening unlikely to have so dramatic an effect in limiting population. And it is not hard to imagine, in the years before millers, uses for chemical excretions by males angry at being disturbed while copulating, uses which need not involve suppression of egg laying by their mates. These regulatory devices of cannibalism and conditioning must, in a sense, be properties of the experimental system. Knowing about them does little to help us understand how a balance in flour-beetle numbers might be established in nature. There should be other regulatory devices at work.

A direct hint that there are other subtle, but very potent, regulatory devices unleashed in crowded *Tribolium* populations came from the studies of Janet Boyce (1946), then a doctoral candidate at the University of California. Boyce wished to find out what would happen to egg production, if the flour was always fresh but the beetles were still forced to crowd, not just eggs surviving cannibalism as Chapman measured, but actual egg production as number of eggs laid. She changed her flour so often that no effects of conditioning could be detected in control populations. She also took all of the eggs out of her cultures every five days and replaced them with the same number of fresh eggs, a procedure that made sure that no eggs ever hatched. There were never any larvae in her cultures, and the numbers of adults were kept constant by replacing any that died. The only population events that were possible were the increase of eggs through laying and the decrease of eggs through cannibalism. In the first few days of the history of any of these cultures, the eggs present at the daily count rose, but there always came a time when the

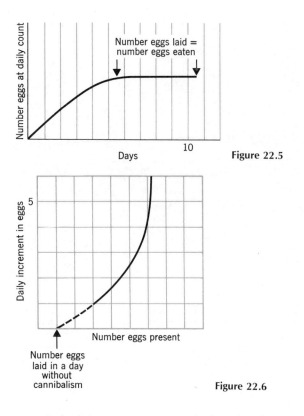

Figure 22.5

Figure 22.6

cannibals took as many as were laid so that the egg content was constant as shown in Figure 22.5. Boyce then plotted the daily increment of eggs in early days when the egg population was rising, against the number of eggs present. Extrapolating such curves to day-zero gave her an estimate of the number of eggs that would have been laid in a day without canni- balism in any particular culture as shown in Figure 22.6. Boyce per- formed similar extrapolations for a series of ever more crowded cultures, finding the rate at which females laid eggs in each. Her results strongly suggested that egg-laying rates did, indeed, go down as the animals were more crowded (Figure 22.7). It seemed clear that the lowered yield of eggs per female beetle which had been apparent in everybody's daily egg counts of crowded populations was only in part a result of the eggs being eaten almost as fast as they were laid. Many less eggs were, in fact, laid to be eaten. This conclusion of Boyce's has since been confirmed by one of Park's students, Rich (1956), who invented a way of marking beetle eggs for future reference, without hurting them. He kept gravid females in tinted flour, which so stuck to the gluey cases of the eggs they laid that he was provided with a supply of tinted eggs. He could add his marked eggs to active cultures at constant rates, then record the rates of cannibalism in those cultures by recording the disappearance of tinted eggs. The same rate of cannibalism could be assumed for the unmarked eggs actually laid

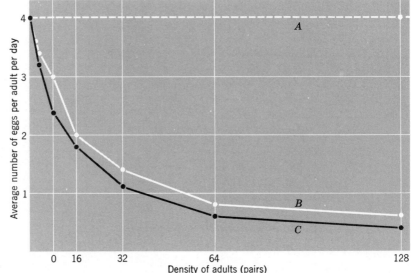

Figure 22.7

Daily egg production of *Tribolium confusum*. *A*, average number of eggs per day that would appear at different densities of adults if there were no cannibalism and no reduction in fecundity. *B*, average real fecundity per adult per day. *C*, average apparent fecundity per adult per day. *A* minus *B* at each density will give an approximation of the reduction in real fecundity for each density. *B* minus *C* at each density will give an approximation of the egg reduction due to cannibalism for each density.

within the culture, and so a number equivalent to those taken by cannibalism could be added to the recorded yield of the culture to give him the actual egg production. Like Boyce, Rich showed that egg production per female fell in crowded cultures, even when they were kept in fresh flour.

Some idea of how crowding in clean flour can affect egg laying has come from further studies in Park's laboratories and elsewhere. Beetles may mate too little, or mate too much; the blundering passage of too many beetles through the flour may interrupt each other's mating; there may be specially favored places for egg laying even in a simple container, and these may be competed for, and so on. The list of ways in which too many beetles are known to restrict egg production or egg survival is thus long; it is certainly by no means complete.

But all the other stages in the *Tribolium* life cycle must also be subject to similar mechanisms. There is evidence that larvae eat pupae; that old adults eat young adults; that larvae eat eggs; that male beetles often so injure females with their thrusting male organ that the females die; that crowded males try to copulate with each other, sometimes inflicting serious injury; and so on (Park et al., 1965). Any one, and any combination, of these density-dependent control mechanisms can act to check the population growth as the beetles become more crowded. It is quite certain that there are many other possibilities not yet explored, even in the lifetimes of work which have been devoted to the subject. But the studies do provide clear evidence that almost any kind of mechanism which you can imagine for the density-dependent control of a population can, in fact, work with real animals. And they can work to limit populations well below the level at which competition for food becomes serious.

The work with these various experimental animals, with *Paramecium*, fruit flies, and flour beetles, as well as with many others, has shown convincingly that populations of real animals can be controlled by density-dependent factors in the manner required by the logistic model. They may be controlled by starvation, which is to say by simple competition for dwindling food, or they may be controlled by other density-dependent mechanisms. It is very clear, however, that the logistic model can only be a very simplified analogy of what actually happens. In the equation

$$R = rN \left(1 - \frac{N}{K}\right)$$

the carrying capacity of the environment "K" is set by whatever limits the density of the animals, and may usually be expected to have many components. In some simple systems, as perhaps in *Paramecium* cultures or the egg laying of fruit flies, K may be set by direct competition for food; but more usually a complex of many factors must set K, as the *Tribolium* studies so clearly show. The matter is made more complex still by the fact that different mechanisms may be at work in different stages of the lives of the animals. In the holometabolous insects so widely used for laboratory studies, this difficulty is readily apparent. Eggs, larvae, pupae, and adults are, in a sense, different organisms, depending for their survival on different mechanisms. Any long-term population balance for the whole species must be an integral function of the production and survival of each stage in the life history, so that a realistic mathematical model of the growth of such a population must be very complicated. I believe it safe to say that even computer simulation has not yet produced a reasonably complete model of the population growth of flour beetles in flour. And this difficulty does not only apply to animals with complex metamorphoses; consider only the changing foods and habits involved in growing a big fish from a tiny egg.

But the real justification for undertaking so much painstaking work with laboratory populations has always been that it might be possible to establish, by experiment, generalizations that should help our understanding of nature; thus have physicists so beautifully proceeded from observation, to model, to experimental test, to satisfying generalization. It is necessary to look at the work with laboratory animals with this general justification in mind, and to see in what ways we have been rewarded. We have started by making a model, the logistic model, of a population that grows rapidly, is damped, and is finally brought into a state of balance. This model apparently represents what actually happens in simple cultures, but it is by no means certain that the balance of nature, which we seem to see around us, has come about through such an history. We

SOME AMBIGUITIES OF
LABORATORY STUDIES

really have to face the fact that our model describes not natural events exactly but the results of earlier experiments in the laboratory. It should not be surprising that subsequent similar experiment finds the conditions of the model to be substantially correct.

And then, the design of any laboratory experiment in which animals are kept alive automatically ensures that the conditions of the logistic model are met; for we must limit the space assigned to the animals, and we must feed them to stop them dying. We divorce our animals from those complex interactions of nature which make up the parameters of their real niche, and force them to find a simpler niche within limits of our own contriving. If we make animals compete for food, we should not draw too much conclusion from the fact that then they do compete for food. G. E. Hutchinson has an elegant way of expressing this dilemma, telling his students that he who designs a population experiment with laboratory animals is really making himself a simple analog computer. The behavioral responses of the animals are like the pulse inputs of computers using an electric input analog. The animal experiment can thus be designed to yield any desired result. When you wire an analog computer to an X-Y plotter, the instrument draws graphs depending on the equations which you direct it to solve. The growth curves coming out of your population experiments are likewise dependent on the "data" which you feed to the system.

These difficulties mean that we cannot rely very much on the results of simple population-growth experiments for an understanding of the processes leading to the apparent balance in nature. For all that the many sigmoid histories of such populations now on record can show, it remains but an hypothesis that the logistic model is applicable to wild populations. That laboratory animals compete, or are regulated by density-dependent factors, is not evidence that wild animals so compete or are regulated.

The hypothesis of competition leading to natural balance is, however, interesting and plausible, and much effort has gone into finding more valid tests for it. There have been two general approaches to the problem. Some have studied wild populations, seeking evidence for competition and density-dependent regulation devices; all the field studies of population problems can be considered to serve this purpose. And others have tried to manipulate the logistic model to predict natural consequences. This last method has led to some of the most penetrating of the understandings of modern ecology.

When animals of the same kind compete together for limited resources, the effect is apparently to restrict the size of their population so that the species persists, but what happens if individuals of different species compete for the same resources? Is there a desperate struggle in which one kind of animal is totally eliminated, or does the battle go to a draw so that each persist with reduced and restricted populations? These alternatives could be investigated by the logistic hypothesis, through developing a pair of logistic equations to describe the growth of each separate population in the presence of the other, and then solving the equations simultaneously to find the relative sizes of the two populations when both had ceased to grow. Such pairs of equations, known as the Lotka-Volterra equations, predict that coexistence between strongly competing species is impossible, one of them always being eliminated in the struggle. Only if they compete very weakly is coexistence apparently possible. These predictions were tested by Gause, in perhaps the most important experiments of ecology, when he cultured together various pairs of species of small animals in conditions carefully contrived so that the animals must compete. In such conditions one species was always eliminated, as the predictions said it must be. Only if the animals could avoid competition in some way did both species survive in the experimental chambers. These results allow the stating of a general working principle, that of *competitive exclusion,* or *one species: one niche.* The suggested, and grand, implication of this principle is that the results of the great struggle for existence of which Darwin wrote are not more competition but the avoidance of competition. A fit animal is one that avoids competition, living free from the necessity to struggle

A MODEL OF
COMPETITION
BETWEEN SPECIES
LEADS TO THE
EXCLUSION OR
GAUSE PRINCIPLE

with others in its own specialised niche. But it can be argued that the principle is not strongly established by experiments with animals in laboratories, because the competitions which we observe in our experiments are contrived, having been arranged by the experimenter himself. It is thus necessary to test the principle of competitive exclusion against evidence of how animals live in wild nature.

CHAPTER 23

A MODEL OF COMPETITION BETWEEN SPECIES LEADS TO THE EXCLUSION OR GAUSE PRINCIPLE

If two kinds of animal are forced to eat the same food, they must compete for it. But what should be the result of such competition? Should it be normal for some conquering species to utterly suppress the other, as the goats suppressed the Abingdon tortoises? Or should competition between adversaries better matched than goats and tortoises lead to coexistence, a balance of power? There are many species in nature, but only one ultimate food source. Are all the kinds of animals about us struggling for the same food in the same way, so that they reach a straining balance like titans wrestling? Or is the outcome of any conflict a local annihilation, so that the sounds of struggle are muted as contestants avoid each other? To distinguish between these two propositions is a prime task of ecological thought.

Competition between species should be different only in degree from competition between individuals of the same species. If the logistic model truly describes intraspecific competition, it should be possible to expand it to describe interspecific competition as well. More, it should be possible to predict the outcome of such competition between species, enabling ecology for once to proceed by the classical scientific method of devising experiments to test mathematical predictions.

A model for predicting the outcome of competition between species living in the same space and on the same food was developed from the logistic equation by two mathematicians working independently in the early part of this century, Vittorio Volterra in Italy (Chapman, 1931) and Alfred Lotka (1925) at Johns Hopkins University.

If we call the two competing species 1 and 2, then the population growth curve of species 1 living alone is

$$R_1 = r_1 N_1 \left(1 - \frac{N_1}{K_1}\right)$$

and the population growth curve of species 2 living alone is

$$R_2 = r_2 N_2 \left(1 - \frac{N_2}{K_2}\right)$$

We must now derive a pair of equations to describe the growth patterns of the two populations when grown together. At first, the growth rate of neither population will be influenced by the presence of a few individuals of the other species, because there is still lots of space and food. But when the animals are numerous, such that N_1 approaches K_1 or N_2 approaches K_2, the effects of competition should be significant. Competition must be dependent on numbers and the strength with which each individual is able to compete. The growth curve of a population of species 1 in the presence of species 2 is, therefore, given by

$$R_1 = r_1 N_1 \left(1 - \frac{N_1}{K_1} - \frac{\alpha N_2}{K_1}\right)$$

where α is the coefficient of competition of population 2. If population 2 does not compete, contrary to our expectations, then α will be zero, αN_2 will also be zero, and the history of population 1 will be quite unaffected. But if there is competition, then αN_2 will be positive, will combine with N_1, and will reduce the term

$$\left(1 - \frac{N_1}{K_1} - \frac{\alpha N_2}{K_1}\right)$$

to zero before the saturation value of population 1 is reached. Assigning the coefficient β to represent the competition of the other species we may write a pair of equations as follows:

$$R_1 = r_1 N_1 \left(1 - \frac{N_1}{K_1} - \frac{\alpha N_2}{K_1}\right)$$

$$R_2 = r_2 N_2 \left(1 - \frac{N_2}{K_2} - \frac{\beta N_1}{K_2}\right)$$

The outcome of the competition between the species would be revealed when finally the system came to equilibrium. The growth rate of neither population must then change, and R_1 and R_2 must both equal zero:

$$R_1 = R_2 = 0$$

and $\quad r_1 N_1 \left(1 - \dfrac{N_1}{K_1} - \dfrac{\alpha N_2}{K_1}\right) = r_2 N_2 \left(1 - \dfrac{N_2}{K_2} - \dfrac{\beta N_1}{K_2}\right) = 0$

These equations may be written in differential form as follows:

$$\begin{cases} \dfrac{dN_1}{dt} = r_1 N_1 \left(1 - \dfrac{N_1}{K_1} - \dfrac{\alpha N_2}{K_1}\right) \\[2mm] \dfrac{dN_2}{dt} = r_2 N_2 \left(1 - \dfrac{N_2}{K_2} - \dfrac{\beta N_1}{K_2}\right) \end{cases}$$

when they may be solved simultaneously to show that coexistence of strongly competing populations is not possible. If they are forced to share food and space, one population will always displace the other. The titans do not long hold the strain; rather, one scores total victory. This conclusion is so important that it is best to try to understand clearly why it is a necessary consequence of the model.

There are four possible ways the competition may take:

Model 1   Species 1 may compete much more strongly than species 2 (β is large compared with α).
Model 2   Species 2 may compete much more strongly than species 1 (α is large compared with β).
Model 3   Both species compete strongly (α and β are both large).
Model 4   Both species compete weakly (α and β are both small).

The simplest model to consider is one that allows only one species to compete strongly. Let us first suppose that species 1 is the strong competitor, making β large and α small (Model 1). The inhibitory effect of any individual of species 1 on its own population is $1/K_1$ and on the population of species 2 is β/K_2. If an individual of species 1 treats an individual of species 2 as one of its own kind $\beta = K_2/K_1 = 1$ and it is clear that for species 1 to compete more strongly with species 2, $\beta > K_2/K_1$. If the reciprocal competition is weak, we have

$$\alpha < \frac{K_1}{K_2} \qquad \text{and} \qquad \beta > \frac{K_2}{K_1}$$

To understand the necessary outcome of such competition, it is convenient to express the possibilities on a diagram. This can be done by plotting all possible combinations of the two species as number of species 1 against number of species 2. If alone, species 1 can have no more than K_1 individuals. If it lives with K_2 individuals of species 2, there must be K_1/α individuals. The line K_1, K_1/α therefore represents all possible numbers of species 1. Conversely, the line K_2, K_2/β represents all possible numbers of species 2. When species 1 is the stronger competitor, as above, we have the graph shown in Figure 23.1. In any mixture of the two populations whose plot falls in the white hatched area, species 1 can still produce more individuals while species 2, already above its saturation value, must decline. The plot of the species composition will move steadily to the right as the relative success of species 1 proceeds, until finally, at k_1 only species 1 is left.

In the second of the possible models, species 2 is the stronger competitor such that $\alpha > K_1/K_2$ and $\beta < K_2/K_1$. The outcome of this competition is illustrated in Figure 23.2. Now the area between the lines of maximum abundance is shaded black to show that species 2 may increase in it whereas species 1 must decline. The competition ends at K_2 with only population 2 surviving, and there is an equilibrium population of K_2 individuals of species 2 having the habitat to themselves.

In these models of competition between the strong and the weak, there can be no surprise that the strong is predicted to achieve an annihilating victory. That victory should be just as complete when the strong meets the strong is not, however, so intuitively obvious. When the strong meets the strong $\alpha > K_1/K_2$ and $\beta > K_2/K_1$

then $$\frac{K_1}{\alpha} < K_2 \qquad \text{and} \qquad \frac{K_2}{\beta} < K_1$$

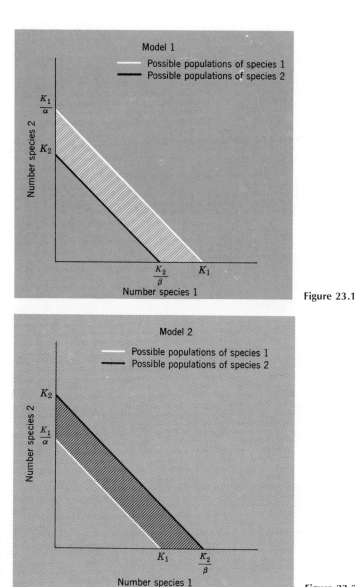

Figure 23.1

Figure 23.2

so that the lines describing possible populations of each cross as illustrated in Figure 23.3. In the black triangle, species 2 can expand but not species 1, and a plot of the relative populations will move to the left with time until only species 2 remains. But in the white triangle, species 1 is favored, and the competition proceeds until K_1 individuals of species 1 have the habitat to themselves. The outcome of the competition between mutually strong contenders thus depends on the initial concentrations of the two populations, but the final victory is always absolute. There is also a theoretical possibility of equilibrium at point S, where the two popula-

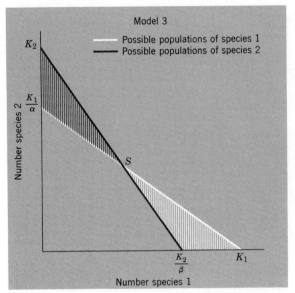

Figure 23.3

tions are beautifully balanced so that both persist. This is so unstable an equilibrium, however, that with populations of real animals the chance of its being maintained are vanishingly small.

There remains the fourth model, that of populations which compete only weakly. When the weak meets the weak $\alpha < K_1/K_2$ and $\beta < K_2/K_1$ so that $K_1/\alpha > K_2$ and $K_2/\beta > K_1$ when a plot of competing populations is as shown in Figure 23.4. The white triangle in which species 1 has the advantage is now at the left of the diagram, and the plot of a population mixture in this region must move to the right with time. But when point E is passed, the plot enters a black region where species 2 should be favored. The tendency would be for the relative proportions to move back to the left. Thus, there should be a stable equilibrium at point E so that both populations persist indefinitely. For weakly competing populations, then, the Lotka-Volterra equations predict that both should persist indefinitely, their populations fluctuating only gently about equilibrium levels.

If the assumptions on which the Lotka-Volterra model is based are true, species can coexist in confinement only if they scarcely compete. Whenever strongly competing animals meet, there should be a struggle for existance which must result in the total victory of one of them. This is a prediction so definite that ecology in the 1930s seemed for once to be really in the position of classical physics, to be on the verge of erecting a body of knowledge based on hypothesis tested by experiment, which experiment should yield a refined hypothesis to be tested by further experiment, and so on.

The crucial experiments were performed not by Lotka or Volterra, but by the Russian, G. F. Gause, working at the University of Moscow. Gause (1934) used as his experimental populations yeast plants and protozoans

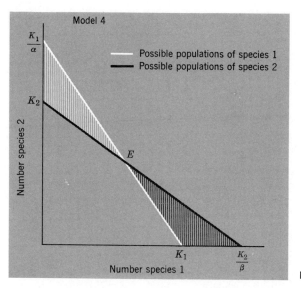

Figure 23.4

of various kinds, plants and animals which should be small, simple, and fast growing. Their simplicity was particularly important to him, because simple animals may have simple wants, which means that you can choose species that are virtually certain to compete in the simple systems you construct for them. Gause's work with different species of *Paramecium* provides, perhaps, the most striking of his tests of the Lotka-Volterra predictions. I have described already (Chapter 22) how he grew *Paramecium caudatum* in centrifuge tubes, spinning the animals down daily to change their water and add a fresh supply of oatmeal medium. Bacteria fed on the oatmeal and the paramecium fed on the bacteria. Gause found that all the species of *Paramecium* which he tried thrived in such tubes of medium. The resulting habitats were so small that he could keep many duplicates and controls, meticulously meeting the require-ments of correct experimental procedure. And he found that whatever species of *Paramecium* he grew alone in these tubes, the population growth was of the expected sigmoid form described by the logistic equa-tion. In different species, inocula of 40 animals developed into stable populations of from 4000 to 20,000 in each of his tubes.

The species of *Paramecium* are so similar that they are hard to tell apart. They thus might be expected to have similar requirements and to compete strongly. It must have been with considerable excitement that Gause chose two cultures to grow together in a crucial test of the Lotka-Volterra prediction. He chose *Paramecium caudatum,* a relatively large and slow-growing species, and *P. aurelia,* a smaller and faster-growing species, for the great adventure, and inoculated a set of his tubes with a few individuals of each. In the early days, both species flourished and both populations grew. This was well, for the crunch should only come

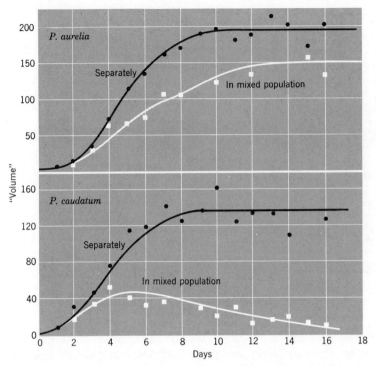

Figure 23.5

Competition between species of *Paramecuim*. Example of data used by Gause to confirm the Lotka-Volterra prediction that closely related species could not coexist when forced to share the same niche. Under the conditions of this experiment (daily changes of water and constant inputs of food) *Paramecium aurelia* always persited while *Paramecium caudatum* died out. Gause was able to alter conditions so that *P. caudatum* could inevitably win the competition instead. Notice that the size of each population is expressed in volumes, a device used by Gause to eliminate the effect of different sizes of the two species from his graphs. (After Gause, 1934.)

when the animals become so numerous that there would be many hungry mouths ready for the daily input of food. In each tube, the population of *P. caudatum* leveled off first, then began to decline, while that of *P. aurelia* went on growing. Eventually, Gause could find few or none of the once flourishing *P. caudatum* populations. In every tube *P. aurelia* thrived alone, occupying the space as if it had started there alone (Figure 23.5). The prediction of the Lotka-Volterra equations was fulfilled. This was inevitable victory by the stronger, suggesting that competition followed Models 1 or 2.

Elton's concept of the niche provides a convenient term for describing the events in Gause's tubes, and we say that the niche requirements of *P. aurelia* were more closely met by the conditions of life that Gause imposed than were those of the vanquished *P. caudatum*. But *P. caudatum* is a good species that thrives in nature. It must have slightly different niche requirements to those of *P. aurelia,* and if we knew what they were, we should be able to rig the conditions of life in the tube so that *P. caudatum* became the stronger, favored competitor. Then *P. caudatum* should triumph, and *P. aurelia* should die out according to the Lotka-Volterra prediction. Gause already knew, from Woodruff's (1913) studies on succession in protozoa (Chapter 8), that some protozoa were able to compete by secreting substances which suppressed rivals. His daily changing of the water should remove such substances, thus removing the advantage of the species which relied on them. He decided to alter his experimental method, leaving the old water in and adding the daily dose of food energy as a concentrate. Any secretions of the animals should be preserved in the tubes. Then Gause tried *P. aurelia* against *P. caudatum* again, but this time *P. caudatum* thrived while the old victor, *P. aurelia,* declined to extinction.

Gause made many other experimental tests of the Lotka-Volterra predictions, using yeasts as well as protozoa. His systems were always simple: small closed containers with food energy supplied in one simple form. The possibilities of different ways of life in these containers were deliberately small so that the animals should compete. There was little chance of the animals avoiding the competition by living different lives, by occupying different niches. And then one of Gause's *Paramecium* experiments had an unexpected outcome; both populations persisted. When *Paramecium aurelia* and *P. bursera* were grown together, neither became extinct, but the population of each leveled off at about half the number it would attain when alone. Gause (1936) soon saw why this was so. One species took to living in the top half of the tubes, while the other took to living in the bottoms. It was possible to find two niches even in so simple a habitat. *P. aurelia* and *P. bursera* avoided competition by living in different ways so that neither became extinct. This was the Model 4 of the Lotka-Volterra formulation, that which allowed coexistence if competition was substantially avoided.

Gause could now conclude that the Lotka-Volterra predictions were verified, so that their predictions could be stated as a working principle. It is known as the **Gause principle,** or the **Principle of Competitive Exclusion,** and may be stated as

"Stable populations of two or more species cannot continuously occupy the same niche."

or, more simply, as

"One species: one niche."

Others have followed Gause's example, chosing their favorite animals and matching species against species. Provided the systems were simple enough, the results have always been as expected; one population of strongly competing pairs has always died out; until very recently. In December of 1969, F. J. Ayala of Rockefeller University reported that strains of two species of fruit fly, *Drosophila pseudoobscura* and *D. serrata*, were apparently able to coexist in culture bottles indefinitely. There seemed no obvious way in which the flies could have divided up their culture bottle habitat as had *Paramecium bursera* and *P. aurelia* in Gause's experiments. The adult flies flew about, laid eggs, and fed together; the larvae burrowed through the agar and pupated together; every seven days an anaesthetic quieted the adults of both species together so that they could be moved to fresh bottles together. These animals must surely be competing strongly, apparently in defiance of the principle of competitive exclusion. Must we abandon it after all? Ayala considered that the only way in which his results were compatible with the principle was for the two species to be really occupying subtly different niches, so that, against all evidence to the contrary, the history of his fruit flies was, indeed, comparable to that of *P. bursera* and *P. aurelia*. He put forward the hypothesis that this must be so, that

$$\alpha < \frac{K_1}{K_2} \quad \text{and} \quad \beta < \frac{K_2}{K_1} \quad \text{(Model 4)}$$

and set out to determine α, β, K_1, and K_2 for his animals by direct measurement. He could calculate the saturation values by growing N_1 and N_2 in his coexisting cultures. From these figures he was able to calculate α and β. His values showed that the conditions of Model 4 were not met. Apparently, the predictions of the Lotka-Volterra equations were not verified for populations of fruit flies, and Ayala concluded that competitive exclusion, as a general principle, must be discarded.

But there is a much simpler explanation of Ayala's results. Fruit flies during their life cycle occupy a series of different ecological niches; the adults fly and feed on yeast; the larvae burrow through agar; the pupae rest and must not suffer interference. It is possible for fly to compete with fly, and species 1 to be the victor. It is also possible, at the same time, for larva to compete with larva so that species 2 is the victor. All that is needed for coexistence is for these two processes to be equal and opposite, perhaps each fluctuating with slight variations in temperature or other environmental parameter. Ayala's work provides a beautiful example of the difficulty of applying simple models to real animals. Fruit flies, like all holometabolous insects, change their ways of life at intervals, occupying a series of niches, and the exclusion principle must be applied niche by niche if it is to have any validity.

There is, however, a more serious general criticism of Gause's work. Like the attempts to study competition in single species populations, his results are open to the objection that the experimental populations were no more than analog computers which obediently followed the programs made for them. The exclusion principle may fairly be likened to a mathematical theorem, reflecting the mathematical formulation on which it is based, and it is indeed sometimes referred to as the Gause theorem. As a theorem it is true in the same sense that Pythagoras' theorem is true. If data required by a true theorem are fed into a computer, we should expect that the predictions of the theorem are fulfilled. Gause's experimental results can be held to reflect no more than this.

And yet the possibility that the exclusion principle really is sound, that Gause's results reflected more than obedient computing by his experimental animals, has far-reaching implications for biology. It shows us how we might understand the most central of biological problems, the diversity of species. Why are there so many different kinds of plants and animals? Darwin had shown us how species had their origin in common stocks from which they were separated by natural selection, a process long known as "survival of the fittest." But what was the "fittest" and why should there be so very many different kinds of these fit animals and plants? The idea of competitive exclusion offers an answer to these questions. A fit animal or plant is one that avoids competition by adopting some private way of life, its niche, which is all its own. Species result not so much from a struggle with others for existence, but from a process of avoiding such struggles. This truly ecological way of looking at evolution and the species problem has many exciting possibilities. It has led to much study of ways in which the principle of competitive exclusion can be tested against the real world and the experience of naturalists.

A practical test of the exclusion principle may be found in the experience of taxonomy, for the niches of animals and plants are reflected in
their shapes. Each unique species, with its unique shape, thus has a
unique niche, making the "one species: one niche" idea easy to accept.
But, more than just being unique, species could now be seen as devices
to avoid competition. The carrying capacity of any place was divided up
into discrete portions, each the undisputed property of the animals or
plants occupying the appropriate niche. All populations living in one
place must thus be significantly different from their neighbors, a conclusion which gives particular interest to instances of close relatives living
together; the so-called closely related sympatric species. Studies of such
sympatric pairs have, in fact, always shown them to be avoiding competition in some special way. Sympatric pairs of cormorants, weaver birds,
big game ungulates, and cone shells all eat different food. American
warblers which eat the same caterpillars, from the same trees, nevertheless hunt in such different ways that each has a private portion of the
crop. Animals and plants of the plankton may use the open water in
turn, and plants may be unpalatable to all but specialist herbivores. The
many studies of related wild animals and plants living together which
have now been undertaken leave very little doubt that direct competition is very rare in nature. But we must then ask how such specialized
feeding habits were developed, and we have gained insight into this
question from a number of studies which suggest what happens when
slightly different populations, long isolated by geographical accident,
are brought together. It seems that extreme types from each population,
those who might compete least, are favored, a phenomenon called *char-*

acter displacement. **When potential competitors accidentally come together to share the same living place, therefore, it seems that two possible things may happen. Either there is direct competition, which results in the total elimination of one population by** *competitive exclusion,* **or there are individuals in each population sufficiently different to avoid competition, and the offspring of these individuals are favoured by selection, a displacement of character that results in two new noncompeting species.**

CHAPTER 24

AN ECOLOGIST'S
VIEW OF SPECIES

The story of Lotka, Volterra, and Gause sounds like a connected series of isolated events; of brilliant men on their own stepping forward with one new insight, each of which triggered some new light in another. But in truth it was not quite like that. Ecologists everywhere brooded long on the implications of competition, and very many had in their minds an idea of how the concept of niche must aid an understanding of the species problem. Gause's experiments made clearer the implications of niche, and led to more formal statement of the exclusion principle, but a general test of the principle was already at hand, showing how it could be received with such satisfaction by biologists. The test lay in the experience of classical taxonomy.

Species are classified by shapes of their bodies, but these shapes reflect function so closely that you can usually deduce much about the niches animals occupy merely by looking at a corpse; talons mean a carnivore, hooves means fleetness of foot, opposing thumbs mean climbing trees, and so on. The deduction of niche from shape is what palaeontologists do every time they reconstruct the life of an extinct animal. To a remarkable extent, therefore, museum taxonomy had already tested the exclusion principle; indeed, taxonomists may be said to have found it on their own, for there are groups of organisms whose shapes are so similar that they have been classified by function. Pathogenic bacteria may be classified by testing them against a host, and parasitic nematodes by the plant hosts in which they are found. This experience of classical taxonomy explains why the exclusion principle could gain ready acceptance as a working tool. A species could be thought of as a morphological expression of the animal's way of life, of its niche. Animal and plant species were unique; they reflected unique niches; one species, one niche.

But the exclusion principle implied something still more fundamental than this, for it implied that niches were developed to avoid competition for energy. When in its own niche, an animal triumphed; it was not then engaged in an endless struggle with other animals. Speciation was not so much a struggle for existence as a device for avoiding such struggles, even though competition was used to decide the boundaries of niches. This is a profitable way of looking at evolution. When we seek to discover the role of an animal in the community, that collection of responses and activities which Shelford once called its mores and which Elton later called its niche, we must particularly look for the ways in which the animal avoids competing with its neighbors.

Within a few years of the publication of Gause's main work in English, in 1934, many scientists were looking at animals with the exclusion principle in mind. The implications of the principle became particularly interesting for pairs of closely related species which lived close together, species which are called **sympatric** (literally "same country"), because such species might be expected to have similar niches and, therfore, be in danger of competing. How might pairs of sympatric species avoid competition? In the 1930s, David Lack, now director of the Edward Grey Institute of Ornithology at Oxford University and the doyen of British bird men, set himself the task of testing the exclusion principle against all the pairs of closely related sympatric species in the British list. He could not find enough data on feeding habits to be sure about many, but for all those for which there were good data he was able to show that they fed differently (Lack, 1944, 1945). One of nicest sets of data was for the two species of British cormorant, the common cormorant *Phalacrocorax carbo* and the shag *Phalacrocorax aristotelis*. These two birds look so alike that they can often fool an amateur (Figure 24.1). They live on the same stretches of shore; they both feed by swimming underwater after fish; they both nest on cliffs overlooking the sea; they are both common; they were both hated by fishermen for stealing their livelihood. This last provided the data that David Lack could use, because the fishermen had once been so vehement in their denunciations that local councils had put a price on cormorants' heads so that they were shot by the thousands. When it became apparent even to the fishermen that the slaughter made not a bit of difference to the fishing, the councils decided that their money might be better spent paying fisheries biologists to report on the diet of the birds. The Plymouth marine laboratory undertook the investigation, examined stomach contents, and made field studies. Shags, by far the most abundant of the two species, ate mostly sand eels and sprats, which were not commercial fish. The common cormorants ate various things, particularly shrimps, and including a few small flatfish but no sand eels or sprats. The flatfish were commercial species, but the take was negligibly small and the fishermen's wrath quite unfounded. This was no surprise to ecologists, but the data were very much to the point for Lack's study. The food of the two species was obviously quite different, so that they avoided competition and the exclusion principle was upheld. The fisheries study had shown further how the catching of different fish was ensured, because the shags did their fishing in shallow estuaries while the common cormorants went further out to sea. Lack was also able to show that the nesting requirements of the two birds were different even though they did nest on the same cliffs. The shags nested low among boulders or on narrow ledges, whereas the common cormorant nested on the high

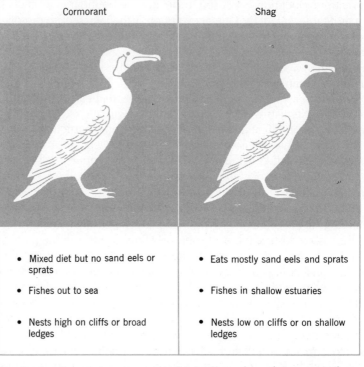

Cormorant	Shag
• Mixed diet but no sand eels or sprats	• Eats mostly sand eels and sprats
• Fishes out to sea	• Fishes in shallow estuaries
• Nests high on cliffs or broad ledges	• Nests low on cliffs or on shallow ledges

Figure 24.1

Closely related sympatric cormorants in Britain. These close relatives, so similar to look at, get their livings in quite different ways and so do not compete. We infer that their ancestors were separated by character displacement and that the species were then preserved by competitive exclusion as genetic-isolating mechanisms were evolved.

tops or on broad ledges. In short, these closely related birds, so similar to look at, had niches which were quite distinct. In their normal lives they were unlikely ever to come into competition.

This kind of study has now been repeated many times, for the discovery of a pair of related animals living together is sure to set an ecologist to finding out how; the exclusion principle is one of the few firm anchors in his diffuse science, and he can use it as a physicist uses his general principle of the conservation of mass; as a base from which he can make his assault on complexity.

Three yellow weaver birds of the genus *Ploceus* bred side by side in one colony stretching nearly 200 yards along the shore of Lake Mweru in Central Africa, and the man who found them promptly shot a few to see what they were eating (White, 1951). The stomachs of one species had hard black seeds in them, those of the second soft green seeds, whereas those of the third held nothing but insects.

Rapacious gastropods of the genus *Conus*, the pretty cone shells of the collector, live many species together on Hawaiian reefs; but careful mapping of their distribution shows that they have divided the sub-littoral zone into six or more narrow strips so that their ranges scarcely overlap (Figure 24.2). They have divided up the feeding grounds in much the way that Gause's *Paramecium aurelia* and *P. bursera* had divided up his culture tubes; there was a simple physical division of the available space. But the *Conus* species also specialized in food, as Alan Kohn (1959) was able to show by keeping them in aquaria or by cutting open wild *Conus* which he saw feeding. Some ate small worms, some ate large worms, some ate different sorts of snail, and the deep water species used their poisoned darts to catch small fish. The whole series of *Conus*, so apparently alike, apart from differences in size and pretty markings, had found a set of specialized ways in which they could each carry on the trade of poisoned dart-wielding carnivorous snail on the same stretch of coast without coming into competition.

Herbivorous animals are provided with easy chances for food specialization by the great variety of plants making up vegetation, the effects of such food specialization being particularly obvious among herbivorous insects. Any amateur lepidopterist knows that you must get the right food plant if you are to successfully rear caterpillars, implying that the closely related butterflies which flitter together in clouds over a meadow are avoiding competition by doing their growing as larvae on different kinds of meadow plant. The plants may be said to force this specialization, with specialized chemistry that makes them inedible to all but the specialist herbivore (Whittaker and Feeny, 1971).

The engaging little birds called warblers are mostly so similar, even in coloring, that learning to tell all except breeding males apart is one of the trials of a beginning bird watcher. In the eastern United States there appears each spring a particularly frustrating assortment of them, flying north from their Caribbean winters, along common flightways, to their common breeding grounds in the woods of New England and eastern Canada. Five species, in particular, nest in the spruce forests of Maine and Vermont. The five birds are closely related. The vegetation in which they breed is without obvious variety, just ranks of spruce trees which are all alike. The beaks of the birds are all the same size, and alike, suggesting they can eat the same food. Investigations of enormous numbers of stomach contents by forestry people, looking for enemies of the spruce budworm, have shown that their food is, indeed, roughly the same. The food of closely related sympatric cormorants and weaver birds had been found to be different, but these little warblers even had tastes for food which were alike. How then can they occupy different niches? How can there be more than one species of them? One of the foremost ecological theorists of our day, Robert MacArthur, earned his doctorate by answering these questions (MacArthur 1957).

MacArthur spent many long hours in the springs of several years watching the little birds. Each time he saw one he noted exactly where it was; on top of a tree, at the side of a tree, on the ground, flying about; and he started a stopwatch going so that he could measure in seconds just

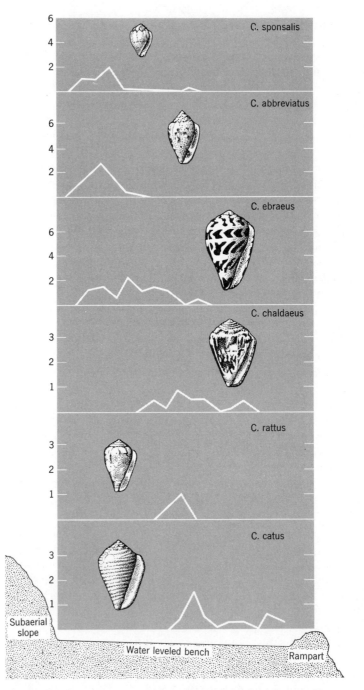

Figure 24.2

Closely related sympatric cone shells in Hawaii. Animals of the genus *Conus* are carnivorous snails that hunt with poisoned darts. These sympatric Hawaiian species live in parallel strips along the shore where each catches a unique array of prey animals (From Hutchinson, 1965).

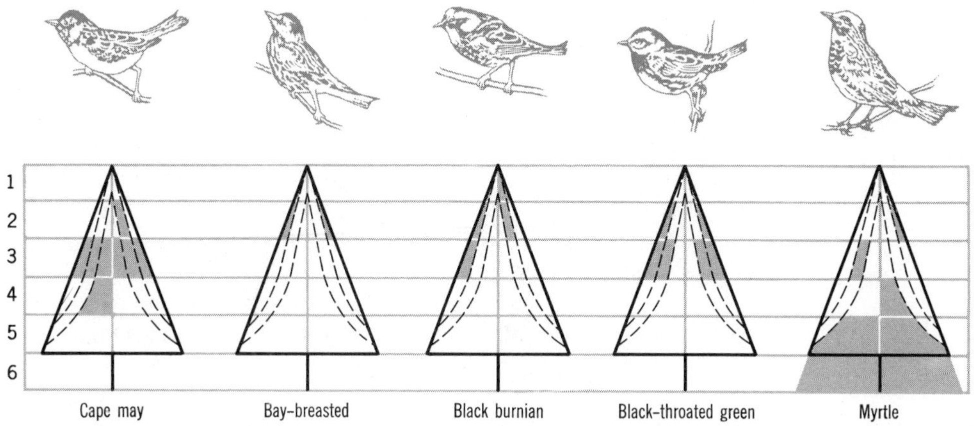

Separation of niches by warblers eating the same kinds of caterpillars. The five kinds of warbler feed on bud worms on the same spruce trees. These diagrams illustrate MacArthur's study which showed that the birds hunt in different parts of the trees, so that each kind has a private crop. MacArthur divided the trees into 5 layers, and then considered each branch as having a top, a middle, and a bottom, which let him divide each tree into fifteen compartments. A sixteenth compartment was provided by the ground underneath. MacArthur then noted in which compartment was each warbler he saw, timing some of them to see how they divided their time between compartments. The stippled areas show where they spent more than 50% of their time or where they were on more than 50% of his observations. The results of timing the birds are on the left of each tree; the results of other observations are on the right. It is evident that the birds go some way to avoiding competition by hunting in different parts of the trees. Different methods of hunting further separate them. (From MacArthur, 1958 and Hutchinson, 1965)

Figure 24.3

how long the warbler spent where it was. This was a tedious and time-consuming undertaking, for the little birds are hard to see in the dense spruce forest, and they never remain in sight for more than a few seconds. But eventually MacArthur had watched for so long that he could be certain of where each kind of warbler spent most of its time. And it was clear that the warblers worked in substantially different parts of the trees (Figure 24.3). One kind spent nearly all of its time on the pointed spruce tops, another one lower down, a third on the ground, and so on. The spruce budworms, the most abundant food for all the warblers, lived all over the spruce trees, but the warblers hunted in their own special preserves. Being such mobile creatures, the birds did poach each others space somewhat, but MacArthur was able to show that other behavioral traits stopped them from poaching many of each other's caterpillars even then. He timed the motions of each kind of warbler, noting how long each spent hovering, running along branches, or slowly plodding, and was able to show that each species had a characteristic pattern of doing things. One was more active than others; another was more deliberate. There seemed little doubt that these different activities reflected different hunting methods. One kind of warbler got caterpillars on tops of needles, another kind got caterpillars hidden under needles, and so on. Even though a warbler might poach another warbler's space, and hunt the same kind of caterpillar there, the two might still not be competing because the hunting method of each caught a different portion of the total crop.

It has long been recognized that the many grazing animals of the African game herds must be specializing in food, and students of the herds are now beginning to show how the niches of each species in those moving masses of ungulates differ. In Chapter 8, I described how recent work had shown that zebras, wildebeest, and Thompson's gazelle grazed in succession. First the zebras took the long dry stems, an action for which their horsy incisor teeth clearly adapted them. Then the wildebeest took off the side-shoots, gathering with their tongues in the bovine way and tearing off the food against their single set of incisors. Finally the Thomson's gazelles came along and were able to pick out ground-hugging plants and other tidbits which the feeding methods of the others had both overlooked and left in view. Completely sympatric though these and other animals of game herds might be, they clearly avoid competition by specializing in the kinds of food energy which they take. And on the flanks of the herds move the specialized carnivores, the cats of different sizes, the hyenas on the watch for the suitably weak, the pack dogs which panic herds to cut out calves. All the exciting list of animals in the great herds can be seen by an ecologist in terms of the exclusion principle: a set of species which represent a set of niches, each one of which is a way of life conditioned to avoid competition with the other ways of life around it.

Short-lived animals and plants may avoid competition if they stagger their activities so that they appear at different seasons of the year; and such seasonal successions as those of insects in woods can obviously be thought of in this way. The changing seasons themselves provide cues for the animals to make their appearance at the right time when the habitat is not occupied by some other animal which might compete, but even when the impact of the seasons is slight, such time-separation of life histories is probably common. Something of the sort seems to be needed to explain the diversity of plant species in the phytoplankton of lakes and the oceans. Planktonic plants all acquire their energy in the same way, by photosynthesis; they all presumably require the same essential dissolved nutrients, such as the limiting phosphorus; they live all stirred up together in a transparent medium which they can hardly divide up into private spaces; and yet there are many species of them. G. E. Hutchinson (1961) has called this "the paradox of the plankton." In lakes, at least, it is known that many of the plant species bloom at different times, and that they possess resting stages in cysts or bottom-living forms which sink to the surface of the mud. It seems likely that they avoid competing, in part, by occupying the open water in turns. Doubtless there are also other subtle separations of phytoplanktonic niches; such as harmful secretions like those used by protozoa and fungi; special nutrient requirements like the vitamin B_{12} that some must receive from solution in the water; and special relationships with animals of the plankton, such as size or taste affecting the extent to which they are grazed.

The filter-feeding animals of the zooplankton which eat the tiny plank-tonic plants might be expected to compete strongly also, and yet there may be several species of cladocera or copepods living together in the same lake. They apparently swim through each other's water, filtering with their legs and mouthparts as they go. How can they possibly avoid competing for the supply of plants? Hutchinson (1951) has examined much evidence on the subject and concluded that even filtering can be selective. Some animals reject large particles from their filters, some reject small particles, and some are able to so direct their filtering devices that only particular kinds of plants are taken. So even filter-feeding animals can chose their food. And their doing so gives another lead to the problem of the paradox of the phytoplankton; since, if animals concentrate on one kind of plant over another, this immediately produces specialized niches for the plants. The plants are so grazed down by their specialized herbivores that they never directly compete for space, and the important parts of their niches become their relations to things that eat them. One plant can occupy the niche of not-being-eaten-by-herbivore-A while another plant occupies the niche of not-being-eaten-by-herbivore-B.

For the longest-lived of organisms, forest trees, there are still some quite general problems of niche to which we do not know the answers. The niches of plants of the succession stages are easy enough to see, as are the competitions which occur as the habitat is changed by its first users to provide for the niches of those plants which come later. The final niche of the dominant tree of a temperate forest is also apparent, and the vagaries of soil and exposure suggest how niches for a few other species of tree may also be developed. But in what ways do the niches of the hosts of species of large trees in some rain forests differ? Rain forests may have 100 species of tree to the acre (Richards, 1952), and it seems unlikely that there are 100 distinct ways in which an evergreen forest giant can get its livelihood in an acre of equatorial lowland. In maturity, the giant trees must be expected to hold space too effectively to be displaced by the subtle pressures of other roots and leaves. But for the species to be maintained in the forest there must come a time when seeds are scattered and seedlings are established; and it is then that the bite of competition should be applied. I have seen a place in the Nigerian rain forest where one of the canopy trees had died, and its corpse swiftly torn down by decomposing fungi and termites. Light streamed into the gloomy forest floor through the gap which its canopy once filled, and numerous tree seedlings were reaching up toward the disc of light; it was like a stilled frame from a motion picture in which the race of competing trees for that single place in the canopy was preserved with all its sense of movement. There may well be 100 ways of spreading seed through the forest and poising a seedling for that race for the canopy, and it is, at least, possible that rain-forest trees must avoid competition most during establishment and seedling growth. Recent studies in Costa Rica (Jantzen, 1970, Chapter 34) suggest that avoidance of seed predation by insects has also been one of the drives leading to many tropical tree species. This explanation is akin to that offered for the paradox of the plankton; the plants, or for forest trees particularly their seeds, are chemically and physically different to escape attack by generalist herbivores. But the matter needs more investigation.

CHARACTER
DISPLACEMENT IN
SPECIATION

Nuthatches (family Sittidae) are small birds with long woodpeckerlike bills who search for food by climbing up and down trees or rocks on their short powerful feet, hunting in cracks and crannies. This way of life is specialized, the progress of the birds being odd enough to attract the attention even of laymen. The physical adaptations for this way of life (long bills, short stubby tails, and big feet) give nuthatches a characteristic appearance, making them all look much alike. Strikingly similar are the common species of Greece and Turkey, *Sitta neumayer* and of central Asia, *Sitta tephronota* (Figure 24.4). A skin of *S. neumayer* from Greece may be so like that of *S. tephronota* from a site in Central Asia more than 1000 miles away as to deceive even a competent·museum taxonomist. The nuthatches of these distant sites, so closely related and so similar, should be occupying closely similar niches. What differences there are between them may be explained in terms of the geographical isolation of their breeding populations. Where their ranges come into contact, as they do south of the Caspian Sea in Iran, we should expect the species to come into competition, and perhaps to find evidence that one was triumphing over the other. What we in fact find is that throughout Iran both species coexist, but that they are here so strikingly different that there is never any difficulty of telling them apart (Vaurie, 1951). Their bills are of different sizes and one bird has a thick black stripe from eye to shoulder, whereas the other has almost lost its eye-stripe. The different bill lengths suggest different feeding habits; the different markings suggest distinctive patterns which could be used in recognition of suitable mates. The species differences have been accentuated, and there is no doubt that the nuthatches occupy different niches where their ranges overlap. Competition has been best avoided by the individuals that were most different in the two populations, and these were favored by selection.

When Charles Darwin first pondered the mechanism of natural selection, it was already evident to him that animal characters must diverge when they lived in the same country, and he outlines his conclusions in a section of *On the Origin of Species* entitled "Divergence of Character." Modern ecology has confirmed and made more definite Darwin's conclusions. The different environments of different places lead to populations differing slightly, as individuals finding niches most suited to local conditions are favored. Then some accident of history makes the ranges of the distant populations merge again. They must compete; and the Gause principle predicts the outcome. If there are individuals present who can occupy different niches, they will be favored, and their differences accentuated. If not, variability may well be suppressed as only one kind triumphs, being able to find only one niche. Vaurie's observations on the Greek and Asiatic nuthatches of Iran, showed that the two populations must already have been diverging before they were brought

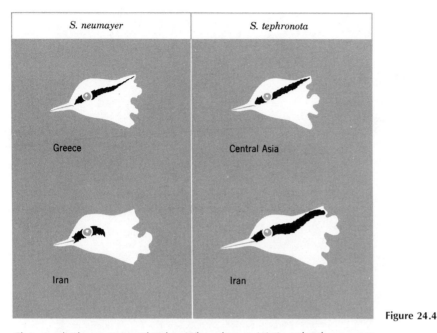

Figure 24.4

Character displacement in nuthatches. Where the two Asiatic nuthatches are sympatric, in Iran (bottom), they may be easily distinguished by bill size and eye stripes, but individuals from far parts of their ranges (top) are very similar. Evidently selection has favored individuals who are most different in the region of overlap. (Redrawn from Vaurie, 1951)

together again. Then avoidance of competition, in keeping with the exclusion principle, brought out the differences in the populations.

Many times since Darwin's day naturalists have noted and reported the exaggerated differences that are apparent in countries where closely related animals coexist, but the phenomenon became more generally talked about after 1956 when W. L. Brown of Cornell and E. O. Wilson of Harvard collected together many examples and invented a name for the phenomenon, **character displacement.** They started with Vaurie's nuthatches, and supplemented this with examples from other birds, fishes, frogs, beetles, ants, and crabs.

Since Brown and Wilson's paper, the outcome of competitions such as Gause engineered have been thought of conveniently in two terms, **character displacement** and **competitive exclusion.** The experiment in which *P. aurelia* and *P. bursaria* lived on in different parts of the tubes showed that **character displacement** had occurred, with the bottom-living individuals of one species and the top-living individuals of the other being favored. The outcome of all other experiments, in which only one survived, could be called the result of **competitive exclusion.** In nature displacement should be more common than exclusion because the real world must always offer many ways in which niches can alter, thus letting

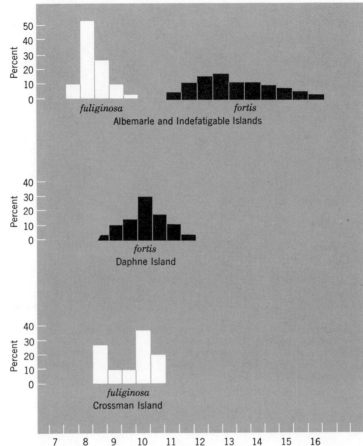

Figure 24.5

Character displacement in Darwin's finches. Histograms of beak-depth in *Geospiza* species (Darwin's finches). Measurements in millimeters are placed horizontally, and the percentage of specimens of each size vertically. If the beak-depths on Daphne and Crossman islands are indicative of optima in the absence of close competitors, the displacement on islands with competitors is most easily interpreted as character displacement. (Lack, 1947.)

every chance for diversity to give advantage. In Gause's tubes the chances of finding other niches were kept deliberately small.

The Galapagos finches of the genus *Geospiza*, whose diverse forms provided Darwin with key evidence for the theory of evolution, provide, fittingly enough, one of the best examples of character displacement. The birds occupy many niches not occupied by finches in other parts of the world, no doubt because other passeriform families are lacking in the Galapagos. The finches eat a great variety of seeds, insects, and foliage. Their beaks are of many different sizes. David Lack (1947), the Oxford ornithologist who had tested the Gause principle against the related pairs of birds in the British fauna, worked extensively on Darwin's finches. He found that where two species *G. fulginosa* and *G. fortis,* lived together, as they do on two of the larger Galapagos Islands, their beak sizes are quite separate with no overlap. But on the tiny island of Daphne there was only *G. fortis,* and these birds had beaks covering the range of sizes of both species on the larger islands. On the equally tiny Crossman Island there was only *G. fulginosa,* and these birds also have beaks of the wider size-range. On the large islands competition has been avoided by character displacement. On the tiny islands, there is apparently only room for one population, which may hold the island against invasion by the mechanism of competitive exclusion (Figure 24.5).

Water ferns of the genus *Azolla* on the Galapagos Islands reveal a different history, showing how exclusion may operate to preserve a range in nature. *Azolla* grows floating on small ponds like duckweed, completely covering the water (Figure 24.6). There are few ponds suitable for *Azolla* on the whole Galapagos Archipelago, only a few dozen at most (Colinvaux, 1968). These now support populations of only one species, *Azolla microphylla.* The plants are apparently spread from pond to pond on the feet of birds, and we must imagine that some lucky flight brought a living fragment some 600 miles from the South American mainland to a suitable pond. Once established, the local Galapagos ducks should quickly spread the ferns to all other Galapagos ponds, after which there should be little chance of a second species ever getting established following another chance transoceanic flight, since the niche of *Azolla* floating on a pond must be as about as restrictive and simple as those which Gause contrived for his *Paramecium.* It is, therfore, not surprising that only one species of living *Azolla* has been found on the islands. Recently, however, we have been looking for *Azolla* spores in drill cores of mud from a Galapagos lake (Schofield and Colinvaux, 1969). *Azolla microphylla* lives there now and we find plenty of the spores, which are released into the water, in surface mud. But in mud deep down in our drill cores, shown by radiocarbon dating to be more than 48,000 years old, we find only spores of another species, *Azolla filiculoides,* a plant known from South America but not now growing anywhere in the Galapagos. How had such a changeover occurred? The ancient mud is separated from the mud of the last 10,000 years by red clay deposits, suggesting that the lake had been dry for a long period. Drought had exterminated that ancient population of *A. filiculoides.* When the lake again filled with water many thousands of years later, chance brought a migrating bird with a piece of the other species on its foot, and a second species held the lake against invasion. This, at least, seems the most likely explanation of the fossil history. It suggests, what is also implicit in the Lotka-Volterra equations, that being there in strength first can give a competitive edge to one species over another which might win if the two started out with equal numbers.

Speciation can be seen as a two-part process requiring, first, geographical isolation and, second, the merger of once isolated populations. As animals and plants, who are the actors in the evolutionary drama, spread

THE THEATRE AND
THE PLAY

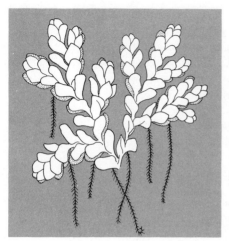

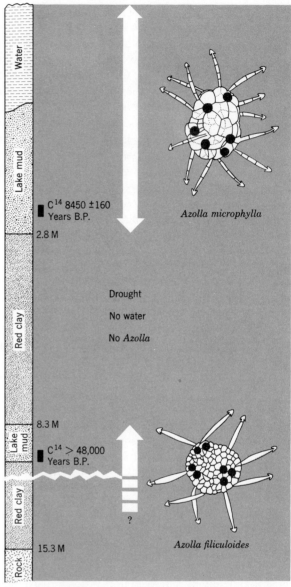

Figure 24.6   History of water ferns in a Galapagos lake. Water ferns of the genus *Azolla* loose their spores into the water incased in structures armed with hooks and called massulae. The shapes of these structures are different in the different species. The lake dried up for many thousands of years in its middle history, after which the new lake was occupied by a different species of *Azolla*, and the only one now known from the Galapagos Islands. This history probably illustrates competitive exclusion, the triumphant species each time being that which reached the Galapagos first by chance transport from mainland South America in each of the wet epochs.

over the land, they come to live in different environments. The offspring of some individuals are favored over those of others in these new places so that distant populations tend to vary; the ecological stage has directed the progress of the players. The niches of distant populations, although having many things in common, are extended in different directions. When accident brings the players back into the same habitat, they are likely to compete and their differences will become important. Now the actors are on a different ecological stage, that produced by the interactions of actor with actor in the common quest for energy. Competitive exclusion and character displacement determine the outcome of the drama, and differences are either buried or so enhanced that new species are formed ready for the start of the next act. The steps of the actors must often be directed by chance; accidents of climatic history or chances of dispersal which allow one population into a habitat before another; but the unfolding of the drama is restricted by the dimensions of the stage; by the range of environments, the chance of dispersal, the common need for energy, and the principle of "one species-one niche." This is the process so aptly described by G. E. Hutchinson's (1965) phrase, "The Ecological Theatre and the Evolutionary Play." It gives an intuitive understanding of the mechanisms which must lead to diversity in nature. But it still leaves ecologists with the task of explaining the actual diversity they observe, of determining why plants and animals of any place are common or rare, closepacked or spread out; and of why the process of speciation has resulted in just the number of species we see rather than in more or less.

Virtually all the agencies of natural death can be expected to work in density-dependent ways, but there is no physical law which says that animals and plants must be under density-dependent control; all that is required is that their numbers are curbed. Periodic catastrophe should do just as well, although this might not make for such predictable annual populations as we actually observe. Strongly in favor of the density-dependent thesis, however, is that the logistic hypothesis, of which it is a part, predicts the existence of discrete species. If discrete species should be expected, on the assumption that populations grow until competition checks them, then the existence of discrete species is strong support that natural populations are, indeed, controlled by competition, or density-dependent pressures acting in a like manner. The animal populations for which we have the best historical records are those of birds, and what we know of bird populations is consistent with the density-dependent hypothesis. The number of birds breeding each spring seems remarkably constant, suggesting that any surplus individuals produced in years particularly favorable for breeding are eliminated in the following winter by density-dependent death. The hypothesis thus requires that regulation is achieved by changes in the death rate, not by changes in the birth rate, and is thus consistent with the theory of evolution by natural selection which requires that animals must always breed to their utmost ability. Convincing although the hypothesis is for birds, the practical difficulty remains that the necessary density-dependent winter death has never been demonstrated. Supporting evidence for other kinds of animals is even less good, but there is nevertheless a general feeling

357

among naturalists that populations of everything in the wild, from butterflies to fish, are very commonly controlled by changes in death rates acting in density-dependent ways.

CHAPTER 25

BIRDS AND THE
DENSITY-DEPENDENT
HYPOTHESIS

Long experience with keeping small animals in laboratory culture has seemed to give ecologists a general feel for the processes of control which should be going on in nature. Laboratory populations come into balance, and this balance can be readily understood to result from density-dependent pressures caused by crowding. Actual mechanisms of control by density-dependent factors can be shown to be operating in these simple culture systems, even though the mechanisms could be decidedly devious, as the long history of *Tribolium* studies shows. There seems to be a similar balance in nature; surely this natural balance must also be set by the mechanism of the crowd, by density-dependent factors.

It takes no great effort of the imagination to see how many of the hazards of the natural world could act in a density-dependent way. Starvation in the wild, like starvation in the laboratory, can readily be seen to be a curb of ever-increasing effect as individuals crowd, for the quest for food will become harder and harder as the population grows. Then it is well known that the dangers of disease go up as crowding increases, both for men and animals, so that disease as well as starvation can be density dependent. The chances of being eaten by a predator may also go up for those who are abundant, for predators commonly eat several kinds of prey and are likely to concentrate on the easiest to catch, which is to say the commonest. Many animals require special places for building nests or homes, and these must necessarily be harder to find as the populations become denser. And many animals must seek special shelter from catastrophes such as cold or flooding, and these refuges, too, will become less available as the populations become dense. It is, in short, the normal experience of the naturalist that animals are commonly subject to adversities that must act in density-dependent ways. These forces for control are always present, and must have effects. The only question of doubt is over how strong these effects really are.

But yet there is no scientific principle which states that populations must be controlled as a function of density. Density effects will always be working in crowded populations, but it is always possible that actual control in the rough-and-tumble life of the real world is exercised on growing populations before they become crowded. All that can be said on general scientific grounds is that populations must be controlled somehow, that they cannot go on growing forever without violating fundamental physical laws.

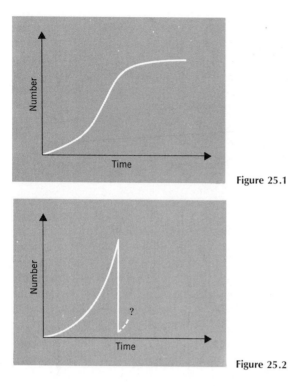

Figure 25.1

Figure 25.2

Preoccupation with the idea of balance can give the impression that a population growth curve must look sigmoid as shown in Figure 25.1 but the requirements of physics are just as completely met if the growth curve looks like that shown in Figure 25.2. "Control" on this second model is brutal and catastrophic; it choses no particular individual over another but acts independently of density. It is, in the colorful language of a Washington reporter describing a conservative senator's assault on a foreign aid bill, the "meat axe" approach to regulation. And it solves the problem of overcrowding just as completely, if less comfortably, than a gentle density-dependent device. This is something that should be pondered well by those who look with equanimity on the rapidly growing human population of the present.

Any control agents that are not dependent on density must take the form of sudden events that are catastrophic to animal populations, but the natural world is well supplied with catastrophic events. None can doubt that things such as hurricanes and volcanic explosions destroy whole populations, although these events are, perhaps, too scattered and local to be used for a general theory of population control. More applicable are the common catastrophies which we know as changes in the weather. Nearly all places on the earth suffer seasonal changes from summer to winter, from warm to cold, from wet to dry. Each change in these yearly cycles must represent hazards of varying severity to the animals of each

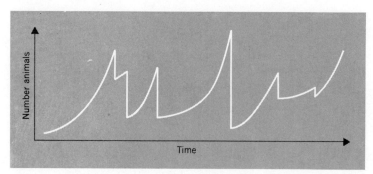

Figure 25.3

place. Even over much of the tropics there are seasonal changes, those of the monsoons for instance. It is thus easy to imagine how growing populations can be frequently cut back, making their normal lives a race to reproduce so quickly that there shall be at least some survivors of the next expected catastrophe. Unless something happens to prevent the catastrophe from arriving on time, such a population might never grow large enough to suffer the effects of crowding. The population would be under some sort of control, in that numbers fluctuated within wide limits, but it could scarcely be described as "in balance." A graph of its population history would be a jagged affair, which might look something like that shown in Figure 25.3. Even such a population history as this gives some hint of density-dependent effects, however, for otherwise we should expect the population to become extinct. The death rate is apparently relaxed in the troughs to let the population increase again, so that death rate is, indeed, influenced by density. But this is not "balance," although it may yet be allowed for in our subjective impression of the "balance" that we see in nature. Our idea of the normalcy of our surroundings may be no more than a subconscious averaging of many local fluctuations. I think again of Lindeman's several year's records of the populations in one lake, when he found that different types of animals were commonest each year (Chapter 11). This is probably true for most temperate lakes, but they all give the impression of being normal, of illustrating, as we say, "the balance of nature."

Ecology in the last two decades has struggled with this great and general question: Is there a true population balance in nature, with animal numbers regulated by density-dependent factors? Or are animal numbers really controlled in a catastrophic and density-independent way by weather, so that the apparent balance in nature is just the result of a general averaging? From its onset, the debate has been influenced by the apparent intellectual triumphs that had come from studying crowds by theory and experiment. "If populations are controlled in a density-dependent way by crowding, we can describe them with the logistic model; if we develop the logistic model to predict the results of competitions between species, we arrive at the principle of competitive exclusion; and the exclusion principle gives us an understanding of the distinctness of all the earth's species." This is a formidable line of argument. If the niches of all species are set by competition according to the exclusion principle, then their own populations must be set by competition also and the density-dependent theory of population control prevails.

Many, perhaps most, ecologists have been well content with this argument, considering that density-dependent mechanisms do exercise control over all wild populations. The effects of severe catastrophies have been acknowledged, but have been considered to introduce but temporary perturbations into systems normally in balance. This viewpoint was summarized in an important paper by the Australian entomologist, A. J. Nicholson (1954). Nicholson worked in the way that Gause worked, by refining mathematical models and testing them with experimental populations. He chose experimental animals of his own, bluebottles or flesh flies of the genus *Lucilia*, which he could conveniently rear in gauze cages supplied with blobs of meat. His animals lived more naturally than paramecia on oatmeal medium or fruit flies on agar, because wild flesh flies also laid their eggs on discrete blobs of meat (corpses) in which their larvae were forced to live and grow. From considering, and mimicking, as many natural systems as possible, Nicholson was able to develop a general hypothesis of the density-dependent control of all natural populations.

But Nicholson's work was still subject to the general criticism that most animals in nature live in conditions quite unlike those of the confined laboratory populations for which mathematical models were so successfully descriptive. And wild populations must be affected by the vagaries of the weather to some extent. To support a general theory of density-dependent control in nature, it became important to show that wild animal populations were truly in balance in spite of the weather. The catastrophes of life must be shown to produce only transient changes in animal numbers, with a natural balance speedily restored when the catastrophe passed. Evidence for natural balance has been perhaps most successfully sought by zoologists working with vertebrates, particularly with birds.

The common European heron (*Ardea cinerea*) is both conspicuous itself and builds its large nests high in the trees of breeding colonies. Some heronries in England are known to have existed for centuries, so that we know that the birds have maintained themselves for long periods. Since 1928, amateur naturalist groups have kept count of the number of occupied nests in many heronries, so that we are provided with a remarkably complete annual heron census for large parts of England. David Lack (1954) collected the census data from the two best studied districts to produce the following heron histories (Figure 25.4). Several things about these heron histories are revealing. The numbers do fluctuate from year to year but, except after the winter of 1947, not by very much. The record is, in fact, just the sort of thing that should be expected if the populations were in balance and controlled as a function of density. That there should be a marked fall after the winter of 1947, was

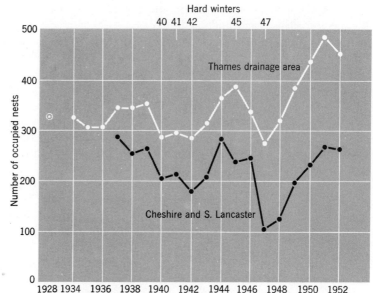

Figure 25.4

History of breeding populations of herons in England. The herons were censused by counting occupied nests in ancestral heronries each spring. The population is remarkably constant from year to year, strongly suggesting density-dependent control. The particularly hard winter of 1947 reduced the numbers of herons breeding next spring, but the population quickly regained its old level in subsequent years. (Redrawn from Lack, 1954.)

also striking, for the English winter of 1947 was notoriously bad. Usual English winters are so mild that lakes and ponds are free of ice for much of the time, and rivers and estuaries nearly all of the time, but in 1947 there was a long freeze of the kind common in continental countries. This was particularly hard on herons, and dead herons were found by frozen lakes that winter. In the spring of 1948, there were many empty nests in the ancestral heronries, but the populations were quickly built up in subsequent years. A catastrophic event had, indeed, severely reduced population, but in doing so the pressures of crowding on the survivors appeared to have been released so that heron numbers could expand back to their old levels.

Lack used this history of the herons as his first example in a book that was to become one of the signal works of ecology, his "Natural Regulation of Animal Numbers," published in 1954, the same year as Nicholson's general density-dependent hypothesis. Lack had early been influenced by Gause's studies on competition, and had been quick to see that his exclusion principle was beautifully upheld by the different habits of closely related sympatric bird species. More perhaps than anyone else, Lack had called attention to the importance of exclusion as an ecological concept. His studies with British birds such as cormorants are still some of

Stork nest on a village roof. Storks which build nests on the rooves of houses are easy to census during the breeding season (Figure 25.6). We have good census data from only the few species whose habits make them conspicuous like these storks.

Figure 25.5

the best demonstrations of the working of the principle, and his study of speciation in the Galapagos Darwin's finches is a classic (Chapter 24). If all bird niches were set by competition, as the exclusion principle suggested, then it must follow that the numbers of the birds were also set by the mechanism of crowding. It could have been no surprise to Lack to assemble the heron census data and to find so strong an indication that heron numbers were in balance, were apparently regulated by the density dependent mechanisms of the crowd.

Lack found that data similar to the heron census existed for the white stork (*Ciconia ciconia*) in Germany. Storks build huge nests on the roofs of houses in German villages and are, like the herons, very faithful to the homes of their ancestors. No elaborate sampling system was needed to take the census of these birds, and the counts could be relied on. Once again the results suggested a population in balance (Figures 25.5 and 25.6).

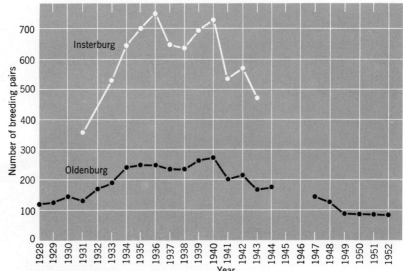

Figure 25.6

History of breeding populations of the White Stork (*Ciconia ciconia*) in parts of Germany. (Redrawn from Lack, 1954.)

Most birds are not as easy to census as storks and herons, and only for a very few do we have reliable long-term census data, but an exception is for the great titmouse (chicadee) *Parus major* in a Dutch wood. The bird naturally nests in holes in trees, but it prefers nest boxes. All that need be done to count the breeding titmice in a wood is to place many more nest boxes than there are breeding pairs there, and to count how many are occupied. This novel method of census has provided a long history of a local Dutch population of great tits going back to 1812 (Kluijver, 1951). The results (Figure 25.7) again show fluctuation about a mean, although a rather wide one in which the densest population if four times the sparsest. But there is about this history nothing of that steady rise followed by unexpected fall which should denote the fluctuations caused by random catastrophies, nor is there a rhythm that might reflect recurrent seasonal crises. The tits lay six or more eggs at a time, showing that fluctuations could be very wide indeed, if there was no density-dependent control; for a population might multiply 36 times in two years, if not checked.

These census data, coupled with much less complete but nevertheless suggestive data about other birds, let Lack conclude that bird populations must be normally in balance. His records were largely based on census of breeding pairs in the spring. Census in midsummer would, of course, show much larger populations since the year's fledglings would be included. The stork or heron population would be half as much again or, perhaps, twice the breeding population, and the great tit population several times larger. And yet only about the usual number of birds would turn up next spring to breed. The effects of density must be applied

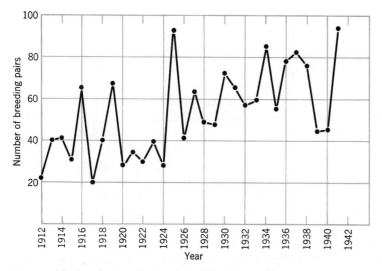

Figure 25.7

History of the breeding population of the Great Titmouse (*Parus major*) in a wood
in Holland. The population of the tits fluctuated more than that of the herons and
storks, perhaps reflecting clutch sizes of up to six at a time, but the population still
seems to be regulated within comparitively narrow limits. (Redrawn from Lack,
1954.)

meanwhile, probably in the midst of winter when mortality might well be
expected to be high. It is hard to be sure of when and how birds die.
Census of birds except when they are breeding is so difficult as to be very
unreliable, and we rarely find corpses. A dying bird just disappears from
human sight, usually leaving no trace behind that we can detect. No
doubt the information about what happened to them is held in the guid-
ance systems of numerous carnivores, but we have devised no way of in-
terpreting these data. The evidence for a balanced population assures us
that there should be density-dependent mortality, but it is hard to observe
this directly.

Of one thing Lack felt very certain: the control was not exercised
through changes in reproductive rate, in the number of offspring which
birds reared. The numbers of eggs laid and young reared did vary a little
for most birds, but only a little. Even the lower rates known for any
species should be quite sufficient to permit an exponential increase in the
population if there was no other control. It seemed much more likely that
birds always reared as many young as the season's food supply let them.
Changes in the weather would be reflected in slight changes in the
numbers of the birds at the onset of the next winter, but these changes
would be much less than those introduced by density-dependent mortality as the winter went on. One of the most spectacular bits of evidence for
this viewpoint comes from the study of snowy owls (*Nyctea scandiaca*) in
the Arctic. These large white birds breed in the treeless tundras of the
North, where they build huge shaggy nests on the ground, each quite an
emminence in, for instance, the flat expanse of the Alaskan coastal

regions near Barrow, each with its big white bird perched conspicuously on top. And they feed their young on lemmings. Sometimes in the Arctic, lemmings are immensely common, being a 1000 times more numerous than at other times (Chapter 35), and in these high lemming years the owls lay perhaps 10 eggs, rearing young from most of them, whereas in other years they should rear only one or two young. But the breeding population of owls in the North is roughly the same every year. The bumper crops of the high lemming years wander away in winter, often invading temperate lands to the South to the delight of bird-watching societies. And in those southern regions death presumably awaits most of them, for they never reappear on their breeding grounds. The density-dependent control system in this species is good enough to overcome even the large perturbations, in a sense catastrophic, commonly introduced into the animal's environment by its fluctuating food supply.

Descent of the snowy owls on the inhabited South may be called an **irruption,** a word coined by Lack to describe such sudden increases in the numbers of birds. The annals of ornithology contain many stories of irruptions, each one an exciting event in the lives of bird watchers. Every few winters some bird long unfamiliar, or even confined to the small-print list of rarities in the back of the "field guide," suddenly turns up in everybody's back yard. But next year they are gone again, and we can only talk nostalgically of the year in which we saw crossbills, or waxwings, or snowy owls. A cardinal fact about such irruptions is that, after they are over, the numbers of the birds drop back to roughly the numbers which lived before the irruption. Some unusual event, quite possibly connected with the vagaries of weather, leads to unusual breeding success, but this means that the crowd next winter is much denser than usual, so that whatever density dependent mortality may be operating presses that much more strongly than usual. The population is brought rapidly back to its original level, the level at which the density pressure for death is relaxed by the thinning crowd. A population controlled merely by the vagaries of weather should, of course, also suffer irruptions, indeed they should be common events, but the numbers of survivors after the irruption should be quite unpredictable. The adversity which would reduce this kind of irruption should act in an haphazard way, leaving a band of survivors whose numbers should bear no relationship to the numbers living before the irruption. That bird populations seem to be stabilized at their old levels following irruptions, is thus additional evidence that their populations are normally under density-dependent control.

Lack's collection of data about bird numbers provided a convincing argument that very many bird populations, perhaps most, are controlled by density-dependent factors. Their normal populations are in balance. If some unusual catastrophe does reduce the population below this level, the pressures of crowding are released and the lost numbers are quickly restored. If a different sort of catastrophe raises the numbers unduly, the death rate goes up so that the additional numbers are quickly lost. It seemed that all the known facts about bird populations could probably be explained in terms of just two mechanisms. Birds always rear as many young as they are physiologically able to rear, which means that the population at the end of the breeding season varies widely from year to year, and occasionally that a remarkably good spring produces so many off-spring as to constitute an irruption. Then, during the winter, the population is reduced by density-dependent mortality so that the usual number of birds are ready to breed next spring. Lack could not say for even one species of bird how the excess individuals actually died, although he felt sure that the deaths must occur in winter. But birds could effectively disappear from the population if they failed to breed in the spring just as completely as if they had died in the winter. This possibility has been exploited by those who see much of the control of bird populations as being effected by territorial behavior rather than winter death (Chapter 32), and there has arisen a prolonged controversy between those who think bird populations are almost completely controlled by winter deaths and those who think failure to breed in the spring has something to do with it. Both schools of thought concede that the populations are under density-dependent control, however.

But most animals are not birds, and a general theory of density-dependent control must describe the conditions of all animals. The data on other groups were (and are) much less good, but there were some suggestive lines of argument which Lack could collect. There are, for instance, many known increases in animal numbers following changes in their environments caused by man, which suggest that human interference may have altered the factor which was setting the old limit. There were many upsurges in deer numbers following human settlement in North America, which could have been the result of shooting predators that had previously kept the deer numbers down. Cutting the wilderness and farming also alters the food supply of deer, perhaps actually giving more food to populations that had been held down by winter starvation. The human coming has also caused extinctions everywhere, suggesting that new factors may have become limiting. Many such lines of evidence, although not proving that animal numbers are under density-dependent control, are at least consistent with the hypothesis of limits set by crowding. But above all is that common experience of naturalists that numbers do not change from year to year to any startling extent. Whether a man watches birds, collects butterflies, or catches fish, he knows pretty well what future months will bring. Some bottom is put to the troughs of population numbers; some ceiling is put to increase. If populations are prevented from getting too low and too high, it must be that density itself influences the forces that set the limits. There must be feedback. This the hypothesis of populational control by density-dependent pressures provided. Ecologists came to accept the hypothesis as a working rule, expecting that all wild populations should be shown to be under density-dependent control.

But there were other lines of evidence which could not be so easily reconciled with the density-dependent hypothesis, such things as the histories of wildly fluctuating insect populations. These must be allowed for in a general theory of population control. We must know what effect the weather has on the lives of such animals, and if this agent of death can anticipate density-dependent controls. And in any complete theory we must show how and when animals actually die.

We expect control of numbers to come about through changes in death rates because natural selection requires that birth rates always be as high as physical circumstances permit. An ecologist thus wants to know how animals die, and when, and how often they die also. This is the same kind of knowledge which has been compiled for human populations by actuaries of the insurance industry who organize such data into life tables. A life table for any natural population should allow a clear understanding of how the numbers of individuals in that population are restricted, suggesting whether death is a controlled happening which is proportional to the size of the population, or whether it is an unpredictable or catastrophic event. If we had life tables for a representative series of natural populations we should probably have the evidence required for an irrefutable general theory of natural regulation of numbers, but life tables are based on census, and census of wild animals is very difficult. The few reasonably complete life tables which have so far been compiled, however, reveal much of interest about the lives of animals as diverse as wolves and rotifers, suggesting that the arduous work of census may often be worthwhile. In addition animal census often gives an opportunity for pleasing field work.

THE TABULATION OF LIFE AND DEATH

369

CHAPTER 26

THE TABULATION OF LIFE AND DEATH

A proper study for the student of life is death. All populations must be controlled by the being born or the dying of individuals. In practice, birthrates seem generally to be set by the energy requirements of the parents, and this is true whether the species follows the strategy of producing very many offspring of small size and low energy reserves, or whether it concentrates its efforts into producing fewer but better favored young. I have argued earlier how all species must be expected always to reproduce as fast as they possibly can. Even in hard times for the parents, this tendency to maximum reproduction is likely to result in more than the young needed for the simple replacement of the parents, and our experience with watching wild and captive animals generally confirms this expectation. Control of populations therefore, must be control by death, and this must be true whether one invokes a theory of control by catastrophe or of control regulated by the effects of density. In the one hypothesis death is mass death; in the other, it is selective death; but the complete establishment of either hypothesis requires an understanding of the manners of death.

We must know when animals die, how they die, and how old they are when they die. We must also know the proportion of them that die at different times and different ages, and this means that we must also know the total population at each age, and the numbers of individuals of each age who survive to grow older. This is a problem of census. Lack was able to extract some convincing circumstantial evidence for density-dependent control of some bird populations by simple counts of a few birds who made conspicuous nests (Chapter 25). Death outside the breeding season was merely inferred, and there was no information on how old were the animals who died, how long-lived they were, or what each individual's chance of survival was. Most animals are not so obliging as to present themselves for census in heronries or nesting boxes at breeding time. For most animals the collection of data on numbers and age is a taxing task which consumes much time and labor.

Most compilation of vital statistics has been done on the one species of animal for whom we have voluminous records, man. Typically an actuary, basing his livelihood on those who need to know the duration of human life in order to insure it safely, seeks to predict the expectation of life of individuals of varying ages. To do this he has census data of a population for a single year, which give him the number of people in the population and the age of each. He also knows the number dying in the year, and the age at which each died. For convenience he converts these data to proportions of a standard-size population, usually 100,000, which he calls a **cohort.** The individuals of the cohort can be separated into so many of each age (called an **age class**) and the number dying in each age class can be tabulated. From these data it is merely a matter of complicated arithmetic to calculate, and tabulate, the death rate (or **mortality**) as a proportion of those surviving to each age, the deviation of age of each age group from the mean age of the population, and the expectation of life of each age group. The resulting table is called a **life table** (Table 26.1). It is intended to predict what will happen to the members of the

TABLE 26.1
A life table for men

Age Interval	Of 100,000 Males Born Alive		Mortality Rate	Complete Expectation of Life
	Number Alive at Beginning of Year of Age	Number Dying During Year of Age	Number Dying per 1000 Alive at Beginning of Year	Average Number of Years of Life Remaining at Beginning of Year of Age
x	l_x	d_x	$1000q_x$	e_x
0– 1..........	100,000	6232	62.32	59.12
1– 2..........	93,768	931	9.93	62.04
2– 3..........	92,837	483	5.20	61.65
3– 4..........	92,354	331	3.59	60.97
4– 5..........	92,023	285	3.09	60.19
5– 6..........	91,738	243	2.66	59.38
6– 7..........	91,495	208	2.27	58.53
7– 8..........	91,287	179	1.96	57.67
8– 9..........	91,108	156	1.72	56.78
9– 10..........	90,952	142	1.55	55.87
14– 15..........	90,246	172	1.90	51.29
19– 20..........	89,172	268	3.01	46.88
24– 25..........	87,692	321	3.66	42.62
29– 30..........	86,053	346	4.02	38.39
34– 35..........	84,222	410	4.86	34.17
39– 40..........	81,979	522	6.36	30.03
44– 45..........	79,036	691	8.74	26.05
49– 50..........	75,188	900	11.98	22.25
54– 55..........	70,165	1184	16.87	18.66
59– 60..........	63,496	1563	24.61	15.34
64– 65..........	54,924	1960	35.68	12.33
69– 70..........	44,253	2373	53.62	9.68
74– 75..........	31,986	2515	78.61	7.43
79– 80..........	19,565	2344	119.83	5.57
84– 85..........	9,159	1587	173.33	4.21
89– 90..........	3,068	712	232.11	3.21
94– 95..........	672	211	313.32	2.35
99–100..........	72	32	438.79	1.62
104–105..........	2	1	621.87	1.05
105–106	1	1	666.56	.96

Abridged life table for white males in the continental United States, 1929 to 1931. This is the standard form of life table developed by the insurance industry. The symbols x, lx, and the like usually appear without the verbal descriptions, and have been adopted by ecologists for animal life tables as well. (From Allee et al. 1949.)

population from what has happened, since that is the desire of the insurers, but it will not do this very accurately for it assumes that the future will repeat the past; that the mortality will not change with time, and that there will be neither immigration nor emmigration. But, as long as human populations change so fast, these inaccuracies cannot be avoided. An ecologist wants the sorts of data that appear in an actuary's life table, but he would like to use the observation of a long time span to describe the actual populations he studies rather than to base his life table on the condition at one census. He is under no compulsion to use the latest records of ever-changing conditions of life to speculate about future trends, and in this he has the advantage over an actuary. But he also must work with animals that do not fill out census forms with their dates of birth, nor whose passing is noted by a death certificate.

Various sampling and census techniques have been developed to estimate the number of inconspicuous or elusive animals (and most animals are one or the other), although the best always leave some uncertainty of the true number. Very numerous small things, such as insects or plankton, may often be sifted out of representative samples of their habitats, killed, counted, and the total population calculated by extrapolation. Less numerous or more motile things must be counted alive. You must catch them, mark them, and let them go again, estimating the number present from the success of trapping. One way of using trapping data is the **mark and recapture** method (or **Lincoln index**). After setting many traps, and marking and releasing the catch, there will be a definite number of marked animals mixed in the wild population. If traps are again set, some of the animals caught the second time will have been caught before, and hence will be marked. The proportion of marked to unmarked in the trapped sample will be the same as the proportion of marked to unmarked in the whole population, if the traps take a truly random proportion of the population. Since the total number of marked animals is known, an estimate of the total wild population can then be calculated from the following simple equation:

$$\frac{\text{Size of sample when recapturing}}{\text{Number recaptured}} \times \text{number originally marked} = \text{total number in population}$$

Unfortunately, the assumptions of randomness of the trapping do not hold true. Animals do not mix at random, nor is one individual as trap-prone as another. Much ingenuity in designing trapping programs is required to minimize these difficulties, and several ingenious statistical methods have been devised to make the most of trapping records (Overton and David, 1969). Wildlife biologists spend much time on these problems. With enough work it is usually possible to arrive at a tolerable minimum estimate of population size from trapping samples.

Aging animals is usually more difficult than counting them, and age is a vital datum for constucting a life table. Some animals obligingly carry indexes of age with them; such as wear on teeth and rings on scales, but most do not. Trapping and marking can be used as a guide sometimes, for at least you know how long a recaptured animal has lived since it was first marked. This method is attractive for work with birds because you can band the young in their nests, but bird studies suffer the difficulty that there are usually very few recaptures, most of the banded birds simply disappearing.

Marking of all the individuals of a population at one time was managed for sessile rotifers (*Floscularia*) in a pond by Edmondson at the University of Washington. The tiny animals live in tubes which they make from particles around them, and to which they constantly add. Edmondson (1945) dusted their habitat with particles of carmine, some of which the animals promptly incorporated into their tubes. In subsequent visits, Edmondson was able to spot the individuals who had lived at the "carmine" time, to determine how much they had grown since, and how many had been replaced; all the information he needed to describe the population. A similar datum point for sessile animals is provided by the natural habit of barnacles, which start their lives as pelagic larvae but soon anchor to bare rock there to be fastened and to grow for the rest of their lives, providing an ecological actuary with the same sort of advantage that Edmondson contrived with his carmine. All one has to do is to scrape a piece of intertidal rock clean, and then to watch the developments of barnacles on it in subsequent years. There have been several intriguing studies of this sort with barnacles (Moore, 1934). An obvious disadvantage is that no record is left of the mortality of the pelagic larvae, doubtless the greatest mortality suffered by the population.

When estimates of age can be combined with an accurate census, it is possible to construct life tables for natural populations of any animal. In 1947 Edward S. Deevey of Yale University drew the attention of biologists to the possibilities in a lucid, and now famous, review article. He compared the problems of ecologist and actuary, and discussed the early work with bird-banding, barnacles, and mark and recapture. And he used some published data on a collection of skulls of Dall mountain sheep to construct the first life table for a big game animal, a beautiful exercise in armchair ecology. Adolf Murie (1944) had spent years around Mt. McKinley in Alaska, studying wolves. The wolves lived largely on sheep, so a collection of sheep remains, mostly skulls, was of interest as a record of the activities of the wolves. Murie also saw nothing to suggest to him that sheep commonly died in any other way than being eaten by wolves. Annual rings on the horns of the skulls recorded the age at which the animals died. All sheep skulls lay about for years on the mountain, so that all dead sheep left their record as ageable skulls. Murie (1944) collected 608 of them, and published their ages. These data provided Deevey with just the information an actuary got from his book of government statistics: the ages at death of a large representative sample of the population. Taking a thousand individuals as his cohort, Deevey was able to construct a now much-quoted life table for Dall mountain sheep (Table 26.2).

The "number dying" tabulation in Deevey's Dall sheep life table showed that many died young and many died old, but that the animals of middle age were relatively immune from death. Deevey concluded that fit healthy adults must be able to escape from wolves either by running or by the group action of herding, and that only the feeble young or the feeble old could be cut out or run down by the wolves. If predation by wolves was the sole check on the Dall sheep population, it was exercised largely through culling the very young before they were old enough to breed. The extent of this predation, however, must depend on the

TABLE 26.2

Life Table for Dall Mountain Sheep

x	x'	d_x	l_x	1000 q_x	e_x
Age (years)	Age as percent deviation from mean length of life	Number dying in age interval out of 1000 born	Number surviving at beginning of age interval out of 1000 born	Mortality rate per thousand alive at beginning of age interval	Expectation of life, or mean life-time remaining to those attaining age interval (years)
0–0.5	− 100	54	1000	54 0	7.06
0.5–1	−93.0	145	946	153.0	−
1–2	−85.9	12	801	15.0	7.7
2–3	−71.8	13	789	16.5	6.8
3–4	−57.7	12	776	15.5	5.9
4–5	−43.5	30	764	39.3	5.0
5–6	−29.5	46	734	62.6	4.2
6–7	−15.4	48	688	69.9	3.4
7–8	− 1.1	69	640	108.0	2.6
8–9	+13.0	132	571	231.0	1.9
9–10	+27.0	187	439	426.0	1.3
10–11	+41.0	156	252	619.0	0.9
11–12	+55.0	90	96	937.0	0.6
12–13	+69.0	3	6	500.0	1.2
13–14	+84.0	3	3	10000	0.7

* A small number of skulls without horns, but judged by their osteology to belong to sheep nine years old or older, have been apportioned *pro rata* among the older age classes.

This table was constructed solely from examination of a collection of sheep skulls from Mount McKinley. It reveals that most mortality was suffered by the very young and the very old, which was as expected because wolves were the main source of mortality and they were known to hunt the weaker animals. (From Deevey, 1947.)

number of wolves, and this number must be set by the energy that the wolf population could win by eating old sheep as well as young sheep. If the wolves killed too many young sheep, there would be fewer old sheep for the wolves to eat some years later, with consequent repercussions for their own population. There must therefore be a rather complex relationship between the number of wolves and the number of sheep, an equation of energy balance whose solution yields a death rate for sheep.

A very different structure of mortality is revealed for another mammal, the gray squirrel (*Sciurus carolinensis*), recently studied by Mosby (1969) in a Virginia woodlot. Mosby trapped and marked to get his data, setting enough traps for each squirrel to have a chance of going in more than one trap. He handled a large proportion of the squirrel population in his wood every year for six years. His record of markings gave him some data on age as animals were recaught, and the disappearance of marked animals gave him rough estimates of ages at death. Squirrels do not carry obliging indexes of their ages, like the rings on the horns of sheep, but Mosby was able to assess roughly the age of each animal when first caught (and every

TABLE 26.3
Life Table for American Gray Squirrels in a Woodlot

x	l'_x	d_x	l_x	1000_{qx}	L_x	e_x
Age	Number surviving in age-class x	Number dying in age interval per 1000	Number surviving beginning of age-class per 1000	Mortality rate per 1000 at beginning of age interval	Average number living between two age intervals	Mean expectation of life remaining
0.5 to 1	93	204	1000	204	898	1.86
1 to 2	74	452	796	568	570	1.21
2 to 3	32	193	344	561	248	1.13
3 to 4	14	97	151	642	103	0.94
4 to 5	5	43	54	796	33	0.72
5 to 6	1	11	11	1000	6	0.50

The data for this table were collected by mark and recapture techniques spread over six years. Unlike Dall sheep, the squirrels suffer roughly constant mortality throughout life. (from Mosby, 1967).

subsequent time he caught it to check his estimates) from its size, weight, development of genitals, and state of fur. After his six years of work he was able to draw up a life table that reflected the actual development of a population through time (Table 26.3). The number-surviving column of this table showed that animals reaching the age of six months (at which age they appeared outside the nests and could be trapped) were dying quickly, and that they continued to suffer heavy mortality throughout what one would expect to be the years of their prime. Few reached old age. This is quite different from the survival pattern of the Dall sheep, which died young and old, as do humans, but were safe in their middle years. Apparently the population of gray squirrels was exposed to forms of death against which there was no special defense in being prime and fit. It would be interesting to know what agent strikes down the weak and the fit with an even hand. It would also be interesting to see a life table for gray squirrels which were preyed on by martens, carnivores which run them down as wolves run down sheep. Would the fit middle years of such a population seem relatively safe times?

The different survivorship data of these sheep and squirrels thus give suggestive ideas of the ways of life of each. Such survivorship data can most clearly be presented graphically as survivorships curves, that is, as number of survivors plotted against time intervals. Different kinds of animals live widely different spans of time, so the scales of individual graphs can be widely different, but if age is plotted as *percent* deviation from the mean life of the population, instead of in actual age, the survivorship curves of different populations can be plotted on the same graph. Figure 26.1 shows Deevey's (1947) computed survivorship curve for the Dall mountain sheep compared with survivorships of Edmondson's carmine-labeled rotifers and Canadian Herring Gulls (*Larus argentatus*). Strangely enough, life expectancy of a tube-living rotifer seems to

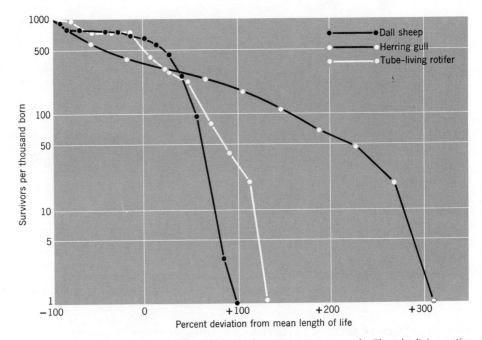

Figure 26.1 Survivorship curves of different animals drawn to a common scale. The tube-living rotifer has secure middle age, like the Dall sheep, but Herring gulls suffer constant mortality like the woodlot squirrels (redrawn from Deevey, 1947.)

parallel that of sheep hunted by wolves, with long expectancy of life once the early years are safely behind it. The comparison is not really fair, how-ever, because the rotifer life is taken, for the purposes of the analysis, to start when a tube is first formed, when in fact there is a highly vulnerable larval stage to be passed before the animal settles to form its tube. The herring gull data, like that from many other birds, suggest a pattern more like that of Mosby's squirrels, a steady hazard throughout life. It is hard, on the basis of this, to think that birds are commonly hunted by predators who must seek the weak and infirm in the way that wolves do. Herring gulls are probably little afflicted by predators anyway, and small birds may be as conveniently taken by a hawk when they are fit as when they are poorly, so superior is the "firepower" which a hawk can bring to bear.

The collection of data for the construction of life tables and sur-vivorship curves must always be very time-consuming labor, particularly when some unusual quirk of circumstance does not come to your aid, like the settling of barnacles on rocks or the preservation of sheep skulls on the tundras of an Alaskan mountain. It is seldom that a man can spend six years to describe the population structure of the squirrels of a woodlot, and few such studies have been attempted. Neither are the immediate intellectual rewards often worthy of the labor, for there may be little that is revealed by a life table that cannot be suspected from less time-con-

suming observations. Nevertheless, the knowledge of population structure embodied in life tables can be very useful for understanding the controls setting limits to the population. When wolves take young and old,
we can expect a population of prime animals which changes only slowly
with time, which is in a sense stable. The life tables of herbivores which
are preyed on by wolves suggest density dependent control like that
postulated for birds by Lack. On the other hand, life tables which show
devastating mortality at some time in the lives of the animals studied
leave much more room for catastrophic explanations of their population
histories. There can be little doubt that if we had long term life tables for a
significant proportion of the animals of our planet, we should be able to
draw up a much more embracing theory of population regulation than is
possible at present. The biologist man-hours necessary for such a task are
not likely to be forthcoming for quite a while, however.

But there is a pleasant practical side to some of the work of collecting
population data, particularly of big animals, which sometimes allows a
professional ecologist to make the most of his naturalist inclinations.
Recently a student at Purdue University, David Mech, (1966) earned a
doctorate by studying the wolves and moose of an isolated island community, using his skills at following the animals on foot and with a light
aircraft in quest of an understanding of their relationship. His study
reflects not only the pleasures of field work but also the remarkable effort
and cost needed for compiling simple statistics for even the biggest and
most obvious animals.

Within Lake Superior, the largest sheet of freshwater in the world, lies
Isle Royale, an island 45 miles long and up to 9 miles wide. The island is
now a reserve of the United States National Parks Service, where hunting
is prohibited, and it supports a healthy population of moose and the
timber wolves that prey on them. The animals of Isle Royale are effectively isolated from other herds and packs by the waters of the lake.
Moose can swim as far as the mainland, and sometimes do, but so seldom
that such contacts do not seriously affect the normal state of the island
herds. Wolves can cross over the ice to the mainland during unusually
severe winters; indeed, the island pack certainly reached the island
across the ice during the last few decades. But Lake Superior is so large
that ice bridges strong enough to permit migrations by the cautious
wolves do not exist in most years, and the wolves are effectively restricted
to the island. Moose are the largest animals native to North America,
animals that should be simple to find and count in the restricted space of
an island. Timber wolves are pretty big, too, and their habit of living in
packs aids detection. It should be possible to make whole population
counts, not just trapping samples; and the counts of successive years
should not be affected by immigration.

Mech counted wolves from his aircraft, finding the main pack of them
as it wandered the island in its endless hunting, and counting its 16
members as they strung out along their trails. Other wolves, living alone
or as two or three made up the island population to about 20, and the

TABLE 26.4   **Life Table for Moose on Isle Royale**

Wear class	Number of moose remains found	Percent of total mortality	Mortality	Population
I	1	2.56	2.17	85.00
II	2	5.12	4.34	82.83
III	1	2.56	2.17	78.49
IV	4	10.24	8.68	76.32
V	8	20.48	17.36	67.64
VI	8	20.48	17.36	50.28
VII	6	15.36	13.02	32.92
VIII	5	12.80	10.85	19.90
IX	4	10.24	8.68	9.05
IXA				.37
Total	39	99.84	84.83	492.80

These moose are preyed upon by wolves like the Dall mountain sheep and, like the sheep, suffer most mortality when young and old (from Mech, 1966). The data have not been converted to a standard cohort, and wear class are used instead of years, since the correlation between the two is not precise.

number did not change in the three years that Mech sought them out. Mech came to know them well, following the main pack at their hunting no less than 69 times, and nine times he circled close overhead all the way from the find to the kill. Twice his pilot landed him quickly enough in the vicinity of a kill for him to run to the victim, chase off the hungry wolves as they fed, and examine the prey before it was much eaten away. He found the remains of many other kills; and this study of the wolves' eating habit told him much about the lives of the moose who were being eaten. All the moose that died were killed by wolves, who apparently singled out the ailing and the feeble. The young and the old were killed, and the fit of middle age survived. This was the same conclusion suggested for predation of Dall mountain sheep by Deevey's study of their skulls, but for the Isle Royale moose Mech was able to confirm it by observation. Many times he watched fit adult moose run completely away from wolves or, more often, stand and successfully defy their assault so that they soon gave up. This observation alone would be enough to suggest that the survivorship curve should be sigmoid. To the observations of the hunting wolves, Mech was able to add direct counts of the moose. A complete census by airplane was possible, although not so simple as might be thought from a first consideration of the size of the animals. Mech closely quartered the island at 400 feet, coming down and "buzzing" every moose or sign of moose to make anything there run. Buzzing a lone moose once set seven more to running which Mech and his pilot had missed until they started to move. But he was thorough, and was finally able to conclude confidently that there were about 600 moose on the island at the start of each winter of his study. Six hundred moose to 20 wolves seems a nice sort of relationship, one in keeping with the energy-flow restrictions relating the numbers of predators to prey. In the spring, Mech could tell adult males from females by their antlers, and he could tell what proportion of the cows had calves because he saw the little animals run by their mothers' sides when he "buzzed" them. From these data of kills, births, and sexes, Mech calculated the birthrate, the recruitment rate, the death rate, and population size, the sex ratio, and the rough age distribution of the population of moose on Isle Royale (Table 26.4).

If populations of some kinds of animals could be checked by recurrent catastrophe, we should look for hints of it in the harsh places where natural catastrophe might well be expected. Ecologists working in one such harsh place, the edge of the South Australian desert, did find hints of recurrent disaster, and developed from them a general theory of population control through the recurrent catastrophes represented by changes in the weather. Andrewartha and Birch were able to show that the densest populations of grasshoppers and thrips in their area were in no way influenced by the pressures of their own crowding, but that the populations were always still rising quickly when there was widespread massacre due to a change in the weather. It was possible for the catastrophe to be so complete that the very existence of the species depended on the chance survival of scattered individuals. The animals whose lives were regulated by weather in this way were *opportunistic species* having great powers of dispersal, and with life cycles carefully synchronized to the seasons. Their way of life was subject to catastrophe, but they could live virtually free from competition between catastrophes. When such animals dispersed to places of more reliable weather they were unable to compete with the regular inhabitants. This could be explained by saying that they had evolved to use much of their available energy for the mechanisms of dispersal, leaving relatively less for the business of competition. Such opportunistic animals have also been called *fugitive species*, because they must always flee competition in comparatively equable places. Their species are able to persist, however, because there are always enough suitable habitats scattered about for local populations, freed from competition, to repair the damages done by   381 catastrophe elsewhere. The pioneer plants of the classical successions may be considered as fugitive opportunistic species in this way. The existence of opportunistic species, whose peak populations apparently seldom crowd, does not conflict with the idea that species are formed through character displacement in competing populations, because even opportunists may crowd sometimes, or because opportunism itself is the result of character displacement of one of two competing *equilibrium species*. But it is interesting to see if speciation is possible without displacement all the same. Models for this have been put forward, the crucial parts of which are "either, or," choices faced by animals at some times in their lives.

CHAPTER 27

INSECTS, WEATHER, AND OPPORTUNISM

Much of southern Australia has a Mediterranean climate, with moist winters and dry summers. Winter and spring rains flush a short-lived green pasture, and allow a hasty crop of wheat to be grown. The winters are so mild that water rarely freezes, but the summer may be hot and parching. And on top of this pattern of sharply alternating seasons there is much uncertainty of how much rain the spring will bring, since the wettest year of a 20-year period may have more than three times the rainfall of the driest. This is an environment of much uncertainty for its natural animals, one in which the vagaries of weather might be expected to bear heavily on the fortunes of populations. Insect pests have sometimes reached devastating numbers in southern Australia, and their study has produced perhaps the best evidence we have of how dramatically weather may in fact act as a agent of population control.

One of the pests is a grasshopper *Austroicetes cruciata,* which has sometimes become such a nuisance that even an official account by an entomologist could describe the swarms as "too obvious to need counting" (Figure 27.1). From 1935 to 1939 the swarms were present every spring, closely watched by entomologists who then could do little more than watch. But in 1940 the swarms went away, fortunately for ecological theory with the entomologists still closely watching. They saw the animals die.

Austroicetes cruciata seems specially adapted to the sharply alternating seasons of southern Australia. There is never more than one generation in a year. The animal passes the long succession of inclement months of the dry summer and the cool winter as dormant eggs, eggs in that quiescent durable state of some insect stages which is called **diapause.** The rising temperature of spring breaks this sleep of diapause, and the young nymphs hatch ready to feed on the green herbiage of spring. Within 50 days the baby grasshoppers have become adults and are ready to mate and lay eggs. But the drought of summer is coming. They must have

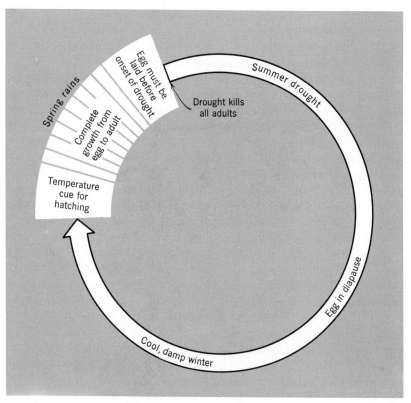

Spring rains

Eggs must be
laid before
onset of drought

Complete
growth from
egg to adult

Summer drought

Drought kills
all adults

Temperature
cue for
hatching

Egg in diapause

Cool, damp winter

Figure 27.1

Life cycle of *Austroicites cruciata*.

enough food to stock their developing ovaries with fertile eggs, and the
success of reproduction becomes a race against time. If the spring is wet
enough, the grass stays green a long time and many batches of eggs are
laid safely in the soil for the long sleep of diapause. But if the rains fail, the
adults die before they can lay their eggs. The result is a disaster for the
population, a catastrophe that could well threaten the existence of the
species. L. C. Birch and H. G. Andrewartha (1941) watched such a mass
death of grasshoppers in the spring of 1940, noting first the omminous
low rains of the previous winter and early spring, rains only a third of
those usually expected, then seeing the young hatch in great numbers
from the eggs as the coming warmth broke diapause, and then seeing the
animals vanish as they starved to death. Very few were left when the mature swarms should have been around, although there were pockets of
survival, particularly near the irrigated wheat fields of the farmers. Here,
however, flocks of birds harried them, and cleaned up nearly every one. It
seemed not unreasonable to suppose that the bird population of large
areas was feeling the insect shortage of the drought and, being able to
search country, sought out the last populations of grasshoppers and

hungrily ate them all. The grasshopper plague was over, for this year and for next year too.

There can be no doubt at all that the grasshopper population was cut down by the weather in a catastrophic way. There was no effect of density in this mass death. The animals were not short of food because of competition, either with their own kind or with other animals. They were short of food because there was no food for anyone. Theirs was not the sort of mass death suffered by overwintering birds after one of Lack's **irruptions.** After an irruption the pressure of crowding increased mortality only until a normal breeding population was left, but from the grasshopper swarms there were practically no survivors. The species itself could only continue to exist because over the huge extent of its range there were likely to be a few isolated favored places where an adult could lay a purse of eggs before hunger swept it away. Replacing the swarm with another swarm must take years of good fortune in breeding.

Austroicetes cruciata has obviously been able to last as a species in spite of the catastrophies to which its homeland is prone, for the sort of mass death of grasshoppers which Andrewartha and Birch witnessed must be repeated, often, whenever the green time does not last for the 50 or so days required for rearing adults from eggs. The grasshoppers are beautifully adapted to exploit this ephemeral season, dodging most of the year in the security of diapaused eggs from which they hatch only on the temperature cue of the rising warmth of spring. The species has found a niche that few others have found, so that it has the fresh annual growth almost to itself. But it must meet the occasional catastrophe, and this it apparently does by dispersal. The adults can fly long distances, and do so, spreading their kind over hundreds of square miles. When disaster strikes, there is a high probability that some grasshoppers somewhere will survive. And in good times the relict populations can breed quickly, for theirs is a niche that gives them almost uncontested access to a food supply which they normally can scarcely impair. They are opportunists, living exposed to catastrophe, but free from competition.

Another insect of southern Australia which sometimes becomes a plague of farmers is a species of thrips, *Thrips imaginis*. Like most of its kind it lives in flowers, where it feeds on pollen and soft tissue. The thrips are tiny black slivers of insects, whose general appearance is familiar to anyone who has sniffed a rose. One or two of them in a flower do no damage, since their food supply is then far in excess of their needs so that their feeding is scarcely noticed. But 40 or more thrips in one apple flower can so damage the ovary that no fruit is set, and *Thrips imaginis* has sometimes descended on the orchards of southern Australia in densities of 40 thrips to a flower so that the apple crop failed. When this happens, the fruit growers of whole regions of southern Australia suffer together, showing that whatever causes the thrips outbreaks must be operating over a very wide area. From the start of the investigations into the cause and cure of thrips epidemics, it seemed that the controlling agent must be the weather.

Thrips imaginis, like the grasshopper *Austroicetes cruciata*, is a native to southern Australia, but it copes with changing seasons and unpredictable weather in quite different ways. At no stage in the thrips life history is there a diapause like that which safely brings the eggs of *Austroicetes* through the drought of summer, so that the animals must be active in all seasons. Most of the lives of the thrips are spent in flowers, which young adults find by wandering flight. The eggs are laid in the flowers, and the young grow to their full size in the same flowers. But then they must walk down from the plant, bury themselves in the ground, and pupate; after which each young adult must fly away from its perch on the ground in quest of a flower in which to pass the rest of its days. The search for a flower appears to be the one desperate enterprise in the life of this animal. If flowers are scarce the chances of finding one must be poor, and then the thrips must die without laying eggs. But, if it finds a flower, then its own old age and the growth of its offspring are virtually assured, since there are apparently very few serious thrips eaters at large in southern Australian flowers. In the cool winter there are apparently enough flowers left to tide the animals over. Summer is the really bad time, because then flowers are very scarce indeed. The species must depend on a few individuals finding the surviving flowers of sheltered places; but the strategy of the species accepts that every summer is a disaster summer, a time when most of the thrips who were reared in the spring must die because they fail to find any food. The few that do find flowers, however, have all the food they need, and so leave descendents to start the population booming again when next the rains provide many flowers for wandering thrips to find. Every year the population must rise in the spring, crash in the summer, and hold its own from scattered bastions of flowers throughout the rest of the year. The monthly population of thrips is thus determined by the number of flowers, which is in turn determined by the weather.

Davidson and Andrewartha (1948), with the facilities of an entomological field station at their disposal, set out to monitor the population size of *Thrips imaginis* over the changing seasons, and at the same time to record the weather. Could the population size be so closely correlated with the weather that the times when thrips became a pest were explained as the consequences of unusual weather? They needed an easy method of census, and found it close at hand in the station's rose garden. Every day they plucked a rose, and counted the thrips on it. There were at least a few roses in the garden all year long in the benign climate of Adelaide. The counts of thrips on the chosen rose varied little from day to day, suggesting that this novel census did provide representative samples, but there was a great increase in the numbers of thrips found in each rose in the spring. This was quite independent of the fact that there were more roses. With the onset of summer drought, the numbers of thrips on each of the remaining roses fell sharply, as had been expected. Davidson and Andrewartha counted their rose-full of thrips almost every day for 14 years, during which time they had counted more than 6 million thrips.

Every spring the rose count showed that the thrips population was at its

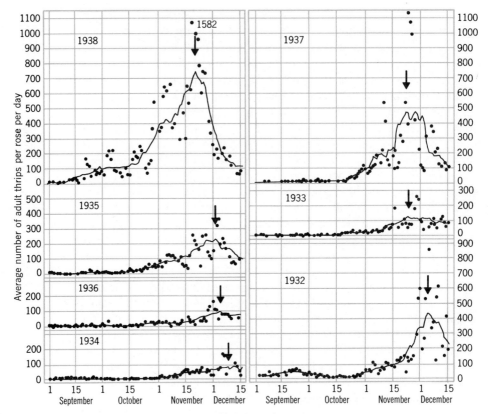

Figure 27.2 The numbers of *Thrips imaginis* during spring each year for seven consecutive years. The numbers in late winter were the same each year, suggesting control at this low plateau by density-dependent factors. But the blooming of many flowers in the spring always caused an irruption of thrips. The size of the irruption was dependent on the length of the flowering season, so that the maximum size of the irrupted population was independent of density. (Notice that the Australian spring comes in December). (From Andrewartha and Birch, 1954.)

maximum, but the populations of different springs varied widely (Figure 27.2). Apparently population expansion was continuous until the drought of summer came, and this could be early or late. Sometimes the spring increase was modest, sometimes large, and just a few times it was very large indeed. Among these times were years in which the apple crop failed because of thrips attack.

Davidson and Andrewartha chose four easily measured parameters of weather which their knowledge of the thrips life history told them should be important to the animal, parameters describing the warmth and wetness of the summer preceding the growing season and of the growing season itself. Then they summed these effects by the method of regression analysis to predict the numbers of thrips that should be expected if these parameters, and these parameters alone, were controlling the thrips population. The match (Figure 27.3) between the expected numbers of thrips each spring and the observed numbers is very close indeed. From this,

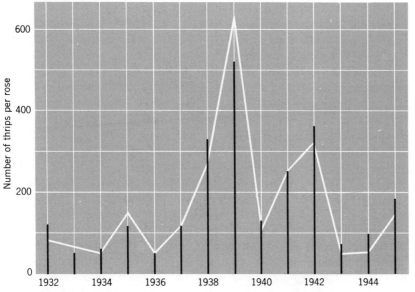

Figure 27.3

Correlation of size of thrips' irruptions with predictions based on meteorological data. Four parameters of weather were used to predict the relative size of thrips' outbreaks in successive springs (continuous line). The bars describe the actual size of outbreaks as recorded in the roses of the garden of the experiment station at Adelaide.

Davidson and Andrewartha could conclude that the population of *Thrips imaginis* directly reflected the weather of the preceeding months. This was useful from the practical standpoint, of course, because it should be possible to predict future outbreaks by watching the changing weather. But it was also important for population theory, because it showed that thrips numbers in the spring were controlled by weather in a catastrophic way. The long-term history of thrips populations was a series of exponential growths followed by mass deaths, and the size to which any expanding population could grow was set directly by the date at which weather struck with its axe of drought. Like the grasshoppers, the thrips of peak populations were killed in ways which were neither caused by, nor triggered by, the density of their own populations. This is not density-dependent death as envisaged by the logistic model. It can be argued that there is a density-dependent quality about the chances of any *individual* thrips' dying, because the choice of survivors who find refuge must be affected by the number seeking refuge, but this is not density dependence as understood in the logistic model. The actual number of survivors is set by the number of refuges available, and this is not a property of the population.

And yet the history of these thrips populations includes times of density dependence, since the populations in winter are nearly constant and the same each year. It rather looks as if the thrips population is kept at a low plateau in winter by some controlling device, which must be density dependent, but that it escapes this plateau for the rest of the year to face catastrophe. Each spring is an irruption year for the thrips. Looking at the little animals in this way makes the problem of their abundance comparable to that of the abundance of birds whose populations sometimes irrupt.

Andrewartha and Birch (1954) had together performed many of the studies which unraveled the histories of the Australian grasshoppers and thrips. They took note that the peak populations of both were limited only by the weather, and that the peaks apparently never rose so high that numbers began to be suppressed by the effects of crowding. The fates of the grasshoppers, indeed, seemed unaffected by density even in the bad years, since the very survival of the species was staked on chance allowing a few individuals to survive each drought somewhere in the vastness of southern Australia. Eventually the species was going to be unlucky so that there should come a drought which allowed no survivors, when the species would become extinct. But species were always becoming extinct; always had been; always would be. So Andrewartha and Birch argued that the animals they knew best were not normally regulated in density-dependent ways, but by the weather. But if these animals, why not other animals also? All countries had weather; all animals must accept an environment that kept changing, often in unexpected ways. Bird men may have convinced themselves that the numbers of birds were the same every year, but most animals were not birds. To a student of insects the morrow was not predictable, for it was well-known that some years were good for insects, and some years were bad. Thinking on these lines, Andrewartha and Birch countered the prevailing density-dependence hypothesis of population control with their own general hypothesis of control by accident of weather. They published this as a book, *The Distribution and Abundance of Animals,* in 1954, the very year that the formal statements of the density-dependent hypothesis by Nicholson and Lack appeared (Chapter 25).

Much of the argument for density dependence, of course, had been based on the apparent success of the logistic equation as a model for the growth of experimental populations and for prediciting the general applicability of the exclusion principle. But this argument was vulnerable (Chapter 22) to the criticism that only populations of simple animals in simple systems conformed to the logistic model, and that all such simple animals conformed to the model because they were constrained to do so by the conditions of the experiment. This is Hutchinson's analog computer argument. In *The Distribution and Abundance of Animals* can be found perhaps the most thorough and formal examination of the deficiencies of the logistic model which the literature of ecology affords. In the course of it, the authors also critically examine the use of the word "competition" in ecology, for the logistic hypothesis had required that animals did compete for limited resources when they became crowded. Andrewartha and Birch were quick to find that ecologists often talked of competition when they had no evidence for it, that although a gray squirrel replaced a red squirrel in England there was no evidence that they had actually competed, and so on. They concluded that neither the logistic hypothesis, nor the underlying idea of competition, had really been tested against wild populations, leaving the way open for their rival hypothesis of control by weather. But they needed more supporting evidence for a general theory, and they found it in the claim that the earth was green.

The **green-earth argument** goes like this: The earth is carpeted green with plants, which means that there should always be plenty of plant food available at the surface of the earth. But most animals are herbivores, a truism summarized by the concept of the Eltonian pyramid and explained in terms of energy flow. If these herbivores were limited by food energy, and if they competed for food, then we should expect the food supply of herbivores to be constantly used and restricted. In which event the earth should not be green but eaten piebald by the herbivores. Therefore, since the earth is green, herbivore numbers are kept down in some way. Andrewartha and Birch then postulated that what kept the herbivore numbers down was weather, just as it kept down the numbers of the two herbivores they themselves had studied; the Australian thrips and grasshoppers. They realized that some insect plagues do overeat their plant food supplies, and that large mammals such as goats do sometimes destroy their range, but they claimed the supposed greenness of the earth as evidence that this is not usual. Thus did their theory of density-independent control by weather gain generality. Some insects are controlled by weather; most animals are herbivores; the herbivores eat so little that the earth is green; therefore, most animals are probably controlled by weather.

But there are now a number of criticisms of the green-earth argument which together are sufficient reason for discarding it. For one thing, views from space have now shown us that the land surface of the globe is not green; it is brown. The impression that it is green comes from the human habit of living in the few well-watered valleys of the earth where it is green, but even places like these have plenty of brown in them. I recently flew over the rain forests of Ecuador and saw that the emergent trees of the canopy were brown, even though the under-story plants were green. But even if you accept that the most fertile regions are really always green, the green-earth argument can still be shown to be unsound. One approach to it is to accept that greenness means that herbivore populations must be kept low by factors other than food, but to deny that these factors have anything to do with weather. Predators are the most likely alternative explanation. If predators compete to eat herbivores, then it is likely that they will eat so many that there will be few escaping. Viewed in this way, the greenness of the earth becomes due to the activities of predators rather than weather, and Andrewartha and Birch are again left with the task of showing that weather is the prime cause of death and control.

A second objection comes from the realization that plants commonly have chemical or physical protection against being eaten, limiting consumption to few chemically adapted animal species and protecting part of the crop from being eaten by herbivores at all. The tannin in oak leaves is a now well-known example of this kind of thing, a development that requires oak-leaf feeders to restrict their activities to the first weeks of spring before the tannin is deposited. After this the oak leaves are inedible and must eventually go to decomposers. It follows that herbivores of such plants can be strongly food-limited even though the tree is green throughout the growing season.

But the third objection is perhaps more potent still. There is growing evidence that many plant populations are, in fact, controlled by the herbivores which eat them, showing, incidentally, that these herbivores are, in turn, limited by their plant food. The earth remains as green as it does inspite of this herbivore pressure because plants have adapted to put out fresh leaves, or fill holes in the canopy left by neighbors, as fast as they are eaten. And they can do this because of the enormous and constant energy supply which they receive from the sun. The replacement of loss is so prompt and effective that we do not notice it. This conception is discussed in Chapter 34.

But if the greenness of the earth argument could not be used to suggest that most herbivore populations are held down by inclement weather, it was yet true that Andrewartha and Birch had shown that some of them could be. The interesting question thus becomes, "How widespread is the phenomenon of control by weather?" Are many animals controlled by weather, as Andrewartha and Birch claim? Or is it only a few and, if a few, what kinds of animals?

Further consideration of the history of *Austroicetes cruciata* populations shows how these questions may probably be answered. The area in which swarms of the grasshoppers had been common formed a belt running parallel with the coast but some distance inland (Figure 27.4). The country on the seaward side of this line was comparatively moist, but that to the north was dry grassland bordering the central Australian desert. The grasshoppers could probably not swarm further inland than they did because, there, every summer would be a disaster summer in which grasshoppers would usually starve without laying eggs. Immigrant adults do often start colonies to the north, but they always die out. To test this hypothesis, Davidson and Andrewartha (1948) mapped the regions with ratios between precipitation and evaporation of 0.25, an index known to describe roughly the edge of desert vegetation, and the resulting isopleth convincingly defines the northern limit of the grasshopper swarms (Figure 27.4). Beautifully adapted though the grasshoppers were to the vicissitudes of southern Australian seasons, they were not able to thrive beyond this line because it was too dry there.

To the south, however, it was not too dry; and yet the grasshoppers never managed to swarm. Davidson and Andrewartha thought it must be too wet, and were able to match another precipitation evaporation isopleth, one suggesting the edge of wetter times, to the southern boundary (Figure 27.4). But this does not mean that wetness *per se* sets a limit to spread of the grasshoppers. Dryness sets a northern limit all right, this we know because Andrewartha and his colleagues actually observed herbs and grasshoppers dying in the drought, of the drought. But nobody has

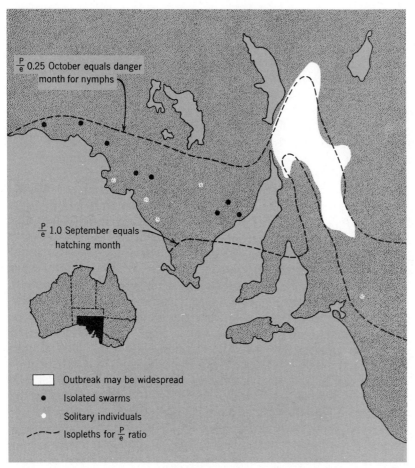

Figure 27.4

The area of South Australia where *Austroicetes cruciata* may, during a run of favorable years, maintain a dense population. The isopleth (precipitation/evaporation ratio) lines show that the distribution parallels a climatic distribution. The northern limit may be set by the shortness of spring, but the southern limit is probably set by other animals (see text). (After Andrewartha and Birch 1954.)

seen an *Austroicetes* die of wetness. The moister climate of the coast is not so moist that grasshoppers are drowned or their eggs flooded. What the comparitive moistness probably does do is to allow other animals inimical to the grasshoppers to thrive there: predators, competitors, or animals which eat its eggs, or pathogens. These agents must certainly be dependent on less dry weather; but they should act in density-dependent ways when they do act. In moister climes, then, *Austroicetes cruciata* is likely to be limited to populations well below swarm numbers by density-dependent effects of the environment, but in the dryer climate of the north it is freed from those restraints. It wins this freedom with its splendid adaptations of diapausing eggs and a life cycle synchronized with the seasons, but it pays the penalty of being subject to the catastrophes of its dry environment. It overcomes the dangers of this penalty, in turn, by having great powers of dispersal which ensure that some individuals somewhere always survive catastrophes. The animal is an **opportunist.** Its general qualities are suited to life in unstable environments subjected to sudden change, and it has notable powers of dispersion. It is opportunist animals such as this which, exposed to hazard of weather in their normal lives, are most likely to have their numbers curbed by accident of weather. Population lows may be controlled by density-dependent factors, but population highs seldom are. And the history of their population growth is hardly ever sigmoid or logistic.

EQUILIBRIUM AND FUGITIVE SPECIES

There must be many kinds of opportunist animal, and of plants also. The earth affords many unstable habitats scattered round the edges of places with more reliably favorable climate. In a lucky year, the unstable places are tolerable, but if the seasons are unfavorable the colonists of other years may die. It is thus good strategy to be opportunist, to be able to disperse far and fast so as to be able to occupy such places in good years, and to have a large scattered population which will ensure that some survive the inevitable decimations. If you are small and short-lived, opportunism is probably the only satisfactory strategy for life in unstable places, but if you are big, and with a low metabolic rate, an alternative strategy is open to you; you may just sit tight through the bad times, living on your reserves and reducing your life processes to the minimum. In this way your population may survive the bad time, continue growing in subsequent good times, and finally reach numbers where the effects of crowding are important so that a population equilibrium is attained. The large plants of deserts follow this strategy. They live in the same sorts of unstable place which forced opportunism onto *Austroicetes cruciata,* but their populations are relatively stable. They are **equilibrium species.** Such an equilibrium tactic in unstable places is easier for plants than for animals, although large animals may manage by devices such as hibernating and aestivating. But small plants adopt opportunist strategies in unstable places also. The annuals of deserts are living lives akin to that of *Austroicetes,* managing a brief flowering in the wet times and making many seeds which shall tide over the dry times.

Opportunists who disperse so well to desolate places presumably also disperse to more desirable ones. But they do not persist there. The reason for this is likely to be that the equable places allow all populations to rise until the effects of crowding force a population equilibrium. Equilibrium strategies are favored in such predictably favorable places. But opportunist species have diverted relatively large amounts of energy into an apparatus of dispersion which is now largely redundant, and they have foregone specializations which would let them make the best use of the resources of a more stable environment. When they come back, they cannot compete with stay-at-homes and win. They are compelled to be permanent opportunists, taking advantage of their ability to disperse by living in marginal places all to themselves, but always having to yield when they find themselves in good places. Hutchinson (1951) has called them **fugitive species;** they are forced to flee the competition of **equilibrium species,** and they persist only because they can exploit fresh opportunities faster than the equilibrium species. It was such animals that Andrewartha and Birch studied, and whose populations they found to be largely controlled by weather.

Hutchinson documented his discussion of **fugitive** and **equilibrium** species with copepod examples, using particularly the story of *Eurytemora* and *Diaptomus* in Britain which Elton told (Chapter 21). *Eurytemora* was a fugitive, occupying fresh formed sewage ponds quickly and enjoying a brief abundance. The equilibrium animal, *Diaptomus,* dispersed slowly, but took the ponds for itself when it did arrive.

Copepods are small and short-lived like insects, but bigger more permanent things can be fugitives, too. MacArthur (1958) found evidence that warblers can be. Two of the five sympatric species which he studied (Chapter 24) laid many more eggs in years in which spruce budworms were common than in other years, but the remaining three species laid roughly the same number of eggs every year. The two who took advantage of the budworms so were being opportunistic. Their populations went up steeply with every budworm outbreak. But they were the two species whose niches, as determined by places of feeding and methods of hunting, were least specialized. In years when budworms were not particularly abundant, their numbers fell quickly. MacArthur concluded that they were true fugitive species. There were probably often local extinctions caused in lean years by their not being able to compete well for food with the other three equilibrium species. Massive increases in numbers during budworm outbreaks then enabled them to recolonize these lost places. Their total population fluctuated widely. And the final agent of control was whatever caused the budworm outbreaks; which may, or may not, have been the weather.

The distinction between an unstable habitat and an equable one is relative; so also must be the separation between opportunistic fugitive species and equilibrium species. There clearly must be halfway stages and halfway strategies that reflect occasional catastrophes and occasional equilibrium conditions. There are gradients of climatic stability, often running across continents, and they must be reflected in changing degrees of opportunism. This is perhaps most clearly revealed by plants, as when the proportion of annuals (opportunists) in a flora increases along a line from forest to desert. Such a cline of increasing opportunism is nicely shown in Raunkiaer's old system of life-forms (Chapter 2).

Even in generally equable places there will be local habitats which are unstable. The ever-changing landforms of the earth's surface are always creating fresh unoccupied places which are potential homes for opportunists. Wherever streams cut, rocks falt, shorelines emerge, hurricanes uproot, droughts kill, floods drown, or lava cools, there is fresh ground ready to be colonized. Opportunists, being specialists at dispersion, are likely to get there first but, since the fresh bare ground may be in a place of generally stable climate, the equilibrium species will not be far behind. The opportunists enjoy the new land only briefly, after which they are eliminated by competition. Some fugitive propagules must meanwhile have reached a fresh patch of bare ground.

In this process of opportunists being replaced by equilibrium species we have an explanation of ecological succession in plants. Many of the first plants to reach any bare spot must be fugitive species. They are replaced in succession by slower moving but more competitive equilibrium plants which follow. We can have little doubt of the reality of this competition for the succeeding plants physically overtop and dominate the pioneers. Successions are thus necessary consequences of the fact that bare ground is always being produced by physical processes, providing niches for opportunistic species.

As the pioneer vegetation becomes established, so the animals associated with it arrive. These, too, are opportunists who must be able to disperse well. It is likely that their populations may be subject to catastrophe, as were the opportunist insects studied by Andrewartha and Birch, but this perhaps may be dependent on the actual size of the bare ground being colonized. If their physical separation from the pool of equilibrium species is slight, some pioneer animals may yet be in reach of other animals inimical to them. Predators travel, and are sometimes of catholic taste, so that control from such agents is still possible for the herbivores of successional vegetation. Yet the universality of the phenomenon of plant succession does suggest that there are places for opportunistic animal species to live in all parts of the world. It may be, therefore, that some populations in all regions are sometimes controlled by accidents of weather in the ways suggested by Andrewartha and Birch. It is also perhaps more likely that in places of reasonably mild climate most animals are equilibrium species whose populations are checked or depressed by density-dependent factors. There are, however, huge areas on earth with unstable, unpredictable climates where control by weather may be normal: the tundras, the edges of deserts, and the lands afflicted by monsoons.

OPPORTUNISTS AND
COMPETITIVE
EXCLUSION

There is yet one question we must ask ourselves, if we accept that some opportunistic animals are normally controlled by weather. What of the exclusion principle? These animals are good species; but we have established an ecological view of speciation which includes the phenomenon of character displacement to avoid competition according to the exclusion principle. If the populations of these animals are kept well below the level that their resources should allow, how can the populations ever come into competition? A simple way of answering this is to say that sometime they did escape catastrophe long enough to establish a short-lived equilibrium population. During this time the populations of similar opportunistic species merged, and were separated by character displacement. Short-lived states of equilibrium are possible for even opportunistic species like the Australian grasshoppers and thrips, and even have some inherent probability among the history of random catastrophes that afflict them. Perhaps even more likely is that character displacement can occur in the unfavorable seasons when even opportunistic species are likely to be under density-dependent control. The Australian thrips of Andrewartha and Birch were clearly regulated to a plateau in the unfavorable season, suggesting that their lives at these times were essentially those of equilibrium animals and thus ripe for character displacement. Another alternative is that the opportunistic habit is itself the result of displacement between two equilibrium species; that one of the original pair avoided competition by developing great powers of dispersal and other trappings of opportunism. From arguments such as these we may say that the exclusion principle is not disputed as a result of the evidence that opportunistic species are sometimes controlled by weather.

But there is another, perhaps more interesting, way of looking at the problem of speciation suggested by these findings. This is to deny that the exclusion principle is necessary to explain speciation. This argument starts by noting, as Andrewartha and Birch note in their book, that Gause's models of competitive exclusion are based on logistic models that are themselves suspect. This is the "analog computer" argument again. But, goes the response, all species do occupy separate niches from which other kinds are barred; how can this be if you deny exclusion? To answer this retort, it is necessary to show how species occupying distinct niches can be evolved without the aid of competition, and there are now models being put forward which do allow for this. A recent example is that of Mitchell (1969) who discusses speciation in parasitic water mites. Three different damsel fly hosts live in distinctly different parts of ponds, providing an element of spatial separation from which the evolving mite parasites can "choose." Two species of water mite live on all three kinds of damsel fly, but one mite fastens always to the abdomen of the emerging fly and the other always to the thorax. Here are two closely related sympatric species whose niches manifestly do not overlap; how did this come about? Mitchell produces a convincing model to show that it is inherently unlikely that a water mite should fasten to any part of the host except these two extreme positions. The mites reach the adult host by lurking on the pupa case and waiting for the fly to emerge. When it does so, it comes partly out, then rests awhile with the thorax exposed. A mite can get aboard or not. Then the fly jerks the rest of the way out, and rests a second time but with the tip of the abdomen near the case. The mite then has a second chance to grapple and board. This behavior of the host ensures that all mites will be on one place or the other. Mitchell shows how two species of such animals could evolve without any competition between the populations. A mite with a thorax-grabbing tendency arrives in one part of a miteless pond, and breeds thorax-grabbing offspring on the local species of damsel fly. Meanwhile an abdomen-grabbing mite breeds abdomen-grabbing offspring on another of the isolated damsel fly populations. The argument is one of probabilities but, given the "either, or" choice with which the emerging habit of the damsel flies presents the mites, it can be shown that two separate, noninterbreeding populations can arise without competition. We know so little about most animals, that it is quite possible that many are sometimes presented with critical "either, or" choices during their lives. If they are, then there is always a definite probability that separate species will evolve without populations having to compete. A rather similar argument has recently been used to explain the many related fish species found in East African lakes. This line of argument does not suggest that character displacement does not occur. It merely suggests that species can separate in more ways than one. And it aids the suggestion that populations can be controlled in more ways than one also, by the effects of crowding and by the vagaries of weather.

From the very first, ecologists have had at the backs of their minds the idea that predators must represent one of the strong controlling forces of nature. Those who have recently developed the idea that complex ecosystems should be more stable than simple ecosystems have relied on the assumption that predators are always a controlling influence on the numbers of their prey, suggesting that the more kinds of predator and prey there are the more perfect the control, and those who have sought to explain a general balance of nature with the hypothesis of density-dependent control have often looked to predators as prime agents of that control. Yet, for large animals, at least, it has been hard to find evidence, even circumstantial evidence, that predators do in fact exert such control. Much of the supposed evidence is anecdotal, being accounts of deer numbers rising after predators were shot during the settlement of North America, but such a rise in deer numbers could have resulted from changes in the vegetation, whether the predators were shot or no. The one apparently well-documented story, that of the Kaibab Plateau, is now known to have been embroidered to the point at which it must be considered to be mostly fiction. There is good evidence that wolves confined on small islands can control the numbers of their ungulate prey, but this does not mean that such control could be attained in the open spaces of a continent. A number of studies now suggest that large mammalian predators commonly take only the sick and the old, thus culling animals which were due to die anyway and not acting as prime agents of control. Some control may be effected when young animals are taken, but that control may be no more than one among many density-dependent pressures. Recent studies of African

DO LARGE PREDATORS CONTROL THEIR PREY?

397

game herds actually suggest that predation is one of the lesser controlling influences. It is thus possible to conclude, if rather cautiously, that large predators are not often the primary agents in controlling their prey populations and that predation is only one of a set of density-dependent factors at work. Many large predators may properly be considered to be scavengers rather than killers of prime animals.

CHAPTER 28

DO LARGE PREDATORS CONTROL THEIR PREY?

Predators kill; it follows that they reduce the numbers of their prey, restricting or even controlling them. This simple equation has dominated man's thinking about nature since before recorded history. Probably the first man who ever kept sheep lost some to wolves, and cursed the wolves accordingly, passing on his opinions to his offspring so that wolves finally became ogres in fairy tales. Such caricatures of predators are with us yet. In Alaska, agents of the government are still killing wolves from airplanes, apparently under the general belief that it must be "good" for the "game" to do this. In the landed estates of Europe, managed for the last 200 years as places where you shoot gallinaceous birds, everything with hooked beak or claws has been hounded down for bounty, even down to sparrow hawks, fish hawks, and pine martens; and on little more than the belief that, "It eats flesh so it must be bad." Curiously spiders, which seem physically repugnant to men, are "good" because they eat "flies," so we must preserve spiders. These attitudes to nature are wonderful materials for the student of the human mind, but they are also germane to a serious discussion of predation because they have affected our attitudes to the problem.

Predators kill prey, but if they kill it all they must themselves starve. On the other hand, they will obviously kill as much as they can get. There must be a "balance" between their efforts to kill and the escape of the prey, a balance that controls both the numbers of predators and prey. Thus the simple explanation of the "balance of nature" which comes from our ancient impression of the role of carnivores. But is it really true? Is the greenness of the well-watered parts of the earth largely due to the vigilance of carnivores, who cull the plant eaters and keep their populations low? Or is being killed for meat a relatively unimportant hazard in the lives of most herbivores, as apparently it was for Andrewartha's grasshoppers and thrips? And how, in turn, are the numbers of the predators themselves really restricted? Simple those these questions seem, it has not been easy to answer them with certainty.

Indirect evidence that large predators may have a hand in controlling the numbers of their prey has come from the American experience of slaughtering predators as a first task of civilization. The history of the frontier was also a history of deliberate campaigns of extermination against the large native carnivores. Wolves, bobcats, coyotes, and pumas have been completely exterminated over most of their original ranges, even though these ranges were sometimes left as forest or prairie. This passing of the frontier has often seemed to be followed some decades later with plagues of deer, suggesting the obvious conclusion that the predators had controlled the deer herds in primitive times and that our killing of the predators had released the deer populations from this, their natural restraint. This is a generally compelling argument, but it suffers from the poor quality of the "experimental" evidence. Killing predators was not the only task of the pioneers, who also burned forests, cleared land, farmed, released livestock, built roads, and so on. They changed the environment in many ways, so why should we decide that the key change was the removal of predators?

Many professional opinions on the importance of predators have been based on the supposed history of the Kaibab deer herd. The Kaibab Plateau is a wild area bordering the Grand Canyon, and it was declared a national park in the conservationist era of Teddy Roosevelt at the turn of the century. All the big predators of the park were then shot out by government hunters while, at the same time, a complete embargo was placed on the hunting of deer. Then, in the 1920s and 1930s came reports of there being too many deer in the park, and there does seem no doubt that the numbers had risen over what were present when the place was first declared a reserve. Unfortunately, we have no good census figures for deer present in the park at any time in its early history, only various visitors' and warden's "estimates". These are perhaps good enough to suggest that the herd tripled or even quadrupled its size in 20 years, although they by no means prove that it did. Such an increase need not be surpising for any place as disturbed as the Kaibab had been, for the frontier had passed through 20 years before it had been declared a park. Ranchers had herded 200,000 sheep, 20,000 cattle, and many horses there, and they had doubtless set their usual fires (Rasmussen, 1941). There can be no doubt that they had shot many deer, too. That the removal of such exploitation as this should result in quadrupling the deer herd need surprise no one. There is no need to invoke "shooting predators" to explain it. Such are the facts of the matter as well as they can be gleaned (Caughley, 1970). But this true story became embroidered by a series of misrepresentations until it became a fictional tale of a plague of deer and the classical illustration of the controlling effects of mammalian predators. It is instructive to follow the Kaibab fiction, for much ecological writing and thinking cannot be properly interpreted without knowing how people have been led astray.

The fateful progress of the Kaibab embroidery is shown in Figure 28.1, as it is revealed in a recent study by a New Zealand student of game animals, Graeme Caughley (1970). Diagram A appeared in a monograph on the Kaibab by Rasmussen (1941), and gives his private assessment of what was known about the history of the deer herd. Remember that there had never been any real census; but there had been forest supervisors in

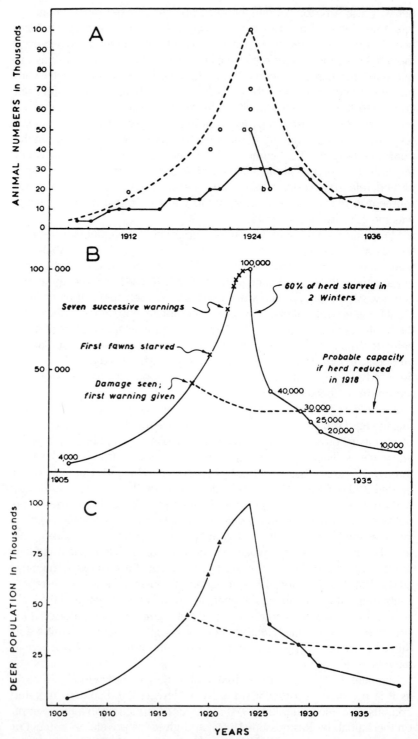

Figure 28.1

The Kaibab deer herd fiction; a history of embroidered data. (A) Population estimate of the Kaibab deer herd, copied from Rasmussen (1941). Linked solid circles are the forest supervisor's estimates; circles give estimates of other persons, and the dashed line is Rasmussen's own estimate of trend. (B) A copy of Leopold's (1943) interpretation of trend. (C) A copy of trend given by Davis and Golley (1963), after Allee et al. (1949), after Leopold (1943) from Rasmussen (1941). (After Caughley, 1970.)

the park who have left us the best professional estimates of deer numbers we can glean. The supervisor's estimates are the solid lines and circles, a history of increase, indeed, but one easily understood in the light of the past history of the plateau. But in 1924 various people had gone to the park and remarked the "great numbers of deer," and there was valid discussion about whether it was time to cull the herd. Visitors estimates are the open circles. But Rasmussen was inclined to accept the largest of them, a nice round figure of 100,000 deer. On the basis of this estimate he sketched in the dotted line as a likely history. It really does make a dramatic curve, much more exciting than that produced by the professionals who lived in the park (the black curve below it). Then came the real mischief. The noted game management specialist, Aldo Leopold (1943) drew diagram B on the basis of Rasmussen's diagram A and published it in a local bulletin on wildlife management in Wisconsin. Notice that some ecological ideas have been superimposed on Rasmussen's flight of fancy, for the left-hand side has developed the nice sigmoid shape of a population undergoing logistic growth and the right-hand side has the neat fall of catastrophy, until it meets the supervisor's estimates again! A number of comments have been added reflecting the "to cull or not to cull" controversy. Then there was more mischief. Leopold's version, exercise in imagination and artistry though it was, was reprinted in the standard ecology text by Allee et al. (1949). This was enough for the history to be taken as true, although there were more imaginery refinements to come, as in diagram C.

In learned articles and textbooks over the last 20 years, the supposed history of the Kaibab deer herd has been used as the standard example of what happens when the controlling influence of large predators is removed. It has now even reached the pages of general biology texts, where Leopold's striking drawing can so easily be reproduced. But we have no real evidence that the deer herd at Kaibab did suffer a population explosion, nor that it crashed. All we have is evidence of a not unreasonable fluctuation following disturbance. And since the population explosion was a nonevent we cannot use it as evidence that American deer populations were controlled by predators.

The collapse of the Kaibab example puts us back in the position of sensing that the settlement of North America resulted in an upsurge of deer numbers, and of wondering if the removal of predators was the reason behind it. Even if we did have a "good Kaibab" on file, a true record of exponential growth of deer numbers following settlement and eradication of predators, it would still prove nothing. Treated as an experiment, the elimination of predators by settlers is valueless. It would be without before and after surveys, without duplication, and without control. A laboratory study carried on in such a way would be unpublishable in the scientific literature as meaningless. Experience may yet suggest that shooting predators is a good way to bring on a plague of deer, but every time the frontiersmen passed through they not only shot predators but also interfered with the vegetation. Commonly they replaced forest with second growth; climax with early succession stages. Very likely such changes provided more food for deer, a sufficient reason for their increase even if the predators had still been there.

There is a second general difficulty in accepting predators as the controlling agents of large vertebrates, and this is their common habit of mostly taking the weak or sick prey animals that might be expected to die anyway. This aspect of predation strongly influenced one of the foremost students of the subject, Paul Errington (1946, 1963).

Errington devoted most of his life time to the study of muskrats *Ondatra zibethicus* in Iowa; from 13 years of his youth spent as a fur trapper, to most of his later years at the Iowa State University. Muskrats are particularly hard to observe, passing most of the daylight hours underwater or hidden in burrows, but Errington learned to discover their presence and activities from "reading of sign," which he describes as "studying the meaning of tracks and trails, of diggings and cuttings and heapings, of food debris and droppings, or miscellaneous traces, of blood, fur, wounds, and carcasses." These qualitative bits of evidence became quantitative when Errington used them to identify the presence of individuals and to map the members of populations. He trapped muskrats on a large scale, marked them, and released them. He was given sample carcasses by the local fur trappers for autopsy, particularly for estimating the numbers of offspring which had been born to females from the number of scars of old placentas in their ovaries. He studied the possible predators of muskrats, notably minks, *Lutreola vison,* in the same way, and was able to detect when they were feeding on muskrats from the remains in their droppings. The thoroughness of his work is revealed in many of the minor observations in his writings, as when he reported that one thin old male traveled 2800 yards in half a day, a statistic that Errington gathered by following the animal for that half day without being detected. Thirty years of such devoted work has provided us with perhaps the best-documented history of a wild mammalian population which we posses. And Errington came to the conclusion that nearly all the muskrats eaten by predators in his part of Iowa were individuals who were doomed to die from some other cause, such as disease, if the predators had not eaten them first.

The lives of Iowa muskrats were strongly influenced by the marked seasonal changes of the local climate, by the alternation of warm continental summers and bitter winters, by the chances of floods and the chances of drought. Muskrats established in fine summer weather in their home ranges seemed almost immune to predation, probably because they had somewhere to run to when approached by a mink or because they faced the potential agressor so confidently that they made the risks of combat unacceptable to it. But if they were flooded out, or left exposed by drought, or suffered epidemic disease, the minks killed very many of them. Sometimes the onset of some calamity attracted predators to such an extent that whole populations of muskrats were almost wiped out. Many muskrats were also killed in the spring as they wandered from their winter quarters without the security of a home range, but Errington concluded that most of these animals stood no chance of finding a decent place to live anyway, and were thus doomed. He argued that virtually all these deaths were ordained once the animal was wandering or the calamity had struck, and that their actually being eaten by predators was without real significance to the population. The predators were in effect acting as scavengers, carrion eaters without the patience to wait for their victims to formally die. It was true that the minks did kill a large number of healthy baby muskrats in the spring, but Errington thought that even this predation was of little significance because the bereaved mothers usually produced second litters. Errington can perhaps be criticized for playing down the effects of predation on babies, but he was able to argue that predation of muskrats had a relatively minor restrictive effect on the population, not a controlling effect.

Errington's account of the muskrats has much in common with the accounts of Australian grasshoppers whose catastrophic population history so influenced the thinking of Andrewartha and Birch (Chapter 27). Like the grasshoppers, there is much in the life of the muskrats which suggests that they must be partly opportunists. Every spring they must sally forth from winter quarters in search of places in which to live, and they must be adapted to disperse so well that the loss of many local populations from catastrophe does not spell calamity for the species as a whole. It seems reasonable that the main chances for the predators of such opportunistic prey should be to exploit the evil fortunes of local populations, to take merely the sick and the lost as Errington concluded. But Errington also noted that large predators generally did take the weak and the sick members of whatever prey they hunted, so that perhaps predation was always a minor influence toward population control, that a predator was usually no more than the giver of a coup de grace to animals suffering other and controlling mortality.

There is, of course, much evidence that vertebrate predators do commonly take the weak and the sick. Murie (1944) had shown that wolves on Mt. McKinley took young, sick, or old sheep, and that sheep in their prime were almost immune from attack. Mech had found that the Isle Royale wolves took only young, old, or sick moose in a like manner (Chapter 26). For pack-hunting animals like wolves, it seems obvious that the weakest animals will be the ones to fall, for they will be the easiest to catch. It seems likely, for instance, that this fact has influenced the evolution of parasites which have two hosts, one an herbivore and the other a member of the dog family, as do many such parasites as tapeworms. The tapeworm cyst waiting in the muscles of its first host does not take too long a chance on not being eaten by the necessary second host, if a pack of hungry wolves is following the herd waiting for a sick animal to drop out.

It is not so obvious that solitary hunters who kill by stealth, like the large cats, should take only the sick. Schaller (1967) has suggested that tigers may take the fit as commonly as the unfit, but his accounts of how tigers killed tethered buffalos does not suggest that killing is safe or easy even for a tiger. The tigers ran at their prey, half climbed on their backs, wrestled them to the ground, then dodged the flailing hooves to seize the buffalo by the neck. The buffalos took some 10 minutes to die, suggesting

that killing sick animals should be a safer undertaking. It is not improbable that a lurking tiger might normally let the more formidable animals
pass unmolested, just as Errington's mink apparently refrained from attacking confident muskrats, because no predator may risk a serious hurt
winning one day's meal. A recent study of the puma (*Felis concolor*) in
Idaho by Hornocker (1969) (Chapter 31) suggests that much of puma
behavior can be explained as directing the animals to achieve quick kills
without getting hurt. Nevertheless Schaller's view needs to be given due
weight, and it may be that big cats, as opposed to canids, commonly kill
the fit and are thus primary agents of death. Introducing big cats might
therefore do more to control a deer herd than introducing wolves.

But the weakest of all animals are the very young. They may be
defended by their parents, or the parent herd, but otherwise they must
rank as safe meals for such predators as can find them. Even Errington's
opportunistic muskrats lost large numbers of babies to minks, and the
compensation of second litters could not entirely efface the effects, for the
loss of one set of babies must slow the rate of population growth. The
controlling effect of losing babies may have been minor in muskrats (Errington is not very convincing on this point), but it might well be decisive
for large herbivores who can have only one family each year. For this
reason, the hypothesis that large predators often control the populations
of their prey can survive the demonstration that very many of the animals
which they eat would have died from other causes anyway.

Inconclusive but suggestive evidence for the influence of predators
comes from our releasing herbivores on small islands without them, for
often the population thrives, undergoing exponential growth for a while.
This happened when pheasants were released on a small island off the
coast of Washington State (Figure 28.2). The experiment was terminated
when the United States Army occupied the island and shot all the pheasants, but we have more complete records of reindeer populations on
Bering Sea islands, whose fortunes have been scarcely affected by human
activity. The islands all supported virgin tundra, a climax vegetation presumably not the normal home of opportunists, and yet the reindeer populations all grew quickly and exponentially until they apparently destroyed
their range and suffered mass death. The most dramatic of these histories
is that of the reindeer herd on St. Matthew Island (Klein, 1968). Twenty-
nine reindeer were put ashore there in 1944 by members of the United
States Coast Guard. A cenus in 1957 found 1350 reindeer, and a second
census in 1963 found 6000. The island is 128 square miles in extent, but
6000 reindeer were enough to destroy their own winter forage, reindeer
moss, which almost vanished. In 1966 there were only 42 reindeer left,
all but one of them females. Let us hope that the remaining male is impo-
tent, so that the mischief of this experiment will be brought to an end and
the unique native vegetation given some chance to recover. But the
duplication of this history on other Bering Sea islands does suggest that
reindeer will normally increase until their range is destroyed unless some
definite check to their numbers is provided. It is commonly claimed that

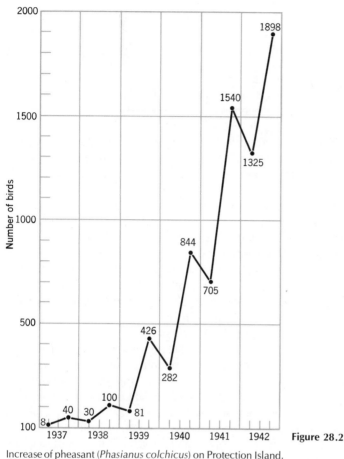

Figure 28.2

Increase of pheasant (*Phasianus colchicus*) on Protection Island,
Washington. The birds were censused each spring and fall. In
1942 the island was occupied by the military and the pheasants
shot. (From Lack, 1954, after Einarsen.)

this check could not be anything but predators, but it is also possible that
reindeer would eventually establish some balance with their "prey," the
lichen plants.

If island deer herds destroy their range and die because there are no
predators to control them, it should be possible to check the growths of
the herds by introducing predators. This has never been done inten-
tionally, but it has been partly done for us once by an historical accident, on Isle Royale. When Mech (Chapter 26) worked on the island,
he found both wolves and moose there; but once there were neither. Up
to about 1900 there were caribou, coyotes, and other animals, but apparently no moose or wolves. The earlier inhabitats became extinct, no
doubt helped on their way by men, but moose arrived to replace them
during 1913 and 1914. Moose swim well, so there is no mystery about

how they arrived (Murie, 1934). In spite of the difficulties of census discussed by Mech, it is certain that the moose herd grew fast, damaged its
range, crashed, and continued fluctuating violently until a pack of wolves
arrived from the mainland sometime in the late 1940s. Since the advent
of the wolves, however, the moose numbers have remained roughly
stable; and it now seems that the numbers of the wolf pack are roughly
stable too.

The wolves of Isle Royale live by hunting the weakest of the moose
herd, by taking the old and the sick as Errington claimed they should, but
also by taking many of the calves. They live on an island with strong
seasonal changes of climate, and on prey who have their vulnerable
calves at only one season of the year, and yet they are able to maintain
themselves throughout the year. Perhaps they are helped to do this
because it takes two or more years before a moose is strong enough to escape their attack, so that there is always a pool of attackable young as
well as of the sick old for them to exploit. But their depradations on the
young are apparently enough to control the moose herd, allowing us to
conclude that releasing predators on other islands, like those of the
Bering Sea, should be sufficient to control the deer herds. Large predators
can indeed control their prey, and it remains only to decide if they normally would do so in undisturbed natural habitats.

A general criticism of island studies must be that island communities
are in limited systems under constraint. Confining moose on an island 40
miles long is not very different from confining paramecia in a bottle, or
flour beetles in a box, so that conclusions from such studies are alike vulnerable to the analog computer argument (Chapter 22). When moose and
wolves are shut up together on an island, with no other likely prey and no
other likely predators, an equilibrium can be reached. It is good to know
this. But does it mean that the populations should be balanced if both can
get away, if both have alternative plans of feeding or escape, or if both are
subjected to the stresses of life in the bigger world of a continent? Fortunately in the plains of East Africa, we still have one last vestige of the ancient world of big animals where we can study vast wild herds in quest of
the answers to these questions.

The Serengeti Plains of Tanzania and Kenya are as large as Ireland, or
the states of Vermont, New Hampshire, and Delaware combined. They
contain almost virgin herds of much of the complex fauna of Africa,
preserved from the exploiters to our day perhaps by the ready spears of
the Masai warriors who not so long ago conquered the other peoples who
lived there. The commonest ungulates on the plains are the wildebeests
(*Gorgon taurinus*), of which there are more than 200,000. Lee and Martha
Talbot, now of the Smithsonian Institution, have spent many years studying the wildebeest herds, and have particularly addressed themselves
to the question: Are the numbers of wildebeest regulated, and, if so, how?
(Talbot and Talbot, 1963).

It is not easy to census the animals on 20,000 square miles, but the task
is made possible by the openness of the plains. Talbot records that from a

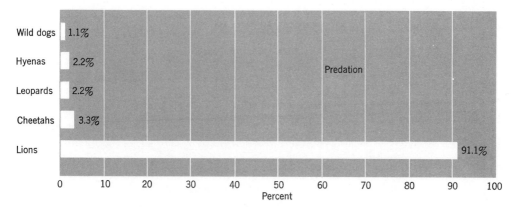

Predation on western white-bearded wildebeests. Lions were the only serious predators of Wildebeests in the Serengeti, but it only took a relatively small fraction of the wildebeests dying every year to support the known lion population. More important causes of death in wildebeests were disease and young animals getting detached from their mothers in crowded herds. (After Talbot and Talbot, 1963.)

Figure 28.3

hilltop one day he could see herds of game stretching before him for 50 miles until they vanished in the haze of the horizon. By driving transects across the plains, and by counting from aircraft, it has been possible to make good enough censuses to show that the numbers of wildebeest do not change much from year to year. The herd is therefore under some sort of control, and the wildebeest seems to be an equilibrium species. How are wildebeest populations controlled?

Seven hundred lions hunt the plains, feeding almost entirely on wildebeest, and there are also leopards, cheetahs, hyenas, and hunting dogs taking their separate tolls in their different ways. Lions live almost entirely on adult wildebeest, probably because only large wildebeest can assuage the appetites of a family of lions. Many studies have shown that each lion must kill about a wildebeest a week, which allows one to calculate the lion take on the Serengetti to be between about 12,000 and 18,000 wildebeest a year. It is not known how many of these animals might have been going to die from other causes if the lions had not taken them, but even if they all were prime animals, their loss represents only a small portion of the annual crop of animals that must die if the population is to be kept in check. The best estimates of the take of the other predators show them to be negligible compared to that of lions (Figure 28.3), so we must seek some other source of mortality. Many wildebeest die from disease, particularly the young and the yearlings. It is very hard to estimate these losses because corpses are always totally eaten, even down to the bones, by the vultures and hyenas, but the Talbots concluded that many more animals died of disease than of outright predation. But even the best estimates of disease and predation combined still seemed too small to regulate the numbers of wildebeest, and the Talbots considered that the remaining, and major, cause of death was simply that of babies getting lost. The wildebeest mothers always seemed to give birth to live calves successfully, but the calves then had to follow very close to their mothers as the herds moved. In large crowded herds, the young, quite simply and literally, often got lost. Many fell to hyenas and jackals, no doubt, but it was unlikely that there should be enough of such predators to clean them all up in the short season when lost calves were available. Most probably starved and were picked up by vultures. Losing of calves was density dependent, since the babies seemed safe by their mother's sides in small herds. Only in the larger crowded herds was this loss significant. Thus it was getting lost and getting sick which were the main causes of mortality in wildebeest herds. Besides these, being cut down in the prime of your health by a lion or a hyena were unlikely fates.

Predators must always have some regulatory influence on the numbers of their prey, because they always kill at least some animals that would not have died otherwise, particularly the young. In the simple moose-wolf system confined within the boundaries of Isle Royale, predation is almost certainly the principal controlling force. It seems likely that wolves provided potent checks on the numbers of other deer in primeval North America, although we have no way of really knowing. The evidence so far collected from the African game herds suggests that predation is for some species not the most important cause of deaths, or means of population control.

But, even if not the main controlling agents, predators will be one factor in the stability of the prey population, particularly when they hunt the young. This may have been true even for Errington's muskrats, for if all the other calamities of a muskrat's life failed to act for some reason, and an unusually large population of muskrats began to breed, the minks would presumably have to concentrate their efforts on the food that was available, a bumper crop of baby muskrats. Populations may be checked by many different density-dependent pressures acting at once, only one of which might be the predators. But it may be prudent to think of some of the largest predators primarily as superscavengers, who cull out the sick and the old but slightly before their time.

Unlike large predators, small predators may commonly exert effective, and even stringent, control over the numbers of their prey. Part of the evidence for this comes from our experience with agricultural pests. In America, where devastating outbreaks have sometimes been traced to the importing of pest species from abroad without their natural enemies, some outbreaks have been countered by bringing over suitable predators that have then reduced the numbers of the pest species to low and acceptable levels. Other evidence comes from the unwise use of long-lasting insecticides like DDT, which have sometimes actually promoted the pests that they were supposed to subdue because they exterminated the active, and therefore vulnerable, insect predators of those pests. Such evidence is most clear-cut for tropical regions where it seems that populations of insect and other small herbivores are often kept at remarkably low levels by the attacks of their predators. In places of seasonal climate the control exerted by small predators is more erratic because life cycles must be synchronized with the seasons, so that whole generations of herbivores may escape the ravages of predators whose own life cycles are not perfectly in phase with them. In the harshest or most seasonal places, where opportunistic species live, even small predators may seldom exert much restrictive effect on the numbers of their prey. That small predators may be much more effective than large predators may be a consequence of their superior killing ability. A tiger which kills a buffalo performs a dangerous and difficult feat of arms, but a wasp assaulting a caterpillar faces no such hazard and may kill with certainty all the caterpillars which it can find.

409

CHAPTER 29

PREDATION ON
SMALL ANIMALS

When European agriculture was brought to America, the various imported crops were followed, often after a lapse of some decades, by European pests. This should not have been surprising because the same
human ingenuity that cleared the land to make it suitable for alien crops
must also have made the land suitable for the insects that ate those crops.
But what was a surprise, and an unpleasant one, was that the insects did
much more damage in America than they had ever done in their native
countries. American farmers often faced ruin from insect pests which had
been known as little more than irritants to their European ancestors. A
possible explanation of this unwelcome phenomenon was that the pests
had been imported but the enemies of those pests left behind. Insect herbivores were able to eat their way to huge populations in a New World
free from the specialized predators which had fed on them in the Old
World. The answer to the pest problem seemed, therefore, to be to seek
out the natural enemies of the pest and to introduce them to American
farmlands. The consequent experiments in biological control of pests
have given us some of the most striking examples of how strongly small
predators can affect their prey.

In the late 1880s the California citrus industry faced complete ruin from
a plague of scale insects, called the cottony-cushion scale, *Icerya pur-
chasi*. These small fluffy insects suck the sap of leaves and twigs, generally resting immobile with their mouthparts buried in the phloem. A massive infestation causes the leaves first to curl and then to fall, so the
orchards of California, on which the settlers had lavished their whole
lives, looked gaunt and dying. There were no oranges to pluck or ship.

When *Icerya* struck California, entomologists already knew that many
herbivorous insects were attacked by parasitic hymenopterans that
hunted down their particular prey to pierce the larva skin with their
ovipositors and lay eggs in the living flesh. This was a specialized form of
predation which directed the aggressor to attack only the herbivorous
species that it was "programmed" to hunt. If such a parasite of *Icerya*
could be found, perhaps it could be reared in numbers and released in
the orchards to the salvation of an industry. After the original home of
Icerya was traced to Australia, $1500 were wangled out of a federal
appropriation voted to equip a trade pavilion (De Bach, 1964), and a
man, Albert Koebele, was sent to Australia to look for likely parasites of
Icerya. He found some, but he also found something, as it turned out,
better; ladybirds. The pretty little red ladybird *Rodolia cardinalis*, now
known commonly as "the vedalia," apparently had a voracious appetite
for the little scales. All stages of the vedalia beetle were carnivorous, the
males, the females, and every instar of the young; and all stages could apparently thrive on the passive cottony-scales. Koebele shipped three consignments home, and there arrived in California a total of 129 live vedalia
beetles.

The vedalia beetles were received by Koebele's colleague, D. W.
Coquillet, who proceeded to introduce them to the orange orchards. He
placed them on the branches of an infested tree, then wrapped a muslin

tent round the whole tree so that the beetles could not get out. By January of 1889 the last of the three consignments had arrived and were put onto the tented tree, and they began breeding rapidly. The vedalia can complete their life cycle in only 26 days, suggesting that a massive exponential increase is possible in a single summer. By April 1889 the tented orange tree was free from scales, but rich in ladybirds. Coquillet opened one side of the tent to allow the animals out, and he began shipping colonies to other orchards about the state. By July the whole orchard of 75 trees was free from the pest. The news spread, and planters journeyed far to collect the precious beetles for their own estates. Within a year the whole of Southern California was rid of the plague of cottony-scale, and began to grow rich again. The grateful planters bought a pair of diamond earrings for Koebele to give his wife, and it is said, probably with regrettable accuracy, that he was the only entomologist who has ever been able to so bedeck his woman with diamonds. Professional jealousy forced Koebele out of his job though (DeBach, 1964).

This destruction of *Icerya* by its predator, *Rodolia,* is a clear demonstration that small carnivores can have a devastating effect on their prey. And the general outlines of the story have now been duplicated many times. Usually the chosen predator is a hymenopteran parasite of the kind that Koebele had in mind when he was first sent to Australia. When the control measures have been successful, the typical result is for the predator or parasite to quickly build up its numbers at the expense of the prey, until its destruction is so great that the remaining members of the prey species are too few to do serious damage to the crops. But the prey usually does persist. Once the first dramatic slaughter is over, there usually remain low populations of both predator or parasite and prey, which are apparently in some sort of balance.

We thus have many demonstrations that small predators can control small prey to an extent that seems almost never to be achieved by large predators working on large prey. These different effects may probably reflect the different styles of hunting which prey size forces on predators. Large vertebrate predators, for all their strength, must manage something of a feat of arms every time they catch a meal, and they must achieve these feats without ever getting hurt. These requirements force on them the habit of catching the sick and the weak. They may also require undisturbed and unaware animals to stalk, which means that there cannot be too many predators about. As Hornocker (1969) argues for his Idaho pumas, (Chapter 31) there cannot be so many cats on the prowl that the deer are always wary, or no cat would catch anything. For these reasons the pressure of predation from large carnivores is nearly always light. But small predators usually face no such problems in hunting or killing. The vedalia beetle, for instance, can walk up to the passive scale insects, chew them up at its leisure, and move on to the next one which sits there waiting to be eaten. (Figure 29.1). Vedalia beetles can thus scarcely be said to hunt, they first search, then they butcher. The hymenopteran parasites more commonly used in biological control work do hunt, however;

Figure 29.1

Vedalia beetle and prey.

but they go about the business of killing in such formidable armor, and with such formidable weapons, that the only escape possible for slow-moving prey is to hide. When the prey caterpillar or other larva is found, the wasp merely climbs aboard and stings it with her ovipositor. In such an encounter there is no hope for the prey and no danger to the predator. Likewise, when a web spider snares a fly, or a wolf spider stalks within springing distance of its crawling prey, there is little doubt about the outcome of the attack or the safety of the attacker. Unlike large predators, small predators can confidently kill on encounter, and it is this special ability that can best explain their effectiveness as agents of control (Figure 29.2).

But introducing predators in attempts at biological control of pests does not always work; indeed, it usually does not. Often the chosen predator or parasite becomes established, apparently living well, but does not make a significant reduction in the numbers of the prey. It lives off the pest, and possibly off other herbivores of the neighborhood as well, but without exercising noticeable control. The evidence from experiments at biological control is thus ambiguous, suggesting that sometimes small predators control their prey, and sometimes that they do not. Evidence that control by predators* can be normal, however, comes from another activity of agriculture, the misuse of pesticides.

In Malaysia there are very large plantations of tree crops, such as rubber and oil palms. The climate is that of the wet tropics and almost without seasonal differences, a place where the virgin vegetation was usually rain forest. There is to be found in Malaysia the tremendous variety of insect life normal for the equatorial tropics, and among this array of insects are very many herbivores which can eat rubber trees and oil palms. A farmer, reflecting on his great troubles with pests in temperate orchards, might well be expected to quail before the prospect of planting trees in such an insect-ridden place; and yet tropical planters have generally had less trouble with plagues of pests than have orchard growers in the north. The trees of the wet tropics, whether of plantations or virgin forest, rarely have badly damaged foliage in spite of the variety of insect life that they support. This must surely mean that the local insect populations are kept down in some way, but we cannot appeal to the Andrewartha and Birch hypothesis of insect control by recurrent calamity (Chapter 27) because there never are climatic catastrophies to disturb the everlasting sameness of the local weather. The most likely alternative explanation is that all those different kinds of local herbivores are kept in check by a comparable variety of local predators; and a practical demonstration of this was to come when the unwise use of DDT and its long-lived broad spectrum relatives killed off so many insect carnivores that they produced the very outbreaks of pests which they were designed to prevent.

In the late 1950s many estates growing oil palms in Malaysia suffered plagues of defoliating caterpillars, and the estate managers called in entomologists to advise them. One of the entomologists, Brian Wood, was able to work out a detailed history of the insect plagues, and to explain how they came about (Wood, 1971). The start of the story seems to have been a small alarm, the appearance of sufficient cockchafers on one plantation to worry the local manager. The cockchafers were not doing much damage, but the manager thought he ought to spray against them as an insurance, so he sprayed the plantation with DDT, and noted the fact in his monthly report. Shortly afterward there were outbreaks of leaf-eating caterpillars on that part of the estate, so the manager sprayed again, and also sprayed the neighboring areas as well "to contain the pests." The next thing that happened was that caterpillars became a serious trouble over all the sprayed area. Other planters learned of it, became convinced that they were faced with a general outbreak of insect pests, and began spraying in real earnest. DDT was followed by dieldrin and endrin, the very long-lasting and toxic relatives of DDT, and these were sprayed often and over all plantations. The caterpillar plague then grew to such proportions that whole oil palms were defoliated, standing out gaunt like rows of discarded giant's fly wisks. The planters faced heavy losses.

Wood examined this history of the pest outbreaks, and concluded that

* In the ensuing discussion the word "predator" is taken to include hymenopteran parasites, or "parasitoids" as they are sometimes called. That these kill by slowly eating their hosts rather than by giving them a swift death makes no difference to their effects on numbers.

FIGURE 29.2
The deadliness of
small predators.

A

B

In each of these encounters the armor, the weaponry, or the masterly manouver of the attacker makes the fate of the prey certain, the safety of the predator assured. Conquest by
large predators, such as lions, of large prey, such as buffalos, cannot be so certain or safe.
These differences must be taken into account when assessing the relative importance of
large and small predators as agents of population control. (a) A stinkbug (Pentatomidae) attacks an insect larva; (b) a devil's coach horse (Staphylinidae) attacks a pillbug (Isopoda);
(c) a water beetle (Dytiscidae) attacks a tadpole; (d) a web spider bites a bound and helpless
fly.

C

D

the only change in the environment or management of the estates which was synchronous with the pest outbreaks was the fact of spraying with insecticide itself. But could insecticides encourage pests? — they could if they killed the predators which had been eating the pests more effectively than they killed the pests themselves.

The worst of the defoliators was a bagworm, a species of moth whose caterpillars lived in shared cocoons, and this cocoon-living might afford them some protection from the spray. Wood found that more than 20 species of hymenopteran parasites commonly laid their eggs in the bagworm larvae, feeling for them inside their cocoons, with their long ovipositors. These hymenopterans would often be more exposed to the sprays than their prey and would be selectively killed. It might, in fact, be generally true that caterpillars and other herbivores, curled up in the foliage, journeying little, might be more protected from insecticides than the active wasps, so that the sudden appearance of large numbers of many defoliating species might be a direct consequence of killing their enemies with insecticides. The test of this hypothesis was, of course, to stop spraying and see what happened; and this Wood recommended. To a large extent his stratagem worked. The plagues went when they stopped spraying, and Wood could conclude that all those myriads of potential pests had indeed been kept down in the days before DDT by their natural enemies. In tropical Malaysia, then, it seems that small predators usually do control their prey.

But Wood's remedy of stopping spraying was not always completely successful. Plenty of the hymenopteran parasites soon reappeared, but the bagworm populations sometimes remained at a disturbingly high level. How could this be? Wood concluded that the regular application of sprays had synchronized the life cycles of pests so that now they were all of a common age. The life cycles of predators would also have been synchronized. There might be periods, for instance, when there were large caterpillars everywhere but no flying adult wasps to hunt them, so that whole generations were able to partially escape predation. Only when the many local populations got their generation times back out of phase would control be as complete as it was before spraying was started. Wood's practical answer was to attack the secure generations of caterpillars with stomach poisons, which is to say with sprays that had to be eaten to kill and which were thus only deadly to herbivores. By this means he was able to undo the mischief that contact poisons had done and leave the plantations in the security of the prechemical age.

A GENERAL
CONCLUSION ON
CONTROL OF PREY BY
SMALL PREDATORS

Our experience with successful introductions of predators to control insect pests in temperate lands, and with the disastrous consequences of spraying general insecticides in the tropics, suggests most powerfully that small predators commonly do control their prey. In climates without strong seasons or intermittent catastrophes, control of foliage-eating insects by predators is probably almost universal. In such places, plagues of herbivorous insects or other small herbivores are likely to be exceedingly rare. An important ecological result of this tight control of herbivores is that most of the energy transformed bv tropical vegetation does not enter animal food chains but is degraded by decomposers. Most of the non-photosynthetic life of such places must be decomposing life, and these decomposers owe the largeness of their plenty to the small carnivores of the forest.

In temperate lands, control by predators may also be effective, but there are clear indications that the control is not so unremitting. There is first the fact that most attempts at pest control by introducing predators do not work; the predator persists without significantly reducing the numbers of its prey. And there is secondly our general experience that insect herbivores commonly do become pests of temperate agriculture. Among entomologists there has been a widespread belief that these pests of temperate agriculture are nearly always introduced animals, but this assertion does not seem to have been documented. It is the experience of Wood and others working in the wet tropics that pests there are certainly local and native, and it seems likely that the pests of temperate lands are local animals too. More often than not, pests are native animals that have "got out of hand," not just herbivores that have crossed the oceans in ships unaccompanied by their predators. So it seems that local herbivores do manage to escape control by local predators and become a nuisance more commonly in temperate lands than in the tropics. And for all the unpleasant side effects of DDT, its use in temperate agriculture did result in drastic reductions of pests. There have been, it is true, many examples of how pesticide sprays have brought on troubles to northern farmers, as for instance the common result of promoting plagues of red spiders (acarine mites) in orchards by killing off their hymenopteran enemies, and the fact that *Icerya* has returned to be a nuisance in California, suggesting that the vedalia beetles have been killed off by the sprays. But generally DDT and its allies have not brought such retribution to Western agriculture as they have to the tropics.

Killing the local predators is apparently not so serious in temperate lands, which suggests that they may not be such reliable agents of control there. An explanation for this is not hard to find. Temperate climates are seasonal climates, and seasonal climates must synchronize the generations of insects. Wood had found that regular spraying of his oil palms had synchronized generations of caterpillars so that sometimes whole generations were only lightly preyed on. But such synchrony is normal in seasonal climates. The ending of winter is the trigger for pupae to open, for eggs to hatch, and for the sleep of diapause to be broken; which events necessarily result in synchronized generations. To this is added the fact that there are less species in temperate climates, suggesting less complex ecosystems, and hence less stability (Chapter 17). We should not be surprised to find that control by predators is more erratic in seasonal climates, that in some years they severely restrict the numbers of their prey but that in other years they seem to have little effect.

In extremely seasonal and marginal climates, like those of southern

Australia where plagues of the grasshopper *Austroicetes cruciata* come and go (Chapter 28), the synchroneity effect might so couple with the shortness of the living season that predation is always negligible. Herbivore populations might then well expand unchecked by predators until climatic catastrophe strikes them down, just as Andrewartha and Birch claim that it does. The effectiveness of small animal predation is thus likely to be a function of the seasonality of climate. In regions of humid tropics without noticeable changes of climate from month to month, predation may be a dominant controlling influence that sets the populations of herbivores at low levels. In more seasonal places, both temperate lands and tropics afflicted with dry seasons or monsoons, control by small predators is likely to be more erratic. And on the marginal zones at the edges of deserts and the Arctic, predators may be largely ineffective, so that animals must be adapted to survive the catastrophes which climate or unchecked population growth must bring on them.

Clashes of predators with their prey can be described in mathematical terms. The simplest algebraic models tend to predict inevitable extinction of the prey, a conclusion that seems unacceptable, but more realistic calculus models, essentially modified versions of the Lotka-Volterra equations of competition, predict coupled oscillations. Such coupled oscillations have been demonstrated in various laboratory microcosms, but the oscillations will only develop when provision is made for the escape of some of the prey during times when the predators are at their maximum abundance. In nature such escape may well be achieved by dispersal, suggesting that patchy distributions of insect pests may result from local populations undergoing coupled oscillations. It is the common experience of entomologists that pests are frequently distributed in patches, but this could as readily be explained as the result of local extinctions, as predicted by simple algebraic models, or from some quality of the environment remote from the activities of predators and prey. We have to conclude that the distributions and abundance of wild animals may be influenced in so many ways that they cannot be used as tests for models of predation. More fruitful has been the approach of simulating the ways in which individual kinds of predators operate, so that their activities may be imitated on a computer. This approach requires that the computer program be constantly tested against the performance of real animals so that the simulation may be true to life and truly descriptive. It may eventually help our general understanding of the workings of nature by letting us see what part of the apparent balance which we see around us is due to the interaction of predators with their prey and how much must be ascribed to other forces.

THE ANALYSIS OF PREDATION

419

THE ANALYSIS OF
PREDATION

When predators meet prey, it is numbers that count. If predators are rare, their effects may be negligible. But if they are abundant, they may devastate or control the populations of their prey. This importance of numbers makes the phenomena of predation unusually suitable for investigation by mathematical analysis.

Very simple mathematical models can be made to describe the effects of ideal predators on ideal prey, to simulate systems in which every hunt is successful and every escape rewarded by the successful rearing of young. This ideal comes closest to being realized when an hymenopteran parasite hunts easy-to-hunt prey. When such an hymenopteran finds its prey animal, it attacks, lays an egg in the host, and has effectively accounted for one individual of prey; and its total killing power has an absolute limit set to it by the number of eggs that it can lay. If you assume that the fecundity, that is the ability to lay eggs, of all the individuals of the wasp population is the same and constant, and further that every hunt is successful, then the fecundity of the wasps is a direct measure of their effect as predators. If the fecundity of all the members of the prey population is also uniform, and if all eggs successfully develop into adult prey animals, then the outcome of any clash between the two populations is directly predictable from their initial densities alone. If the fecundity of host and parasite are equal, there will always come a time t_g when the last generation of prey is eliminated and both populations perish. This is given by

$$t_g = \frac{n}{p}$$

where n is the initial density of the host and p the initial density of the parasite.

If the host can outreproduce the parasite, its population can escape the disastrous effects of being attacked provided that

$$\frac{p}{n} > \frac{(h-s)}{s}$$

where h is the fecundity of the host and s the fecundity of the parasite.

This very simple model allows effectively only one ending to the contest, extermination (DeBach, 1964). If the prey is not quickly killed off as a consequence of not being able to outbreed its attacker, its population will grow until some nonpredation curb is applied, when the predators will quickly catch up, attain the critical density necessary for them to exterminate their prey, and proceed to do so. Such models suggest that all predators should long since have exterminated their prey, and that the world by now should be populated only with plants. The models can be made more true to life by allowing for the fact that not all hunts are successful, that fecundity is not constant, that sex differences are important, and so on; but they remain unreal because they do not allow the force of these factors to change as the populations change. The true relationship between predator and prey must be dynamic, with all the parameters that measure success at killing or escaping constantly changing as the ratios of the populations change. Such changing relationships cannot be described in simple algebraic terms, but require modeling by calculus or similar mathematical methods.

The results of eating or being eaten should reveal themselves as changes in the rates of population growth of predator and prey, and it is convenient to seek to model their relationship by equations describing the growth of each population in the presence of the other. This is a process analogous to that used by Lotka and Volterra to investigate the effects of competition. Both men, again independently, adapted their logistic equations to predict what happens when predator meets prey (Lotka, 1925, Chapman, 1931). The growth of each population is again seen to depend on the intrinsic rate of increase and the population density, but the limiting term to the growth of each is a function of the numbers of the other population thus:

$$\begin{pmatrix} \text{Rate of growth} \\ \text{of prey} \\ \dfrac{dN_1}{dt} \end{pmatrix} = \begin{pmatrix} \text{rate of growth} \\ \text{when without check} \\ r_1 N_1 \end{pmatrix} - \begin{pmatrix} \text{rate of fatal} \\ \text{encounters} \\ \gamma_2 N_1 N_2 \end{pmatrix}$$

$$\begin{pmatrix} \text{Rate of growth} \\ \text{of predator} \\ \dfrac{dN_2}{dt} \end{pmatrix} = \begin{pmatrix} \text{rate of successful} \\ \text{encounters promoting} \\ \text{growth} \\ \gamma_2 N_1 N_2 \end{pmatrix} - \begin{pmatrix} \text{rate of death} \\ \text{of predators} \\ d_2 N_2 \end{pmatrix}$$

where

r_1 = intrinsic rate of increase of prey
N_1 = number of prey
γ_1 = fraction of contacts which prove fatal to prey
d_2 = rate of death of predators in absence of prey
γ_2 = rate of growth allowed predator per unit contact with prey

The simultaneous solution of such a pair of equations shows that the populations should fluctuate cyclicly. When predators are scarce but prey numerous, the predators should be able to build their population quickly, inevitably reducing the population of their prey. Eventually the predators should be numerous and the prey scarce, at which time the predators should compete with each other so vigorously for the prey food that they would suffer enormous mortality from starvation. The predation pressure on the prey would then be relaxed, and the prey population should expand, thus completing the cycle. There should thus be oscillations in the populations of predator and prey which would be coupled (Figure 30.1). There should always be "too many" predators or "too many" prey.

The idea that oscillations in related processes should be coupled is a general one with many examples in studies far removed from ecology. Economists are familiar with the idea as the theory of supply and demand,

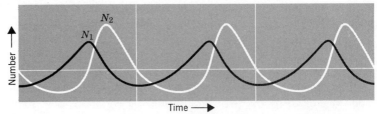

Figure 30.1

Coupled oscillations predicted by the Lotka-Volterra equations for predation. (After Volterra, from Chapman, 1931.)

and a form of it was well and bitterly known in the American Midwest of the depression years as the "corn-hog" cycle. When corn sold well, farmers planned next year to keep more corn for sale even if it meant rearing less hogs. This resulted in a glut and falling prices. Many farmers responded by keeping their corn back next year to fatten hogs instead; which resulted in a glut of hogs and a shortage of corn. Recurrently there would be a glut of hogs or a glut of corn, each glut being attended with a crop of farmer bankruptcies, each glut being accompanied with a shortage of one of the coupled commodities. In engineering, such coupled relationships form the foundation of servomechanism theory, and they are discussed in physics texts under the heading, "The theory of coupled oscillations."

Essentially the contribution that Lotka and Volterra made to predation theory was to show that coupled oscillations were a quite plausible outcome of meetings between predators and their prey. As when discussing competition, Lotka and Volterra put forward as a basis for their work the hypothesis that population growth could be truly described by logistic equations. They added to this the second hypothesis that there is a rather simple and direct relationship between the numbers of the predator and of its prey. Reducing the numbers of predators must immediately allow the prey to recuperate, and increasing the numbers of predators must immediately depress the prey. It was clear from the first that these conditions were much too simple to be met by many populations of real animals, but it was also decidedly interesting to test the models to see how close they did come to reality. Did the populations of real animals fluctuate with some resonance of the elegant rhythm that the equations suggested?—if they did, we might expect the numbers of even equilibrium species to fluctuate within wide limits. Perhaps the changes in animal numbers that most men noted within their own lifetimes were reflections of the relationships between predators and their prey.

Gause (1934) tested these predictions in his laboratory in Moscow, just as he had been the first to test the predictions made by Lotka and Volterra on the outcome of competition (Chapter 23). For prey animals he used populations of *Paramecium* grown in centrifuge tubes and supplied with his oatmeal medium, just as he had for his competition experiments. He

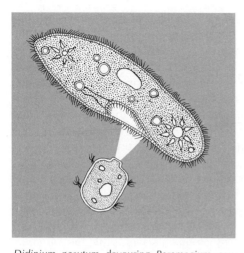

Figure 30.2

Didinium nasutum devouring *Paramecium cau-datum*. *Didinium* is so effective a predator of *Paramecium* that introducing a few into a *Paramecium* culture results in the complete exter-mination of the paramecia. A *Didinium-Paramecium* system can be made to oscillate by adding "immigrant" paramecia at the time when the *Didinium* population is reduced through star-vation.

now needed a predator that should live under the same conditions, and he found it in the protozoan carnivore *Didinium nasutum*. This animal is about the same size as a paramecium, and it seizes its quarry one at a time to suck out their contents with a tubular proboscis (Figure 30.2). *Didinium* reproduces by binary fission in the same way as *Paramecium*, and its feeding habits suggest that there should be a rather simple numerical relationship between its numbers and those of the paramecia that it catches. In Gause's tubes, then, he had introduced a system of predator and prey so simple that even the unrealistic conditions of the Lotka-Volterra models might be met; he had, as Hutchinson might have said, built himself an animal analog computer that might well be set to correctly solve a pair of equations describing coupled oscillations. But there were no oscillations between the numbers of *Didinium* and its prey in Gause's bottles. Whenever he introduced *Didinium* into flourishing *Paramecium* cultures, the *Didinium* quickly reproduced at the expense of its prey until every *Paramecium* was killed and eaten, and only *Didinium* remained. Every *Didinium* then died of starvation.

This result was probably not unexpected to Gause, for it must seem obvious that a small centrifuge tube of water which was crowded with *Didinium* should be so deadly a place for paramecia that none might es-cape. The coupled oscillation hypothesis must allow the effectiveness of predation in some way to be lowered when the prey is very scarce, so that some individuals escape being eaten for as long as it takes most of the predators to die of starvation. An obvious weakness of the hypothesis was that it required prey to escape when they were outnumbered by desperate predators seeking food. Gause's experiments confirmed this weakness, showing that the hopes for survival of outnumbered paramecia in a centrifuge tube were negligible. But Gause could argue that in the real world there might be places in which the survivors could hide, that chances of escape by hiding would, in fact, go up as the survivors became few because then hiding places should be easy to find.

Gause "doctored" his experiment to simulate the successful escape of a few survivors. He waited until nearly every *Didinium* had died of starvation, then he put in a few fresh paramecia which he called "immigrants." By choosing the right moment at which to introduce his "immigrants," Gause could push his system past the critical stage. While the *Didinium* was still rare, the paramecia reproduced quickly to build another flourishing population, but then *Didinium* caught up and wiped them out again. Then, while *Didinium* was next starving, Gause would add a few more "immigrants" to start the cycle all over again. In this way the oscillations could be kept going indefinitely, and Gause had shown how real animal populations might be made to behave in the way predicted for them by the Lotka-Volterra equations. Or we might say that he had trouble-shot his animal analog computer so that it correctly solved the equations and produced the required coupled oscillations.

Much perseverence by others has led to the production of a few laboratory microcosms in which the numbers of predator and prey will oscillate without such interference as the manipulation of "immigrants." The most striking is perhaps that constructed from weevils, which burrow into beans, and their hymenopteran predator by the Japanese Syunro Utida (1950, 1957). Weevils are holometabolous insects, whose larvae are soft maggots but whose adults are beetles with hard exoskeletons. The azuki bean weevil, *Callosobruchus chinensis,* can subsist when provided with no more than piles of dry beans in a box, and complete their life cycle in about three weeks. Utida kept them in petri dishes 1.8 centimeters high and 8.5 centimeters across, the little flat glass dishes with covers which are standard equipment in any biology laboratory. Ten grams of beans (some 50 of them) were quite sufficient to allow the weevil population in each dish to rise to several hundred, provided that the beans were replaced every three weeks. And the space in the small petri dish was also quite large enough to maintain a large population of the tiny parasitic wasp *Neocatolaccus mamezophagus.* These wasps hunt the fat final instar larvae, feeling for them in their burrows within the beans with their long ovipositors, and laying eggs beneath their skins. The parasitic baby wasps then live on the larvae even as they feed and pupate, so that finally there emerge from the pupae not weevils but wasps. Utida found that stock cultures of wasps could be maintained on the weevil cultures in the dishes, that a petri dish provided a sufficient universe for the complete life cycles of both these complex animals. And the universes were so simple and small that it was easy to maintain sufficient replicates to meet the requirements of proper experimental procedure. It should be possible to watch closely the effect of one population on the other.

Utida started his experiment by placing eight pairs of freshly emerged beetles in a dish, and letting the population grow by itself for 10 weeks. Three generations had then passed, and there were between 300 and 400 weevils where once there were only 16. Then Utida released on them four pairs of the freshly emerged wasps. Every three weeks he used ether to quiet the inmates of the dishes, gently disentangled them from the remnants of the old beans, put in 10 grams of new beans, and tallied the numbers of the animals. He persevered with this work for more than six years. And at the end of it he had the most dramatic records of oscillating populations that we posses. Oscillations between predators and prey did appear, and were apparently coupled as Lotka and Volterra said they should be. The fluctuations were not perfectly regular. Sometimes a population would rise quickly; sometimes both populations seemed to stagnate for a while; but the cycles never disappeared, were only temporarily damped, and continued for more than 110 generations in a way which promised they might go on for ever (Figure 30.3).

The coupled oscillation hypothesis requires that some prey escape even when a horde of predators is competing for their bodies. Such escape was not possible for a few paramecia in tubes full of *Didinium,* but it apparently is possible for the last weevils in petri dishes buzzing with wasps. There are several possible reasons for this. It may be that the physical habit of burrowing into a bean gives the weevil victim a reasonable chance of dodging attack, although this is perhaps doubtful. Parasitic wasps are generally skillful at finding their targets, even if they must feel down the tunnels through a bean with their long ovipositors. A perhaps more likely cause of escape is that the generations come to overlap slightly so that a few weevil larvae are too young to be attacked while the wasp plague is at its height. This is inherently likely because the mean wasp generation is a few days less than the mean weevil generation, and yet the wasp must stay synchronized to the weevils by its necessity of attacking only last instar larvae. Another possibility which must act to some extent is the random element in wasp hunting which means that some wasps must attack larvae that other wasps have already attacked. Wasp parasitism, although completely lethal, does not kill at once so that it is possible for the same prey to be attacked by a number of different predators, each of which delivers a potentially fatal blow. When many wasps hunt a few larvae by a process of random search, some larvae will be hit only once, but others will be hit many times. A few may not be hit at all, and these may provide the adult weevils which can rebuild the prey population after the wasps have died. Probably the complete explanation of the survival of the weevils in the wasp-filled dishes involves some combination of processes such as these.

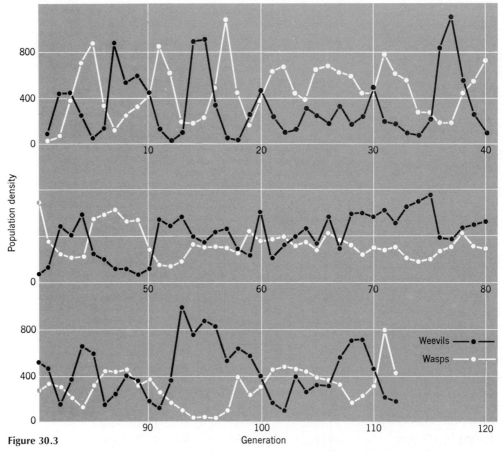

Figure 30.3

Coupled oscillations in Utida's wasp-weevil system. The system consists of beans (renewed at intervals) weevils and a parasitic hymenopteran in closed petri dishes. The populations oscillate without further manipulation, apparently because some weevils escape during dense predator populations because the wasp life sycle is not perfectly synchronized with the weevils or because some larvae are overlooked by the wasps, which sting others repeatedly instead of spreading their attacks evenly. (From Utida, 1957.)

Utida's petri dish microcosms are closer to real life than those used in other experiments, such as paramecia in tubes or *Tribolium* in flour. Wild wasps do hunt wild weevils in wild beans, and many hymenopteran parasites are host specific so that their own numbers are directly controlled by their success at hunting one sort of prey. But wild weevils are offered the extra chance of escape which lies in fleeing, and wild wasps have the additional problem of dispersing after their quarry. Dispersal is an obvious way in which prey might survive local concentrations of predators, as dispersal is also a way in which starving predators may find new concentrations of food. The fact that wild animals can disperse might thus make local coupled oscillations between predators and their prey seem more

The Huffaker universe of oranges and rubber balls across which predator mites hunt herbivore mites. Herbivorous mites infest oranges. When a predator mite finds a colony, the
predator population increases until the herbivores are exterminated. This universe, in which
travel from orange to orange is impeded by rubber balls and moats of grease provides just
enough opportunity for some herbivore mites to disperse away from the focus of predation
and to build a new population on another orange before the predators again find them. The
result is a perpetual game of hide-and-seek among the oranges in which neither population
becomes extinct within the universe.

Figure 30.4

likely than oscillations over large areas. Dispersal was really what Gause
imitated when he provided his collapsing systems with "immigrant"
paramecia, but a more realistic way of modeling the effects of dispersal in
a laboratory microcosm has recently been devised by C. B. Huffaker
(1958) of the Division of Biological Control, University of California at
Berkley.

Huffaker loosed a carnivorous mite, *Typhlodromus occidentalis*
on an herbivouous mite, *Eotetranychus sexmaculatus*. The herbivore
ate oranges, and could massively infest their surfaces. When the predator
was loosed on such infesting mites, it killed them voraciously, expanding
its population until its descendents totally consumed the prey; after
which the predators died of starvation. This was Gause's *Didinium* story
all over again, an encounter from which some of the prey could survive
only if they could manage to get away. Huffaker set out to devise a universe in which it was possible for some of the persecuted prey to flee to a
place of safety where they had enough peace to build up a new population before wandering predators found them again. He solved the
problem by interspersing oranges with rubber balls of the same size on
trays, and separating them with barriers of vaseline or oil which were
more or less difficult for the mites to cross (Figure 30.4). He made many
experiments with different combinations of oranges and balls, and with
different barriers between them, learning the while much of the dispersal
powers of the little mites. And he found one configuration in which the
populations of predator and prey in the universe of his trays fluctuated in
the desired rhythm of coupled oscillations (Figure 30.5); and without his
doing more than to replace oranges as they went bad.

These various experiments show that coupled oscillations are indeed
possible for real animals if the conditions are right. But are the conditions

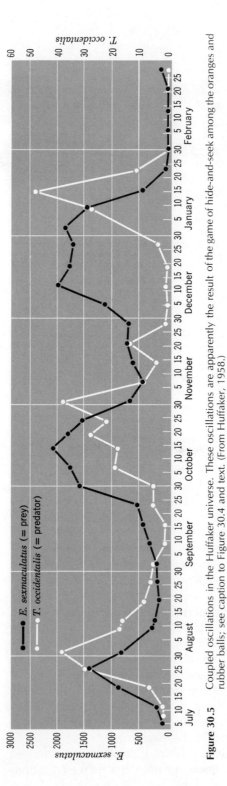

Figure 30.5 Coupled oscillations in the Huffaker universe. These oscillations are apparently the result of the game of hide-and-seek among the oranges and rubber balls; see caption to Figure 30.4 and text. (From Huffaker, 1958.)

commonly right in nature, so that the fluctuations in animal numbers we see around us do truly reflect an endless rhythm of kill and escape? This has been a difficult question to answer. The obvious approach was to collect evidence about any animals whose populations were known to change by large amounts, to look for rhythm in the fluctuations, and to see if this was coupled with a corresponding rhythm in the populations of likely predators. Interesting fluctuations have long been known in rodent populations, and various historical records have suggested that some of these fluctuations are rhythmic or cyclic. Most famous are the fluctuations in arctic animals, such as hares and lemmings, whose numbers change by orders of magnitude with rhythms of a few generations. These animals are beset by many predators, and the numbers of the predators do fluctuate sharply also, but many careful studies have shown that the populations of such predators and their prey fluctuate synchronously. There are many predators when there are many prey, but they have only a small depressive effect on the prey. Then the prey dies from other causes, and the predators die too. The oscillations do not alternate, but are in phase. The evidence for this is discussed more fully in Chapter 35.

But it is among insects that we should expect to find evidence for coupled oscillations in the wild. Insect predators do seem to have a general controlling influence on their prey, and many insect predators are specialized feeders, often being limited to single species of their prey. This is an important requirement if coupled oscillations are to be realized, for a predator of catholic taste might not eat down local populations of one kind of prey so closely. Census of insects on a large scale is often dauntingly difficult, and yet a proper demonstration of coupled oscillations requires simultaneous census of both predators and prey. Even when these data are obtained, there is still the great difficulty of unscrambling the effects of other checks to population, such as weather and disease. Then, over how large an area are you to look for your evidence of oscillating populations? If the animals constantly move perhaps you may end up by sampling different parts of different populations each time you make a census, so that your figures are useless for your purpose. For these reasons, the actual demonstration of the existence of coupled oscillations in wild insects is next to impossible.

But entomologists have commonly noted one significant fact: massive infestations of pests are often distributed in local patches. One tree in an orchard, or one branch, may be sore beset by mites or caterpillars or bugs, while other trees or other branches have almost none. At other times, census of the same orchard may show that different trees and branches carry the infestations. The total population of pests in the orchard may well be the same, but they are differently distributed in dense patches. This pattern makes sense if there are local oscillations between predators and prey in progress, so that census time finds one pest population at a maximum and another at a minimum.

The distribution of insects in patches becomes even more suggestive of the existence of coupled oscillations when we take into account the more

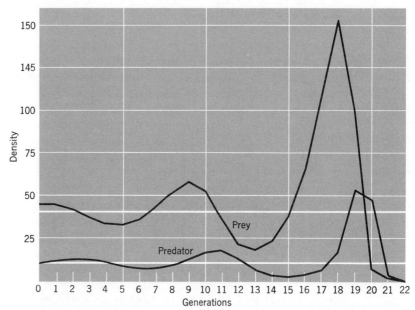

Figure 30.6

Nicholson and Bailey model of predator-prey system. This model is a development of the Lotka-Volterra model, but one that allows for various lag responses. The prediction is that the populations oscillate into extinction, the same conclusion suggested by simple algebraic models. Patchy distributions of insect pests may reflect the operation of such processes in nature.

complete form of oscillation theory developed by Nicholson and Bailey (1935). Nicholson and Bailey modified the Lolka-Volterra equations to provide the time lags which the qualities of real animals should inevitably introduce into the system. They put forward their conclusions in a monographic mathematical work in 1935, a work well beyond the mathematical competence of the biologists of their day so that it was more quoted than read. But one of their principal findings is well-known; it is that the meeting of predator with prey should produce coupled oscillations, certainly, but that the oscillations should proceed with the progressive diminution of the prey until it became extinct (Figure 30.6).

If the assault of a predator on a prey population should always end with local extinction, then the common discovery of patches of herbivorous insects about an orchard or elsewhere is even easier to understand. The local clashes have brought local extinctions, but meanwhile fugitive herbivores have started colonies elsewhere which will flourish until the wandering predators find them again. This is the pattern of Huffaker's oranges and mites, rather than that in Utida's petri dishes. Nicholson and Bailey in effect concluded that most wild insect predators can overeat their prey, although it takes them several generations of oscillating fortunes to do so. They commonly will unless something stops them. This something may frequently be dispersal, which should lead to the patchy distributions of insects commonly observed.

There is, however, a simpler explanation still of patchy distributions.

When insect predators are devastatingly efficient, as say were the vedalia beetles in California orchards or the predacious mites in Huffaker's trays, they should be expected to produce patchy distributions through speedy local annihilations without any bother of a few oscillations on the way. Patchy distributions could be no more than the result of hide-and-seek between dispersing predators and dispersing prey. A fugitive starts a colony which thrives until a hunting predator finds it. The colony is then wiped out, but another colony has been started elsewhere meanwhile, and so on. There are many studies in the entomological literature which suggest that this sort of thing does happen in the wild. The process could be quite well described by the simple algebraic models of Thompson, with which this chapter began. An overall regulating effect is achieved by the predators by processes of extinction and escape, without there being any need to allow for the subtle dynamic relationships of one population on the other.

But the subtle dynamic relationships are real. They show clearly in experiments like those of Utida with his weevils and wasps, and their omnipresence is felt by any biologist who studies populations in the field. There may be a few devastating killers like the vedalia beetles loose from time to time, but most predators, perhaps even most hymenopteran parasites also, seem much less ruthless. Attempts at controlling pests by introducing their enemies usually do not work; many herbivores are not noticeably patchily distributed; many predators attack many kinds of prey; and many predators apparently do not control their prey at all. A proper analysis of predation must guide us to an understanding of all these subtle relationships. It should be obvious that all predators do not act in the same ways, and that realistic models of predation must start by describing the responses of individual sorts of predators to their prey.

When prey is plentiful, it must be easy for predators to get enough to eat. A result of this should be that many of their offspring should survive to breed themselves, so that the predator population grows. The predators will bear down heavily on the prey simply because there are so many of them, and it is this **numerical response** of the predators which is described by the models of predation based on logistic equations describing

HOW PREDATORS FUNCTION AT DIFFERENT DENSITIES OF PREY

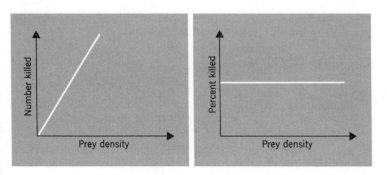

Figure 30.7

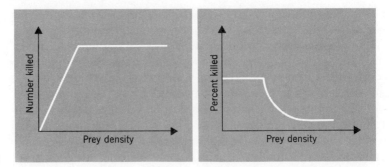

Figure 30.8

Type 1 functional response curves.

population growth rates. But there is another way in which predators may respond to changing abundance of their prey; each individual predator may kill more or less animals as the prey becomes easier or harder to find. The predator's **function** may change as the ratio of predator and prey individuals changes, and its effectiveness as an agent of control will be given by adding the **functional** and **numerical responses** of the predator together (Solomon, 1949).

The general models of Lotka, Volterra, and Nicholson and Bailey assume that the numbers of animals killed by predators are directly proportional to the predator or prey density. The predators are assumed to act in a way both mindless and without regard to the difficulty of catching and handling more and more prey animals as the prey density increases as shown in Figure 30.7.

It is readily apparent that this model is unreal. At the very least there must be some limit to the killing ability of predators so that they become fully occupied and can kill no more however dense the prey may be. In addition, a naturalist must doubt that complex animals like predators behave in such simple and mechanistic ways. We must learn how predators really function if we are to understand their full effect on their prey, which brings us back to the role originally assigned to predators in the general theories of density dependence. Do predators kill more prey when it is numerous, so that the pressure of predation acts as a check on increasing numbers? General models like those of Nicholson and Bailey (1935) seek to answer this question in terms of the numerical response, in terms of increasing numbers of predators, but might not part of the effect be due to the changing functions of predators? This possibility has received much attention in the last decade, especially in laboratories working on the biological control of insect pests. Particularly well-known are the studies of C. S. Holling (1967; 1970) and his colleagues in Canada, who have sought to describe the various **functional responses** of different predatory animals and to isolate these several responses into components of behavior. These components should then be built into mathematical models that describe the activities of a predator in a realistic manner, allowing its behavior to be simulated by computer. Once

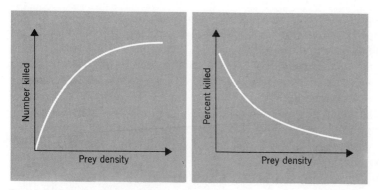

Figure 30.9

Type 2 functional response curves.

this was done, computer simulation could be used as a tool to help design practical control measures.

Holling (1965) begins his analysis by examining the functional responses of different sorts of predator. The simplest theoretical model is one that allows the predator to take a constant proportion of prey as in the numerical response models, but meets the physical requirements of the system by setting a plateau to represent the saturation of the predator's killing ability, as shown in Figure 30.8. Holling found that the eating habits of some types of small predator did let them function in this simple way. These were mostly filtering animals, like brine shrimps and water fleas, which was not surprising since their filters should take a constant proportion of what flowed past them up to the limit of their filtering capacity. It may be argued that such animals are not real predators in that most of their food may be single-celled plants, but hunting a free-swimming plant is functionally similar to true predation. It would be interesting to know if large filter-feeding animals of the sea, like whales and basking sharks, show the same simple functional response; it seems probable that they do.

Holling called this simplest of functional responses a **type 1 response.** Such a response does not provide a density-dependent control device, since the proportion of prey animals taken does not go up as the prey becomes denser. Indeed, once the saturation value is reached, the carnivores take an ever-decreasing proportion of the prey so that the predation curb is eased. Such predators can only control their prey through a numerical response, through increasing the numbers of their kind.

But most predators are not filter feeders. Holling studied water insects, corixid bugs, and dytiscid beetles, animals well-known for voracity, and tried exposing them to various concentrations of things that they ate, of tadpoles and mosquito larve. These voracious animals did eat more and more as the prey became denser until a plateau of saturation was reached, but they approached the plateau gradually, in a curvilinear manner (see Figure 30.9). A search of the literature discovered work by the Russian fisheries biologist Ivlev (the same who had so ingeniously measured the energy intake of tubificid worms; see Chapter 11) on the feeding of freshwater fishes, which showed that they too might show this curvilinear functional response.

Studies of a Canadian colleague of Holling's, T. Burnett, (1958) found that there was a similar response by wild parasitic wasps searching for sawfly cocoons in the litter under trees, a finding of much importance since so many effective insect killers are hymenopterans of this kind. Holling called the curvilinear response a **type 2 functional response,** and it was clear that it was widespread and important in nature as the type 1 response was not.

This type 2 response is easy to understand when you reflect that a predator has only a set amount of time in which to do its killing. Some of this time must always go in hunting, in searching and preparing the assault, but the time available for hunting is progressively reduced as more and more time is consumed by eating. It is true that abundant prey needs less hunting, but this saving in hunting time is apparently more than made up for by the increased time taken up in handling many victims. Holling put it forward as an hypothesis that the curvilinear functional response was indeed caused by the pressure of events on the predator's time (a familiar sort of complaint), and set out to test this hypothesis. He needed a simple model to examine, and he found it by using a blindfolded girl as his predator and a set of sandpaper disks as his prey. The disks were scattered at varying densities on a table top, and the blindfolded girl was asked to hunt for them by dabbing with a finger and picking up all she found. Her hunting and killing were timed. When she had tried her prowess on various densities of disks, her scores could be recorded as a graph of kills against density; which showed that she functioned as a type 2 predator should (Figure 30.10).

In this simple model, the identification of the quarry (feeling the sandpaper under a finger) was virtually instantaneous, once the finger had completed its search. The time taken to pick up a disk was always the same, since all the disks were the same. Which meant that the only variable was the time spent searching. It seemed inescapable that it was, indeed, the compression of this searching time that caused the proportional decline in hunting success, as the hypothesis claimed. As a further test, Holling altered his model to introduce another time-consuming activity; he asked his blindfolded girl to search for the disks by tapping with the blunt end of a pencil instead of her finger. She now hunted by sound, which introduced a slight time-consuming uncertainty into her attack procedure. Her response was still curvilinear, but the slope of her "number killed" curve was lowered a little as predicted.

Like the type 1 response, the curvilinear type 2 response does not provide density-dependent control. The proportional take, in fact, falls steadily as the prey population grows, so that all control exerted by predators of this kind must be through their numerical responses. Since most of the predators known to control their prey are insects, particularly

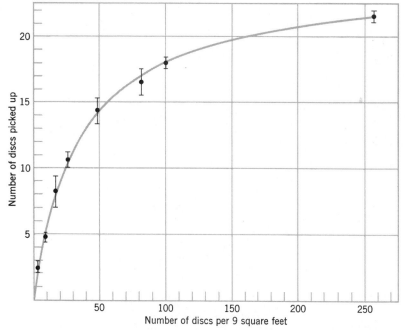

Figure 30.10

Functional response of a blindfolded girl predator hunting sandpaper disks with her finger. The number of disks that could be found in a set time increased with density but at a decreasing rate. This result suggests the time available for each hunt was being compressed as the number of kills went up. Insect predators hunting insect prey were found to respond to density in a like manner. (After Holling, 1959.)

hymenopterans, responding to prey densities in a type 2 way, it is clear that it really is numbers which count. It is reasonable to disregard the functional response when making very general models like those of Nicholson and Bailey (1935). The changes in populations are what we must principally study.

But Holling had found from his own work that vertebrate predators may respond in a different way to the densities of their prey. He was fortunate in having at his disposal Canadian pine plantations that were being used to test various control measures against an infestation of pine sawflies *Neodiprion sertifer*. The sawfly caterpillars eat pine needles, but they pupate inside cocoons in the leaf litter at the base of the trees, where they are hunted by various small mammals. Campaigns of spraying selected lots with various doses of insect viruses, provided Holling with cocoon densities of from 39,000 cocoons per acre to over a million cocoons per acre. Two kinds of shrew and a mouse lived in the woods, apparently largely subsisting on sawfly cocoons when they were available. Holling could keep count of the cocoons in the woods from standard soil samples, and he could keep count of the take of each animal meanwhile because they left the empty cocoons behind, each marked by

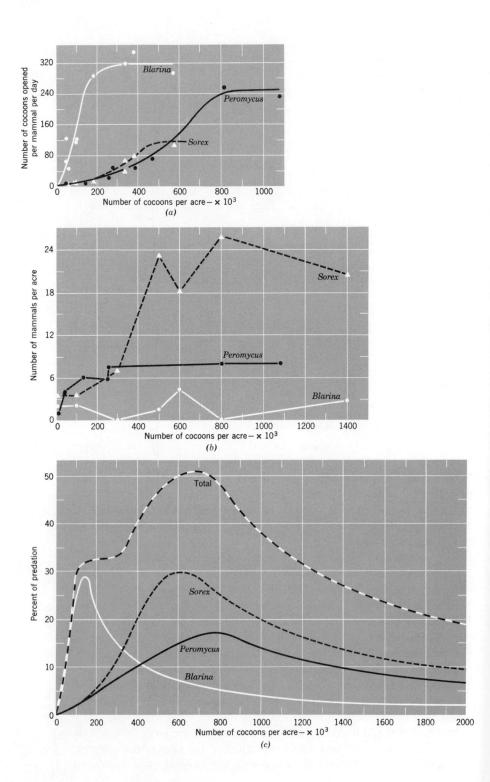

(a)

(b)

(c)

(a) Functional responses of three small mammalian predators. These shrews and a mouse were hunting the same prey, sawfly cocoons on the floor of a wood. There was a range of prey density over which each of these predators took not only an increasing number of prey as density rose but did so at an increasing rate. This is apparently because they learned to concentrate their efforts on hunting the common prey. Eventually their efforts meet the same time limitations that controlled the hunting of the blindfolded girl, so that this functional response, the type III, is sigmoid. (b) Numerical responses. These are the same animals and prey as in (a). Only the population of the shrew *Sorex* showed a positive correlation with the density of the prey. (c) Functional and numerical responses of the three mammals combined. The two effects are additive, and their overall shape suggests that control by the predators is likely to be effective up to a critical prey density. But above this density, the ability of the prey to reproduce should exceed the rate of predation and control would be lost.

Figure 30.11
**Functional and numerical
responses of mammalian
predators.**

an opening characteristic of one of the three predators. Mark and recapture techniques let Holling (1959) monitor the mammal populations, so that he had all the data needed to calculate the functional responses of these animals. His results are given in Figure 30.11a. The functional responses of these animals are not curvilinear but sigmoid, as shown in Figure 30.12. Holling checked these results by keeping mice and shrews in cages and feeding them different densities of cocoons. The functional responses of these caged animals was always sigmoid, just as it was for the wild animals. Holling called this sigmoid response a **type 3 functional response.**

The sigmoid trace comes about because the animals do not respond readily to changes in density when the prey is scarce; it is only when the prey is relatively abundant that the kill rate of the mammals goes up quickly, and then the response is a curvilinear rise to a plateau. At high prey densities the mammals apparently kill as rapidly as time allows, just as if they were mindless insects, but at low prey densities they respond only slightly to changes in abundance of their prey. An explanation for this probably lies in the learning power and lack of specialization of the mammals. They normally probably eat many different kinds of food, and are able to maintain relatively large populations because of this. A slight change in the density of one kind of food will lead to little extra killing by the predator, because it still feeds in many ways. But if one sort of food becomes suddenly much more abundant, the marauding mouse or shrew encounters it often and learns from this frequent contact to hunt specially for this one kind of food. From a generalist he becomes a specialist, eating all that his skill and time allow. It seems reasonable to expect such behavior of all higher vertebrates that are capable of learning.

The hypothesis that the mouse and the shrews were learning to concentrate on the common food could be tested by providing captive mice and shrews with various mixtures of foods as well as with different densities of sawfly cocoons. Holling (1959) tried experiments of this kind, and was able to show that the caged animals did quickly learn to discriminate between foods, not just by ease of hunting but by palatability also. There have, of course, been very many behavioral studies of vertebrates that lead to the same conclusion. A **type 3 functional response** is, therefore, probably usual among vertebrate predators.

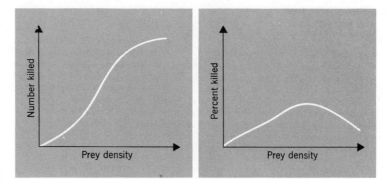

Figure 30.12

Type 3 functional response curves.

Of particular importance is the fact that the percentage take of vertebrate predators goes up during the initial increase of the prey population. The predator becomes aware that one of its food sources is
becoming abundant and concentrates its activities on that source. This
functional response, therefore, acts in a density-dependent way as a
curb on the prey population. Insect predators could only control
their prey through the numerical response, but vertebrate control should
be the sum of both functional and numerical responses.

Holling's several years of study in the Canadian pine plantations
provided him with estimates of the increased populations of his three
small mammals as well as estimates of their functional responses, so that
he could add the two together to arrive at the total effect of mice and
shrews on increasing numbers of sawfly cocoons. His results are given as
Figures 30.11b and c. The shapes of the functional response curves
are still apparent in the summed curve for percent total predation, and
indeed have been amplified, showing the potential that such predators
have for controlling their prey.

True control is exercised when death from predation exactly balances
live births of prey animals. If x is the replacement rate of the prey, there
must be x percent predation to control its numbers. Figure 30.13 shows a
functional and numerical response curve for an hypothetical group of
predators whose combined response was equivalent to the mice and
shrews of Figure 30.11c. The thickness of the response curve is taken to
represent variation in the effectiveness of the predation. The controlling
predation level x, which just balances the birth rate, is drawn in as the line
x percent. The death rate will equal the birth rate for all prey densities
between A and B, and between C and D, when the prey will be under
control. If the prey density is less than A, the prey will increase until it
comes under control at AB. If the prey density is between B and C, the
death rate will exceed the birth rate and the prey will be suppressed until
it again comes under control at AB. Only if the prey density contrives to
get beyond D will its numbers get out of control so that its population
continues to expand. To do this it must be able to reproduce so quickly
that its numbers can jump from A to D in one generation, to suddenly
outbreed its predator with enough in hand to swamp the predator's func-

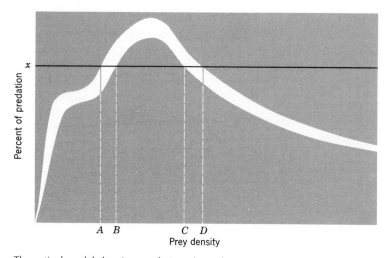

Figure 30.13

Theoretical model showing regulation of prey by predators.

tional response as well as its numerical response. Its chances of doing this are good if A and D are close together, and this in turn is possible if there are not many kinds of predator hunting it. A glance at Figure 30.8c shows how the wide separation of A and D is, in fact, the result of summing separate steeper curves; for the prey to suddenly outbreed one predator may not be hard, but outbreeding an association of them is not so easy. This functional analysis of Holling's gives us an extra understanding of why control by predators seems much less perfect in seasonal and fluctuating climates, since such climates support fewer kinds of predators; the prey can play the game of getting its population past theoritical density D with much more hope of success.

But it is also necessary for a prudent man to remember that the maximum predation that the system can produce may yet be less than the birth rate of the prey. If this is so, predation will never control the prey, and the prey population will increase until some other check is provided. This appears to be what generally happens when large predators meet large prey (Chapter 28). But even for such animals a total predation curve should be extremely interesting, since it will show just how much control needs to be applied by other agencies to explain the observed density of the prey. It is an indictment of modern biology that we do not possess such information for a single species or assembly of large predators.

Long ago Aldo Leopold, in his classic work on Game Management (1936), split predation into five components, as follows:

COMPUTER
SIMULATION OF
PREDATION

1 The density of the game population.
2 The density of the predator population.
3 The prediliction of the predator, that is, his natural food preferences.
4 The physical condition of the game and the escape facilities available to it.
5 The abundance of "buffers" or alternative foods for the predator.

Holling's analysis so far has considered the influence of the first three of these components and the last. The fourth could be avoided by using immobile prey like sawfly larve, although it obviously must be allowed for when considering most prey animals. Using the terminology of Solomon (1949) and Holling, the five components became

1 Prey density.
2 Predator density.
3 The functional response of the predator.
4 The functional response of the prey.
5 Availability of alternate foods.

And to make a complete model of changing events with time we must add

6 Numerical response of the prey.
7 Numerical response of the predator.

Many of these seven components may be readily subdivided into still smaller components. Holling and his colleagues have set about doing this, a process they call "experimental component analysis." They have concentrated first on the functional responses of insect predators showing the type 2 functional response, since these are likely to be basic to the responses of all animals. Time, the predator's limiting resource, must be taken up by eating and digesting prey, as well as in catching it, and these are components that must be allowed for. Then hunger may be important also, since there may be time lost before a satiated animal gets hungry enough to go after more food. Holling (1965) proceeds by estimating the time taken by each of these subcomponents in experiments with suitable animals (preying mantids seem particular favorites of his) and then substituting these values in equations of the general form:

$$N^a = a\,(T_t - t_h N_a)\,N_0$$

N_a = number of prey attacked
a = rate of successful search
T_t = time the predator is exposed to prey
T_h = time spent in handling prey
N_o = prey density

Expansions of this equation to describe all the components of the responses of real animals become very complicated, but the solution of many-termed equations is now simple, thanks to digital computers. Holling made a model of the functional responses of preying mantids which contained 22 components. He measured these individually, and then used the data to program a computer to generate the functional response curve of a preying mantid. The result was a proper curvilinear type 2 response curve. He then measured a true functional response curve for a preying mantid in a cage, and the result is given in Figure 30.14. The computer-generated response curve looks as if it has been

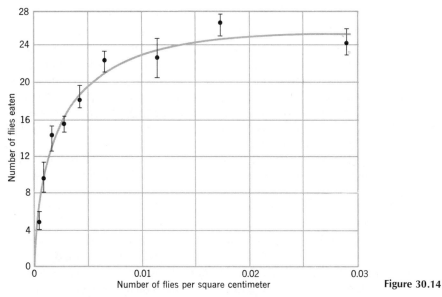

Figure 30.14

Functional responses of individual preying mantids (*Hierodula crassa*). The mantids were fed adult houseflies. Their catching prowess increased with density but at a decreasing rate, a result similar to that of the blindfolded girl experiment. (After Holling, 1965.)

drawn through the experimentally determined points, but it hasn't. The computer program had behaved like a true-to-life mantid.

This success of Holling's, and others like it, offer the first real hope of analyzing and understanding complex affairs like predation. If you can measure and simulate 22 components of a mantid's functional response, you can also subdivide and measure the rest of Leopold's five main components of predation. You can then build into your computer program the numerical responses of both predator and prey, so that you have in your computer's memory the correct set of responses which have been measured for individual sorts of predators and prey. But you can do more than this. One of the great difficulties of studying population dynamics in the field is that problem of dispersal, the constant worry lest the population changes you measure are the results of immigration and emmigration rather than of numerical responses of the populations themselves. Events in real life are going on at different rates in different places, or are out of phase. This problem is dealt with on a computer by seeing the area as divided by a grind into discrete subsections, simulating the events in each subsection separately, and then synthesizing the results from all subsections. This is just the sort of bookkeeping chore that businesses buy computers to do for them, and a program for handling biological data in this way has been written by Watt (1964). Finally, a computer can be made to simulate history. It can take events as elements and compute them in sequence, and it can do it at the famed speed of computers.

Once the components of predation are so well measured that a computer can be programed to simulate all the activities of real predators and real prey, it will at last be possible to see exactly what does happen when predator meets prey, and continues to do so generation after generation. We shall be able to reproduce the events of epochs in a short time and seek to understand just why we are confronted with the particular numbers of predators and prey of our own day. Which of these sets of numbers are in rough states of balance? and which are still responding to catastrophes now distant in time?

Holling (1968) and his colleagues are fond of saying that their models are "holistic, precise, realistic, and general." This statement always impresses me like a passage of Browning's poetry, it gives a feel of uplift but I am not sure that I understand. What they mean is that their models are true to life. The models give us a chance to study nature which we have never had before, to study whole systems at once and to examine events so complex that we have hitherto looked at them in manners artistic rather than scientific. The group of Canadian entomologists who have been mainly responsible for these insights have been guided by the need to understand the way predators work so that they can manipulate those predators as agents to control the pests of Canadian crops. But a disinterested biologist can see the more glittering goal of using such models to understand the numbers and doings in the great game herds of the earth in the last few decades that can pass before they vanish forever. Let us undertake component analysis of predation and escape in herds of wildebeest and prides of lions of the Serengeti, so that we can really know why there are just so many of them there. We need field biologists with a working knowledge of FORTRAN to guide their watchings and their counts.

Songbirds in the spring become aggressive toward their neighbors, finally living and rearing their young as spaced-out pairs. This behavior apparently promotes durable pair-bonds without which mobile animals such as birds might not be able to breed successfully. A second effect of the behavior is that each pair comes to have a private piece of land in which to hunt for food, a landholding that has come to be called the birds' *territory*. Territoriality is sufficiently explained as a device to promote the breeding success of individual birds, but if there is a minimum size to a territory the behavior might have the consequence of population control. A ceiling would be set to bird numbers by the total area of land available, and any birds in excess of the ceiling could gain no livelihood and must perish. This hypothesis is attractive because it explains the remarkable constancy of many populations of large animals more completely than does the hypothesis of control by density-dependent winter death, but it faces the grave objection that it is apparently a mechanism of birth control which should not be allowed by a mechanism of natural selection favoring individuals who leave most offspring. This objection can be met, however, by supposing that the population regulation is an accident, a mere side effect of behavior giving such overriding breeding advantages to individuals as durable pair-bonds. The behavior has not been evolved as a population-checking device, although it does sometimes have that effect. A number of studies now suggest that the surplus nonbreeding populations required by the hypothesis do, in fact, commonly exist. It is likely that any behavior which confers its advantages through spacing-out of animals may have population effects. This applies both when solitary hunters, like mountain

443

lions, keep themselves apart in their favorite preserves and when social animals, like chickens or field mice, drive social outcasts from the favored living places.

CHAPTER 31

THE INFLUENCE OF TERRITORY AND SOCIAL DOMINANCE

At first light on spring days, birds begin to sing. They are joined by others and still others until the air is a-twitter with their song, a splendid sound which we call the dawn chorus. It ends when the sun is fully risen, but individual birds go on singing at intervals throughout the day. Frequently, the singer seems to be on some vantage point, on a post, or a tall tree, or hovering high in the air over a favored field as the skylarks do. Coupled with this beautiful noise are other signs of spring excitement; birds display, they chase one another, they meet in symbolic struggles, and sometimes they fight in deadly earnest. All this noisy ritual clearly has something to do with breeding; but what? Romantic analogy long gave naturalists the idea that they were witnessing gallant sex struggles, in which males sang their challenges from the treetops, then met in desperate combat before the blushing eyes of their brides to be. Only the fit should breed, the fit should be chosen in battle, and the female rewarded with her love the biggest thug. This is a view of the struggle for existence that has been the apology for some of the ugliest doctrines of human tyranny that the military bullies of this century have produced. Fortunately, the hypothesis is unsound and we do not have to accept so depressing a view of nature. We were provided with a general understanding of the real significance of the spring behavior of songbirds in 1920, when Eliot Howard published conclusions from 20 years of inspired watching of English birds in a book called *Territory in Bird Life*.

The tenor of Howard's many observations of many kinds of birds can be given by following his thoughts about just one of them. A common English resident is a sparrow-sized finch called a yellow bunting (*Emberiza citrinella*), or more popularly the yellowhammer. The male has a yellow head, which makes him conspicuous, and he sits by the roadside singing a song that anybody can recognize as "a-little-bit-of-bread-and-NO-cheese." In winter, yellowhammers live in flocks, males, females, and other kinds of birds as well, all living together. But in spring, males begin to leave the flocks for short periods, tending round a favorite perch and experimenting with song. At first they often go back to the flock, but soon they spend longer and longer near their perch, leaving it for short times but then hurrying back. Many times Howard would see a cock yellowhammer plummeting to his perch in a fast, straight flight, as if he had forgotten something. Once back he would sing, "a-little-bit-of-bread-and-NO-cheese."

Eventually, each male took to foraging only near his perch, and alone. Where once the males were parts of gregarious flocks, each was now by itself, each spaced apart from the others, each advertising its presence with song. And the separated males became irascible, rushing angrily at any other male that came near the favored spot. There sometimes ensued one of those fights that had often been interpreted as combat for a mate. But Howard noted that there was usually no presumptive mate to be seen watching the gladiators. In other birds which he watched, migratory birds who wintered out of England, the males were at their quarreling sometimes days and weeks before the later-arriving females were in England at all. The aloofness, irascibility, and the associated singing must serve some other function.

The "little-bit-of-bread-and-NO-cheese" song did seem to attract a mate, for each male was joined by a female in the fullness of time. In migratory birds, the hypothesis that the song served as a beacon for females seemed particularly strong, for the late-arriving hens had to home in on cocks scattered far and wide before their arrival. In stay-at-homes like the yellowhammers, male singing probably served the same function. But it must be something more as well, because the males went on singing even after they acquired mates. Each pair built a nest and reared young near the vantage point from which the male had made his first experiments with song. They foraged nearby in that same local space which he had come to treat as home. And they both became irascible and intolerant of other birds. Now both would fly at an intruding male or at an intruding female; and sometimes the matronly hen would fly at an intruding hen, attacking her in real earnest; so much for the idea of the blushing female watching the combats of cavaliers. And each male, when his household duties would let him, went on singing from his favorite perch until the young were fledged and off his hands.

To Howard, the advantage to the birds of this package of habits was clear. The singing and isolation of the males in early spring resulted, when they were joined by their mates, in isolated, separate pairs. Each hen became used to the same surroundings that had evidently become fixed in her cock's senses as home, and the pair were kept together by this mutual sense of home. And it is vital that the pair should stay together after they had mated. They had to build a nest, incubate eggs, and rear completely helpless offspring. Without durable marriage the birds could not breed at all. Viewed against this necessity, their surly treatment of visitors of either sex was an equal necessity, because there must be no chance taken with the completeness of the pair-bond. The mindless brains of the birds could not be expected to maintain their couples through affection; selection had produced for birds this mechanism of joint selection and defense of a favored place to serve in its stead. In these terms, Howard was able to construct an hypothesis that completely explained the wonderful singing of birds on a spring morning, a time clearly resulting from the synchronous habits forced on birds by the cycle of day and night. His hypothesis also explained the celebrated fights and displays that the birds made in spring. Both were parts of patterns of behavior necessary to keep such free-ranging animals as birds together while they reared their helpless young. Such behavioral patterns must be enforced by natural selection with extreme rigor, for any deviation must result in failure to breed and consequent hereditary oblivion.

But there were other consequences of these habits, secondary consequences perhaps, yet important. The birds fed in restricted areas into much of which they would not allow birds of the same kind to enter. This meant that the special food supplies of their species in the immediate vicinity of their nests were reserved for their sole use. Each pair could feed on its own doorstep, so to speak, without having to undertake wasteful journeys in quest of food. Energy was conserved, and could be diverted from the energy of unnecessary flight to rearing as many young as possible. This consequence of the behavior should thus also be favored by natural selection. A final consequence of the behavior was that birds held the space in which they fed, that they dominated an area, even to the extent of defending its boundaries against other birds of their own kind. To call such a space a bird's **territory** was an almost inescapable English usage. Birds sang and fought to gain **territory,** the favorable consequences of which were durable marriage and economy of energy for the rearing of young. Such was the Howard hypothesis of territorial behavior in birds, which is now so substantiated as to have the status of a general theory. Howard used the word "territory" with misgiving, knowing how it might be abused by those wishing human sentiments on animals, but feeling also that it expressed his general hypothesis so clearly that he could not avoid using it. It is perhaps fair to say that territory was a name for a symptom expressing a whole syndrome of behavior. The symptom was the spacing out in defended areas, but the vital parts of the syndrome were the sets of habits which ensured that pairs should stay together in a place where they could rear their young.

Most songbirds sang for territory, but birds other than songbirds also showed territorial behavior. Howard particularly noted how ledge-nesting seabirds like razor bills (an auk *Alca torda*) also had their favorite nesting space, though it be but a square foot of ledge. They, too, were attached to their home, displayed on it, and would brook no interference with it. They did not hold space in which to feed, but the other essentials of the behavior were there and the function of maintaining the home while the young were reared was met. "Territorial behavior" could become a name for this behavioral syndrome, also.

Behavior that can be called territorial is now known to be widespread in many kinds of vertebrate animals, although the matter has been most studied in birds. There is a growing literature illustrating examples of it among fishes. European sticklebacks, *Gasterosteus aculeatus*, small fish that live well in aquaria, are well-known for the nests they build at the bottom of the water, and for the way the male fish swims on guard, rushing at visitors with a display of his red breast as robins do. Many mammals seem to defend their homes, also. There seems, indeed, to be a wonderful array of home-holding activities among larger animals, all of which have the effect of spacing animals out, giving to each individual, pair, or sometimes group, territory of its own.

All these forms of territorial behavior can be fully and sufficiently explained as devices to help individual animals attain high breeding success. Animals are helped to find mates, to preserve the union while rearing young, and to collect food from nearby, as Howard so clearly showed for the birds he studied. Property rights may also mean that time and energy need not be wasted in competition for food with other individuals, thus enhancing the chances of individual survival. Predators may be best avoided when all travels are in a limited space that the owner knows very well. Spacing can reduce the spread of epidemic disease, and so on (Klopfer, 1969). With so many advantages conferred on individuals, it is not hard to see that selection for the behavior should be very strong.

If survival and success at breeding are enhanced by territorial behavior, as its widespread incidence suggests they must be, the effect on population should be to force it up. Territorial behavior becomes one more gambit in the race for maximum population, one more pressure that must be curbed by Chapman's "environmental resistance" (Chapter 22), or by the mechanisms of weather and density dependence. But there is a possible corollary of the behavior, which Howard pointed out in his book and which has been a source of hot argument among biologists ever since. *There cannot be more pairs of breeding animals than there are territories in which they can breed.* At once this introduces the idea of population control. Territory is defined in space, and the total space available to animals is limited. If, as Klopfer (1969) puts it, "the size of the territory cannot be reduced beyond a certain point, and if successful reproduction requires that the bird possess a territory, the regulatory function of territories becomes a function that is beyond dispute." Populations might be expected to increase in size until they reach a limit set by the number of possible breeding territories, after which the population is kept constant from year to year. Surplus animals become territorial outcasts. Denied a home, they perish. Only the normal population, set by the available territories, persists from year to year, unchanged except by chance catastrophe. There thus develops the possibility of a general mechanism setting limits to the numbers of many of the larger animals, a mechanism more certain than predation, more precise than weather, triggered by density yet not proportional to density, a mechanism that might uniquely account for the remarkable constancy in the numbers of many birds and other vertebrates.

This hypothesis has received, and continues to receive, formidable opposition from some of the most articulate and scholarly of zoologists. David Lack (1966), whose review of population regulation set the tone for much modern work (Chapter 25), has several times set down his reasons for rejecting the hypothesis, making himself spokesman for the

THE IMPLICATIONS OF TERRITORIALITY FOR POPULATION REGULATION opposition. The rejections are organized round the difficulty which the regulation hypothesis apparently raises for genetic theory. We are asked, so goes the argument, to accept that it is normal for individuals bested in territorial display to be outcasts who are denied progeny. Why, therefore, are these individuals not selected against so that their genotypes disappear? Only if the loser in ritual combat can also breed successfully will the genetic basis of the behavior be preserved. It seems that there must be territory for everybody, even if some are on inferior ground, so that none are denied the chance to breed. But this argument may be readily answered by supposing that the victors in ritual display are not genetically preordained. Bird encounters, ritual or real, like human encounters, ritual or real, must be influenced by circumstance. The times of arrival, physiological condition and health, the history of the individual birds and their learning from past experience, their mood, their individual suitability to local peculiarities of the habitat contested, the fortunes of the combat itself; all these will influence the outcome of territorial struggles. We must assume that a bird which triumphs in an encounter might well have been bested by his victim if they had met on a different day.

The critical arguments are more difficult to meet if the position is taken that the territorial mechanism has been evolved expressly as a population regulatory device. If this was true, then the advantage to the population would be that some individuals should fail to breed so that the survivors should live on in comfort and safety. But the altruists who sacrificed themselves would leave no progeny, and the genes for altruism must quickly vanish from the population. This raises the whole issue of group selection as proposed by Wynne-Edwards (Chapter 32) and for which there is so little evidence. But there is no need to assume that territoriality has evolved as a device for restricting populations in order to believe that it might have that effect. Regulation, if it occurs, is just one of the many effects of behavior that has other consequences of overwhelming importance. The behavior forges marriage bonds, conserves energy, and protects individuals from predators and disease. These are formidable evolutionary reasons for its existence. If a consequence of the behavior is also that population size is restricted, this may be no more than an extra effect, one of indifference to the evolutionary process promoting survival, an effect analogous to those which a physician calls "side effects." Proof that the behavior sometimes led to population regulation should only imply that this was an extra effect, not that it was the prime reason for which territorial behavior was selected.

There are thus no theoretical arguments which say that territoriality cannot regulate populations, and we are left with the question: "Does it?"

For territorial behavior to limit population size there must be surplus birds that fail to breed. The most famous evidence for the existence of such a surplus has come to us by accident, evidence gathered initially for a quite different study.

The state of Maine, like the portions of the boreal forest in Eastern Canada, has been much afflicted by spruce budworm, a caterpillar that

sometimes becomes so abundant as to destroy whole forests. Many insectivorous birds eat the caterpillar, and it became important to know how
effective the birds were at controlling the pest. Accordingly members of
the United States Department of Agriculture and the Fish and Wildlife
Service experimented with removing birds from a small plot of forest with
the intent of comparing the local budworm population with those of areas
possessing their normal compliment of birds. But first they made a careful
count of the birds that arrived in the spring to take up territories in their
chosen plot. Two weeks of field work in their 40-acre section revealed
that there were 148 pairs of various species present. The investigators
then went back with shotguns, intending to kill every bird in the wood.
They spent two weeks at this endeavor, at the end of which time they had
shot 302 adult male birds; and still there were territorial birds singing all
over the wood, 148 to start with; 302 shot; many still there at the end.
This was rather convincing evidence that there was a large surplus of
birds in the neighborhood, who filled territories as fast as they became
vacant (Stewart and Aldrich, 1951). Next year a team went back to the
same 40-acre lot to see if they could confirm these results (Hensley and
Cope, 1951). They made their initial census with even greater care, and
they found 154 breeding pairs. This was remarkably close to the 148 pairs
of the preceding spring. Then the team took to shooting again, this time
getting 352 males before they quit.

This evidence of the Maine gunners has seemed decisive to many, and
is commonly quoted as demonstrating that territoriality does, indeed, regulate populations. J. L. Brown (1969) in a recent review has made the
only serious criticism of it, noting that the shooters bagged hardly any
females. Of the 10 commonest species in the 40-acre wood not a single
female was replaced; or, at any rate, not one female more than could be
accounted for by the original census was collected. Was there a surplus
of males only? (something odd enough to require explaining in its own
right), or was it too late for the females to come in and fill territories made
vacant by death? The latter seems more likely, suggesting that mating
time for wallflowers was over, however much the males were still
prepared for a chance at sexual display. But it would be nice to know. On
the whole, this evidence must still be thought of as backing the regulation
hypothesis.

Perhaps clearer still is the evidence from a long and thorough study of
Australian Magpies (*Gymnorhina tibicen*) by Carrik (1963). These birds
live together, but they do so in two kinds of groups. One of these kinds is
composed of birds which breed, which nest together in a home tree, and
which collectively defend a mutual territory. They live completely in the
territory, where they seem to thrive. Carrick called these groups "tribes."
The second kind of group, which Carrick called "flocks," were of
nonbreeding wanderers. The flocks had no territory, they suffered many
losses from disease and other hazards, their numbers fluctuated markedly, and they never seemed to come into breeding condition. Occasionally a flock bird would be recruited into a territory, or a new territorial tribe would be forged out of a flock, but otherwise the flock birds had

no hereditary future. They were a floating population, just as required by the regulation hypothesis. The magpies are now immensely common birds of farmlands, being clearly one of those species that has expanded its population into habitats made by men. It is likely that the prominent "flocks" are able to persist only because of the species' preadaptation to agricultural life, but in showing us a persistent floating population they show clearly how territoriality may limit populations even though there was no selection for the behavior which caused the effect. This example seems incontrovertibly to answer the question: Does territorial behavior sometimes regulate populations? with a yes!

Vital to the hypothesis of population regulation is the thesis that territorial size cannot be reduced beyond a certain minimum. It is well-known, and generally accepted, that territorial sizes can be varied very widely. Some of the most striking variations are those in the territories of arctic birds feeding on lemmings (Pitelka et al., 1955 and Chapter 35), which may be huge when lemmings are rare but are closely packed during lemming highs so that the tundra seems dotted with nests. This variation seems a logical consequence of one of the apparent functions of territory about which there is no dispute, the advantage breeding birds derive from feeding close to home. If food is very plentiful, birds can feed very close to home indeed, journeying little and establishing boundaries near the nest. This mechanism should allow territory to fluctuate with the food supply; but it is not necessary that territory size fluctuate with the demand to explain the observed variation. The fact that territories do fluctuate from year to year is thus not evidence against the regulation hypothesis. All the hypothesis requires is that there is a minimum size for any abundance of food beyond which territories cannot be further compressed.

For most animals we have as yet no evidence confirming that there are minimum territory sizes, but such information is hard to get and must always require careful observation over many years. Studies with various tits (Parulidae) and the Scottish red grouse (*Lagopus scoticus*) (Klopfer, 1969; Lack, 1966), however, have rather convincingly suggested that there are lower limits beyond which some territories cannot be shrunk. Surplus birds are forced to live outside the favored habitats, where they commonly become the food of predators or scavengers. This is the same fate that overtook many of the muskrats studied by Errington (Chapter 28); most of those that could not find a place in which to live were surplus to the population and were eaten by minks. There is general belief among field ecologists that such histories are typical of the local populations of many animals.

The general likelihood of the territorial mechanism acting as a population ceiling is, for many ecologists, enhanced by the difficulty of explaining the common constancy of bird numbers without its aid. Bird numbers do fluctuate from year to year, but not by much. In particular, occasional gluts produced by odd favorable springs, and which appear as irruptions, are completely removed so that a normal number breed the following spring. If the territorial mechanism plays no part in this surgery of the gluts, then we must place our faith in other density-dependent mechanisms, presumably acting in winter as Lack's general theory requires (Chapter 25). The importance of winter death cannot be doubted, and there can also be little doubt that there is a density-dependent component in this death. But this does not mean that the same number of birds should emerge from each winter's ordeal. The numbers of birds entering the winnowing floors of winter varies with the success of the nesting season. The severity of the winters varies also. Winter death is presumably ultimately death from malnutrition, whether this be outright starvation or one of the various secondary consequences of being poorly fed. But starvation in vertebrates must not be expected to act in simple density-dependent ways, for often there will be mass death rather than a subtle killing which relieves the pressure on survivors. For this reason, the hypothesis which expects winter death to regulate with such elegant precision has not seemed so persuasive to all ecologists as it has to Lack. If the effects of winter death can be supplemented by the further reduction of surplus survivors through their being unable to find territories, the observed stability of bird populations is more satisfyingly explained.

All animals, except perhaps the drifting creatures of open water, must have their own special places in which to live. Commonly an animal may have a resting place, where it spends much time sleeping, digesting, or rearing young, and a larger area of familiar ground in which it finds its food. It is natural to call this tract of familiar ground its **home range,** a term that scarcely needs definition. It would be as natural to call this tract the animal's "territory," if it were not for the fact that "territory" had already been used by Howard in the somewhat different context of a defended possession. It has been a common practice of textbook writers to stress the difference between the two concepts by defining "territory" as *the defended portion of the home range.* For songbirds in the spring, territory and home range were identical pieces of ground, but it would be easy to imagine wide-ranging animals like mammalian carnivores hunting over much larger areas than they could attempt to defend. If such animals had territories, these presumably should be smaller areas near their lairs. So much seems reasonable, but it is now clear that this separation between the concepts of territory and home range is misleading.

THE TWIN CONCEPTS OF TERRITORY AND HOME RANGE

Laying stress on defense in a definition of territoriality warps the thrust of Howard's observations, and distorts even more the conclusions of the many students of the subject who have followed him. Territorial behavior is essentially behavior that leads animals to keep to themselves. One pattern of behavior that produces this result in songbirds is aggressive display to neighbors, but the same result would be achieved if neighbors were merely aloof so that they mutually avoided each other. It is the common experience of naturalists that vertebrates of many kinds do seem to live spaced out, but this does not mean that their spacing must be achieved by battle, that they carve up the map of the earth like nations of warring men. To assume that they do so is to forget the individual advantages which we

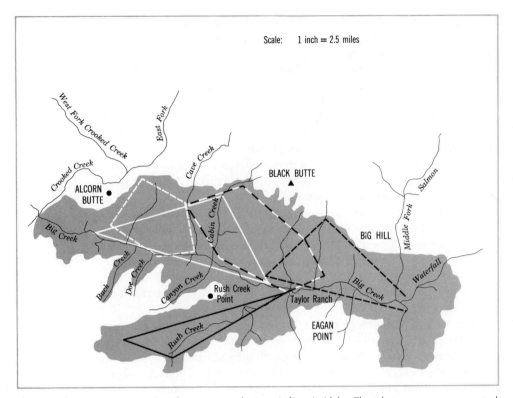

Figure 31.1 Minimum winter home ranges of mountain lions in Idaho. These home ranges were mapped by following the lions tracks in the snow. The ranges were fairly discrete, and it seemed that the animals were seldom in the overlapping portions at the same time. One home range was by two male lions, and another by a male and three females, but the other ranges had solitary occupants. The spacing of the lions in winter can be explained as a mechanism for ensuring that each is able to hunt an area where the deer have not been disturbed by other lions (After Hornocker, 1969.)

see for animals in their territorial behavior. Any actions which space them so that they breed in peace will give them these advantages. Fighting may have nothing to do with it. The combat rituals of some birds are distinctive, but they must not be allowed to influence our thinking about how all animals manage their affairs.

The unsuitability of the traditional separation between territory and home range is nicely illustrated by a recent study on the mountain lions (*Felis concolor*) of Idaho by M. G. Hornocker (1969).

Few men have ever seen a mountain lion, let alone settled down to study them. They are secretive animals, wandering over large tracts of country, killing in one place and next killing many miles away. *Felis concolor*, under its various local names of mountain lion, puma, cougar, and panther, has been shot into oblivion over most of America, and there is still a price on the head of this graceful and thrilling beast in many states of the Union. But in the wilderness of Idaho the mountain lions still thrive, and there Hornocker learned to track them in the deep snow of the mountain winter. He would patiently pursue their tracks, day after day, having with him trained dogs to corner the lions at the end, while he subdued them with anaesthetic darts. Each animal he caught was identified and marked. In four years of work he handled 43 different individuals, catching many of them many times over. Nine he got to know well, catching them a total of 59 times. He mapped their trails, and matched the ground each covered with the individual lion as he caught up with it. In time, his maps showed him the extent of each lion's wanderings, showing him the limits of the home range of each. And the maps also showed him where ranges and individuals met.

The ranges of resident lions were remarkably distinct. Each prowled the many square miles of country which it knew well (Figure 31.1), and which it had largely to itself (surely, it is reasonable to say its "territory!"). But there were lions, particularly young animals, who wandered more and who crossed and hunted inside the territories of even mature and established males. But the established lions offered no resistance to this poaching; instead, as tracks and spoor clearly showed, they went out of their way to avoid the youthful intruders. Once Hornocker actually caught a young animal at its kill in the territory of its virile elder, and then found the tracks of the owner first approaching, then turning away from where the younger and weaker beast was feeding.

The lions marked their passage with scrapes and excrement, just as do domestic cats. The trails revealed that all the lions, young or old, males or females with cubs, turned away from the presence of others. The territorial spacing of individuals, apparently so similar to those of songbirds in the spring, was achieved by mutual avoidance. Established animals had home ranges that were as distinct as the territories of songbirds; but they did not defend them.

The survival value of these discrete home ranges to animals of the lion's habit seemed clear to Hornocker. A large population of deer is needed to support one lion, which requires that the lions be scarce. Furthermore, a mountain lion is not so large and terrible a hunter that it can casually harvest its deer. It must hunt by stealth, getting very close to an unsuspecting, and perhaps weak, deer before delivering its attack. The quarry must not be nervous, which means that it must not be hunted too often. The survival of individual lions necessitates that each be able to wander over country that has not lately been hunted by other lions, and this requires that it have it to itself. A private hunting ground is a necessary consequence of the hunting habits of mountain lions. The home range becomes perforce a territory, its boundaries set by the twin needs of the animals to wander on familiar ground and to be alone.

It is likely that many animals establish something very like territories by avoidance, as do the mountain lions. These private spaces are very different from the territories of songbirds, both in means of establishment and in evolutionary purpose. But both result in spacing out individuals and pairs. And both are likely to result in there being individuals who never succeed in finding a permanent home in which to breed or feed. Both result, therefore, in ceilings being set to the sizes of populations.

Rather similar effects on population density and individual survival may come from the phenomenon of social dominance, the now well-known idea of the peck order. Gregarious animals often spar together or fight, just as do territorial birds, but may go on living together afterward all the same. After the first combat, however, they adopt perpetual positions of victor and vanquished, with one of the pair becoming a low-status animal that always gives way to the other. A series of pair encounters may establish a complete hierarchy for the group, with one individual being socially dominant overall, and the status of the others being arranged in descending order. Such social structures in animals were first clearly recognized in domestic hens by the German biologist Schjelderup-Ebbe (1935) in the 1920s, and it is from his observations on the pecking of hens by each other that the term "peck-order" has come into common use. There seem to be real advantages to individuals in living in socially stratified societies, since many experimental and field studies show that the individuals of settled societies feed better and are more successful at rearing young than those in groups afflicted by the constant strife which accompanies the reception of a stream of outsiders (Allee et al., 1949). Probably this is because the animals who live in a socially-settled state neither waste time on strife nor suffer the effects of too much stress.

Social dominance, like territoriality, can be understood as aiding the survival of individuals. But, also like territoriality, it can be expected to have effects on the size of populations as well. Many field studies have shown that vertebrates choose habitats, preferring special sorts of places in which to live even though other nearby places look satisfactory enough to the human eye. This is, of course, true for territorial birds, but it has been shown to be true for many small mammals as well, particularly rodents like the muskrats studied by Errington (Chapter 28) and various field mice and voles (Klopfer, 1969). Commonly, parts of fields will be crowded with animals while nearby areas are nearly empty. Within the favored place a social hierarchy may be established. When crowding becomes intense, it is easy to see how individuals at the bottom of the hierarchy may be so oppressed as to be driven out all together. Such animals should be wanderers from home, easy meat for predators, and liable to swift death. A hierarchy in a crowded place might thus force density-dependent death on the population, tending to stabilize the numbers of animals.

This possibility has recently been reviewed by J. C. Christian (1970), one of the foremost students of small mammal populations. Christian reasons that the outcome of social meetings may be partly determined by circumstance, just as the outcome of territorial encounters must be in birds. He cites experiments in which the same pair of animals have been brought together at different times and in which first one has become the dominant animal, and then the other. There are also intriguing studies on record (Allee et al., 1949) in which the hierarchy seems to be circular, so that A is dominant to B is dominant to C is dominant to D is dominant to A. It is thus possible for individuals with a variety of genetic components to be social outcasts, and denied a chance to breed. Social ostracism can thus serve as a check on numbers.

Many individuals may be cast out from crowded populations, and the fate of the wanderers is then probably death, again like the homeless muskrats in Errington's studies (Chapter 28). But if the population is low, some wanderers might succeed in finding new preferred living places, and the casting out of low-status individuals should then serve the species as a whole in good stead. The habit then provides the species with a dispersal mechanism, something essential to animals whose preferred homes are ephemeral. All animals which live in plant communities of the early successional stages face this problem of dispersal. They must be opportunists, ready always with colonists when fresh successions start. And social intolerance may well serve the small rodents of such places, the rats, mice, and voles so well-known to students of behavior, with the required drive to disperse. But for all the social outcasts who become the founders of new populations, there must be many more who die. These social habits may thus apply powerful checks to the growths of the populations of opportunistic mammals. They may also provide a ready source of food to many sorts of predators, which are then able to act as scavengers rather than potent controlling influences in the prey populations.

The hormone systems of men react to many different forms of stress in the same way, which results in a set of symptoms called the *general adaptation syndrome*. This syndrome presumably represents a mobilizing of the body's resources to meet the stress. We know that much the same sort of response is present in rats and suspect that it is general in all vertebrates. Extreme or unusual stress can cause an overreaction of the hormone system leading to shock disease and even death. Enlarged, or otherwise abnormal, adrenal glands are common symptoms of death by shock disease and they have been detected in a few populations of crowded wild animals and in overcrowded laboratory rats and mice. It seems probable that wild populations which have escaped the normal checks on their numbers so that they are forced to try to live in densities which the environment cannot indefinitely support, and to which their social mechanisms are not adapted, may suffer shock disease even as they are facing other inevitable agencies of their destruction.

THE POPULATION
CONSEQUENCES
OF STRESS

457

CHAPTER 32

THE POPULATION
CONSEQUENCES
OF STRESS

Men under stress have more or different hormones circulating in their bloodstreams than men not so stressed, and this changed circulation of hormones seems clearly related to physical changes in the endocrine glands themselves, notably in the adrenal cortex. It seems that the body adapts to fatigue, hardship, or anxiety with a change in hormone balance. Since it has long been clear that many of the body's responses are in fact regulated by hormones, it is not surprising to learn that changes in the stresses of life do lead to adaptive changes in the hormone regulating system. But it is surprising to find that a whole variety of stresses, from physical exhaustion to disease or worry, results in essentially the same set of hormone responses. Much work, particularly by the physiologist Hans Selye (1950) in the 1940s, has shown that the bodies of men, and their laboratory analogs, rats, do respond to different stresses in a common way. Different stresses produce similar effects on vital organs of the body, including of course those of the reproductive system. Selye called this common pattern of response to stress **"The General Adaptation Syndrome"** (which often appears in scientific writing as GAS). We now know that stress produces the general adaptation syndrome in many kinds of vertebrate, it may well be in all of them. It seems likely that something of the sort happens in other groups of animals as well, even though we yet have little evidence for it.

But if stresses of all kinds can invoke the general adaptation syndrome, might we not find that breeding is sometimes reduced by stresses of life that may have nothing obvious to do with reproduction. Reduction of breeding effort because there was not enough food for the young is one thing, but reduction of breeding because an animal was frightened too often or engaged in too much social activity in a crowded population is quite another. And yet there are common hormone responses to these varied forms of stress. Perhaps they do all lead to reduction in breeding effort. This is an attractive idea for population theory because we should expect stress to rise as populations get more crowded. If this extra stress invoked the general adaptation syndrome, and if the syndrome then reduced the breeding effort, there would be a beautifully designed density-dependent control device. But there should also be the possibility that local stresses could become so extreme that the general adaptation syndrome became overloaded, causing shock disease. Then populations might face diasterous loss. These possibilities were first clearly pointed out to ecologists by the work of J. J. Christian (1950).

Christian found a simple way of assessing the general adaptation syndrome in rats when he found that stressed animals had heavier adrenal glands than unstressed animals; apparently the larger output of the adrenal cortex meant that the glands had to be physically bigger, and an experimenter could abandon hormone chemistry for the simpler techniques of the analytical balance. This has sometimes proved a misleading and innaccurate measure in practice but it did serve well enough for some suggestive early studies. Equipped with this simple method of estimating the presence of the syndrome, Christian proceeded to examine rats which led placid lives and to compare them with those who suffered stress. He found that wild rats living in sewers had adrenal glands that generally weighed half as much again as the adrenals of well fed, docile laboratory rats of the same size. Apparently the rough and tumble of sewer life represented stresses in the rats' lives which were met by the general adaptation syndrome. The sewer rats were fine healthy specimens. For them the general adaptation syndrome was truly adaptive, helping them to breed well and achieve a high population in spite of the stresses of sewer life.

Christian then designed experiments to see what would happen if stresses on rats were really severe, and the easiest way to contrive this was to let the animals become overcrowded. He kept colonies in comfortable pens, giving them plenty of bedding, food, and water. Colonists in these pens showed no signs of stress, being in every way normal. But, being normal rats, they began to breed like rats, which is to say quickly and often. In a very few short rat generations the pens became crowded. The rats still got all the bedding, food, and water they could use, so that their physical wants were met, but they were clearly "falling over each other." There were frequent fights, so frequent that males seemed to have little opportunity to mate. And the females who were mated resorbed their embryos, ate their young, or simply neglected them. Breeding came to almost a complete standstill, and the consequences of crowding had acted as an almost total check on population increase. The adrenal glands of these overstressed rats on autopsy proved to be twice as heavy as those of normal rats, which suggested, reasonably enough, that the general adaptation syndrome was rather heavily invoked.

If rats under such crowding were kept penned up indefinitely, clearly a catastrophic population decline would follow. In spite of the fact that there was plenty of food, the population would suffer a massive loss, and this loss would be due to hormonal response to stress. But this hormonal response can still be considered adaptive, because conditions were such that breeding could well be impossible for all that there was plenty of food. A good breeding strategy for a rat under such impossible conditions for the rearing of the young would be to save its energies for breeding when things get better. So if the rats survive crowding, even though they do not breed when the crowd is at its worst, it is still reasonable to consider the adaptation syndrome to be truly adaptive, to be increasing the overall chance of every individual leaving as many viable offspring as possible. But it is also possible for the general adaptation syndrome to be overloaded so that the stressed individuals actually die.

It has often been noted in nature how some animal populations grow very large only to "crash," that is to suffer catastrophic death. Plagues of rodents, for instance, very often end in this way, and have excited speculation throughout recorded history. Christian postulated that these crashes were due to failure to breed or shock disease brought on by stresses of crowding.

Christian (1950) could find ready support for his hypothesis from the works of others. Several times men had collected the dead when populations of mice or voles had crashed, seeking by autopsy to explain the cause of death. They had usually failed to do so. There was rarely any evidence of infectious disease (although this had been found in hares) or starvation, and the animals might even seem fat and in apparently good condition. And yet they were dead. It was a mystery. But if they had died from hormone imbalance interfering with vital functions, as Christian suggested, the mystery was cleared away. There was a series of investigations in the literature in which symptoms described as "shock disease" had been explicitly blamed for the deaths. This was the study of Green (1930) and others on the periodic sudden deaths of snowshoe hares. Hares both young and old would die suddenly and in convulsions. Disturbing a quiet hare, such as catching it, was apt to bring on the seizure. Green's autopsies had revealed signs of physiological abnormality, such as too little sugar in the blood and fatty degeneration of the liver, symptoms that fitted perfectly with Christian's hypothesis. The hares had been crowded, suffering much stress. The hormonal responses of the general adaptation syndrome had been pushed as far as they could go. One extra piece of stress, such as being grabbed, could not be accommodated, adrenal exhaustion ensued, the body's systems became disorganized, and rapid death followed.

There was also the evidence from crashing populations of deer. There were many known histories of deer populations that had grown very large over the years, only to lose more than half their number in a single winter. These deaths were usually attributed to starvation, often for the very good reason that the animals seemed in poor condition in the weeks before they died. But it was seldom that thorough autopsies had been performed. Perhaps, if they were, evidence for hormonal abnormalities might be found in addition to the evidence for starvation. After all, starvation might be viewed as just another form of stress. Eventually, Christian and his colleagues (1960) had a chance to test this hypothesis by studying the adrenal glands of a population of deer as the numbers first grew to a peak and then passed through a catastrophic decline.

A herd of the sika deer (*Cervus nippon*) had lived on little James Island in Chesapeake Bay, Maryland, since their introduction in 1916. The population had grown steadily, particularly in the 1940s and 1950s, until there was a density of one deer to the acre in 1955. Deer have been thought to starve in Michigan at a density of 30 to the square mile, so it seemed likely that something would happen to the James Island herd before long. It was a cue for Christian and his colleagues to collect some deer for study. The animals seemed fit and healthy, but Christian recorded their general state all the same, and he weighed their adrenal glands. In the winter of 1957 to 1958 60 percent of the herd died. There was no sign of epidemic disease or an abnormal load of parasites on the bodies. The corpses were fat, and their coats sleek. The range was not yet so damaged that there should be any reason to suspect sudden starvation, nor was there any record of climatic catastrophe to blame. But sections of the adrenal glands revealed abnormal tissues. When survivors of the disaster were collected next spring, they were found to have adrenal glands that weighed only half as much as those of the striken animals, or of any animals collected in the two years of crowding. This circumstantial evidence for death resulting from overloading of the general adaptive response to stress does seem rather strong.

These studies of Christian and others have convincingly demonstrated both that stress can kill or prevent breeding, and that killing stresses can be induced by crowding. We are provided with an answer to the question; How can mass death be produced in nature when massive infection and starvation are absent? The phenomenon of mass death through shock is revealed as one more of the formidable checks to population increase which the natural world affords. It seems likely that populations who sometimes suffer it are normally kept in check by other means and that it is only when they have escaped these other controls by some accident that their populations grow so large as to crowd themselves into a state of shock. Opportunistic species should be particularly liable to this form of death because they characteristically reproduce quickly and experience high mortality of their wandering colonists. If their population falls on unusually good times, as when left on an island like the sika deer or able to thrive in the permanent succession stage of agriculture like many mice and voles, they experience crowding to which they are not adapted. Shock disease ensues, and there is catastrophic death.

The possibilities of shock disease are bound to interest students of the human condition. Our cities are notorious for their problems of crime and mental disease; might not these problems have their roots in the general adaptation syndrome? Does crowding men into cities cause them to suffer stress so great that the output of their adrenal glands becomes abnormal? This opens the way to an attractive hypothesis purporting to explain the ills of our day. Let us blame the miseries of the inner city on the stresses of crowded city life, onto purely physiological responses of the general adaptation syndrome.

IMPLICATIONS FOR
THE STUDY OF MAN

There must be grave doubts of whether this hypothesis has any substance. Men are remarkably adaptable animals; and they go to cities of their own choice. The history books of Western societies are cluttered with stories of farmers decrying "the drift from the land" as their laborers headed for the "better life" of the city. Most civilized (the word means "accustomed to living in a city") men prefer living in cities to living anywhere else. There are undoubtedly many extra stresses in city life, but these are probably met by the truly adaptive qualities of the general adaptation syndrome. Men in cities should probably be compared with healthy sewer rats; they like it there.

When stresses become very bad, men are still able to overcome them by mechanisms of mental discipline denied to other animals. The many tales of cheerful endurance in beleagured fortresses is testimony to this. Physiological pressures tending to shock disease are always present at such times, and some people do indeed succumb, but more conquer their own glands, avoid disease, and survive. When discipline fails, however, men do sometimes act like animals, as many nasty incidents of military excesses attest. Perhaps the leaders of such soldiers fail them, so that they react to the stresses of fright and pain with hormones rather than with steady minds. It would be interesting to weigh a few of their adrenals to see if the poorness of their military discipline had unleashed the uglier side of the general adaptation syndrome.

The population consequences of occupying territories, of social hierarchies, and of other forms of behavior are perhaps best thought of as side effects of habits which conferred selective advantage on individuals. Yet the population restraints which they imply might, in theory, serve to keep the numbers of animals which possess them comfortably below the carrying capacity of their habitat and so insure the species against the dangers of damaging its own resource by pressing on it too closely. Wynne-Edwards put forward the hypothesis that behavior which might restrict the size of populations had been evolved not as side effects but as the prime advantage which selection had worked to preserve. A requirement of the hypothesis is that it must be possible for natural selection to choose between groups, or societies, rather than just between individuals. Some forms of group selection in which individual survival is enhanced by membership of the group may be possible, but the hypothesis faces the very difficult fact that natural selection will always choose individuals who leave most offspring in preference to those who practice birth control. Behavior that has a limiting effect on population is only likely to persist if it gives overriding breeding advantage in other directions, as does the territorial habit in birds. In social vertebrates the setting of a modest limit to the population might well have a long-term advantage for the group, if it could be preserved by selection. But insects and many other sorts of animals which show complex group behavior are likely to be subjected to such other restraints that any population consequences of that behavior are likely to be unimportant. The general hypothesis of control by group behavior must be considered unnecessary to explain the habits we observe.

CHAPTER 33

THE HYPOTHESIS OF
CONTROL BY
GROUP BEHAVIOR

The animals whose numbers seem to us to fluctuate least are mostly conspicuous vertebrates, the birds and mammals whose regular presence does much to give a sense of normalcy to the world around us. These animals must be under some sort of control, on conventional thinking, control by density-dependent death, but they are the very animals for whom density-dependent death is hardest to demonstrate. Some of them, like hawks and most mammalian carnivores, live at the ends of food chains, habits that make them immune from predation. There is also doubt (Chapter 28) that even the herbivores and insectivorous birds among them are controlled by predators. Both herbivores and carnivores suffer particularly from epidemics, but there is little evidence to suggest that many vertebrate populations are actually regulated by disease. This leaves starvation as the only other plausible density-dependent factor, and there is very little evidence that starvation is a good regulating agent. In summer, most animals seem to get plenty to eat. In winter, many do sometimes die when the food supply fails, but these are sudden losses, not regulating losses. Many ecologists have, nevertheless, placed their faith in winter starvation as the controling device (Chapter 25), but they have been unable to support the belief with much hard evidence. It is thus not unreasonable to put forward the hypothesis that neither predation nor disease nor starvation normally regulate vertebrate populations, but it then becomes necessary to put forward an alternative hypothesis to explain the constant numbers. The Aberdeen zoologist, V. C. Wynne-Edwards, has pondered this possibility of an alternative hypothesis, and has put forward his views in a massive book called *Animal Dispersion in Relation to Social Behavior*.

Wynne-Edwards was struck by the fact that many of the things that animals did seemed to affect their breeding success. The territorial behavior of birds was one such form of activity. The prevailing view was that this behavior had been evolved to ensure durable marriage and to conserve energy in the rearing of young (Chapter 30), but some ecologists also believed that the behavior inevitably set limits to numbers, even though this was no more than an unavoidable side effect of the behavior. Such is the view taken in this book. But, Wynne-Edwards went further and postulated that regulation was the prime function of territorial behavior, that territoriality had evolved as a device for population control. The evolutionary advantage as he saw it was that bird populations should never get so large as 'o endanger the food supply and expose the population to the danger of extinction through mass starvation. Species without such a built-in control system should inevitably overeat their food sometime, become extinct, and be replaced by others who were able to manage their affairs better.

But territoriality was not the only form of behavior which seemed to have population consequences. Social dominance, the phenomena of peck orders, also seemed to influence the family success of animals. Such effects could also be thought of merely as side effects of advantages conferred on individuals by the behavior (Chapter 28), but again Wynne-

Edwards took the argument a stage further and postulated that the real evolutionary significance of much of social dominance was the control of populations.

If the very widespread phenomena of territoriality and social dominance were to be explained as devices for population control, then the evolutionary advantage of being able to keep your population in check must be great indeed. Other behavioral mechanisms might well have been evolved to serve the same purpose. Wynne-Edwards reflected on what was known of the behavior of animals generally, particularly seeking patterns of social behavior for which there seemed as yet no adequate explanation. And he found what he sought in a variety of communal displays.

Many fish live in shoals, great pods of fish which swim together through the ocean. Why do they do this? Many birds "flight" together, like those great flights of starlings which swoop in wonderful formations tens of thousands strong across the dusk skies of London, a twittering grey cloud which disports itself for 10 minutes or so before plunging to the communal roost. What is the use of this to the birds? Deer and elk congregate briefly for the rut in the spring. Fur seals go to breed together in herds numbering 100,000 or so on tiny congested Pribilof beaches while miles of similar beaches are left unattended. Many cock gamebirds visit ancestral display grounds in the spring, there to strut and crow at each other before mating. Seabirds sometimes nest on traditional cliffs, while those roosting on nearby cliffs apparently cannot breed. Why should all these things be?

For Wynne-Edwards these events all had something in common; they all arranged that the animals should be crowded together at some important times in their lives. He was looking for behavior which should so affect the reproduction of animals that their populations should be kept within safe limits; and he found widespread habits which made animals crowd themselves, particularly during the breeding season. He thought of these various forms of behavior as devices for counting heads and termed them **epideictic display.** There must be physiological feedbacks which report the amount of noise, or other social contact, to the reproductive systems during the crowd displays. A dense crowd should result in a lowered reproductive effort, a sparse crowd in numerous offspring. Evolution had selected these various curious forms of behavior through their effects on reproduction; and their evolutionary advantage was that populations were kept so low that the food supply was never in danger.

For this hypothesis to stand, it was necessary for Wynne-Edwards to suggest how the feedback mechanism might work, and for this he could draw on the work of Christian and others on the general adaptation syndrome (Chapter 32). Christian had shown that crowded rats and mice failed to breed, and he had further shown that this failure was associated with changes in the adreno-pituitary system brought on by crowding. Christian had to forcibly crowd his rats in cages, but if the crowding had come about through some compulsion of behavior it was not unreasonable to suppose that similar changes in the adreno-pituitary system might be produced. Wynne-Edwards was thus able to suggest that the many forms of group behavior which were known in vertebrates were devices to crowd the animals together at key moments in their lives, that the density of the resulting crowds was a measure of the local populations, that there were proportional adjustments in the adreno-pituitary systems of individuals, and that these hormonal adjustments regulated reproductive success. The known stability of vertebrate populations was thus not caused by density-dependent death, as was generally supposed, but by birth control.

This is a grand hypothesis. It seems to have generality, offering a unifying explanation for an enormous number of the odd things which different animals do. It couples all these oddities with the one vital function of reproduction. It suggests a beautifully tuned feedback system between the pressures of the environment and the responses of the organism. It grants evolution the success of having made animals which live without damaging the sources of their lives, in elegant and satisfactory balance. And it was set forth in a compelling book, lucidly written, documented in a truly Darwinian manner with forcible arrays of examples. But there are grave reasons for doubting the general validity of the hypothesis all the same.

The most serious difficulty is the way the hypothesis conflicts with accepted evolutionary theory. Natural selection has always been thought to act on individuals, always promoting the survival of the most fit individual at the expense of others less fit. A fit individual must be one that leaves most surviving offspring, and all animals must be adapted to reproduce as fast as their ways of life and the conditions of their times permit. This is the reasoning that led Lack to argue that all birds rear the maximum number of eggs which their food supply allows (Chapter 25). But Wynne-Edwards is suggesting that there may be evolutionary advantage in sometimes rearing less young than food and circumstances permit. He suggests that all individuals of a local interbreeding group may, when crowded, produce less young than the food supply allows. They are to forego some of the food available to them and have a small family so that the species may benefit in some future generation. Why is it, then, that some individuals of the group do not go on producing all the young they can, use up the surplus food, leave more offspring than those who practice self-denial, and in time completely replace them so that the genes of the self-denying ones vanish from the population?

And there is perhaps a more serious difficulty yet. Even if a local group practices birth control in the interests of its descendants, it still has to compete successfully with other groups of the same species that might compete for food more aggressively, gain more energy, and leave more offspring by its policy of rapid breeding while the going is good. Wynne-Edwards has to infer that reproductively abstemious groups will replace the profligate ones in times of food shortage, even though there will be a larger number of profligate individuals whom chance might help over the bad times. Somehow abstemious individuals in the original group must finally leave more offspring than do those with more aggressive reproductive drives. These are difficult requirements for classical selection theory to meet. As a result, the Wynne-Edwards hypothesis has come under continuing and indignant attack by many evolutionary theorists and ecologists (Lack, 1968). The possibilities remain intriguing, however, and it seems not impossible that theories of group selection based on the phenotypic expression of interacting genotypes (Griffing, 1967; Wiens, 1966) might yet show how some of the effects of group selection postulated by Wynne-Edwards can be reconciled with classical theory.

One recent study offers some indication that a group regulation of population might actually be occurring in a species of socially active animal, the Dall mountain sheep (*Ovi dalli stonei*). These animals perform an elaborate rut in the breeding season during which the males fight with their big horns. Valerius Geist (1966) spent a total of 35 months watching the mountain sheep in British Columbia, setting himself the task of determining the real significance of the large horns in the lives of the animals. That the horns could be used as weapons, as shields, and to absorb the shock of combat was easy to verify, but the same functions were served by the much smaller horns of many ungulates. There must be some special purpose to explain the very bigness of horns in this one species. Geist, through numerous observations of males approaching each other, and masterly statistical handling of his data, was able to show that horn size conferred social dominance in this species. A male with small horns automatically treated a male with much larger horns as his social superior, and yielded to him. A result of this was that the males with the largest horns mated nearly all the ewes, and males with very small horns mated none at all. A selection pressure towards largeness in horns is immediately revealed, but this leads to the question: Why has the genotype for small horns not disappeared entirely? Geist found that although the biggest-horned males got most of the ewes, they usually only survived a single reproductive season. The largest-horned males were short-lived animals. Males with middle-sized horns, on the other hand, always mated a few ewes and had a mating life of many years. In spite of the social superiority of big horns, it seemed likely that males with moderate horns each left more surviving offspring in the long run, and the tendency towards a population in which all the males had giant horns was checked. This is evidence that leaving less than the maximum number of offspring in any one year need not be fatal to an individual's genotype. It may also be that in years of dense populations, the intense activities of the rut may mean that the socially dominant animals do all the mating, and that only their short-lived offspring are left. In subsequent years the population should tend to decline as a result, and the longer-lived males with middle-sized horns would do more of the mating. This seems to imply the sort of mechanism that the Wynne-Edwards hypothesis demands. It is not surprising to find this evidence in socially organized and evolutionary advanced animals like mountain sheep. These are also animals which, failing other checks such as predators (Chapter 28), may be in special need of regulation devices to avoid population explosions which destroy their range.

The story of the mountain sheep must give encouragement to the belief that population checks can result from rutting activities, but this need not be more than yet another example of a population check applied by habits evolved for quite other purposes. Social dominance is aided by large horns, but there must be some limit to the practicable size of horns, and we find an evolutionary mechanism, coupling horn size to longevity, which ensures that horns do not get too big. A side effect of this is that checks are applied to population growth. This story is analogous to those of other forms of social dominance and territoriality described in Chapter 31. These widespread forms of behavior do all seem to apply checks to populations, but this does not mean that they have been evolved as regulation devices. Critical examination of very many of Wynne-Edwards' examples shows that the population consequences he illustrates are of these territoriality and social-dominance types. It may be, as I incline to believe, that these activities nearly all restrict population size, but this does not mean that they have evolved for that purpose as Wynne-Edwards claims.

Wynne-Edwards' arguments have been further weakened by his own thorough demonstration that much group behavior also occurs among invertebrates. He cites the singing of crickets and cicadas as being possible devices for estimating population size, and dwells on the group behavior of whirligig beetles spinning round together in their restless way over small patches of pond water. It is one thing to imagine that the noise of a starling roosting-flight produces hormonal feedback, but quite another to suggest that clouds of swimming beetles do the same thing. There is little difficulty in accounting for the control of insect numbers through predation, disease, and weather, agents of control so formidable that insects are highly unlikely to need to apply additional restraints on themselves. And yet many insects show group behavior analogous to that of vertebrates. If the insect groups owe their existence to matters other than birth control, may not the vertebrate groups likewise be concerned with other things?

That many hypotheses can be put forward for any one syndrome of group behavior is nicely illustrated by the story of the diurnal migration of plankton and some fishes. In temperate lakes at high noon the animals of the plankton may be concentrated in a layer about a meter below the surface, but in the evening they gradually rise to the top, then scatter. There is thus a diurnal rhythm, with the tiny animals rising and falling through a meter of water as day follows night. Much the same thing happens in the sea (Figure 33.1), but there some planktonic animals such as copepods may migrate up and down a full 100 meters with the diurnal cycle. Larger crustacea and small fish may achieve vertical migrations of as much as 800 meters, appearing by day as a trace on the sonar of ships known as **the deep scattering layer** and swimming upward at night to vanish from the sonar screens as they disperse in the surface waters of the ocean.

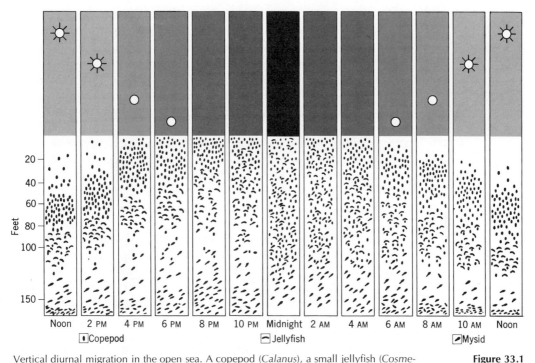

Vertical diurnal migration in the open sea. A copepod (*Calanus*), a small jellyfish (*Cosmetira*), and a mysid crustacean (Leptomysis) are shown in their relative positions throughout a twenty-four hour cycle. A number of hypotheses have been put forward to explain this behavior, none of which seem applicable to the lives of all the different kinds of animals that show it. [From Allee et al. (1949), after Russell and Yonge.]

Figure 33.1

These diurnal vertical migrations have attracted much attention from students of lakes and the sea, and many hypotheses have been put forward to explain the habit. Wynne-Edwards listed this behavior as group behavior since he was at pains to collect as many examples as possible. No doubt he did not seriously think that diurnal migrations of plankton were an advanced form of social behavior himself, for there was no shortage of possible explanations. But having listed such "examples" gave a special advantage to reviewers who could speedily dismiss the hypothesis on the strength of them.

Consider some of the alternative hypotheses put forward to explain the vertical migrations, all of which probably have some validity in fact. The one-meter migrations in lakes probably result from the animals following the tiny flagellate plants of the phytoplankton. These tiny plants probably swim up or down to the optimum light intensity for photosynthesis, sinking as the sun rises and rising as it sets. The animals merely keep pace with their food supply. This explanation probably does not apply to the hundred-meter migrations of ocean copepods, nor the 800-meter migrations of small fish, but there are other explanations for these. One is that there are often lateral currents flowing below the sea surface so that a daily descent will place the animal into water moving sideways, so that it can come up to feed in a different place each night. The habit is thus viewed as serving the purpose of dispersal. Another hypothesis is that the lighted surface waters are too dangerous in daytime because there is no place for small animals to hide from their predators. They hide in the depths by day, only coming up to feed in the safety of darkness. A third hypothesis rests on thermodynamics. The animals have to feed in the warm surface waters, for that is where the plants are, but they go down to cold water below the thermocline in daytime where they lower their body temperatures, slow their metabolisms, and conserve energy while digesting. To most biologists (probably including Wynne-Edwards) any and all of these hypotheses seem more reasonable than that the animals dive each day to count heads. But this is no reason for dismissing the whole hypothesis.

In conclusion, it may be said that many forms of group behavior, such as territoriality and social dominance, probably do exert strong influences on reproductive success and population size. These effects, however, may be no more than side effects of habits evolved because they grant advantages to individuals. The behavior of individuals preadapts the population for survival by restricting its numbers in good breeding years, and the population then avoids the danger of starving in subsequent lean times. But there is little reason to believe that social behavior has evolved specifically to promote birth control, as Wynne-Edwards argues.

The theoretical difficulties with group selection may yield to modern population genetics, suggesting that some forms of behavior in a few groups of animals might have been preserved because they provide insurance against excessive population. The behavior thus preserved by group selection would likely be behavior originally evolved because it served individuals well, but which also preadapted the population to living within its means. Such forms of behavior are likely in gregarious and sociable vertebrates, or in those large predators, like mountain lions and grizzly bears, whose survival may depend on having a private place in which to hunt. When populations of these animals are not provided with other checks, such as predation or disease, the preadaptive advantages of behavioral checks might well be preserved and refined by group selection of some sort.

The numbers of wild plants have, until recently, not seemed as interesting as the sizes or area covered by those plants. Students of vegetation
have tended to concentrate on the differing success of various species of
plants in terms of habitat differences, believing that an understanding of
the physical requirements of plants should lead to an understanding of
their distribution in nature. But this approach would be misleading if the
plants were subjected to strong grazing pressures by herbivores. An
ecologist must think that a stationary item of food, like a plant, is
inherently likely to be attacked by something which eats it, and a variety
of recent studies suggest that such attacks, often very strong attacks, are
virtually universal. The stationary plants have responded over evolutionary time with dispersal mechanisms, or by making themselves inedible in some way. Bad-tasting plants have resulted in the evolution of
specialized herbivores which can eat them, thus accounting for much of
the observed diversity of plants and for the even greater diversity of herbivorous insects. The many species of trees in a tropical forest may
reflect the concentrated herbivore pressures on their seeds and seedlings which allow only widely dispersed individuals to develop. It
seems that looking at plants as the prey of herbivore attackers gives a
more useful insight into the distribution and abundance of plants than
thinking of them as parts of vegetation controlled by the habitat.   471

CHAPTER 34

THE CHECKS ON PLANT NUMBERS

Studies of vegetation have long been influenced by the apparent completeness of the carpet of plants which covers much of the earth. The wetter places have always seemed to be perpetually green with plants, and even the browner places, now revealed by satellite photographs to be typical of most of the earth (Chapter 27), have seemed to be covered with as much vegetation as water or other limiting factors allow. It has been easy to think that, within any formation, numbers of plants are set by dividing the size of a typical individual of the required form and habit into the space available. Plants are always being eaten down, it is true, but they make good these losses so well that much of plant produce is left uneaten by herbivores, going by default to decomposers. Surely, then, the numbers of plants are determined solely by their abilities to win space and nutrients from other plants. This has long been the prevailing view of botanists. The ground is shared by different kinds of plants, it is true, and the numbers of any one kind vary from place to place, but these differing numbers could be explained by slight habitat differences. If you studied plants carefully enough, you would find that their individual tolerances varied, that one needed more phosphorus or wetter ground than another, that a third could not endure late frosts, or shade, or high wind. For several decades this line of reasoning has led to attempts to study the distribution of individual plant species by growing them in environment chambers, in isolation, under rigorous control. There would be no animals present to spoil the measurements, letting the experimenter see just what the plant could do, afterwards attempting to explain its natural distribution in terms of its measured "tolerances." Similar thinking led to field workers identifying microhabitats, which would be covered by the appropriate kinds of plants. Breaks in this pattern caused by disturbance led to successions, as plants suited to the new kind of microhabitat moved in, but the numbers of successional plants could also be understood almost solely in terms of competition with other plants for a limited supply of suitable microhabitat. In practice, numbers were not very important, anyway, it was cover that counted, the area covered by plants of changing shape.

Botanists have, of course, always been aware that grazing animals often alter the forms of plants or affect the species composition of a range, but these activities could be allowed for under a general "controlling influence of habitat" hypothesis. It is usual in plant ecology books to find the influence of animals, or "biotic factors," discussed in turn with water, temperature, and soils as one of the parameters of the habitat.

Such is the thinking which has dominated plant ecology up to the present day. But there is evidence which suggests different explanations for the observed numbers of many plants. Revealing is the famous story of the attempts to control the prickly pear in Australia.

THE PRESSURE OF GRAZING ON HERBS

The prickly pear cactus (*Opuntia* spp) was introduced to Australia from America toward the close of the last century, for the plant is pretty and you can make good fences out of it. But it escaped, spread, and went on spreading until in 1925 it was the dominant plant over 60 million acres of range land. Its numbers had expanded exponentially, and other plants apparently could not resist its coming. Sheep would not eat it; men and horses could not force their way through it; the range was ruined. Botanists apparently had no answer to this problem, but Australian entomologists thought they did (Dodd, 1940; Holloway, 1964). They sent explorers to America to collect likely insects which ate prickly pear, and brought back a number, among them a moth *Cactoblastis cactorum*. This moth they reared in Australia, and then started to put patches of its eggs on wild prickly pear plants. The caterpillers thrived; the moth population expanded rapidly and expontentially. The caterpillers ate up every cactus within crawling distance; as full grown moths they flew on to the next patch. A wave of caterpillers spread across Australia, devastating the cactus, leaving none behind. In a very few years the 60 million acres were back under grass, and the sheep farmers prospered once more (Figure 34.1).

But prosperous sheep farmers were not quite the end of the story. Prickly pear still exists in Australia, although the survivors are few and far between. *Cactoblastis cactorum* still lives there, too, and it still seems to feed only on the cactus. What seems to be happening is a great game of hide-and-seek between the plant and the moth. In odd places the caterpillars have missed a cactus, and it thrives for a few years until a wandering moth finds it. Then it is destroyed, but meanwhile other cactus plants have got away from the moths somewhere else. Plants are always being eliminated, but others are always popping up. Caterpillars often starve, wandering moths often die without finding a plant on which to lay their eggs, but occasionally one of the searchers finds a cactus, its larvae destroy it, and another generation of searchers is produced.

Very clearly the numbers of prickly pear plants in Australia are now set, not by other plants or by microhabitats but by animals that eat prickly pear. And yet the Australian ranges are still as green (or perhaps brown), as ever they were, and the total mass of plant production is probably still much more than is taken by animals, even by the sheep. Other plants filled the place of the slaughtered cactus as fast as it was removed. But why should not these other plants be likewise engaged in escaping from some herbivore, even as the cactus is? There are many species of plants in a range, and there are many more species of herbivorous insects, each a specialist at hunting out particular kinds of plants. We should expect the lethal qualities of many of the hunts to be muted over evolutionary time, as selection favored plants with defences against their animal aggressors, perhaps particularly chemical defences, but enough games of hide-and-seek going on at the same time could provide the basis for that ever-present plant cover. The constant supply of food energy from the sun should do the rest, allowing plants temporarily not under attack to expand as fast as herbivores clear ground for them.

It is thus possible to put forward the hypothesis that many plant populations are, in fact, regulated by animals. Herbivores act like the searching

Figure 34.1

Rangeland in Australia before and after the moth *Cactoblastis cactorum* was introduced. Larvae of *Cactoblastis cactorum* feed only on species of *Opuntia*. Introduction of the moth to rangeland dominated by *Opuntia* led to rapid population growth of the moth whose larvae almost totally destroyed the cacti in every part of the range. Once the wave of caterpillars passed, both the cactus and the moth became rare, with isolated patches of cactus springing up until a wandering moth finds them and lays a clutch of eggs.

predators in Holling's models (Chapter 30), and their anchored quarry, the rooted plants, act much like simple prey. This hypothesis has the special attraction that it seems to give an understanding of the diversity of plants, as well as providing a fresh way of looking at plant numbers. Plants under attack by something intending to eat them have, in an evolutionary sense, only two ways of escape. They can be good at dispersing, either by casting seeds far and wide or by organizing themselves to throw out vegetative shoots while their attackers wander on or change generations. This they plainly all do, in some way or another, making use of that constant, unflagging source of food energy. But they can also manage to taste bad or even be poisonous to their tormentors. This also they plainly do. And herbivores, particularly insects, have responded by evolving specialized abilities to feed on just a few of the strange-tasting array of plants. Here we have an explanation of both the great variety of plants that live in similar places, as in a pasture, and some part of the even greater diversity of the insects which eat them. And this satisfying insight is based on the hypothesis that grazing pressures severely limit the fortunes of individual plant species.

This way of looking at plant numbers and plant species is implicit in Darwin's discussions of plants in *The Origin of Species,* but like so many other things in that wonderful book the hint has been passed by many who succeeded him. The viewpoint has recently been restated by J. L. Harper (1967) of the University of North Wales.

Harper's thoughts have developed from the evidence collected by agricultural workers, from the evidence of attempts to control weeds like the prickly pear, and from the experience of those who have tried to find how best to manage pastures. A famous story of biological control of weeds he finds even more revealing than that of *Cactoblastis* and the prickly pear. This was the successful control of the Klamath weed or St. John's Wort (*Hypericum perforatum*) in California by the beetle *Chrysolina quadrigemina* (Harper, 1969). In its essentials the story runs parallel to the cactus story. An obnoxious weed dominated pastures, the beetle was brought in, it reduced the weed to but a trace in a few seasons, and now there lives on a sparse population of plants playing hide-and-seek with the beetles. But there is this difference; the weed is still common in shaded woody areas where apparently the beetle does not care to live. Harper points out that the microhabitat in the shady woods is quite different from that of the open fields. Anyone coming on the present distribution of *Hypericum perforatum* in California should be excused for thinking that its range was set by its requirement for shady woodland sites. But they would be wrong. It is rare in pastures but common in woods because of a beetle. No amount of growth-chamber study would reveal this fact.

It is perhaps easy to see how insects of specific tastes may single out and control particular species of plants, but how about larger grazing animals? Harper (1969) was able to draw on the many observations and experiments on sheep grazing in Britain to answer this point. Some plants are much more palatable to sheep than others; the animals, although larger than their prey, do pick and choose in the pasture. Heavy grazing of a pasture originally composed of palatable species quickly led to increased diversity; where there were once few species there were now many in the turf. This result was quite in keeping with the beetle and *Cactoblastis* studies. The sheep concentrated on their chosen food, reduced its numbers, and made room for other species to come in. But when a pasture was largely made up of plants known to be unpalatable to sheep, heavy grazing reduced the diversity. Apparently the sheep were hunting out the few plants they really cared for, and got almost everyone as soon as they became established. In this way they reduced the species list. In both series of experiments the pastures remained green all the time, the regulations of populations being hidden as survivors and immigrants constantly expanded to fill vacated land.

The elimination of rabbits from Britain by the virus disease myxomatosis gave botanists a ready-made experiment on the effects of this herbivore in controlling the numbers of some plants. The disease struck swiftly in 1954, swept across the island, and was so completely fatal that its coming killed virtually every rabbit in places where they had been present in the tens of thousands. The British countryside changed from one year to the next as long ungrazed pastures sprang up and new plant successions began. But the first and immediate effect was a sudden increase in the numbers of plant species present in the old rabbit pastures. This happened on the very quadrats that had long been studied by professional ecologists. The plants of a tiny island off Wales had been studied for 300 years. In the year after the rabbits died there appeared 33 species that had never before been seen on the island. Presumably their seeds had blown from the mainland, as they had done every year for 300 years, but this time there were no rabbits and they were allowed to grow big enough for the botanists to find them. Truly the rabbits had been acting as plant population regulators of formidable effect (Lancey, in Harper, 1969).

THE PRESSURE OF
GRAZING ON TREES

These various lines of evidence; the effective biological control of weeds, the quality of so many plants which makes them taste bad to most herbivores, the specialized eating habits of the herbivores themselves, and the effects of large grazing animals; leave little doubt that many herb populations are, in fact, controlled by animals who eat them. But what of woody plants, of shrubs, and particularly trees? Before man came along most of the more verdant parts of the earth seem to have been covered with forests, and trees are not obviously grazed down like grass in a pasture or cactus pursued by a moth. Yet there is clear and immediate evidence that even trees live under relentless attack. The very woodiness of them may itself be partly a defence against being eaten, and one which is very effective. Few animals eat even the cellulose of wood, apparently none without the aid of symbiotic microorganisms to provide the necessary enzymes. This poses the question, "Why can't animals eat cellulose?" which is, as Harper (1969) remarks, one of the greatest mysteries of biology, I think as great as that similar, "Why can't large plants fix nitrogen?" (Chapter 15). Being woody helps trees to be large, but it clearly also defends them by its uneatable qualities, and the trees increase this protection by developing lignin and a variety of substances to impregnate the heartwood. These may best be explained as extra chemical defences, perhaps particularly against fungi.

It may also be argued that the very size of trees also defends against attack, since there are few animals large enough to browse on trees in the forests we know. But in the past there have been many, not only things like elephants and giraffes, but giant ground sloths, various relatives of the elephant, and bovine animals with long necks. In modern rain forests there are still tree sloths, monkeys, and gorillas, all adapted to feeding in the canopy.

Even temperate forests of settled countries are under herbivore attack, mostly by insects and other arthropods. Casually looking at a forest may suggest that the take of these is small, but in fact we are beginning to realize that the grazing pressure of such animals may be crucial to the success of a tree.

G. C. Varley (1970) of Oxford University has for many years studied the effects of a small moth whose caterpillars attack the young leaves of oak trees in the spring. The dates at which the oak leaves open vary from year to year and from tree to tree, and the caterpillars must use some environmental cues to ensure that they hatch from their eggs at the right time. When they synchronize well, they may totally destroy the first crop of leaves, after which they pupate and worry the tree no more that season. The tree quickly grows more leaves, and the casual observer hardly has reason to believe that its fortunes have been affected much. But Varley found that the energy cost of having to grow a second crop of leaves was so great as to represent something like half the energy store the tree was able to accumulate during one season's growth. This is not half the total production of the tree during the season, of course, because much of that goes to maintenance, but half the interest on this capital which it could use for new growth. The spring strike of caterpillars, so quickly made good by the tree, clearly acts as a severe curb on the tree. If the tree is competing for light and space with other trees not so afflicted, the caterpillars could well be the decisive factors in its success, and in the eventual size of its population.

Grazing on tree leaves by caterpillars may fairly be called "parasitism," since the animals live on their host without killing it. But herbivores may also act as predators, killing and eating whole individual plants. The insects used in the biological control of weeds, or the grazing of sheep and rabbits described by Harper, were acting in this way as predators, since they killed the plants they attacked. Animals that attack large trees can also act as predators, if they eat seedlings or seeds. In Chapter 23 it was suggested that the main crunch of competition in tropical forests should come during the establishment of seedlings, but it may be that predatory

PREDATION ON THE
YOUNG OF FOREST
TREES

attacks on seedlings and seeds may be even more important than competition. This possibility has recently been put forward by Jantzen (1970,
1971), and followed up with illuminating results.

Jantzen concentrates his work on tropical forests, attempting to explain
the dispersion of rain-forest trees and the associated large number of
species in terms of seedling and seed predation. He argues that the area
under the canopy of a tree is a deadly place for the tree's offspring
because animals adapted to eat that kind of plant will congregate there.
For one thing the floor under the canopy will be frequently visited by caterpillars that fall or drop down from above, and these parasites of the
canopy can be highly efficient predators of seedlings. A caterpillar may
kill a seedling in a very few minutes, then wander on in quest of another.
A second form of concentrated deadliness comes from insect seed
predators which are programmed to hunt seeds of that particular species.
Seeds falling under the parental canopy are concentrated, making a bold
target for insects and a place where their populations can build up to annihilating proportions. Furthermore, the tree itself may attract such
mobile predators as rodents, acting as a "flag" to show them a good place
to hunt, as the "flag" on the fairway guides a golfer to the hole. These
three agents of seed and seedling death; falling caterpillars, specific seed
predators, and rodents attracted by the flag; are likely to make it impossible for young plants ever to grow in the shade of their parents on the
floor of a tropical forest. The only young plants for which escape is possible are those from seeds dispersed to a safe distance before the
predators find them, probably by being carried away by birds or small
mammals, then being left uneaten. The odds are thus heavily against any
individual seed becoming a tree.

Jantzen's argument is illustrated by the diagram in Figure 34.2. There is
very little chance of a seed being dispersed far from the parent (curve *I*),
but there is no chance at all of its developing into a tree unless it is dispersed beyond a critical minimum distance (curve *P*). The product of the *I*
and *P* curves shows at what distance the chances for growth to adulthood
are best; the place represented by the peak of the population recruitment
curve (*PRC*). Jantzen (1970a) thus concludes that trees in mature
tropical forests are only likely to be established at considerable and definite distances from their parents. He suggests, in effect, that the dispersion, and hence the population size of tropical tree species, is set by
predation on the young; not by competion for space; not by limiting
factors; but by the activities of herbivores.

As described above, Jantzen's views are only to be called an hypothesis, although an attractive one. They are now in the process of being
tested. Jantzen (1970a, 1971) himself presents some suggestive data in
support of the hypothesis. In Costa Rica he has been able to demonstrate
both the deadliness to seedlings of caterpillars falling from above, and the
efficiency and specificity of insect predators of seeds and seedlings. There
really does seem to be a deadly zone under the canopies of Costa Rican
trees as he postulated. But there is still the worry that the Costa Rican

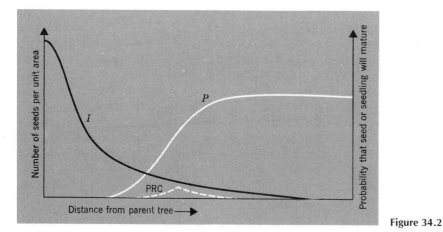

Figure 34.2

landscape is very disturbed by agriculture so that the dispersion of trees on which he worked is abnormal. Modern Costa Rica may be more a place for opportunists than for trees of mature rain forest.

In favor of the hypothesis is the way it seems to explain the much lesser diversity of temperate forests. For in temperate forests there are not many species of climax trees, and individuals of each species do not live widely separated from each other as they do in rain forests. Instead, we find one or two dominant species, the individuals of which live side by side. If seed and seedling predators force tropical trees to be widely separated from their own kind, why do they not do the same in temperate forests? There are certainly plenty of caterpillars, seed-boring beetles, and rodents available in temperate forests. Jantzen's answer is that the populations of insect herbivores in places of strongly seasonal climate must fluctuate widely from year to year, as indeed we know they do. There must, therefore, be some years when the seed crop of temperate forest trees is spared, even under the parental canopy. When this happens a dense stand of young trees will grow under the parent, will soon be tall enough to escape the larger populations of predators in subsequent years, and will exclude other invading tree species by competition. The result is the familiar simple forests of the north with which the early students of plant communities began their studies. The hypothesis "that spacing of individual trees of the same species is largely influenced by seed and seedling predation" thus seems to be widely applicable. It explains not only the species richness of tropical forests but also the comparative poverty of temperate forests. And an essential corollary of the hypothesis is that many populations of forest trees are controlled by predation, not by competition or inanimate qualities of the habitat.

It is well to remember that Jantzen's hypothesis has not yet been widely tested with field data, but the understanding of forests which it seems to give is so attractive that many ecologists suspect that further testing may well substantiate it.

It seems that plants, like animals, have populations that can be checked in many ways. Opportunists must find a place in which to live, and their numbers reflect both the incidence of such places and the effectiveness of their dispersing devices. But they must also compete with other slower dispersing plants, the equilibrium species. The numbers of both will, during this process, be affected by the classical components of density-dependence, the limiting resources of light, space, water, and nutrients. But all will be subject to being eaten. There should be between plant and herbivore all the range of relationships which we are finding between predators and animal prey. Disease and the weather must be working, too. Perhaps soon we shall find evidence for other built-in checks on plant numbers, evidence of the inevitable population consequences of social habits, of growing together or being dispersed. Plants do behave in different ways; shading out one another, inhibiting one another with root secretions, timing their flowering and the germination of their seeds to avoid seed predators. It is becoming respectable to talk of the "behavior" of plants. Recently, in a paper entitled "Reproductive Behavior in the Passifloraceae," Jantzen (1970b), described how some passion flowers, moved their stigmas in the early morning, first letting bees collect pollen, then raising the stigmas so that later bee visitors deposited pollen on them. There is much subtle behavior in plants. Perhaps some latter-day Wynne-Edwards will compile a synthesis of plant behavior pointing out the population consequences, and throwing down some general thesis of plant evolution to infuriate his colleagues and set them trying to prove him wrong. This would be a far cry from traditional plant ecology which has dealt in vegetation rather than numbers, perhaps missing the trees for the forest.

Plagues of mice and other small rodents have been known throughout recorded history. The plagues come with little warning then vanish with less. In the Arctic these plagues are so frequent as to have discernible rhythm, and they provide data against which all hypotheses of population control can be tested. It seems likely that every imaginable mechanism for density-dependent restrictions on population can, in fact, be shown to occur. Predation, disease, damaging of resources, social behavior, adrenal stress; none of them can be ruled out, but also none alone are sufficient to explain the coming and the going of the plagues. A general hypothesis is possible, however, if we assume that although such agents commonly do greatly restrict, and even control, the population, accidents of weather sometimes allow these restraints to be escaped for a time. The population becomes too large to be maintained indefinitely by the local resources and is forced to crowd in ways for which it is not adapted. Some combination of the customary restraints is then reasserted, and the population is reduced to a normal level. Histories of plague animals reveal, what should anyway be intuitively obvious, that no system of density-dependent control is likely to be so perfect that the vagaries of the weather cannot sometimes overturn it.

MOUSE PLAGUES AND THE FLUCTUATING NUMBERS OF ARCTIC ANIMALS: PROBLEMS FOR THE REVIEW OF POPULATION THEORY

481

CHAPTER 35

MOUSE PLAGUES
AND THE
FLUCTUATING
NUMBERS OF ARCTIC
ANIMALS:
PROBLEMS FOR THE
REVIEW OF
POPULATION
THEORY

Throughout recorded history agricultural man has at times been beset by plagues of mice. Our oldest book, the Hebrew bible, tells of such afflictions as visitations of God on the enemies of Israel. Since then there have been many desperate turns in the fates of nations as mouse hordes harried their crops. The mouse plague which ruined much of German agriculture in 1917 and 1918, together with the constricting blockade of the British fleet, probably did more to break the German will to fight than all the valor of the allied armies in France. And before then, too, it is likely that tides of battle have often been turned by lowly mice.

The mice who mastered human events always appeared suddenly, wrought their mischief, and went. Suffering men have always tried to defend themselves. In recent years they have killed by poison, guns, or traps, or, as the Germans did in 1918, by seeking to spread disease among the rodents. In earlier times men could only offer prayer, or witchcraft, or curses. When ancient Philistia was afflicted by mice, the Philistine priests were induced to return the Arc of the Covenant and they placed in it golden images of "the mice that mar the land" [Samuel, I (5)6] as their hope for respite. All these efforts have had one thing in common, as E. S. Deevey (1958) points out. They all worked. Whatever men have done about a rodent plague, that plague has always gone away. It has also gone away if the men did nothing at all, revealing perhaps the most puzzling thing about these outbreaks; they apparently always end in mass death of the animals. The history of a mouse plague tells of a sudden spectacular increase in numbers, often over a huge extent of country, and with at most one or two years' warning that numbers were building up. For a season or two the mice rampage, destroying crops, and running about in the open. Then suddenly, quietly, most of them die. This is a sequence of events which must be explained by any general theory of population checks if that theory is to gain credibility.

The first sudden massive increase in numbers is not hard to understand in simple biological terms. All species must be intrinsically capable of it, since all can leave far more offspring than are required to keep their numbers steady. Mice, which can produce several large litters a year, must be thought of as being always on the brink of stirring expansionist times. But they are usually checked. A mouse plague must start when some normal check on numbers is removed and, whatever the check is, it must be removed from the populations of very large stretches of country at the same time. Changes in weather seem indicated. It is, at least, a good first working hypothesis that some coincidence of favorable seasons allows the mice of large regions to breed over two or more generations with unusual success, and that the resulting hordes of offspring cause the plagues we see. The mice are thus to be thought of as opportunistic species, animals normally held unsteadily down by the almost endless vicissitudes of life. But then we must explain the mass deaths that follow. These must somehow be density dependent, or at least triggered by density, since otherwise the plagues should oscillate away and not just vanish. A working hypothesis must look at all the proposed forms of

density-induced death; of disease, starvation, predators, social intolerance, physiological shock, and perhaps even genetic changes in the
crowded animals.

The absorbing interest and enormous theoretical importance of mouse
plagues to population theory were set down for ecologists by Charles
Elton in one of the classics of scientific literature, a book published in
1942 called *Voles, Mice, and Lemmings*. The book was the product of
more than a decade of work, of years spent agreeably in archives of the
ancient libraries of Europe tracking down eyewitness accounts of the
great plagues of history, and of other years spent, no doubt just as
agreeably, among the mice of harvests and of wilder places such as
Labrador. Elton showed not only that mouse plagues were relatively
common events over much of the world but that in the Arctic they were
regular events possessed of an irruptive rhythm of their own. In the
Arctic, a population theorist must explain not only the sudden onset and
the even more sudden end to a mouse plague but also its cyclic occurrence.

Elton's arctic data came from the records of those who had traded in
the North for furs. The missions of the Moravian church and the Hudson's
Bay Company had been trading with natives in Canada for more than 100
years, and they had kept careful records of their commerce. The main
trade items were skins of predators; but the predators ate rodents so that
the fur yield must give some indication of rodent populations. Two
staples of the fur trade of those parts were skins of arctic fox (*Alopex
lagopus*) and the Canadian lynx (*Lynx lynx*). The sales records of these
skins revealed some extraordinary facts. About every four years there was
a bumper crop of fox skins; and about every ten years there was a bumper
crop of lynx skins. The fur industry was one of boom and bust, with the
booms and busts following each other in regular rhythm ever since the
records started more than 100 years before. The trapping was the same
every year, so it must be the supply of animals which was changing.
There were more foxes alive every fourth year than at other times. And
likewise there were many more lynxes alive every tenth year than at other
times. Why was this so?

The trappers themselves knew the immediate cause of their booms and
busts, and they had left records of their opinions in their diaries and
reports to head office. They blamed the rodents on which the fur-bearing
animals fed. Lynx and fox skins were collected in winter, when the
animals carried their finest fur, and the winters of bumper harvests always
followed summers of rodent plagues. Lynxes ate mostly arctic hares
(*Lepus arcticus*), and reports of this animal being abundant always
preceeded winters in which lynx hunters grew rich. Trappers actually
collected hare pelts as well as lynx pelts, letting estimates of hare
numbers and lynx fur yields to be plotted on a graph (Figure 35.1), an exercise that leaves no doubts in anyone's mind that the fates of the two
populations were linked.

RODENT CYCLES AND
THE ARCTIC FUR
TRADE

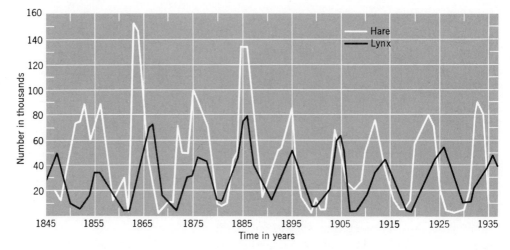

Figure 35.1   Population histories of lynx and snowshoe hare. These histories are compiled from the records of the Hudson's Bay Company trading posts and represent skins traded. Since the trapping effort was the same each year, the fluctuations must mean that numbers of animals fluctuated by a proportionate amount. These are not coupled oscillations in the sense of the Lotka-Volterra models for the peaks of predator and prey numbers are synchronized. Predator populations are helplessly tied to prey populations. (Redrawn from MacLulich, 1937.)

Arctic foxes did not often catch hares, but lived on mouselike animals, on voles and particularly on the various species of lemming. Every good fox winter followed a summer in which trapper's diaries recorded that the tundra was alive with little brown mice. But the trappers noted the presence of the mice only casually, and Elton wanted to be certain of his facts. So he went there himself to talk to the traders. His results left no doubt that the various mice, particularly the lemmings, were to blame. About every fourth year there was a plague of lemmings on the tundra, a population perhaps a thousand times as big as the populations of the lean years in between. After every summer in which the lemmings exploded so, there was a bumper crop of fox skins brought into the trading posts. Elton had shown that in the Arctic there were not only mouse plagues, as there were everywhere else, but that the plagues came and went in cycles. His evidence pointed to those cycles having existed for over 100 years, suggesting that they were virtually perpetual.

When numbers of predators and prey fluctuate sharply there is bound to be a suspicion that what we see is a coupled oscillation between the two populations of the kind predicted by the Lotka-Volterra equations. Gause had tried to generate such oscillations by loosing the predator *Didinium* on his cultures of paramecia with but indifferent success. Might not the Arctic have staged things better? The foxes must surely take a terrible toll of the lemmings in their summer of maximum abundance; might not this toll have been the prime reason for the lemming's sudden disappearance? And the foxes had help, for many other predators also waxed fat at the lemming tables; snowy owls, weasels, jaegers, hawks, and anything else that could eat a lemming. Might not these animals together succeed in overeating the lemmings every four years, only to run out of food, starve to death themselves, and thus give the lemmings respite to catch up? But, as Elton saw, this hypothesis could be quickly discounted. The arctic predator and prey populations were in phase. You either had lots of foxes and lemmings together or both were scarce. It seemed that, far from controlling their prey, the predators were helpless victims of its fluctuating supply.

Elton was able to confirm this failure of the predators to control their prey with evidence from arctic Norway. For about a century the Norwegian government had paid bounties for predators of all kinds, including the foxes, hawks, and owls which preyed on Norwegian lemmings. In places this system of bounties had been unpleasantly effective so that many fine beasts had become rare. Their numbers were always too low for them possibly to control the numbers of their prey, but Elton found that the numbers killed for bounty had gone up every four years just the same. The rare and harried beasts were yet not quite so rare in high lemming years as at other times.

This left the rhythmic lemming plagues still to be explained. In the Norway of bounty hunters, the myriad individuals of a lemming year vanished before next summer, just as they did in lands which still had their predators. Obviously, predation could have some effect on lemming numbers, but it was not decisive. It is proper to look at the arctic predator populations from a different point of view, not as agents of control but as animals cursed with a food supply that fluctuates wildly.

If trappers do not catch the foxes of a bumper year, what happens to them? We do not really know, but probably all biologists who have thought about it assume that most of them die of starvation when they cannot find any more lemmings to eat. But, if so, how are the favored few survivors chosen? And how can they be fed well enough to rear families while other foxes die for want of food? As far as I know, we do not have answers to this.

ADAPTATIONS OF ARCTIC PREDATORS TO FLUCTUATING FOOD SUPPLIES

The responses of two bird predators to these fluctuating crops of food are better known, and their study has well illustrated some of the checks that are constantly applied to natural populations. Snowy owls (*Nyctea nyctea*) respond to a tundra crawling with lemmings with a massive breeding effort (Pitelka et al., 1955) laying clutches of as many as ten eggs and rearing most of the young who hatch. They are prisoners of the evolutionary drive to rear the maximum number of offspring which the food energy supply allows. The result is that there are very many owls left to roam the tundra during the following winter, many too many for the dwindling food supply of vanishing lemmings. They wander far from a North which can no longer sustain them, and we see them in temperate countries as big white oddities. Perhaps most die there, suffering that density-dependent winter death which is relied on by theorists of the Lack school to explain the apparent stability of bird numbers.

Another predatory bird seems to take advantage of the lemmings with greater skill. This is the pomarine jaeger (*Stercorarius pomarinus*). The jaegers are big gulls, sea birds who spend most of their lives far out to sea feeding on the life of the ocean surface. But the pomarine jaegers breed in the Arctic, where they rear their young on lemmings. The adults probably live for many years, perhaps as many as a dozen years, so that they see two or more lemming cycles out. And every year in the spring some of them appear over the tundra. They have long been known to arctic travelers, perhaps principally for their engaging habit of dive-bombing any man who comes near their nests. But our understanding of their breeding behavior, like our understanding of many other things about the Arctic, awaited the construction of the Naval Arctic Research Laboratory at Barrow, Alaska. From there F. A. Pitelka and his students (1955) have studied lemmings, and the animals living with lemmings, for many years. They found that the pomerine jaegers set out and maintained territories on the tundra in ways which were closely comparable to the territories of the songbirds which Howard studied (Chapter 31). Each pair fed on the lemmings of its own chosen space, becoming intolerant of interlopers. In a summer of lemming abundance the jaeger nests would be close together, reflecting the small size of territory over which the birds must journey to catch the lemmings they needed. But if lemmings were scarce, as they usually were, only a few pairs of jaegers would nest, or none at all. The rest of the birds presumably came in from their ocean wanderings, had a look at the tundra, were not stimulated to breed, and departed back to sea. It therefore seems almost certain that many pairs of jaegers are denied the chance to breed in low lemming years by not being able to find territories. In high lemming years, many more breed. But the normal adult jaeger life is so long that most may well have an opportunity to leave viable offspring before they die.

The massive breeding effort which all the lemming predators seem to undertake in high lemming years suggests very strongly that their breeding in other years is checked by want of food for the young. For these animals the food supply in the breeding season is thus a strong regulating force. But the predators apparently do not reciprocate by controlling the numbers of the lemmings which they eat, and we are left with the question: What does? Why do lemming numbers fluctuate rhythmically so that they are a hundred times more numerous in some years than others?

POSSIBLE COUPLINGS
FOR THE LEMMING
OSCILLATIONS

So violent an oscillation looks as if it ought to be coupled with something. Prey is not coupled with predator in the manner of Lotka-Volterra; very well, perhaps it is coupled with something else. Disease seems inherently likely. Perhaps, as lemming populations mount, disease becomes more and more prevalent until it becomes a serious epidemic. Most of the lemmings die. The disease, finding few animals left to infect, becomes rare, and scattered healthy lemmings are left to breed and start the cycle all over again. This was a good hypothesis because it was testable. All you had to do was to visit the Arctic in a high lemming or other rodent year and look for signs of epidemic disease. This has now been done several times, but the results are inconclusive (Deevey, 1958; Elton, 1942). Northern hares in the peak years may often be diseased, but MacLulick (1937), who found perhaps the strongest evidence, noted that different diseases seemed to operate in different years of decline. His finding does encourage the belief that crowded animals should be suspect to disease, but does not support the view that the oscillations in hare numbers were coupled with the spread of a disease organism. Evidence that the lemmings of peak years are diseased is so scanty as to suggest that epidemics may not usually be involved in the mass deaths. One of the most dramatic bits of evidence for this comes from the work of Charles Krebs (1964), who recently saw a whole four-year lemming cycle out in the Canadian North. He examined the bodies of 4000 lemmings that he trapped during the population die-off, generally finding that the animals were not only healthy but actually fat and sleek looking. Probably a prudent man should say that sometimes there may be epidemics among crowded lemmings, but that these are almost certainly not frequent enough to be a prime cause of the lemming cycles.

Perhaps the lemming populations were coupled not with the things that ate them, predators and disease, but with the things they themselves ate. This hypothesis suggests that the lemmings at peak abundance so destroy their tundra food that they die of malnutrition. Only a few survivors get through. The pressure is removed from the tundra which slowly recovers. As it does, the lemmings begin to have large families again, and soon they are so numerous as to destroy the tundra once more. This hypothesis has a very satisfactory and likely sound to it. It conforms well with our ideas of animal numbers in simple systems, escaping control by predators and disease, then building up until they destroy their range. Then they die, wretchedly, in the winter. A first difficulty with this hypothesis is that the lemmings crowd only part of their range. They do indeed eat down the tundra until there seem to be hardly any plants left, but Krebs (1964) found that only some 5 percent of the immediate area was affected. If the lemmings were starving why did they not move over a 100 yards or so to find fresh forage? We know that they are animals that do move because of the famed lemming migrations of Norway. If they can walk miles and miles, why not a 100 yards in search of supper? But more serious still to the starvation hypothesis is the fact that you do not find starving lemmings. The lemmings which Krebs' caught during the crash looked sleek and were quite definitely fat, a fact which led him and others to completely discount the food-supply hypothesis.

All the same, the effect of lemmings at maximum abundance on their food supply is very strong. Anyone who has seen the scorched brown waste they leave behind, with even the roots of the grasses having sometimes been torn up and devoured, cannot help being impressed that this damage must have something to do with the pending decline. I let my own judgement be influenced this way when I was first in the Arctic during a peak lemming year, the summer of 1965 at Barrow, Alaska. The

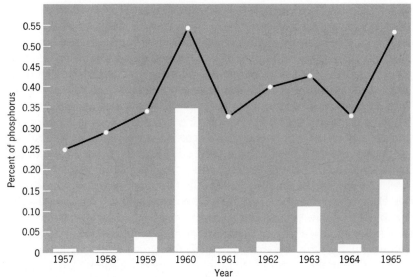

Figure 35.2

Correlation of phosphorus in forage with size of the lemming population at Point Barrow, Alaska. The curve represents phosphorus and the bars represent the relative sizes of the lemming populations in the summers of the several years. The phosphate content of forage apparently drops following a lemming high, suggesting that the mineral is then being cycled back to the soil. (Redrawn from Schultz, 1969.)

tundra, which I remembered from other places and other years as tolerably green, was a bleak brown mat, obviously not capable of supporting much life. I put this viewpoint to Charles Krebs the next year when I was visiting Indiana University. A scholar's evening was ruined by one of those impassioned arguments over Vietnam which recently so afflicted Academe, but I remembered the scorched tundra just as Krebs got up to leave, and called out to him, "I saw the lemming crash at Barrow last year and it was perfectly obvious to me that they had destroyed their range!" Krebs paused with his hand on the doorknob and said, "Paul, you just show me a starving lemming!" And then he was gone.

But there are more ways of dying of malnutrition than outright calorie deficiency. Suppose the damaged range left lemmings short of some mineral or vitamin, or of protein? might not such a diet have led to some subtle symptoms that were overlooked in Krebs' autopsies? or might it not interfere with their later breeding success? For this hypothesis to stand, two things must be shown; the impoverished quality of vegetation (as opposed to its total supply) must be demonstrated, and this failure in quality must then be linked to the deaths of the lemmings. The first of these has now been demonstrated by the studies at Point Barrow (Pitelka, 1957; Mullen, 1968; Schultz, 1969). The overeaten vegetation of a high lemming year at Point Barrow has proportionately less protein, less calcium, and less phosphorus than at other years (Figure 35.2). But there is still no evidence which links this poor quality of the vegatation to either increased mortality of lemmings or to their failure to breed. There seems little doubt that the oscillations in lemming numbers at Barrow are accompanied by oscillations in the nutrient cycles, with a major portion of the nutrient reservoir being held in living things during the lemming highs and this same portion flowing back to the plants via the soil during lemming lows, but there is no evidence at all that this shunting of the nutrients controls or even triggers the rise and fall of lemming numbers. For most ecologists the starvation hypothesis, even in its modified food-quality form, must go the same way as the predator and disease hypotheses. Stresses of predation, disease, and hunger may all act on the population in density-dependent ways. Alone each form of mortality is not decisive, although there must remain a suspicion that in combination they might be. Which is where Elton left the discussion in 1940. There are, however, other matters to be considered.

One particular oddity of the lemming cycles should, on the face of it, be a clue to their ultimate explanation. This is the synchroneity of events over enormous areas. The problem was not only one of why should there be a peak every 4 years but was also one of why should these populations oscillate in phase over such tremendous spaces. The weather has always seemed implicated as a controlling agent. But there was the difficulty of the 4-year cycle and, of course, of the 10-year cycle of the hares which must also be met by a general hypothesis. Did the weather of the Arctic fluctuate with rhythms of 4 and 10 years, so that it either blessed the lemming's conjugal efforts or alternately cut them down in their prime? For a time it did seem as if the hare cycle might be correlated by the famous sunspot cycle, but eventually the sunspots were shown to erupt every 11 years and the hares to show a more irregular period, the average of which was 10 years (Elton, 1942). So these events could not be correlated. Nobody has ever discovered anything like a 4-year weather cycle. And the problem is even worse than that, for the lemming plagues are a little irregular in their appearance. Sometimes they wait 5 years, sometimes even 6, so that the 4 years is only an average. The true mean between plagues is variously quoted as 3.3 or 3.7 years, and we know of no celestial events that might produce weather cycles so curiously unrelated to terrestrial seasons.

THE ROLE OF WEATHER

For many years we had to be content with hypotheses which suggested that vagaries of the seasons had gradually acted to bring cycles generated by other means into step with each other, but recently we have been shown how weather can indeed play a dynamic role in initiating cycles and in making them synchronous over considerable areas. D. A. Mullen (1968) studied the fortunes of the Point Barrow lemmings for 6 years. He concentrated his thoughts on what influenced breeding success. Lemmings breed both in summer and in winter in burrows under the snow where they can live on last year's roots. Pitelka (1957) had earlier

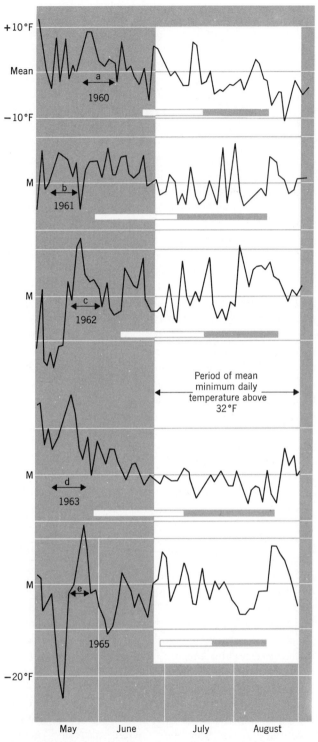

Figure 35.3

Deviations from the mean minimum daily temperatures in the spring at Point Barrow, Alaska. Mullen suggests that the "warm" snaps marked a to e are cues that bring the lemmings into breeding condition. The grey horizontal bars were the observed breeding periods. These ended at about the same time each year, the different lengths being set by the onset of breeding. A broad correlation of the onset of breeding with the spring "warm" snap is suggestive. (Redrawn from Mullen, 1969.)

concluded that much of the population increase could actually take place in winter, but Mullen found that the success of summer breeding was crucial to the rapid growth of a lemming population, as you might expect. In some summers, lemmings bred very well, so that each female reared a large litter to maturity, but in other years few young appeared, or else they died before they were full grown. Why was this? Mullen found what looks like a convincing answer in the history of the local weather.

The Barrow summer is a short uncertain time of 6 or 7 weeks with temperatures above freezing most of the time. The end of this relatively benign period is set at about the same time each year by the frosts of fall, but its beginning may vary with errors measured in weeks. A successful breeding year for lemmings is one in which births are nicely timed for the beginning of a longish summer but, to do this, they must copulate on the right date in early spring. Mullen's records (Figure 35.3) suggest rather strongly that there is usually a week or so in the spring of relatively mild weather, then another cold snap, and then what passes for summer. That first spring mild period seems to be used by the lemmings as an environmental cue, bringing them into physiological condition to breed. If this spring temperature cue comes late, as it often does, then the lemmings get off to a late start and have little chance of rearing their young in what summer is left after their birth. But if the cue is accurate, their large families are ready to face the winter. Two good breeding seasons must be all that is required to make the tundra come alive with lemmings. It seems that the odds are good that there will be two lucky seasons within a span of about 6 years. An averaging of this luck probably accounts for that average time of 4 years taken for the lemming populations to change from rarity to superabundance.

Mullen's weather hypothesis has certainly not been conclusively proved, but his correlations may be called strongly suggestive. The hypothesis seems to explain the curious synchroneity of Arctic events, for all the animals of broad regions could easily be living in the same air mass and so subject to the same environmental cues.

Weather for breeding may account for synchroneity between populations, and for the length of time needed by a population to get very large, but it does not explain the mass deaths that follow. We have tentatively explained how the hordes are born; now we must explain how they vanish. It may be that there is really no mystery here, that predators, disease, and winter starvation can usually combine to cull an excessive lemming crop. It is, for instance, particularly hard to know what is going on under the arctic snow in winter. But there remain the views of men like Krebs who have studied the lemmings and who have convinced themselves that the lemmings never starve, and there is still that awkward evidence collected by Elton to show that numbers can fluctuate even when predation is demonstrably negligible. And yet somehow the stresses of life in dense lemming populations must cause mass death. It is thus tempting to look for the cause in the social pressures of crowding, to call on the Christian hypothesis and the general adaptation syndrome.

LEMMINGS AND THE
GENERAL ADAPTATION
SYNDROME

When John Christian first pointed out the implications of Selye's general adaptation syndrome for population theory, it seemed at once that part of the problem of the lemming cycles was solved. Here was a crowd of animals, many of which died suddenly, mysteriously, and within a short time of each other. Moreover, there was plenty of reason to believe that lemmings behaved oddly in the weeks before they died. In the folklore of Norway are the tales of those intermittent years when the lemmings were "on the march." They appeared in city streets, running about in the broad light of day. They were said to go mad and bite people. They were believed to make suicidal migrations, pouring down the Norwegian mountains like furry rivers, plunging to their deaths in the sea. This is abnormal behavior for any animal, and the Christian hypothesis suggests that crowded animals become first deranged, then die. Those whom the gods would destroy they first make mad. If crowding lemmings to the point where the tundra became alive with them produced more stress than could be met by their hormone systems, then indeed we might expect first madness and then death.

The hypothesis of death from shock disease was testable. Christian had shown that rats and mice under stress had enlarged adrenal glands; therefore one must go North, collect dying lemmings, and examine their adrenals. When Charles Krebs set out on his 4-year study of lemmings in the Canadian North, part of his purpose was to test the hypothesis that mass death of lemmings was due to adrenal shock. He weighed the adrenal glands of 4000 lemmings before he was through. And he found that adrenal weights did *not* vary with crowding. The adrenals of doomed lemmings at the crash seemed like the adrenals of lemmings living in the spaciousness of the intervening years. This was a suggestive finding, but it is well to remember that weighing a whole gland is a dangerously simple and inaccurate measure of hormone function.

Krebs then went on to find more evidence which argued against the Christian hypothesis; he was able to cast doubt on the reality of the lemming madness which was supposed to precede death. In a high lemming year you did see lemmings running about all over the place, and there was no doubt that their populations did spread, even from high mountains to lowlands, as in Norway where they truly do advance down narrow valleys as populations grow on the slopes above. But there was no madness in this. The lemmings at their peak were a hundred or more times as numerous as in other years. When once you had had a chance of seeing one lemming in 3 months, you now had an equal chance of seeing one everyday. This fact alone should account for the press reports of strange behavior. Just imagine what the gentlemen of the fourth estate would do with cat reports if the cat populations of New York or London were to multiply by 100 in the span of a single summer! The tales of suicidal acts were probably fables. Lemmings wandered awhile in spring, in years when they were rare as well as in years when they were common. In crowded years they might wander some more, as they undoubtedly do in Norwegian valleys, but the noticeable thing was that there were 100 times as many lemmings wandering. So you saw them. If a few got drowned, you saw them too. Krebs saw for himself how such tales could be multiplied into mass suicides when he saw altogether some 50 on the ice of a frozen lake during the course of a day's walk. The local press multiplied this to thousands in a mass march. Such a tale, already so multiplied at source, would become the migrations to the sea which we learned in childhood by the time it had passed through the hands of a few more raconteurs. No tale of mass suicide of lemmings had been verified by scientific witness, and Krebs could conclude that the lemming suicides and madness were just fiction.

With madness discounted, and with lemming adrenals always of normal size, Krebs felt he could dismiss the Christian hypothesis like all those other hypotheses that had gone before it. There could not be many hypotheses left, but Krebs produced another alternative, one which had originally been put forward by Chitty (1958) for voles. He suggested that there were two strains of lemmings; good fighters who were poor breeders, and good breeders who were poor fighters. When lemmings were very numerous there should be much fighting, for the animals would be bumping into each other all the time. Perhaps only males who were fighting bullies should be able to win females for themselves and mate. Nearly all families would then be sired by fighters. But the qualities leading to fighting success, whether genetically or culturally derived, should, on the Chitty hypothesis, also lead to young who could not live to adulthood. Hence the mass death that followed crowds. The fighters were at once almost eliminated from the population. Now the meek could mate. These produced fine viable babies, and the population began to build up once more. This was an ingenious hypothesis to explain lemming cycles when all other hypotheses had failed. There was, however, no evidence to support it. Nor was it easy to see how it could be tested; and an untestable hypothesis is a poor vessel in which to trust your faith. Few ecologists were prepared to believe that Krebs was right.

Meanwhile, evidence supporting the Christian hypothesis after all had come to light. R. V. Andrews (1968) had taken advantage of the laboratory at Point Barrow to pursue his own studies on lemmings. He approached the problem not as a population man, but as a physiologist. He, too, tested for signs of the general adaptation syndrome, but by the methods of clinical physiology. He caught live lemmings all through the lemming cycle, then kept them in comfortable cages until they should have recovered from the shock of capture. Then he killed them, immediately removed their adrenals by aseptic surgical procedure, and maintained the glands in physiological culture. He then monitored the daily output of adrenocorticotrophic hormone (ACTH) from these cultured glands. The glands from lemmings killed during the year of maximum abundance produced daily many times the amount of ACTH that the glands of other years produced. This was a practical demonstration that the syndrome of adrenal shock was, indeed, present. Krebs' method of weighing the glands on autopsy had clearly been too crude to detect the increased adrenal output. The Christian hypothesis that crowded lemmings suffered adrenal stress could again be allowed, although there was certainly still no proof that overcompensation for stress led to shock disease in lemmings.

A TENTATIVE
HYPOTHESIS FOR
RODENT CYCLES

A general explanation for lemming cycles may perhaps go like this. The normal state of lemmings is comparative rarity. They live and breed in a harsh environment. The winter is long, and very cold, but the lemmings seem to manage well enough living in burrows under the snow. They breed under the snow, escape predators there, and have food there, provided the range was not too damaged during the previous summer. But the main breeding effort is in summer, and it is in summer, strangely, that they face their greatest hazard. The summer is short; its duration is unpredictable. The lemmings must time their copulation so that the babies are born as near as possible to the first day of summer or there will not be time for them to grow large before winter. A temperature period in the spring is used as a cue for copulation. One effect of this is that breeding is synchronized over very large areas. If the temperature cue comes late, many families perish at the onset of winter, for they are not large enough or fat enough to see the winter through. In addition to this hazard, hawks and owls and foxes harry the lemmings. Local populations can just maintain themselves against these hazards, but the hazards are very real and local extinction may well be common. In all this, lemmings conform to the idea of opportunistic species, and they have that other essential characteristic of opportunists, the ability to disperse. There is wandering in the spring, perhaps particularly as a result of social pressures in populations that are denser than others. Local extinctions are made good by migrations. The lemmings get by from year to year, suffering numerous catastrophes, but surviving as species because they occupy large ranges and can disperse well. There are density-dependent pressures on populations, perhaps particularly introduced by highly mobile and territorial bird predators, but these pressures are unimportant compared with recurrent failures of summer breeding caused by unpredictable weather.

But sometimes the timing of births in the spring may be well done, and the lemmings of huge areas successfully rear large families. Predators thrive, and rear large families themselves, providing a density-dependent check on lemming expansion. But this check is not enough to prevent numerous lemmings entering winter with a good chance of survival. They go on multiplying under the snow, producing a middle-level lemming population for the breeding effort next spring. If the copulation cue again turns out to be good, then a second successful breeding season follows and we see the effects of exponential growth in animals with large families. A poor breeding year may yet let the population enter another winter at middling levels, ready for a second chance at timing its breeding season well. And so on. A couple of good breeding years, not necessarily side by side, are probably all that are required to produce a lemming high. This could happen in just 2 years; or it might take 6 or 7. The long-term average seems to be 3.3 or 3.7 years, or something like that.

As the lemming populations swell to their peak, the predators multiply; some disease may well appear; the lemmings begin to destroy their range; they are intolerant of low status animals, which must wander, down into Norwegian valleys or on the drift ice to drown at sea. All these things represent density-dependent death, and they must collectively take a tremendous toll. Perhaps sometimes they are sufficient to hold back the lemming rise a season or two; this we do not know. But eventually, on an average every 4 years, these density-dependent checks are not sufficient to prevent the lemmings multiplying a hundredfold. Then the lemmings are densely crowded. Particularly dense local populations may so attract predatory birds that they are wiped out, as Pitelka (1957) thinks some-times happens at Barrow. If this does not happen there might yet be an epidemic. And if that does not happen they may so impoverish the range that insufficient food will be left under the winter's snow to see them through to next spring. But one thing is certain; the productivity of the land is such that very large lemming populations cannot be maintained. It follows that lemmings cannot be adapted to persist at these densities. If some normal catastrophe does not carry them off, there must be crowd-ing of a kind which is not allowed for in their social behavior. Not only are the animals harried and hungry, but they suffer social stress as well. They overload the protective devices of the general adaptation syndrome, and they die. There are just a few survivors to overwinter and begin another year of normal lemming rarity. The jaegers and the owls depart. The foxes and the weasels starve. The tundra grows up green again. And the journalists turn to embroidering other tales.

The main parts of this explanation can be used to explain other rodent plagues. A general quality of mice, voles, rats, and other plague rodents is that they all have large families. They are all capable of increasing their populations a hundredfold in time spans easily measured in months. All that is required for them to do so is the lifting of some prevailing check. The animals are all such short-lived creatures that the vagaries of succes-sive winters, or other hard times of the year, are met by different genera-tions. The size of each generation is thus necessarily influenced by the weather at its birth, and the resulting fluctuations may be much too wide for density-dependent factors to contain. Sometimes the population grows so remarkably that we see the results as a plague. But the rodent habitat cannot support plague numbers for long, and so the animals are not adapted to living in such numbers. The physiological basis of their social behavior breaks down, and most of the animals die of stress.

Perhaps the most important general thought to emerge from nearly half a century of study of rodent numbers is that all hypotheses of population checks are applicable. Any imaginable agent of density-dependent death is probably truly at work, killing in its density-dependent way. But when systems are simple, they may be so unstable that density-dependent constraints are evaded. And all behavior has population consequences. These consequences must usually be to maximize the breeding effort, for such will be strongly reinforced by the mechanism of natural selection. But other behavior, such as territoriality, confers its advantages in other ways so that the secondary consequences of checking population growth are preserved.

Over all these pressures hangs the brooding uncertainty of the weather. There are seasonal good times and bad times to which all organisms must adapt. Both efforts to increase, and checks on numbers, change rhythmically with the changing seasons, affording constantly recurring chances for populations to escape their restraints. And as if these regular changes in seasons were not enough, there remains that wonderful element of the unexpected in the weather of any place, the age-old provider of variety in the lives of animals no less than in the affairs of men. This uncertainty means that no number will remain indefinitely in balance.

A CONCLUSION: THE NATURAL REGULATION OF NUMBERS

Much of the understanding of nature which ecologists have reached has come from realizing that the true limit to the activities of living things is set by the supply of energy. It has been tempting, therefore, to conclude that all local populations must grow until all the potential food-energy available to their way of life is being used. But this way of looking at populations implicitly assumes that every kind of animal and plant has been designed only as an energy-getting machine, when in fact evolution works not as a process of design but by selecting from the limited array of designs already available. The chosen design will be that which survives hazard to leave the most offspring. This requirement may be met in many ways, only one of which is that of making the most efficient energy converter. Many social habits, for instance, are preserved by selection because they help individuals breed even though different strategies might seem to allow more food to the species in the short term. Evolution plays for safety, and greed for food-energy, which is also fostered by natural selection, must often take second place. It follows that any general theory of population regulation which sees numbers as being set only by a struggle for food-energy must be false. It is also false to assume that plants, or different kinds of animals feeding in a particular way, must therefore be regulated to some common pattern. Each trophic level is not subjected to a characteristic kind of population restraint. Instead of looking for general mechanisms of population control which should be applicable to particular ways of feeding, we should look for patterns common to habitats presenting similar physical hazards, or to family groups of plants and animals which have perforce had to adapt to hazard from common engineering plans.

497

CHAPTER 36

A CONCLUSION:
THE NATURAL
REGULATION OF
NUMBERS

"A struggle for existence inevitably follows from the high rate at which all organic beings tend to increase. Every being, which during its natural lifetime produces several eggs or seeds, must suffer destruction during some period of its life, and during some season or occasional year, otherwise, on the principle of geometrical increase, its numbers would quickly become so inordinately great that no country could support the product. Hence, as more individuals are produced than can possibly survive, there must in every case be a struggle for existence, either one individual with another of the same species, or with the individuals of distinct species, or with the physical conditions of life."

From Chapter 3, *The Origin of Species.*

In that passage of the *Origin*, Darwin succinctly stated the facts of the regulation of numbers. There must be checks, drastic checks which effect great destruction. So much is decreed by physical laws. But it does not matter to physical laws what these checks may be. Anything that kills will do, whether it is a losing contest for food, the jaws of an enemy, or the hostility of the inanimate world. The studies of a 100 years have amply illustrated Darwin's words. Plants struggle for the food of light, and many of their populations must be curtailed by the success with which the local dominants take that light. The effective working of the exclusion principle, with its accent on feeding structures and feeding behavior of animals tells of the widespread effectiveness of a struggle among allies for food. The work of pest-control men has shown how astonishingly effective some insect predators can be at curbing the expansion of prey populations. Herbivorous insects, like the caterpillars that slew the prickly pears, can likewise be devastatingly efficient, effecting great destruction on the populations of their food plants. And exceptional weather can kill Australian grasshoppers or English herons, and can decide in which winter crowded deer must die.

But the intent of ecology has been more than to provide examples to illustrate Darwin's work. We have wanted to arrive at a general theory, to make a model that should be appropriate for the growths of all populations, and to see that this model is consistent with the mechanism of speciation itself. Many have asked themselves, as did Darwin before them, what is the ultimate thing for which all living beings must struggle? And the answer has seemed surely to be "energy," "food energy." The laws of physics, the concept of the ecosystem, common sense, have all allied in suggesting that populations should normally grow to the limit set by the supply of energy. And what we know of the mechanism of evolution has also encouraged this belief. Selection favors he who leaves most offspring, and he who leaves most offspring is generally he who gets most food. Any habit or adaptation that interferes with an individual's leaving the maximum possible number of offspring must be ruthlessly suppressed by the process of selection. Population regulation, therefore, has come to be seen as a process of sharing out energy. If a population is not itself limited by struggles for food, then it is limited by predators who struggle to eat it, who are themselves struggling for food. Compared with these struggles after energy, even the hazards of the physical conditions of life, of death by storm and death by drought, are to be taken as no more than irregular perturbations of normalcy. That (Andrewartha and Birch notwithstanding) I take to be the prevailing consensus of ecology.

But there is an error in this train of reasoning, a misunderstanding of what is possible for selection. Selection is a process of choice; it choses from among individuals those which are best adapted to avoid the hazards of life at that time and place. But every choice has many effects, and many of the effects will reduce the individual's chances of getting food energy. A better choice would be to select only individuals which survive *and* have an enhanced food-getting ability, but such a superchoice cannot often be available to selection. Natural selection, like a politician, is an architect of the possible. It can only chose from what is available, not what is theoretically desirable. Reconsideration of some examples should make clear the overriding effect that this has had in practice.

The best-studied example is that of territorial behavior in birds. There can be little doubt that the territorial mechanism ensures marriage, that it promotes survival by promoting mating and the rearing of young. Individuals without the necessary territorial traits would never leave offspring at all, so that the selection pressures preserving the behavior must be very strong indeed. But the behavior has many effects, absolutely unavoidable effects. Singing territorial birds, for instance, are heard, and aggressive territorial birds are seen. It would be remarkable if these spring habits did not sometimes make the stalk easier for a predator. Many birds even have conspicuous colors in the spring, colors dangerous enough to their existence apparently for selection to have seen to it that the colors are lost when the breeding season is over. But in the spring these dangerous things must be tolerated. It would be so convenient if a bird could still have its territorial behavior and be as inconspicuous as at other times of the year. But this is not in the realms of the possible. Selection preserves the appurtenances of territorial behavior in order that the birds may breed at all, and accepts the side effect of some being killed because of it. It has made a politician's choice. And another effect of territorial behavior is that populations are not allowed to grow above the minimum territory size needed to preserve the behavior. Limits are set to the population as a second-order effect of traits selected for, as part of the politician's choice, as side effects. And energy is left unused, to become the property of other birds or of decomposers.

Every choice made by selection to fashion an animal or plant must have a population effect. The social hierarchies, so important to things such as mice and muskrats, mean that low-status animals are often cast out to be eaten by predators. This often limits populations; a pity to waste some good muskrat food, but much better than losing that vital social behavior. Baby wildebeests run by their mother's sides within minutes of being born, for if they do not they will be killed. They must then keep running with their mothers or die. Natural selection has made the typical wildebeest baby just as good at running by mother as wildebeest engineering will allow. Yet, as the herds become denser, more and more babies become separated from mother, and are doomed to the vultures. The wildebeest populations are regulated by this density-dependent death of babies, perhaps being kept well below the carrying capacity of the range (this we do not know). But the way of wildebeest life makes such density-dependent death of babies inevitable. It is not selected for. It is just an effect, an accidental parameter of the wildebeest niche. To expect selection to find a superwildebeest which can do everything that an ordinary wildebeest does, and still not lose some babies, is asking the impossible.

In Wynne-Edwards' treatise on animal dispersion is a splendid collection of behavioral traits in animals which seemed to him to have population consequences. I suggest that he is right in this. But this does not mean that the traits have been preserved by selection *because* of their population consequences. Every trait must have a population consequence, and many will involve foregoing food energy in order that a viable population will survive. And this kind of choice, accepting the loss of energy to avoid hazard, can be seen in plants, too. Consider the deciduous tree which sets its leaves late in the spring. It wastes a week or two of good spring sunlight, but it avoids the hazard of frost. It is better to be safe than greedy.

The world is a place of recurrent hazard, a place where many agents are ever ready to wreack great destruction. Since all living beings must be adapted to avoid these hazards as well as may be, they probably all have acquired traits which force them to surrender energy which they might otherwise have. Furthermore, many organisms are bisexual. Not only must they avoid hazards, they must mate. The selection which ensures that they mate without fail must often produce results which would be bizarre if the only function of living things was to pursue food. For these reasons a universal model of population regulation based on a struggle for energy is bound to be false. Some organisms are limited by food, others are not.

One attempt at a general synthesis, by Hairston, Smith, and Slobodkin (1960), has considered animals and plants trophic level by trophic level. The decomposers are considered first. These apparently always consume all the food available to them, because every year organic matter is oxidized at a rate that almost exactly balances the rate at which it is produced by photosynthesis. The decomposers complete the carbon cycle, and therefore use up all their food. It follows that their population *as a whole trophic level* is limited by food energy and must be set by competition for food. This is true, but quite uninteresting. All that it means is that there are in every ecosystem some ultimate scavengers that clean up after the rest and are limited by food. A few species of bacteria alone could competitively do the cleaning up. It is not true to say that all decomposers and scavengers are limited by food, that, because some among them must be, all flesh maggots, fungi, protozoans, vultures, and even bacteria are resource limited. Decomposers, like other organisms, are limited by weather, or things which eat them, or second-order effects of their own behavior, or sometimes, and perhaps even often, food.

The argument then considers plants, accepts the "green earth" argument as evidence that plants are seldom controlled by animals which eat them (Chapter 27), and concludes that the numbers of plants must be set by resources, by space, water, nutrients, and light. So both decomposers and primary producers (plants), *as whole trophic levels,* are thought to be resource limited. Herbivores come next. These cannot be resource limited, because their resources are plants and the claim has already been made, in the green earth argument, that plants are not limited by grazing. But something must keep the herbivores down. Of the two obvious candidates, weather and predators, Hairston, Smith, and Slobodkin prefer predators, because predators have been known to limit herbivores sometimes. So herbivores, *as a whole trophic level,* are limited by predators, and predators, *as a whole trophic level,* are limited by resources (the herbivores they can catch). We now have the comfortable statement that decomposers and plants are limited by resources, that herbivores are limited by things for which they are a resource, and that the highest trophic level, the predators, are again limited by resources.

These conclusions, of course, disregard much evidence that real animals and plants are controlled in ways quite unlike those assigned to them by the scheme. Many herbivorous insects have been shown to be controlled by their food after all (not by predators), as Harper's papers have reminded us and as Darwin noted in the *Origin.* Other insects are indeed controlled by weather. Larger herbivores, like muskrats, lemmings, and wildebeests, are often not controlled by predators either. And predatory animals, such as insectivorous birds, often have populations regulated as inescapable consequences of their behavior, not by some absolute measure of their food resources.

Probably the greatest flaw in the logic that led to these conclusions was the inference drawn from the fact that the earth is apparently always covered with plants. Hairston, Smith, and Slobodkin borrowed the essence of the green earth argument from Andrewartha and Birch (Chapter 27), actually saying that the earth was "green." But it is not green, it is brown. Even granting that the earth's browness yet provides for as much plant cover as simple limiting factors of its habitats allow, this does not mean, as Hairston, Smith, and Slobodkin assumed, and as Andrewartha and Birch assumed before them, that herbivores are unable to run out of food. It simply reflects the fact that much plant tissue is uneatable, and that the remaining parts which are eatable can be speedily replaced. The plant way of life is provided with effectively boundless energy. This gives the plants an unswampable repair force. The herbivores are like an army attacking a fortified enemy who has an unlimited supply of instant reserves. The total position cannot be won. The only effect the attackers have is to promote turnover in the defenders. But this turnover is enormously important for the regulation of life on earth. As one herbivore attack is pressed home, so the species of plant defender may be changed. The response may be swift, as in such games of hide-and-seek as are played across Australia between the prickly pear and the introduced moth which hunts it. Or it may be slower, as when seed and seedling predators deny space near a tropical tree to its own descendents. The many herbivores who live in such ways are limited by their food resources. And their quarries, doubtless contributing a significant portion of what greenness there is to the earth, are, strange to say, limited by a sort of predation.

It perhaps remains true that much terrestial vegetation is resource limited *as a whole trophic level* since the overall greenness of any part of the earth is commonly set by the water supply and other limiting factors. This is as uninteresting as the similar statement that decomposers are resource limited. But it is important to notice that there is no valid reason for saying that the *species populations* of plants are resource limited. And when applied to animal trophic levels the logic fails completely. Herbivores may be resource limited, as described above, or they may not. They may also be limited by predation, as are many tropical insects (Chapter 29) and as Hairston, Smith, and Slobodkin claim. Or they may be controlled by weather, or by losing babies like the Wildebeests, or in other ways. The accent on predation in the Hairston, Smith, and Slobodkin model was partly influenced by the Kaibab deer story, which they had no reason to doubt, and by the suspect belief that all irruptions of insect pests were the result of introductions of herbivores without their predators (Chapter 29).

It should be clear that animals of all trophic levels, and plants as well, may be limited in a great variety of ways, and that the claims for different mechanisms prevailing in different trophic levels are mistaken. The living beings of all trophic levels are broadly limited by available resources, but the sizes of actual species populations in each level may be set in many ways. The same family of checks is applied in every trophic level. It is important to stress this because the paper by Hairston, Smith, and Slobodkin (1960) putting forward the trophic-level idea has been widely reprinted in books of readings, and its appealing, if fallacious, logic followed by many students.

But if a general theory cannot be built up, trophic level by trophic level, what better tactics can we use instead? The answer probably lies in identifying places on the earth in which particular sorts of checks might be prevalent. In seasonal or unpredictable climates, accidents of weather are likely, providing fluctuating resources for all and minimizing controls which act precisely. In parts of the tropics with stable climates, checks by predation will be more important. The moving waters of the oceans force small sizes on plants allowing special relationships with the animals which eat them. And so on. We must begin with each environment, noting the kinds of destruction that it promotes and the life-styles it forces on the local inhabitants. The local agents of destruction and the local life-styles between them then decree what sorts of checks shall apply to populations.

Each kind of environment may perhaps best be thought of in terms of its special hazards. The hazards will not only be killers, invoking their own brutal controls, but they will also determine the most notable adaptations of the living beings of that time and place. These adaptations, in turn, will necessarily have population consequences. And the adaptations possible will also be some function of the kinds of things that live there, for an insect, with its short life cycle must adapt to environmental hazard in ways quite different from a long-lived vertebrate. Likewise, an insect predator, one of a numerous flying horde, searching in patterns of meticulous thoroughness, killing each victim found with virtual complete certainty, does not adapt to hazard, or represent the same hazard to its prey, as a stalking tiger, which may wander far without finding, which may often attack without being able to kill. The actual numbers of all animals of plants of a place must thus depend on the actual chances of disaster that make up the history of local physical conditions of life, and the types of animals and plants available to live there.

A realistic model of the numbers of living things must start with the local habitat, must describe the vagaries in it's structure and the vagaries of events. These qualities of habitats can be mimicked with a computer, treating the area as a grid and the passage of time as a series of slices, as Holling did in his models of predation. Any one kind of animal, any one engineering plan of animal that is to say, can only meet these vagaries of habitat in certain ways. These ways can be mimicked by hypothesis, model, and experimental test, again as Holling has shown us to do. The adaptations themselves will have second-order population consequences, as do the territorial adaptations of birds. These can be represented by additive terms in a computer model. Finally, each successive organism represents extra hazards to those already there, whether as predator, competitor, or blundering obstructionist. Each one of these effects of other animals will evoke counter adaptations, but only if such counter adaptations are engineering possibilities, and only if they are not ruled out by some politician's choice of the possible over the desirable which selection is forced to make. Following on these lines, it should be possible to make both conceptual and practical models of individual ecosystems which should successfully describe the numbers of animals and plants that live there.

The ultimate keys to understanding numbers are the hazards and the physical parameters that delimit and "design" ecosystems. The great air mass boundaries that divide the land surface of the earth into huge areal systems set different rules, each to its own section of the earth. Within each of these great formation-ecosystems different kinds of animals and plants live, or at least different proportions of the commonly available kinds; more reptiles and insects under tropical air masses, high proportions of migratory birds under temperate air masses, and so on. Animals and plants adapt to the conditions of air mass and latitude in the ways possible to them, and these adaptations have their own second-order effects. Within each formation-ecosystem are other ecosystems, the keys to whose understanding are again the local physical conditions of life. The actual numbers of plants and animals on earth is set by the range in the physical conditions of life on earth, and the extent to which it has been possible to adapt a limited variety of life-forms to meet the hazards of these physical conditions of life.

A second part of the understanding is perhaps that different phyla and different sizes of plants and animals are prone to be controlled in particular ways. Vertebrates adapt to hazard in ways that may have profound consequences to their own populations, but insects adapt less well, so that the first-order effects of death by accident, starvation, or being eaten are commoner for them. These phyletic differences are more important than differences of trophic level.

THE CONTROL OF MEN

Men once lived precarious lives as gatherers or top carnivores. They had to find food that wandered and, having found it, kill it. They lived like parties of lions, in wandering bands. They probably starved often enough, or lost their children through hunger, thus being largely food limited. Then they took to agriculture and became more numerous. Their populations grew because they were achieving cultural adaptations that enabled them to avoid hazard at the same time that their herbivorous tastes were granting them more energy. Now they could store grain to avoid winter starvation, and irrigate lands to avoid the hazards of drought. These adaptations did not generally hold second-order consequences that should restrict populations, and their populations went on growing.

But a number of human cultural adaptations did have some population effects. Men took to living in cities, which was adaptive because it helped organize agrarian societies and trade. But city life promoted the spread of disease, and this surely slowed the overall growth of population. As birds tolerate the population effects of territories, so men tolerated the population effects of cities. For both, the adaptive advantage of the behavior outweighed the resulting loss of population.

Some societies of men engaged in vendettas, in tribal warfare, and in headhunting. The advantage of these traits to the societies who had them were probably social cohesion and preserving space for the tribe. These advantages alone should be sufficient for survival of the traits. But tribal warfare and headhunting obviously have the second-order effects of slowing population growth.

And some human societies seem to have developed awareness that their numbers must be kept down for the common good. They often killed some of their babies, particularly female babies. And many more societies seem to have had a semiconscious understanding of the needs for birth control. They made difficulties for young lovers. They engineered elaborate, and sometimes fatal, ordeals that must be undergone at puberty or before marriage. They practiced dangerous circumcisions. Some populations may well have kept their numbers roughly constant for many successive human generations by such means.

These various traits, of tribal warfare, infanticide, and barbarous circumcision have mostly been abolished. And good riddance. Now we have even abolished some of the epidemic diseases that were the second-order consequence of living in cities. And good riddance to them too. But our population, accordingly, is growing fast. If we elect to allow our population to go on growing, as all the nations of the earth have so far chosen, the very best we can hope for is an equilibrium set by the food supply. There would then be a society of the greatest misery for the greatest number, with every wretched couple rebelliously raising the two and a bit children of our replacement rate. But I doubt if anyone who has thought carefully about the implications of crowding people more and more believes that modern civilizations can survive the process until the food-supply limit is reached. A more likely future than the food-supply equilibrium is war and the disintegration of order. Once order and law disappear before the pressures of wretched crowds, the ancient restraints can return. Famine, pestilence, and human brutality will then reimpose their traditional checks.

But more likely still is a world of mixed nations, some returned to war and barbarism and others who have taken the decision of the two-child family, but too late. Anarchy might then confront regimentation. The nations who accept restraints at the eleventh hour will avoid anarchy at the price of disciplining their people to accept small shares of what their country provides. Free societies are then replaced by communes in which the ways of life of all are narrowly restricted by law.

PART 4

AN
EVOLUTIONARY
VIEW OF THE
PATTERN OF
LIFE

One of the outcomes of evolution which an ecologist must explain is that some kinds of animal and plant are common whereas other kinds are rare. That some species are, indeed, much commoner than others has been amply confirmed by census of breeding birds, collecting moths at light traps, counting diatoms that fall on glass slides, and so on. Moreover these census data reveal that there is a pattern to commonness and rarity, for relative abundance seems to follow the mathematical form known as a log-normal distribution. Many data in the botanical literature describing the frequency with which different kinds of plants occur in quadrat samples are also explained if plants are distributed log-normally. Log-normal distributions are known to statisticians to result when a number of random processes are superimposed. The log-normal distribution of species which results in commonness and rarity presumably reflects random qualities in the environment in which the species have evolved. Because the environmental stage is made up of different patches in random array, and because all these patches are subject to accidents that may also be random, the evolutionary play has resulted in the log-normal distribution of commonness and rarity which we see.

509

CHAPTER 37

Some species are common and some species are rare. This is another of those familiar observations of naturalists which is so well-known as to seem unremarkable. But it is a little odd, all the same. Why should some species be so much more common than others? For there is no doubt that, in any one place, a few things are common whereas many more are rare. Specialists in every group of animals and plants the world over have compiled long species lists, noting as they did so the relative abundance of each species, or actually making counts of individuals. And always they have found that a few common species make up the bulk of the collections whereas many more are caught or seen only once or twice. Why should this be?

An analysis of species abundance and diversity has to begin with data on samples. We have accounts of how many birds were seen in a day, or of how many could be counted in a small plot of land; of how many plants there were in 25 one-meter quadrats; of how many insects were caught in a trap set for so many days or weeks or years, and so on. Such data always reveal very clearly that the common things are indeed common and most of the rest disproportionately rare. A light trap was set for moths in Maine and kept running for 4 years (Dirks, 1937). Every moth that came was killed, identified, and counted, giving a final grand total of 56,131 specimens. More than 40,000 of these belonged to just six species, but the total number of species which came to the trap in those 4 years was 349. Forty-two of these species sent only a single individual into the trap. This suggests that common moths may be at least 10,000 times as common as rare moths, truly a startling difference that needs to be explained. But there must be, at the outset, some worry about how these data were collected. A light trap does not sample the population at random; what it samples is just those moths that fly to light traps, and even these presumably in proportion to their predisposition to fly to lights. Perhaps the moths called "common" in that Maine survey were 10,000 times more attracted to light than those called "rare," and not really commoner at all. This sort of objection can be made to all sample data short of absolute counts of every last living individual, which is always impracticable. But it loses its force when the same sorts of results are obtained by varieties of census techniques. Any generalization that can come from a study of many different kinds of animal or plant, sampled by many different techniques, must reflect some underlying pattern of diversity in nature. To understand the one grand generalization which it has been possible to make it is convenient to work with an example.

In 1936 A. A. Saunders published a census of the birds that bred in Quaker Run Valley, of the Allegany State Park in New York. A total of 141 species had been reported from the park at one time or another, although this number certainly included some which were chance visitors that could not be expected to breed. Saunders found that 80 species were breeding when he made his census. He estimated the numbers of nesting pairs of each kind, mainly by the device of counting the singing

Numbers of Birds of Each Species in Quaker Run Valley					TABLE 37.1
1670	1656	1196	868	723	723
675	506	477	389	367	324
311	310	288	282	280	270
220	188	181	179	161	160
158	152	138	111	109	91
90	88	79	60	57	56
50	46	46	43	43	35
34	33	32	32	30	28
28	26	24	23	22	17
15	14	12	10	10	10
10	10	8	8	7	6
6	5	5	4	4	4
4	3	3	3	3	2
1	1	?	?	?	?
?	?	?	?	?	?
?					

males when they made themselves conspicuous. The commonest species had 1670 pairs in the valley, but the tenth commonest had only 389 pairs, and the numbers tailed off sharply after that as can be seen in Table 37.1. This is the familiar pattern of the common being very common while most others are relatively rare.

If you make a plot of species rank against number of individuals, as is done in Figure 37.1 (curve A), you get a sharply descending curve suggesting an exponential function. But the function is not simple, since if you convert the numbers of individuals to a logarithmic scale you get not a straight line but curve B. There might not be anything particularly remarkable in this distribution if it were found only in the birds of the Quaker Run Valley, but in fact the form of the distribution seems to be widespread. Other bird counts yield distributions of commonness and rarity which on a log scale are of the form of curve B, and so do the counts of moths coming to light traps and census data of other insects. The bird and insect counts form the largest bodies of census data on which we can draw, but there are counts of a few other groups of organisms that can be used in the same way. A species-abundance plot of diatoms settling on a glass slide suspended in a stream yields a curve of the form of curve B, for instance. So it seems that this type of distribution of rarity is widespread in nature. If we are to explain it we must first have some formal statement of it, an obvious need that led to a number of attempts to fit the curves with equations for modified harmonic series (Fisher et al., 1943). But the real mathematical distribution that these data reflected was revealed when F. W. Preston (1948), a consulting engineer by profession, turned his ingenious mind to the matter.

Preston started from the observation that commonness and rarity are relative terms. A bird species with 50 nests in Quaker Run Valley only seems uncommon because other species have more than 1000 nests there. It is natural to talk about one species being twice as rare, or four

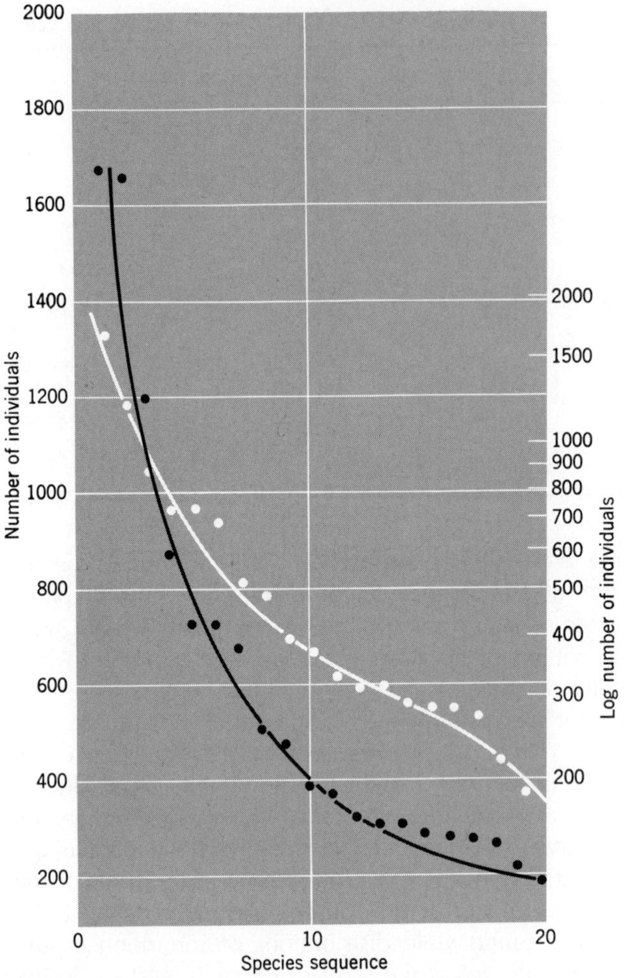

Figure 37.1

Abundance of the 20 commonest bird species at Quaker Run Valley. Curve A is a plot of abundance on a linear scale; curve B, a plot of abundance of the same 20 species on a log scale. Data are those of the 20 commonest species in Table 37.1. Census data of large samples of many kinds of organisms show that such distributions of abundance are widespread. These distributions are log-normal.

times as rare as another; very well! let us order the data in groups of relative abundance. We start with a simple hypothetical series as shown in Table 37.2. Species with four individuals appearing in the census are twice as common as those with only two individuals, which is to say that species of group C are twice as frequent as group B, and so on. The sequence of frequencies can be viewed as a sequence of octaves, which are bounded by the numbers 2, 4, 8, and the like. It is possible to sort real census data into these octaves. If there are 9, 10, 11, 12, 13, 14, or 15

TABLE 37.2

Hypothetical Census Data Ranked by Relative Abundance

Species group	A	B	C	D	E	F	G	H	I	J	K	L
Approximate specimens observed of that species	1	2	4	8	16	32	64	128	256	512	1024	2048

individuals, they belong to octave D. If there are 374, they belong to octave I and so on. Any collection that falls on a line is halved and shared between the octaves on either side. The procedure is well illustrated by Preston's treatment of the Quaker Run Valley bird census (Table 37.3). There are clearly only a few species in the octave at the beginning of the sequence and few species in the octave at the end. Most are in the middle octaves, so that if we plot species per octave against an octave scale we

TABLE 37.3

Species Abundance of Quaker Run Valley Birds, Ranked into Octaves

	<1	1 to 2	2 to 4	4 to 8	8 to 16	16 to 32	32 to 64	64 to 128	128 to 256	256 to 512	512 to 1024	1024 to 2048
Octave												
Species per octave	>1	$1\frac{1}{2}$	$6\frac{1}{2}$	8	9	9	12	6	9	11	4	3

get a humpshaped curve (Figure 37.2). Such a hump-shaped curve is Gaussian, or what is usually referred to as a "normal" shaped curve. But this is not a normal distribution as it is known in statistics because it appears only when plotted on a logarithmic base; for Preston's octave series is the same thing as a logarithmic series to the base 2. The frequencies, in fact, follow the distribution that statisticians have long known as **log-normal.** Preston had thus shown that the frequency of occurrence, which is to say the commonness, of the Quaker Run Valley birds followed a definite, ordered, mathematical distribution. This was a far-reaching conclusion, suggesting that some general process was ordering the relative abundance of bird species.

But what of other kinds of animals? Preston was quickly able to show that the abundance of other species for which good census data were available was also log-normally distributed (Figure 37.3). The pattern was widespread. Perhaps particularly striking was an estimate of the entire bird fauna of North America suggesting that species abundances were log-normally distributed. Since Preston's work appeared, Ruth Patrick (1954) has shown that the species abundance of diatoms caught on glass slides suspended in water is also log-normally distributed, and Whittaker (1965) has shown the same for plants of the Sonoran Desert, suggesting that the distribution can apply to plants as well. But perhaps what was

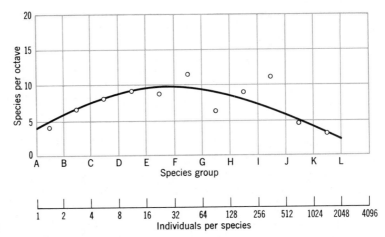

Figure 37.2

Species abundance of Quaker Run Valley birds ranked into octaves. Plot of the census data as organized in Table 37.3. The result is a hump-shaped or Gaussian curve. This is the log-normal distribution which underlies the characteristic shape of curve B in Figure 37.1 (From Preston, 1948.)

even more telling for the general prevalence of the distribution was the way in which it explained a long-standing botanical puzzle, the Raunkiaer Law of Frequency.

Ranukiaer was one of the pioneer plant ecologists of the Scandinavian tradition, a man who sought to describe vegetation from the data of counts in random quadrats. One of the things that he specially studied was the **frequency** with which species occurred in quadrat samples laid out in what was considered a single community of plants. If he threw 25 hoops in one piece of vegetation, and species A was ringed by only five of them, he would say that species A had a frequency of $5/25 \times 100 = 20$ percent, if it occurred in only one hoop then it was 4 percent frequent, if in all of them, 100 percent, and so on. Raunkiaer's **frequency** was a measure of abundance of sorts, but it measured abundance relative to sample size rather than to other species as do the more generally used terms "common" and "rare." Raunkiaer found that a few plants were very "frequent" but that many more kinds were infrequent, as we should expect. Then he divided his species into frequency-classes, not logarithmically into octaves as Preston was to do, but into five equal percentage groups. These would have been 0 to 20 percent, 21 to 40 percent, 41 to 60 percent, 61 to 80 percent, and 81 to 100 percent if he had based his data on 100 quadrat samples. But he standardized on 25 quadrats. For a plant to be 22 percent present in a sample of 25 quadrats it would have to occur in 5 1/2 of them, which is obviously impossible. So Raunkiaer divided his groups at the impossible fractions, making the five groups A, 2 to 22 percent; B, 22 to 42 percent; C, 42 to 62 percent; D, 62 to 82 percent; E, 82 to 100 percent. Raunkiaer found that *if you chose the size of your quadrats correctly* group A had more species than B, which had

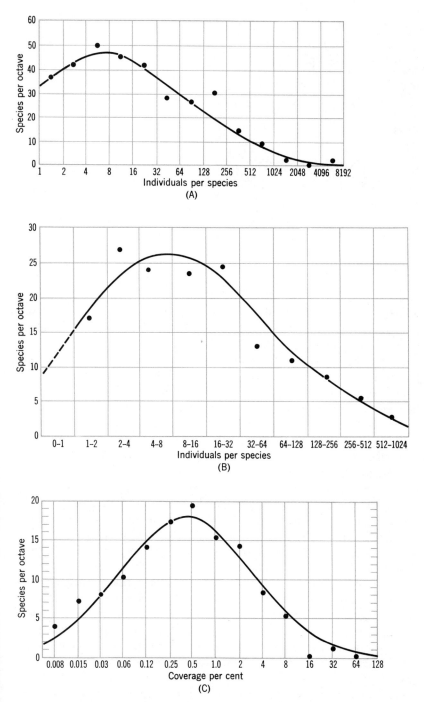

Figure 37.3

Some log-normal distributions. (A) Dirks' data for moths that came to a light trap in Maine (from Preston, 1948). (B) Data for diatoms settling on a glass slide in a stream in Pennsylvania (from Patrick 1954). (C) Data for plants in the Sonoran Desert (from Whittaker, 1965). The curves are all humped, but the left-hand side may apparently be missing, suggesting that rare species were too rare to be included in the sample.

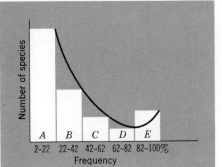

Figure 37.4

Raunkiaer's Frequency Distribution.

more than C, which had more than D, which had *less* than E, or

$$A > B > C > D < E$$

which may be graphed to yield a curve like a back-to-front "J" as shown in Figure 37.4. When Raunkiaer was working more than half a century ago, it was still proper to describe your discoveries as "laws," and this result became known as **Raunkiaer's Law of Frequency.** It was quickly found that it applied to every layer of every piece of vegetation in which it was tried. Demonstrating the frequency distribution, with its "J" curve, became standard in plant descriptive work, and there are enough published demonstrations of the law to leave no doubt about its generality. But why should group E always have more species than group D? It seemed a very odd thing. Those who pondered the matter critically early realized that the answer must have something to do with sample size for, to demonstrate a Raunkiaer frequency distribution, you had to *choose the right quadrat size.* It was left to Preston half a century later to show that if you took the right proportion of plants whose relative abundance was distributed log-normally, the Rankiaer distribution followed automatically. Raunkiaer, without knowing it, had discovered a property of log-normal distributions. A formal proof of this is given in Preston's paper. But the importance of Raunkiaer's work in retrospect is the support it gives for the Preston hypothesis that very many species of plants and animals in nature are log-normally distributed. Apparently the log-normal distribution describes not only birds at census and moths at light traps, but plants in the communities examined by European phytosociologists as well. Whatever process leads to the log-normal distribution seems to be very general.

We started with a question: Why should some species be common and others rare?, which seemed a formidable enough question by itself. Now Preston had presented us with a complication that seemed stranger still; Why should there be an apparent mathematical relationship between common species and rare species so that their relative abundance follows a log-normal distribution?

Log-normal distributions were known to statisticians before Preston alerted biologists to their presence in biological data. They are typically the probable outcome of superimposing a number of random processes on each other so that they undergo "random walks" (Feller, 1951). But must the relative abundance of animals, then, be merely a random thing? For opportunist species, at least, it clearly can. Both the Spanish ecologist Ramon Margalef (1957), and Robert MacArthur (1960), have pointed out that the numbers of opportunists, and hence their relative abundance, must be random. The local population of each opportunist is constantly changing as it takes advantage of temporary habitats, or as it escapes for a while from the pressure of some predator. The fortunes of the many opportunists in nature must fluctuate largely independently of each other. At any one time their relative abundance should be the accumulated integrals of their several rates of increase, and these accumulated integrals should undergo the random walks that lead to log-normal distributions. Preston's findings would thus be fully explained if all species were opportunists whose populations fluctuated widely with environmental accident. But most students of population dynamics would be reluctant to believe that all species are opportunists. All the evidence which shows that some animals maintain population equilibria is against this conclusion. How, then, can the existence of equilibrium species be reconciled with the omnipresent log-normal distributions?

Part of the answer to this riddle may lie in the patchiness of the environment. Equilibrium species may well maintain their equilibria, and determine their relative abundance, by processes which are far from random within a uniform physical space. But if adjacent patches of the environment are different, the animals there might be expected to maintain populations of different sizes and to have different relative abundances. If the census sample is large relative to the size of environmental patches, it will embrace a mixture of patches, each with its different sets of relative abundance. The range of conditions from one patch to the next should vary as much at random as the conditions which determine the sizes of opportunistic populations. The accumulated effects should be the same, and a log-normal distribution should be produced by the random-walk process. The log-normal distribution is thus consistent with there being equilibrium species present, if the sample size is large relative to the size of environmental patches. Casting an eye back on Preston's data at once reveals that his samples did tend to be large. A light trap must collect moths from a very large area. The Quaker Run Valley with its thousands of birds obviously covers many different habitats. And the most spectacular of all Preston's results, the fitting of the distribution to the estimated abundances of all the bird species of North America, becomes very clearly explained indeed.

If log-normal distributions are sometimes dependent on lumping many different microhabitats in the sample, we should expect them to disappear if we adjust sample size to habitat size. R. H. Whittaker (1965) has shown that the Preston hypothesis fails for plant communities in which there is strong dominance, and low diversity (which is to say not many

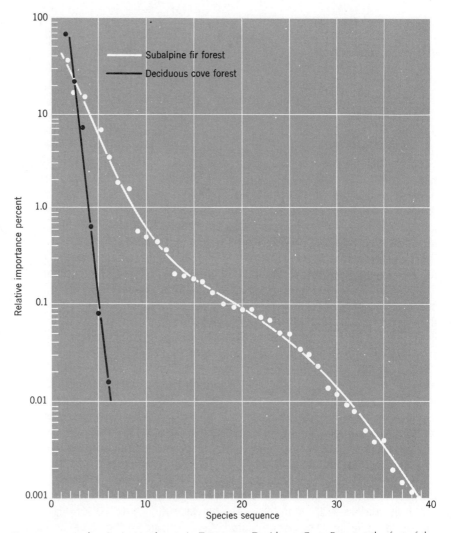

Figure 37.5

Commonness and rarity in two forests in Tennessee. Deciduous Cove Forest at the feet of the Great Smoky Mountains has very many species, being a forest where dominance is hard to detect, like the forests of the tropics. The species importance plot on a log scale looks like that of the Quaker Run Valley birds (Figure 37.1, curve B). The relative abundance of these forest species is log-normally distributed. But the subalpine fir forest, high in the mountains, has few species and strong dominance. The plot of species importance of this forest on a log scale is a straight line. The phenomenon of ecological dominance imposes a special and nonrandom relationship between commonness and rarity in such communities. Notice that relative abundance is plotted as relative importance, which reflects a way of weighting the data to remove the effects of plants' being of different sizes. (From Whittaker, 1970.)

species with one or two of them strongly dominant over the rest). He chose a subalpine fir forest in the Great Smoky Mountains of Tennessee as fitting these requirements, and expressed the relative importance of plants in the community by measuring their net production. Expressing their importance in this way got round that common botanical difficulty of the widely different size of individuals. The relative importance of these plants was not log-normally distributed, but the relative importance of plants in a much richer forest in the foothills of the same mountains was log-normally distributed. The difference is shown in Figure 37.5. The curve for the subalpine forest is a simple logarithmic function, quite different from the curve for the richer forest, whose data would reveal the familiar Gaussian hump if plotted in octaves. This finding is quite consistent with the explanations offered for the log-normal distributions that do occur, for there is nothing random about the establishment of dominance.

But for samples of large areas, some varient on the log-normal distribution does seem to hold good for all communities except those subjected to ecological dominance. It suggests that the general explanation for the commonness and rarity of species lies in the unpredictability of the environment. To living things the environment fluctuates in ways that are hostile, both in time and space. Some species, the opportunists, meet the fluctuations in time merely by fluctuating themselves. By the random-walk process their relative abundance becomes log-normal. Other species can endure the fluctuations through time well enough to establish equilibrium populations. But their overall abundance is set by the fluctuations of the environment through space. Their relative numbers differ from patch to patch, and the products of all these local abundances results in a log-normal distribution also. The answer to the question, "Why are some species common and others rare?" must be, "Because there are many parameters of the environment which are random." The relative numbers of the different kinds of living things directly results from the irregular structure of the physical world, an irregular structure that greets the organisms which endure it as a set of random hostilities.

There is one last intriguing thing about the log-normal distributions, and this is revealed by a study of their mathematical descriptions. The log-normal distribution may be defined as follows:

$$n = n_0 e^{-(aR)^2}$$

where

n_0 is the number of species in the modal octave
n is the number in an octave distant R octaves from the mode
a is an empirical constant which can be calculated from the evidence

The constant a has been calculated for all of Preston's examples, and for other data that have been obtained since. G. E. Hutchinson (1953) pointed out in one of his most famous essays that this constant always seems to have a value close to 0.2. The value itself does not have any direct meaning for us, but it does seem extraordinary that the constant should have the same value no matter the size or reproductive capacity of the organism, whether we are dealing with diatoms, moths, or birds. Perhaps it bears some subtle relationship to the range of variation in the earth's environment, a range set by fundamental properties of the solar system or the elements in the periodic table. But we do not yet know what the value of constant a really does mean.

Within small patches of the environment the relative abundance of equilibrium species need not be log-normal. MacArthur tried to predict the relationship between commonness and rarity of similar sorts of animals living together in the same place. He assumed that the species should have been separated by character displacement, and thus should occupy discrete niches which let them avoid competing. He further supposed that of all the parameters making up the niche hypervolume only one was critical in apportioning the resources of the habitat between the several species. The result was the celebrated broken-stick model. Census data from birds and other animals seemed to show that the relative abundance of many species over small areas was, indeed, as predicted by the model. For a time it seemed to some ecologists that the underlying assumptions of the model, discrete niches through character displacement and critical single niche parameters of overriding importance, would be shown to be generally valid for very many sorts of living things, but it has since been shown that the distribution of commonness and rarity required by the broken-stick model may result from many other sorts of interactions between species, not all of them biologically probable. It is perhaps true to say that the only general statement we can make about commonness and rarity is that it has been guided by random processes.

THE HYPERVOLUME
AND THE BROKEN
STICK

521

CHAPTER 38

THE HYPERVOLUME
AND THE BROKEN
STICK

Preston's way, and the way of those who had gone before him, of seeking to explain the abundance of species had been to try to match observed distributions to curves and equations. When finally a good fit was found, they looked around for some biological meaning in the equation. This approach did produce biological enlightenment, because it revealed the overseeing role of environmental diversity in determining biological diversity. But a more satisfying approach for a biologist is that of starting with known characteristics of living things and then trying to predict their consequences. We want to use mathematics not just to fit biological data, but to predict biological data. It should be possible to predict the relative abundance of species from models based on properties of the species themselves. We make a model, state the hypothesis that the model correctly describes how living things must act on one another, and test the hypothesis by comparing predicted distributions with observed distributions. This is the essence of the scientific method. Its application to ecology by Gause led to some of our most trenchant understandings. It should be possible to follow Gause's example, and to build a model based on the same fundamental concept of niche which should explain the diverse abundance of species, just as the Gause model had explained the existence of species themselves. This, Robert MacArthur, then of Yale University, was the first to attempt.

MacArthur (1957) began his attempt with a study of bird diversity, because many birds seem to be equilibrium species whose numbers vary little from year to year and whose density may be fairly uniform over large areas. He was able to support this belief with the conclusions of David Lack and with his own studies on American warblers, both described earlier in this book. The populations of such equilibrium bird species over comparatively restricted areas might be immune to the sort of random-walk adjustments which apparently accounted for the log-normal distributions found by Preston for such large samples as the Quaker Run Valley. A different distribution of abundance ought to prevail for smaller samples, and this should be predictable from the theory of the niche.

Species divide up their world into niches, so the problem of the abundance of species becomes one of the relative abundance of niches. But how can you measure a niche so that you can divide the bird's world into niche spaces? Help with this problem came when G. E. Hutchinson produced a formal definition of the niche. He called it "an *n* dimensional hypervolume." There were very many parameters to the niche of any species, as had long been recognized, all the manifold influences that together made up its "profession" (using Elton's analogy) being parameters that could theoretically be measured. Calling the niche a "hypervolume" allowed each of these parameters to be treated individually so that mathematical expressions for the niche were possible. Furthermore, the prevailing idea of "limiting factors" suggested that only one parameter was usually crucial in the competition between a pair of equilibrium species. A given resource might be so divided by a number of species as to determine their niche volumes, and thereby their relative abundance. MacArthur felt that he could choose just one axis of the hypervolume, the one crucial for interspecific competition, and proceed to consider how this one crucial axis could be divided.

The critical part of the environment was considered by MacArthur as being represented by a line, which could be divided into segments representing the portions allotted to the several species, and thus their relative abundance. The phenomena of exclusion and character displacement suggest to us that the separation between niches ought to be discrete so that dividing the line into a number of discrete lengths to represent niches seems a biologically sound proceeding, although not the only conceivable way because the niche spaces might overlap. But MacArthur started with the assumption that niches were discrete. The line could now be thought of as a stick of niches joined end to end. The problem was to break the stick into a series of niche lengths which could be used to predict a series of abundances of birds in nature. MacArthur postulated that dividing up the niche space was a random process. $n - 1$ points are thrown at random onto the stick, and the stick is then broken at each of these points. There are then n segments of stick, the lengths of which MacArthur postulated were equivalent to the abundance of n species of birds. The expected length of the rth shortest interval is given by

$$\frac{1}{n} \sum_{i=1}^{r} \left(\frac{1}{n-i+1} \right)$$

so that the expected abundance (I_r) of the rth rarest species among n species and m individuals is given by

$$I_r = \frac{m}{n} \sum_{i=1}^{r} \left(\frac{1}{n-i+1} \right)$$

The derivation of these equations is not simple, and nonmathematical biologists may be comforted by the knowledge that there is more than one mathematical paper in existence about them (Feller, 1951). But, accepting the math, they do provide a model that purports to predict the relative abundance of equilibrium bird species and that can be tested against field data. MacArthur (1960) proceeded to test the model against the census data from the Quaker Run Valley (Chapter 37).

There were 80 bird species recorded as nesting in the valley, comprising a total of 14,594 individual birds. Using the equation of the broken-stick model it was possible to calculate how many of each species there ought to be. MacArthur did this for a number of species, finding that the abundance of the birds fell along a curve of the form A in Figure 38.1. Notice that this is not a straight line, although it looks nearly straight. MacArthur then plotted the actual abundances of the Quaker Run birds as recorded in the census and got curve B of Figure 38.1. The curves are clearly quite different, so that the model does not correctly predict the relative abundance of the 80 species of birds nesting in the Quaker Run Valley. But this was expected because Preston had already shown that

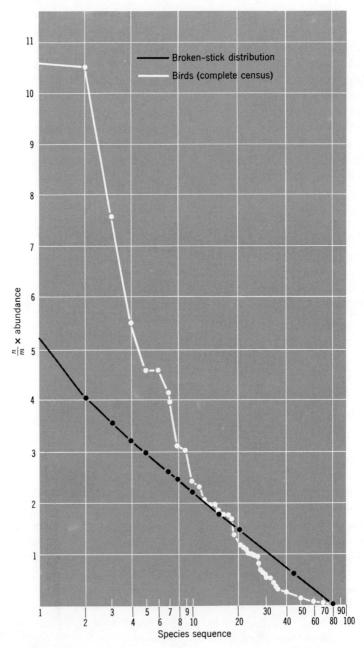

Figure 38.1

The broken stick model and the birds of Quaker Run Valley. The heavy black curve shows the distribution of relative abundance which would be expected if the birds in the valley had divided all its resources for bird life into discrete non-overlapping niches. But the actual distribution of relative abundance is log-normal (Chapter 37), and this distribution does not match the predictions of the broken stick model. Broken stick distributions should only be looked for in populations of equilibrium species occupying small uniform habitats. n = number of species in sample; m = total number of individuals of all species. (From MacArthur 1960.)

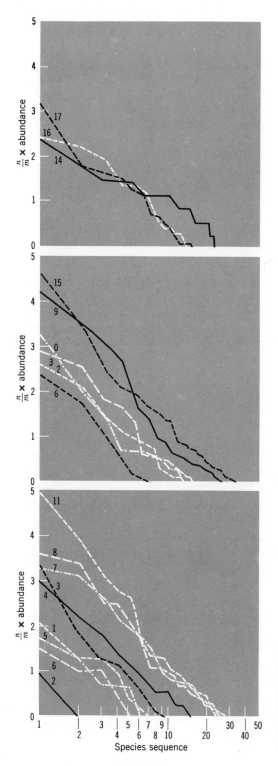

Figure 38.2
Relative abundance of birds in 17 small habitats in Quaker
Run Valley. Each of these curves of abundance looks roughly
parallel to the almost straight line of the broken stick distribution shown in Figure 38.1. It looked as if the broken stick
model survived the test of predicting the distribution of relative abundance of equilibrium species in small habitats, even
though the sum of all these distributions is log-normal. (From
MacArthur, 1960.)

their abundance was log-normally distributed. Even though many of the valley birds might have been equilibrium species whose densities were not set at random, the valley was so large that many different patches were included. Any broken-stick distributions should be masked.

To test adequately the predictions of the broken-stick model we must be able to compare it with census data from a small uniform habitat to which its assumptions could apply. Fortunately this could still be done for the Quaker Run Valley birds, because Saunders had divided his original census data by habitat. There were five uniform stands of virgin timber in the valley as well as various agricultural habitats, and Saunders had counted the birds nesting in each separately. MacArthur plotted the abundance of birds in each of these subcensuses against his ranked species scale, producing the series of graphs shown in Figure 38.2. A glance at these shows that they look like the predicted curve of *A* in Figure 38.1; they seem to have about the same angles to the axes of the graph, and look roughly "parallel" to that almost straight line. They do give the impression that the model works, and if the model works then its assumptions must be sound. And if its assumptions must be sound, then surely it follows that equilibrium species divide up their world into discrete separate niches; which is what many ecological theorists had long wanted them to do anyway. The MacArthur brokenstick model was thus a satisfying thing for ecologists to read about. By following the classical scientific method, ecology seemed to be grasping a new level in its quest for an understanding of natural diversity.

Faith in the broken stick model was enhanced by the failure of attempts to make a successful model of diversity on the assumption that niches were not discrete. If niches overlapped freely and without affecting each others' extents, it should not be so easy to decide where to break the stick, and a more complex mathematical model was required. MacArthur's first attempt to make one yielded species abundance curves that were quite unlike those yielded by his bird census data, suggesting that an overlapping niche model was not the correct one. It later turned out that an overlapping niche model gave even more unreal predictions than MacArthur had thought, for there was an error in his published mathematics (Pielou and Arnason, 1966). The correct curve was even more unlike the census data than MacArthur's. It took 6 years for MacArthur's mathematical error to be discovered, even though his paper was one of the most talked about in biology; which may be taken as a measure of the subtlety of the mathematics involved. In the event, the demonstration of the error served to increase confidence in the original conclusion that discrete nonoverlapping niches was indeed the true state of affairs.

Attempts were soon made to test the broken-stick model against abundance curves for other groups of organisms, and there developed a small literature on the subject. Generally it was found (King, 1964) that you could produce curves for which there was a good "eyeball" fit with the predicted curve of the model if, but only if:

1 The life cycles of the organisms were synchronized.

2 The organisms were fairly closely related, of similar size and form, and
living in similar ways.
3 The area sampled was small and uniform.

These were conclusions that should have surprised nobody because they
were the assumptions on which the broken-stick model was based. The
organisms must be expected to compete, so they must be active at the
same times; competing required similar niches, and such should be occupied by closely related species; and you had to have a small uniform
habitat, or you ended up with the summed effects that led to Preston's
log-normal distribution. But within these restrictions, the model did seem
to work, and our feeling that we had some nebulous grasp on how natural
communities were organized into the activities of different species was
encouraged.

But there were some strong hints of trouble with the model also. N. G.
Hairston (1959) showed that you could make curves of abundance of
even apparently opportunistic species like soil arthropods look roughly
like calculated broken-stick curves if you found the right sample size. Too
big a sample gave you a different shaped curve, so did too small a
sample, but a properly chosen sample size yielded an abundance curve
which looked tolerably like that of the broken-stick model. There is an
echo of the method used to get the Raunkiaer frequency distributions in
this, a suggestion that we are really dealing with some curious statistical
property of an underlying distribution that we do not yet understand.
Added to these doubts was the apparent inapplicability of the model to
plant communities. Whittaker (1965) found, not unexpectedly, that the
phenomenon of dominance produced abundance distributions that were
quite unique to plants, and in no sense consistent with the broken-stick
model. Where dominance was not strong, measures of species importance by productivity or cover measurements (Chapter 37) yielded log-
normal distributions.

But then there appeared what looked like a triumphant vindication of
the broken-stick model. If a community of closely related equilibrium
species was to arrive at the predicted distribution of abundance by
random process, the species clearly must take their time about it. There
should be an interval during the youth of a community when equilibria
were not established and broken-stick distributions should not be found.
But as the community got older, then the relative abundance should
approach closer to equilibrium, and the curves should come closer to that
almost straight line predicted by the broken-stick model. We needed a
history of the changing abundance of members of an animal community
through time, and it was found for us in the record of lake mud
microfossils by Clyde Goulden (1966). Headshields, postabdominal
claws, and other parts of cladocera are preserved in lake muds by the
thousand, and may usually be identified to genus or species. Goulden
took a boring through the sediments of a postglacial lake, extracted the
cladoceran microfossils at successive intervals through the mud column,

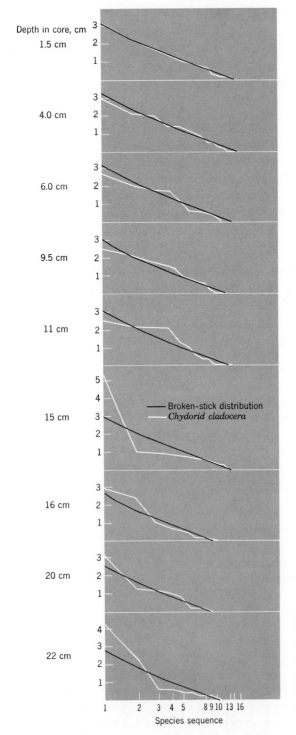

Depth in core, cm

1.5 cm

4.0 cm

6.0 cm

9.5 cm

11 cm

—— Broken-stick distribution
—— *Chydorid cladocera*

15 cm

16 cm

20 cm

22 cm

Species sequence

Figure 38.3
The development of a broken stick distribution through time. A core of sediments from a Guatemalan lake contained remains of chydorid Cladocera, providing a series of censuses of the lake's chydorid inhabitants through time. Notice that the number of species increased progressively from 9 to 16 and that the distribution of relative abundance approached that predicted by the broken stick model as the community became more complex. (From Goulden, 1966.)

worked out the relative abundance of the taxa at each interval, and compared each with the distribution predicted by the broken-stick model. The results (Figure 38.3) seem to be a striking vindication of the model. In the early history of the lake, the abundance curves did not fit very well to the predicted curve. But they got successively closer, until at last the fit was good. And once the fit was good, it stayed good at successive intervals. In Goulden's data we seem to see the establishment of equilibrium numbers through random process, and by closely related organisms in a confined uniform space (the lake) taking place before our eyes.

There was reassurance in Goulden's findings, and these were soon amplified by a similar lake study by Tsukada (1967). There had been two catastrophes in the history of Tsukada's lake, volcanic eruptions which had dumped great quantities of volcanic ash into the water. These catastrophes might be expected to disrupt equilibria, perhaps even exterminating some of the lake's inhabitants, and starting the random process groping toward equilibrium once again. And Tsukada found that agreement of fit between observed and predicted abundances did, indeed, disappear in samples immediately overlying each layer of ash, but that goodness of fit was slowly reestablished in the equable years that followed each catastrophe. Tsukada had, in Deevey's (1969) words, "coaxed history into making an experiment" of disturbing an equilibrium system so that he could witness the process by which a new equilibrium was attained. (Figure 38.4)

But more powerful objections to the broken-stick model were being unearthed to undermine this reassurance. Joel Cohen (1966, 1968) discovered that quite different sets of assumptions could lead to the selfsame mathematical equation that MacArthur had derived from his broken stick.

Cohen's first approach was to consider the small area to be studied as a set of a large number of subniches. There would be many species there, each of which occupied a number of subniches. It should commonly happen that two or more species should share some subniches, but each species would nevertheless have a unique combination of subniches which should make itself unique and allow it to conform to the Gause principle of one species, one niche. This seems a more realistic way of defining the differences between species than that of considering only one critical parameter, one axis of the hypervolume, as was done for the broken-stick model. Cohen likens his subniches to a set of bottomless buckets of unlimited capacity. Species are represented by different colored balls, and the habitat is then filled with species by a set of players (the species) taking turns to throw their balls so that they fall into one of the buckets at random. The rules of the game are rigged by the referee (simulating nature) so that no two ball colors can end up occupying the same set of buckets (which is to say the same niche), nor can any two colors occupy exactly the same number of buckets. The referee goes on allowing turns to fresh players until the last empty bucket receives a ball. The game is now ended, for every subniche is occupied and some sort of equilibrium is established in the habitat. The number of balls of each species is its abundance. This model of filling the habitat by random im-

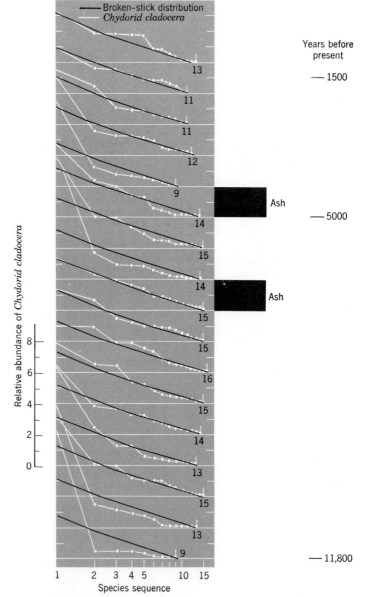

Figure 38.4

The broken stick distribution reestablished following catastrophes. Conditions for life in a Japanese lake were drastically changed by falls of volcanic ash twice during the last 12,000 years. Before the first fall of ash the relative abundance of chydorid Cladocera had slowly changed through several thousand years until it was closely similar to that predicted by the broken stick model, a history which was revealed by analysis of fossil parts of the animals. The distribution of abundance was drastically different after a layer of volcanic ash settled through the water, and it began to approach the broken stick distribution again when a second ash fall was accompanied by a repeat of the history. (After Tsukada, 1967.)

migration into subniches, a species at a time, is at least as realistic as the broken-stick model. Cohen even shows how it can explain some difficult ecological phenomena in ways that the broken stick cannot. And yet he also shows that, if there are n boxes the abundance of the rth species will be given by

$$I_r = n^{-1} \sum_{i=1}^{r} (n + 1 - i)^{-1}$$

which is also the equation of the broken-stick model.

But there was worse to follow yet. Cohen (1968) next postulated something much more unsatisfactory for an ecologist: that abundance had nothing to do with the discreteness or overlapping of niches; indeed, that abundance had nothing to do with the presence of other species at all. Then he postulated that all of n species underwent exponential population growth that was identical but independent of all the others, and that the growth of each population went on until some arbitrary time. The abundance of the rth rarest species after taking part in such an un-biological mess was given by

$$I_r = n^{-1} \sum_{i=1}^{r} (n + 1 - i)^{-1}$$

which equation should by now be familiar. Cohen finally turned the knife in the wound by suggesting that there are probably a number of other more or less plausible models of species interaction that should yield the same equation.

It seems that many sets of assumptions can lead to the distribution of abundance predicted by the broken-stick model. It also seems that distributions of this form can be found in nature, usually when we sample small areas and consider only related species whose life cycles are synchronized, but sometimes with a medley of organisms if we juggle our sample sizes. In the grand diversity of nature it is not improbable that this distribution is produced in more than one of the ways that are theoretically possible. In some circumstances it may be that the resources of a habitat are divided up by species finding discrete niches according to the broken-stick model. But sometimes circumstances result in a process akin to the balls-and-buckets model taking place. And other interactions may lead to like distributions. Cohen suggests that we try to distinguish between these sets of assumptions by devising experiments with animals in the laboratory; but such attempts will always lead to the objection of the analog computer argument. If the math is sound, it ought to be possible to manipulate your animals so that they "solve" any required equation. The experiments show you what animals can do, not what they do in nature.

An alternative seems to be to go back to testing assumptions by direct observations in the wild. This is what we were doing before we called on mathematics to aid us, and it was not good enough. Perhaps the final answer will have to be to build more complex sets of assumptions, leading to more complex equations that shall be unique. This presumably means computer simulation, with models being ever more refined by observation and experiment until they predict very closely the natural distributions of abundance. This seems to take us back to ecosystem studies and to the big teams again.

A generalization that seemed valid after Preston had shown the prevalence of log-normal distributions was that commonness and rarity were ultimately caused by random processes. This generalization seems confirmed by the history of the broken-stick model and its analogs. All these models depend on random process, and yet, in some circumstances, their common equation does describe the diversity observed in nature. It seems that the extremes of commonness and rarity which we find in nature come about because the occupancy of niches, either by immigration or by speciation, is a random process. I suspect that this is not a trivial conclusion.

The result of evolution up to our day is a world supporting a rich variety of species, but with more species in some places than in others. Some of the places with few species, like hot springs or sites inhabited by the pioneer plants of successions, seem to be harsh or difficult places in which to live because they are rare or ephemeral. There has been little time for evolution to forge a new species to occupy any one of them, and the chances for extinction of an opportunistic species attempting to migrate from one to another of them must be high. It is easy to understand why these places may support only a few kinds of plants and animals, but there are other apparently harsh habitats, like the arctic tundras, which support few species even though they are neither rare nor ephemeral. Other apparently harsh places, like the Sonoran Desert, support a rich diversity of species, at least, of plants. We must develop a general hypothesis which explains differences in species diversity such as these. All the hypotheses which have been put forward have had to be directed to perhaps the most striking aspect of the pattern of diversity on the earth; the apparent gradient of species richness which runs from the tropics to the poles. It is not enough to say that the large trees of tropical forests make tropical habitats more physically complex, thus giving opportunities to more kinds of animals, because the number of extra niches which might result from this geometrical complexity is likely to be small compared with the large numbers of extra species which tropical habitats seem to support. Most of the parameters of animal niches are not measured in space, anyway, but reflect the activities of other animals. A tropical habitat certainly seems to hold more niches since it supports more species, but the packing in of more niches in the tropics cannot be used as an explanation for the greater number of species without the argument becoming circular. An early hypothesis of promise supposed that there were less species as the poles were approached because glaciers of the Ice Ages had caused widespread extinction, and there had not yet been enough time for the lost species to be replaced by evolution. There is some validity to this hypothesis. It explains, for instance, the remarkable paucity of tree species in the European forests, as compared with those of America or China, but it fails as a general and sufficient hypothesis for all the observed differences of diversity. The climatic catastrophes which accompanied the Ice Ages were not confined to the North, being felt as drastic changes in climate even at the equator, and fossil evidence shows that there was a gradient of diversity from equator to poles in remote geological periods without Ice Ages. The hypothesis that places with a rich diversity of species are more productive places is even less satisfactory, for some of the most productive habitats have very few species indeed. There should be a theoretical upper limit to the number of species which are possible in any one place set by the energy supply, which is to say the productivity, but it is likely that this limit is very rarely reached. The most satisfactory general explanation is that places with few species are places with the highest probabilities for extinction; being strongly seasonal, unusual or ephemeral, they are places of great stress for life where accident can often lead to extinction. Places of less stress may not acquire new species by evolution any more quickly, but their loss rate through extinction is less and so in time they come to accommodate more.

CHAPTER 39

WHY ARE THERE MORE SPECIES IN SOME PLACES THAN OTHERS?

One of the most striking facts of natural history is the richness of the life of the tropics. There are more sorts of trees, more birds, more insects, more mammals, more snakes, more of almost everything in the tropics than there are in other latitudes. There seems, indeed, to be a gradient of diversity running from the wonderfully diverse life of the equator to the almost lifeless ice of the poles. Parallel changes are met on ascents of high mountains, for not only do formations and life zones change with altitude in the ways that so exercised the early naturalists, but the number of species steadily declines as you go higher also. Other changes of diversity from place to place are well-known, too; a square mile of land supports more species of animals that does a square mile of ocean; an acre of a large island has more species than an acre of a small island; Lake Baikal in the Soviet Union has perhaps 100 times as many species of animals as Lake Superior, although both lakes are freshwater and of comparable area, and so on. Why should species diversity vary so?

At first sight, the places of low diversity usually have characteristics that might seem to explain their impoverishment. They are often, for instance, what we might call harsh places, or rigorous places. Such are the poles and tops of mountains. Such also are extreme deserts, where life is scattered and the species list is low. Salt flats, caves, hot springs, and brackish estuaries all seem likewise impoverished, and each of these may be said to provide rigorous habitats to life as it is lived over most of the earth. But why should a rigorous climate result in fewer species? An estuary, for instance, is a highly productive place supporting hosts of living things; but it happens that there are comparatively few kinds. If some sorts of things can live there, why not many more? And the arctic tundras for hundreds of miles beyond the tree line are a complete green carpet. Vascular plants occupy most of the space available in spite of the rigor, so why should there be only a few kinds of them?

Rigor does not always result in low diversity. The Sonoran Desert, surely a rigorous enough place, is the home of a great variety of perennial plants with its many kinds of catcus and other desert shrubs. These plants alone probably represent as much diversity as is to be found in many a temperate forest with adequate rain, and yet the Sonoran Desert is also the home of many kinds of ephemeral plants, the annuals that spring up in the brief spring rains. And serpentine soils, whose unusual chemical composition excludes many familiar plants, are legendary for the numbers of species which they support (Whittaker, 1954, 1960). So it seems that rigor per se cannot be the general explanation for low diversity. Connell and Orias (1964) make this point clearly in a recent review of the subject by pointing out that the dry land must be thought of as a very rigorous place for organisms who have evolved first in the sea; yet it is the land that is rich in species, not the sea.

But there remains the fact that local rigor and low diversity do commonly go together. Perhaps an extreme example is afforded by hot springs, surely a rigorous sort of place in which to be and which accordingly supports very few forms of life. That something can live in so bizarre a place as a hot spring, shows that life is perfectly possible there. If all the world were a hot spring, might we not expect it to be well populated with living things? But this leads us to what is probably the crux of the matter. All the world is not a hot spring; hot springs are rare. When we say that they are rigorous, what we mean is that living conditions there are unlike those normally encountered by living things. It is their unusualness that matters. And not only are they unusual; they are ephemeral. They exist at a place like Yellowstone for a few thousand years after the volcanic upheaval that gave them birth, then the buried magma cools and the water bubbling from the local spring is no warmer than elsewhere. Animals and plants that would live in such a place must adapt from whatever local species are able to mutate in the required way, or they must be extreme opportunists, able to disperse from one remote place of vulcanism to another. The opportunities for hot-spring life are few, fleeting, and difficult. We should not be surprised that the random process of evolution by natural selection has not been able to make many species suitable for such a life.

Other places that have low diversity, and that appear to us to be unusual or rigorous are the habitats of pioneers in plant successions. The dunes along Lake Michigan, where Cowles saw the dune grasses and cottonwoods start the classic xerarch succession, must surely be thought of as rigorous, and the bared ground on which secondary successions start must be a fleeting and unusual thing, at least until farmers came along. The species lists in both pioneer communities are low. Indeed, it is possible to regard species enrichment as a major part of ecological succession. Throughout most of the early stages of plant successions the species lists grow, and only in the climax is there sometimes a retrenchment. The site begins with almost no species, but then it acquires more and more by immigration and replacement. It should not surprise us that diversity was low when only the first immigrants had arrived. The diversity of opportunists who start successions must therefore always be low. It is not the rigor of the sites that they occupy which causes low diversity but rather the limited time during which the sites are available to the pioneers.

So sites which are unusual or fleeting must be expected to support only species which are unusual or which are being rapidly replaced by immigrants. Many local patches of low diversity can be explained in this way. But we must find different explanations for the grand patterns of diversity, for the change with latitude, and from land to sea. Different latitudes, and the separation of soil and water, are both usual and permanent, implying some more fundamental reason for the difference in diversity.

It is tempting to look first for an explanation in the *structure* of the different places. More species means more niches, and more niches might well be found in places of complex structure, of complex geometry. The presence of forest trees, for instance, immediately suggests many more ways of life, many more niches, for insects than would be present in the same patch of land covered only with grass; you can have canopy-living species, trunk-living species, understory-living species, ground-living species, and so on. This may be true for larger animals also, like birds for which we have much better data. MacArthur (1964, MacArthur and MacArthur, 1961) tested this possibility for birds, measuring at a number of sites both the diversity of the birds and the physical diversity of the foliage. He divided the forest into three layers, the ground cover, which was usually only up to 2 feet high, the middle story of bushes and young trees, and the canopy. Then he worked out the proportion of the total foliage which lay in each of these three layers, an operation that was essentially one of measuring the cover of each by mapping. The proportions of foliage in the three layers gave him a measure of the total diversity of foliage, which he could compare with the diversity of birds encountered. He found that for widely different habitats over different parts of the

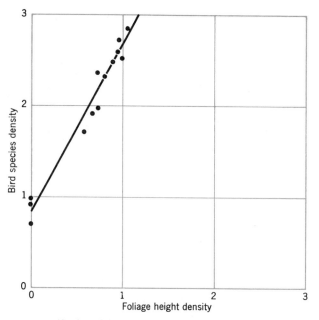

Figure 39.1

Diversity of birds and the diversity of foliage. As trees get larger, and vegetation more complex, there may be more ways in which the niches of birds may be physically separated. This figure shows results of census of birds and foliage-height diversity for a series of habitats each large enough to hold 25 pairs of all species combined. The close correlation suggests that the number of kinds of birds may be close to the maximum that can be allowed for such motile animals, getting their livings in special ways, so that the numbers of species are influenced by matters of simple physical space. (After MacArthur, 1965.)

United States the bird species diversity was proportional to the foliage height diversity (Figure 39.1).

MacArthur's study of the effects of layering in foliage, and other studies like it, have confirmed what seemed intuitively obvious: that complex physical structure should result in more possibilities for specialized living, and hence in more species. Some part of the low diversity of insects in the Arctic may thus be because there are no trees in the Arctic to make a complex world for the insects. The relative poverty of the sea likewise seems to conform to this general concept, since the physical structures of the open sea, although present, are much less elaborate than the structures of the land. If there were large plants in the open sea, there might be more kinds of animals, but large plants are usually absent for the reasons discussed in Chapter 13.

Clearly, geometric complexity does have an important influence on species diversity. But there must be more to the matter than geometry because, if we look closely at the evidence, we find that the decline of diversity with latitude is revealed even when the geometry seems comparable. A good illustration of this is the greater richness of trees in tropical forests. This has been discussed several times in this book. A temperate forest may be dominated by just one or two species whereas the canopy of a rain forest may be shared by 100. Likewise insect species are many times as numerous in tropical forests, a great disparity in species abundance that cannot be explained by the slight increase in structure represented by the larger trees.

The great differences in diversity between things such as insects and trees of tropical and temperate forests is sufficient ground for rejecting the thesis that the decline of diversity with latitude is primarily a consequence of declining physical structure. But there are still more compelling theoretical grounds for rejecting this explanation. The argument leads to supposing that the prime parameters of niches are set in physical space; are in only three dimensions. This is the trap of using the word "niche" as it is used in church architecture, as a three-dimensional slot in which you place a figurine. But the ecological niche is a hypervolume of very many dimensions, only three of which are set in physical space. Most of the rest are set by the neighbors of the animal or plant; by those who would eat it, feed it, compete with it, or associate with it. A complex matrix of ecological niches has biometric rather than geometric complexity. This may well provide more niche spaces, but the complexity results from the presence of many species and is not a cause of them. The argument has become circular. We still face the problem that there is a gradient of diversity from the equator to the poles, and that this can be explained neither by rigor per se nor as the consequence of simple geometry, although the effect may sometimes be amplified by geometry.

An attractive explanation was the one put forward by Alfred Russel Wallace (1878) when biogeography was in its infancy, an explanation long followed by biogeographers (Fischer, 1960). This notes that the polar regions of the earth have been recurrently overrun by glacial ice, and that the last of these afflictions was comparatively recent in a geologic or evolutionary sense. Perhaps the recurrent catastrophes to the north have meant that there has never been time for a diverse northern flora and fauna to evolve. The 10,000 or so years that have elapsed since the end of the last Ice Age might seem a short time in which to evolve a complex biota, or even to repair the losses by extinction suffered when the northern habitat was last overtaken by ice. At the other latitude extreme, so claimed the hypothesis, the tropical rain forest may have been left in undisturbed occupation of its habitats through whole epochs, allowing evolution to create much diversity. This hypothesis would be particularly attractive if the cline of diversity from the poles was a peculiarity of Pleistocene or Recent geologic time, and was not so pronounced in past epochs that were less subject to glaciation. But there is a convincing body of evidence from fossils which suggests that this requirement is not met, which suggests that a similar cline of diversity to the poles has been present in all ages. We have for instance (Stehli et al., 1969) good evidence that there were more species of planktonic foraminifera, small animals that leave their calcareous skeletons behind by the million, near

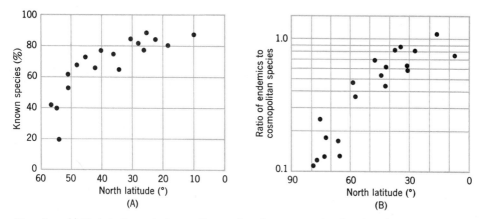

Diversity and latitude in the geologic past. Data such as these suggest that there has always   **Figure 39.2**
been a cline of diversity from the equator to the poles. (From Stehli et al., 1969.)

the equator during the Cretaceous epoch than there were further north. Other animals of which we have splendid records are the bivalves called Brachiopods. In remote Permian time there was a steep decline in diversity of these from the equator northward, also (Figure 39.2). Added to these difficulties is the fact that there is a growing body of evidence which suggests that tropical regions have not been undisturbed for epochs as was formerly supposed. Rain forests may commonly have developed since the last glaciation, and right down to the equator the events of an Ice Age have been felt in such drastic ways as turning lakes to dust bowls or dust bowls to lakes.

For these reasons the Wallace explanation, in its simplest form, must fail, but it can fall back on suggesting that northern regions have always been more liable to catastrophe than have lower latitudes. Whether there were Ice Ages or not, life in the North must always have been subject to an excessive number of catastrophes such as might set the clock of evolution back. But we have no reason for believing that this has been so. Apart from the sweep of ice sheets, there are not many calamaties so great that they can wipe out the biota of whole regions. Hurricanes and droughts are just as likely, perhaps more likely, near the equator as near the poles. And even an ice sheet itself is not so deadly a killer as might be thought, because animals and plants retreat before its slow advance so that most of them survive in the frigid belt before the glacier's front. In Europe the glaciers were able to descend from Scandinavia and squeeze out the forests against the Alps. This is a valid explanation for the low diversity of European forests when compared with temperate America. But elsewhere the species were commonly displaced by the ice, and were prompt to reinvade, as the pollen evidence clearly shows (Chapter 7). Nor is the effect of the ice felt only in the North, since we now know that the climatic pattern of the whole earth is changed by the presence of an ice sheet so that adversity caused by climatic change may be felt at the equator as well as in the North. A reasonable conclusion is that there is some validity in the hypothesis, that some local patterns of reduced diversity in the North, as in European forests, can be explained because the North is disaster prone, but that the hypothesis is not adequate to explain the grand gradient of diversity that we see.

ARE TROPICAL NICHES DIFFERENT IN WAYS THAT ALLOW CLOSER PACKING?

Rejection of the Wallace hypothesis as a sufficient explanation leaves us with the suggestion that the saturation, or equilibrium, number of species that can live in temperate climates is less than at the equator. Remembering that geometry and some ill-defined sense of rigor are not adequate explanations, it is tempting to ask if there is not some special quality of the tropics which allows more niches to be packed into unit space. In short, can tropical niches be different in some general and important way?

An ideal tropical life would not be afflicted by changing seasons. A niche in such a place need not contain parameters requiring adaptation to seasonal change, meaning perhaps that it should be smaller than the niches of northern places with winters. If tropical niches could be smaller, then you could pack more of them in and the answer to the diversity problem was found. But how can you measure a niche to show that it is larger or smaller? Klopfer and MacArthur (1961) attempted to get round this difficulty by considering the implications of packing in lots of small niches as was apparently done in the tropics. The occupier of a small niche must be a specialist, suggesting that he would have the central part of his niche very much to himself; a safe refuge that might make it unnecessary for him to compete fiercely at the periphery. Partial interlopers might be tolerated, and we might say that denizens of the tropics would be more likely to have overlapping niches. The problem seemed to become one of testing the postulate that niches overlapped more in the tropics.

The test tried was based on observations of Hutchinson (1959) that the discreteness or overlap of bird niches could be inferred by comparing the sizes of their bills. Birds choose food with their bills, so that character displacements which set the degree of separateness or overlap of the niches of neighbors is likely to be reflected in the relative sizes of their bills. The downy and hairy woodpeckers, (*Dendrocopus pubescens* and *D. villosus*) of the American East and Midwest, for instance, go woodpecking on the same trees, but they avoid competing for food because one catches one kind of insect with its short stubby bill whereas the other catches a different crop with its more massive chisel. Hutchinson measured the bills of large collections of museum specimens of North American birds, and found that the mean ratio between pairs of closely related species was 1.3. If this was a measure of niche distinctiveness, or the amount of risk of overlap that should be allowed in north temperate regions, then Klopfer and MacArthur could predict that bills of related sympatric birds in a tropical world of overlapping niches should be closer to unity than 1.3. They went to Panama and watched birds, carefully noting which species seemed to feed together. Then they resorted to museums for specimens of the birds they had seen, and measured the lengths of their bills. The mean ratios between their pairs of birds was about 1.1. This suggested that the niches of tropical birds were indeed closer, implying the overlap which they had predicted. Umfortunately, as their friends have often told them (MacArthur, 1965), Klopfer and Mac-Arthur had not shown that the species they had studied were closely related. They had chosen birds that fed together, but there was no reason to believe that these were character displacement pairs, as Hutchinson's formulation required. The "test" of the postulate was thus no test at all.

But, tested or not, there is a graver objection to this approach of seeking to explain increased diversity on the basis of smaller or overlapping niches. It begs the question. If there are more animals and plants in the tropics, then there are more niches. This comes from our original definition of the niche. For a niche exists only when occupied. More animals and plants mean more niches. To explain the numbers of animals and plants by the numbers of niches involves circular reasoning. It might be of interest to know that many niches mean overlapping niches, but a rigorous demonstration that this was so would not help with the problem of explaining diversity.

The mechanisms considered so far are all ones that should increase diversity in particular habitats, what MacArthur (1965) calls **within habitat diversity.** But there is another way in which a country could support more species; by being broken down into smaller habitats. This should lead to enhanced diversity of a kind which MacArthur calls **between habitat diversity.** The animals and plants become more spacially isolated from their relatives, supporting smaller populations but avoiding competition by being strictly "local," as a collector might say. It is easy to see how this could come about. Imagine a new volcanic island being filled with colonists by random immigration. The first comers have the place to themselves, and adopt niches as wide as their physiology and behavior allow. They amass large populations. As more of their old neighbors from the ancestral continent arrive, the pioneers must retreat to smaller niches, although still occupying the whole island. But there will come a time when species have become so numerous that survival will require retreating to only the part of the island to which each is best suited. Increasing **within habitat** diversity has at last been replaced by increasing **between habitat** diversity. An analogous process should occur in an ancient continent, but on a longer time scale, by evolution rather than by immigration. If speciation continues even after habitats can be shared no more, then a pattern of smaller habitats should develop; the within habitat diversity remains roughly constant but the diversity of the whole country continues to increase. A limit should be set to this process by the extinction rate. The fundamental difference between these two sorts of diversity were independently pointed out by Whittaker (1960), who

THE TWO SORTS OF DIVERSITY

called them α (within habitat) and β (between habitat). This nomenclature is not used in this book because of a prejudice against unnecessary Greek.

The realization that increased diversity can be accounted for in two ways opens another chance for comparitive studies. Are both kinds of diversity greater in the tropics? As yet we have little data. Comparisons of species area curves for birds by Preston (1960) suggest that, for birds at least, the big difference is in the between habitat diversity. This is not surprising because such studies as MacArthur's (1964) correlating bird diversity with foliage height diversity suggest that the possibilities for within habitat diversity in birds are saturated in most latitudes. It is therefore easy to accept that the increasing diversity of tropical birds is accounted for by an increase in between habitat diversity. For no other group of organisms do we have such complete data. The richness of tropical forests clearly suggests that for trees the increased diversity of the tropics is, at least, partly accounted for by within habitat diversity, and a subjective view of tropical insect populations suggests the same thing for them.

Even if we had the data which would tell us that the increase in tropical diversity was mostly of the one kind or the other, it would still not help much with the general problem. Whether the extra tropical species are accommodated by subdividing niches or by "moving over" to split habitats, we are still essentially dealing with biometrics, with the creating of more niches by the responses of the animals themselves. The underlying tropical condition that permits the greater expansion of either process remains obscure.

THE HYPOTHESIS THAT HIGH DIVERSITY RESULTS FROM HIGH PRODUCTIVITY

What is different about the tropics then? Their geography and geometry do not seem remarkable. Their lack of "rigor" did not seem to explain much. Their history is only doubtfully free from rejuvenating catastrophies. There may, indeed, be smaller or overlapping niches there, but ascribing this as a cause of increased diversity involves circular arguments that beg the question. Nor would manipulations of niche size help much if increased numbers of species could always be accommodated by splitting habitats to provide between habitat diversity. What is special about the tropics? One answer seems to be that they are more productive.

If a region is more ecologically productive, it should be able to support more plants and more animals; hence there should be more possibility for splitting the grand total into more species. More simply, we might say that a larger cake can be made to yield more slices. The idea that productivity could be behind increased diversity is attractive, since it seems reasonable that the "rich productive tropics" should have more species than anywhere else. There can be no doubt, for instance, that the wet equatorial regions of the earth are much more productive than are the arctic regions of northern Alaska. We can readily see how productivity could set a limit to the process of species enrichment, for this must go on until the rate of speciation is equaled by the rate of extinction. And populations that are too small, that are undernourished, must become increasingly liable to become extinct.

It is possible to erect a scheme of interlocking possibilities on the original base of greater tropical productivity, all of which should require increased diversity. This has been done by Connell and Orias (1964). More energy for plants can mean larger more complex plants; and it can also mean large purely local populations of the kind required to promote between habitat diversity. Diverse plants and complex plants both provide fresh niches for animals, increasing animal diversity. More energy for each animal trophic level means that large local populations of animals are possible, again increasing the chances of between habitat diversity. As similar effects are felt at each trophic level, the effects are compounded many times. Each additional niche which is found becomes part of the boundary for new potential niches. And the lack of seasonality can aid this process of speciation caused by a high energy drive, since animals in a stable climate need not spend so much energy on systems that maintain their internal regulation, thus freeing more energy still for making population. It seems, in short, that if enough time is granted, speciation will proceed until an upper limit is set by productivity.

But there are reasons to doubt that even this attractive scheme provides the practical answer. It is our experience that many of the most productive places on earth are, in fact, places with very few species. Estuaries of great rivers are often enormously productive, fed as they are by the nutrient supplies carried in the flowing water. Yet estuaries are places that classically do not have large species lists; they support large populations of what they do have, but the kinds are few. Salt marshes, which typically have only one important species of plant, are one of the most productive temperate systems known. Redwood forests, which include few species of plants, are likewise immensely productive. It is true that tropical rain forests, so wonderfully diverse, are among the worlds more productive systems, but single species cultures of tropical sugar canes or papyrus swamps are possibly more productive still. These highly productive systems of low diversity can with advantage be compared with the Sonoran Desert, a place of splendid diversity but which is not much more productive than the open ocean. All these examples, and many more like them, make a too heavy load of exceptions for the hypothesis to bear. They are already a sufficient reason for rejecting it, but there is yet a provocative additional argument. If you make a body of water more fertile by polluting it with sewage or phosphates, the productivity increases but the species list goes down. The data to support this are now very numerous, coming from almost all studies of pollution, but the matter has been specially well documented for diatoms by Ruth Patrick (1961).

It seems logical that more productive places should have more species, but it is not true that they do. Whatever mechanism sets the number of species in any place, it works at a level well below that set by the available energy. A valid inference from this seems to be that many more species could be allowed almost everywhere on energetic grounds, but that the accumulation of species has gone further in some places than in others. The grand question then becomes, Why have more species *accumulated* in tropical regions than in higher latitudes? A clue to the answer has recently come from studies of the deep sea.

THE TIME-STABILITY
HYPOTHESIS

The ocean floors beyond the continental shelves and down into the deep abyss are dark, cold, unproductive places. Life down there is dependent on energy supplied as detritus from the lighted zone far above, or carried to the bottom as organic molecules in solution. There is no local primary production at all. There are, of course, no plants. The animals must subsist only on the energy which has been imported after escaping the vigilance of animals and decomposers of more favored places. We may expect the pickings to be lean down there in the abyss. And yet it has just been shown that the deep-sea floors are rich in species; richer by far than the productive continental shelves; and richer still than the even more productive estuaries at the edges of the ocean. This has been demonstrated by examination of dredge hauls carried out by vessels of the Woods Hole Oceanographic Institution, particularly by Howard Sanders (1969). Sanders concentrated only on the animals that lived buried in the surface mud, the **infauna,** ignoring those that lived on top of the mud. Furthermore, he counted only the polychaete worms and the bivalve mollusks of the infauna, but these seemed to comprise about 80 percent of the species in his hauls, suggesting that he was dealing with a large part of the total diversity. These studies show, most convincingly, that as you follow the bottom down to the comparatively unproductive depths the density of the animal populations goes down, as you would expect, but the within habitat diversity goes markedly up. There are fewer animals but more species down there.

The gradient from an estuary, through the continental shelf to the deep-sea floor may be thought of as a gradient of environmental hazard for life. At one end is the estuary, a place whose salinity and temperature may change rythmically with daily tides, with the seasons of the moon, and with the seasons of the year; but also in unpredictable ways as storms or floods may swamp the region with either saltwater or fresh, bringing perhaps mass death. Although highly productive, the estuary is a place of regular stress and unpredictable hazard. The deep-sea floors, on the other hand, are well insulated from environmental shock. Not only is there little change between night and day but even seasonal changes are likely to be slight. The chance of mass catastrophe, unless by turbidity current, must be very small indeed. Clearly it seems reasonable to correlate the gradient of diversity with a gradient of physiological stress and the unpredictability of hazard. And having done this we must surely note that the other great gradient of diversity, from the poles to the equator, is also correlated with a gradient of stress and hazard.

Sanders' work illuminates the fact that the most general and significant things with which the gradients of diversity are correlated are gradients of

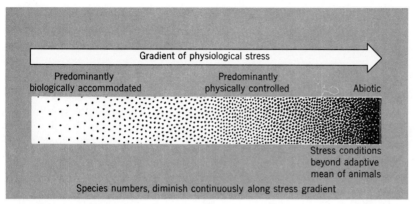

Figure 39.3

The stability-time hypothesis.

stress. If we also accept that shortage of time compounds with physical adversity to increase the stress of an environment, the conclusion embraces all that is valid in the earlier hypotheses. Hot springs were not only physically adverse, but they were also fleeting. Bare ground at the start of secondary successions was available for but a little time. Glaciers did represent something of an extra stress to places in the North. But more striking still is the way the observation is consistent with the anomalies that deny the productivity hypothesis. Salt marshes and estuaries are places of high physical stress, so they have few species for all their high productivity. Pappyrus swamps and sugar cane fields are ephemeral systems shortly to be replaced in succession. And polluting a river stresses it, so that the species list is reduced even though the river is made more productive. Sanders was thus able to postulate that *all places of high diversity would have stable or predictable environments* and that *all places of low diversity would either be places of unpredictable hazard or they would be short-lived.* He called this result the **stability-time** hypothesis.

Sanders and Slobodkin (1969) developed the stability-time hypothesis to suggest the kinds of animals that must live in low and high diversity places. Places of low diversity were stressful or available for limited time, which meant that animals must be opportunists. The bits of Australia which Andrewartha and Birch describe fitted admirably the qualities of a place of low diversity as Sanders and Slobodkin saw it, and their opportunistic animals had the usual qualities for life in such places; they dispersed well, they were capable of very rapid population growth, their numbers were often checked by weather, and they were fugitives from competition. Animals of stable, high-diversity places were, on the other hand, equilibrium species whose numbers were more closely regulated by density-dependent mechanisms. They were more "biologically accommodated." The full scope of the stability-time hypothesis can thus be illustrated by a diagram as in Figure 39.3.

We do not yet have a detailed test of the stability-time hypothesis. Sanders (1969) shows that it fits the data for the infauna over wide parts of the oceans, and it does seem to conform to the general facts about the distribution of diversity. But the suggestion that life of stressful places should always be opportunistic is not so generally satisfying. It may be true for most animals, but it is probably not true for perennial plants. The strategy of many plants of unstable areas is to set aside reserves in the good times and to endure bad times as dormant adults (Chapter 27). This allows an equilibrium population of biologically accommodated individuals to be maintained inspite of unpredictability and environmental stress. On the other hand, annual plants often appear in stressful places, and these adopt the strategies of opportunists. With the reservation that perennial plants may not be opportunistic in places with high stress, however, the stability-time hypothesis seems a satisfactory working postulate. High diversity is brought about by stability; low diversity results from instability or lack of time. Accepting this leads to an understanding of how the extra species of high diversity areas were accumulated.

HOW THE SPECIES OF HIGH DIVERSITY PLACES MAY HAVE ACCUMULATED

In unstable places, or where there is not much time, the dangers of extinction are great. Populations of opportunistic animals must frequently be cut down by weather, and there must always be some definite probability that breeding efforts may fail entirely leading to a year-class failure. The loss of several consecutive year-classes means extinction, even for long-lived animals. But such year-class failure is less likely in stable climates, and a series of consecutive year-class failures is very unlikely. Extinction is thus more likely as environmental stress increases.

But the actual number of species present in any place is a product both of the loss of species by extinction and of their replacement with new species. There is still dispute over whether speciation is more rapid in stable or unstable places. Sanders and Slobodkin (1969) argue that aquisition of species by invasion is easier in stable places because generalists from more rugged places could stand the conditions of life there and thus encounter no difficulty in mastering the environment. On the other hand the physical conditions of life in severe or fluctuating environments must be almost a total barrier to immigration by species from milder places. They suggest, for instance, that a tropical species stands no chance in the winters of the North, but a northern species is already preadapted to tropical weather from the experiences of its own summer. Probably most ecologists would take exception to this, suggesting that competition in equable places is virtually a total bar to the invasion of fugitives from unpredictable climates, whereas the superior competing power of equilibrium species must often let them invade the homes of opportunists. This argument may never be completely resolved for it is hard to see how valid data can be obtained. But it may be reasonable to suggest that speciation rates are not very different in the two sorts of place. What does seem to be true is that the extinction rates are not the same. Stable or ancient places, like some parts of the lowland tropics, go on accumulating species because the extinction rate is low. Unstable or ephemeral places, like higher latitudes or hot springs, do not accumulate species so quickly because the extinction rate is high. This is the likely explanation for the grand diversity gradients.

WHAT ARE THE LIMITS TO THE NUMBERS OF SPECIES?

We must yet ask if there are limits to the numbers of species which can be accumulated, and if these limits are ever reached. The productivity of the earth must certainly set a theoretical upper limit, but we have no evidence that this has ever been attained. Most suggestive is perhaps the evidence of species swarms in ancient lakes, of which Lake Baikal in the Soviet Union is the most striking example. Lake Baikal is deep and cold, noted for the purity of its water. It is accordingly poorly productive. And yet it contains many hundreds of endemic species, having some 300 species of the genus *Gammarus* alone (Kozhov, 1963). The stability-time hypothesis explains this wonderful diversity of the Baikal fauna as the result of species accumulating over a long period of time in the stable conditions of the lake's deep basins. The great number is the product of continuous evolution since the Pliocene or even Miocene, when the lake was first formed. We have no reason to believe that the process would not continue indefinitely, if it was not for the fact that the lake is now being polluted.

Limits other than those of productivity must presumably work by limiting the number of possible niches. There is no way in which we can measure limits to possible niches, since we only know of the existence of a niche when it is occupied. But a few kinds of animals do seem to be telling us that the possibilities are exhausted. The diversities of some groups of birds, for instance, do not seem to vary very much from place to place, suggesting that the earth may be saturated with birds of these kinds. That this should be so seems reasonable, because birds require much space and live in specialized ways. The variations on a theme of "birding" might well be limited. The numbers of ways that it is possible to permute the profession of being a big cat may also be limited to allow not many more than the numbers of big cat species which existed in the recent past. But, apart from evidence such as this, we have no reason to believe that the possibilities for new niches are anyway near exhausted for the majority of species. A large part of the impressive species lists of high diversity areas is made up of herbivorous insects, and the uniqueness of these sorts of animals seems to be due to their having to adapt to the biochemical defences of plants. Perhaps the greatest drive to speciation in plants has been selection for unpalatibility. It has resulted in a parallel selection for animals, particularly insects on land but including zooplankters in the sea, specialized to ignore such biochemical defence. This process of producing ever more specialized plants and ever more specialized predators and parasites of those plants seems to have no limit short of that far away limit set by productivity (Whittaker, 1969).

It seems possible to conclude that diversity varies from place to place because the stability of the environment varies from place to place. The mechanism that links low diversity to low stability is the greater rate of extinction in unpredictable environments. In a few specialized organisms such as birds, a limit to the number of species that can accumulate is set by a restricted number of possible niches. For most other kinds of animals and plants the possible number of niches is much larger than the existing number of species. The actual patterns of diversity which we have inherited are the products of different rates of accumulation in the different environments of the earth.

Attempts to understand the causes of ecological succession have all
along been hindered by the belief that the climax represented an
organized entity. Yet the evidence put forward that such organization
exists has always been very slight. Even the belief that there are climax
communities toward which successions proceed has turned out to be
largely an illusion. The ecological dominance, which seemed so characteristic of the supposed climax communities of temperate regions, is
now seen to be the consequence of things such as erratic grazing pressures in places with seasonal climates. In the tropics there may be no
dominance, and everywhere on earth there is a blending of plant communities that are subtly different. So discrete climax communities do
not exist. Replacement of the plant community as a unit for study by the
ecosystem made the illusory quality of the climax even clearer, for
nowhere can we find discrete ecosystems let alone ecosystems with the
self-organizing properties implied by the concept of the climax society.
The communities for which we have the best histories are those of lakes,
and we now know that the apparently inevitable course of succession
toward the old age of the lake is a result of physical changes in the basin
and not a result of the activities of the lake's community. Yet attempts to
explain succession as an organizational process have continued, so great
is the fascination of the idea. It was once suggested that successions
proceeded toward that community which attained the greatest efficiency of energy conversion. This is now known to be false, that communities which we call climax are actually often less productive than
early successional stages. It has further been suggested that successions
proceed toward communities that efficiently cycle nutrients, but   549

although this is true for some successions it is false for others. Successions do always seem to proceed toward complex physical structure and maximum biomass, but this reflects no more than the increasing bigness of individuals and the layering of communities imposed by gravity from below and the sun from above. Yet the increase of biomass does imply that the store of potential energy represented by a community proceeds toward a maximum; a finding which has led to the latest heresy of an ecosystem directing its evolution through succession. The energy reserves of dead biomass have been equated with information, in the "information theory" sense, and the hypothesis put forward that this increasing information is the self-regulating principle. This hypothesis fails, not only because of the false correlation of biomass with information but also for the reasons that earlier attempts to find organizing principles failed; because there is no evidence that the properties of communities are anything more than the consequences of animals and plants having to live together. A realistic explanation of succession is that any disturbance makes room for fugitive opportunistic species that are always gradually replaced by equilibrium species. The apparent determinism with which these changes occur in any one country reflects no more than the rate at which these replacements take place and the availability of local species to take their places in the succession.

CHAPTER 40

SUCCESSION
REVISITED

From the beginning of ecology there was the problem of succession. Even if you discounted some proposed primary successions as being without sufficient supporting evidence (Chapter 6), there were still undoubted secondary successions. Changes could be seen all around a discerning naturalist, and there seemed to be order in them. Change we could understand and expect, because the physical world itself was always changing. But why should the changes seem so orderly? It was as if there was some grand organizing principle at work, repairing the ravages of man and nature, restoring through successions the natural balance. All the main phenomena of ecology seemed linked to this central phenomenon of succession. Ecological dominance; commonness and rarity; the regulation of number; the distinctness of species; the tuning of lives to share space, nutrients, and energy; all must reflect on that curious orderliness which seemed inseparable from ecological succession. Here was a grand centralizing theme in ecology. Why should there be so much about successional change which seemed deterministic and even predictable?

A first grand generalization was the persuasive philosophy of Clements and his disciples (Chapter 6), who likened the climax stage to some superorganic being. Successions were but the life histories of these beings. As organisms, the climax formations were born, grew, matured, and died. This was exciting stuff; so exciting that it slowed the development of plant ecology as a truly scientific discipline for almost half a century. For it was not only the avowed disciples of Clements who were influenced by it; all of plant sociology, by its very nature, accepted some of its premises. Plant communities were not just chance collections of individuals but were organized according to some socially accepted or ruling plan. Early successional stages were subordinate units that eventually gave way to the better organized societies of the climax. Whether you believed in the monoclimax required by dogmatic interpretations of Clements' writings, or whether you took the pragmatic view that there could be many local climaxes in any climatic region (so-called **polyclimax theory**) you tacitly assumed that succession was a process of social organization; that it led to an organized entity which you called a society. But there was no evidence that this was true. Perhaps the best line of supporting argument which you could offer was that communities undoubtedly became more complex as succession proceeded (up to a point), because they acquired more and more species. You said the society was abuilding. But this argument loses its force when you reflect that succession starting on bare ground begins with no species at all. A progressive increase of species present is then inevitable, and does not require a socially organizing principle.

This was essentially the objection raised by Gleason (Chapter 6). The so-called "society" was no more than the randomly acquired assortment of plants suited to the neighborhood, an assortment that naturally increased with time until all possible local plants were included. This was all that was implied by your "climax."

The ideas of the superorganism and the social entity no doubt acquired much of their plausibility from the prevalence of the phenomenon of ecological dominance in plant communities of the temperate zones. The dominant beeches and maples of a beech-maple forest undoubtedly impose constraints on the community of which they are members, and these constraints might perhaps be construed as organizing principles of sorts. Any other plant that survives where the beeches and maples live must be adapted to tolerate their presence. But this is a conclusion very different from that of detecting grand designs, resulting in superorganisms or organized societies. These concepts are philosophical analogy, not the fruits of deductive science. The proper approach to the phenomenon of ecological dominance is to ask how it can come about as a result of evolution through natural selection, not to see it as some guiding principle around which societies might be organized. The establishment of dominant trees seems, at least in temperate forested lands, to be closely associated with the phenomenon of climax, and thus the event that terminates successions. On the way to understanding succession itself, therefore, we ask the question: why does dominance occur?

Ecological dominance is peculiarly a property of terrestrial vegetation, particularly of forests. The tiny plants of the plankton cannot attain it, nor

in a real sense do animals. It is true that a few species of animals or phytoplankters may be very much more common than others, so that they
may make up most of the biomass in their respective communities. But
this sort of excessive abundance merely reflects the hump of the log-
normal distributions, as Preston has shown (Chapter 37). Such excessive
abundance is predictable from the basic assumption that speciation and
evolutionary success reflect random processes. Commonness is important, but it is not dominance. Real ecological dominance as understood
by the plant sociologist means the physical overshadowing of most plants
by a few. But that the big should squeeze out the small is hardly
surprising. If succession led merely to bigness, we could explain it in the
way in which Gleason explained the complexity of the climax. You start
with no species; which means that you must collect progressively more if
you are to end with many. Likewise, if you start with nothing, you must
expect increasing size if you are to end with bigness. It is not just the fact
that the big overshadow the small which made dominance so important
to the philosophy of succession, but that it was so very few species of big
plants which did the overshadowing. In the beech-maple forest there
might be 20 species of trees, but only the beeches and maples were
numerous enough to be dominant. It was this ability of just one or two
species to unfailingly dominate which gave so much impetus to the idea
of the organized society of plants. The proper question to ask now
becomes: How could natural selection lead to this dominance of the
many by the few, even when subordinate species were also big?

A likely answer to this question has developed from the discovery that
dominance is much attenuated or even absent in tropical forests. A rain
forest with a hundred species of trees to the acre cannot be described in
terms of its dominant trees as can a temperate forest. There would have
been much less temptation for botanists of philosophic mind to find in
plant communities the organized structure of societies if they had lived in
the tropics instead of northern latitudes. But why should the tropical
canopy, with its prize of light, be shared by many species while in temperate lands it is dominated by the few? Jantzen has shown us to look for
the answer to this in the activities of animals which eat seeds (Chapter
34). Seed predation close to parent trees in the tropics is so thorough that
only widely dispersed seeds can develop into adults. But the seasonal
climates of temperate lands forced seed predators to be opportunistic.
Their numbers fluctuated, sometimes allowing large numbers of seeds to
germinate and develop under the parental canopy. The trees that have
once got themselves established are thus able to swamp the temperate
woodland floor with their own offspring, to disbar further immigrants by
their sheer competitive numbers, to establish and maintain their dominance. The form of ecological dominace that so exercised the early plant
ecologists was thus a phenomenon principally of temperate lands, and
explainable as the result of the local insects being opportunists subject to
recurrent catastrophes.

In the event it was not necessary to await the work of Jantzen on seed-eating insects to see that the superorganismic approach as put forward by the plant sociologists was unsound. The very societies that were thought to represent the climaxes became more elusive the more they were studied. Work on gradient analysis, particularly the studies of Whittaker (Chapter 2) revealed the frequencies with which communities blended with one another along environmental gradients, so that the discontinuities we thought we saw were merely blips in the spectra of abundant and conspicuous plants. Real disjunctions between one type of vegetation and another, like those formation boundaries that can be correlated with air-mass fronts, were the exception; disjunctions between smaller blocks than formations were commoner in the mind of the beholder than they were on the ground. There were not so much organized societies of plants as aggregations of independent plant species, each grouped round the sites for which it was best suited. This was essentially what Gleason had maintained all along.

But if the beliefs in superorganisms and directing societies eventually faded away for lack of evidence, fascination with the apparently deterministic process of succession did not. There it remained, the most ubiquitous of all ecological phenomena, the fact that replacements of plants and animals by others always tended to proceed in an orderly and predictable way. What guiding principles directed these changes so?

The plant society gave way to the ecosystem as the central unit of ecological study, but this merely added more intriguing complexity to the problem. Unidirectional, orderly, predictable successions were now seen to lead to essentially stable ecosystems, instead of just to formations or societies of plants.

For a while the study of lakes, the easiest ecosystems to study because bounded by so uncompromising an edge, resurrected the Clementsian superorganism (Chapter 18). Lakes, it was argued, started life as pools of infertile water in raw holes. They then acquired not only more and more species of inhabitants, as would be expected during successional change, but they also grew more fertile. Then the rate of change became less, so that both fertility and the numbers of their inhabitants were stabilized for much of the history of the lake. The lake ecosystem had gone through a sigmoid growth history, like the numbers of *Paramecium* in one of Gause's tubes. Finally the steady rise of sediment in the bottom so shrunk the cold bottom waters, the hypolimnion, that they became anoxic in summer. The oligotrophic lake had taken on some of the qualities of a fertile eutrophic lake. The speed of deposition of bottom mud quickened, old age was accelerated, the hydrarch successions closed over the old basin, and the life of the lake was extinct. As an organism the lake had been born, had enjoyed early rapid growth, a falling growth rate at the close of its youth, a long period of tranquil maturity, and final death. But this resurrection of Clements' creature did not last long. The early changes in production and fertility were shown to be due to inputs of sediment and nutrient from outside. The eutrophy of old age was no more than it had always been seen to be, the result of physical changes in the shape of a basin inexorably filling up in the course of time. Only the successions that responded to these physical changes remained, as ever orderly, roughly predictable, and requiring explanations not based on deterministic philosophy.

THE HYPOTHESIS THAT
SUCCESSIONS PROCEED
TOWARD MAXIMUM
ECOLOGICAL
EFFICIENCY

A new possibility was revealed when Lindeman and Hutchinson introduced the ideas of energy flow and the efficiency of its conversion into the study of the ecosystem (Chapter 11). The possibility is clearly stated in Lindeman's (1942) classic paper on trophic dynamics. The bare ground or water of an unoccupied site should clearly produce nothing at all, having an ecological (Lindeman) efficiency of zero. Production must start with the first pioneers. It must be low at first, as the ground is but partly covered, but then must rise until a maximum is reached. Such is decreed by physical laws. But even though the productivity of the site must rise, the energy flux from the sun remains the same. It follows that the ecological efficiency must also increase as succession proceeds. But productivity and efficiency cannot increase indefinitely; there must be a limit, a maximum. And what more natural limit to expect to find than that of the climax?

Lindeman postulated that successions should proceed toward maximizing the efficiency of energy conversion, suggesting that the climax would be found to maintain maximum efficiency. As succession proceeded we should expect adjustments among the animals and plants of the ecosystem until the highest efficiencies were obtained. Energy was the ultimate limiting resource for all living things. When the best possible use was made of that which was available, then some sort of real maturity would be reached and we should expect change to cease. A climax would have been reached. Now the apparent deterministic growth of Clements' superorganism seemed to be explained, not as some quasi-philosophical property but as the result of a struggle for that most concrete of entities, energy.

Lindeman had no data on which to draw in the test of his hypothesis, and could only illustrate the hypothesis diagramatically (Figure 40.1a). We still have very few estimates of the ecological efficiencies of whole communities, but we do have a growing body of literature of the productivity of communities of various kinds. These leave us with no doubt that early successional stages are not only highly productive, contrary to Lindeman's postulate, but are in fact probably the communities that achieve the highest productivity of all. Old established forests that have been examined seem to produce less (to have lower gross primary productivities) than either young growing forests or the pioneer successional stages of herbs which precede them (Odum, 1969; Kira and Shidei, 1967).

This leads us to conclude that maximum productivity is achieved very early in many successions, that roughly the same high productivity is maintained throughout most of the successional changes, but that there is an actual tailing off as the climax formation develops. On this view, the productivity of a succession starting from a bog mat as postulated by Lindeman should be given by Figure 40.1b.

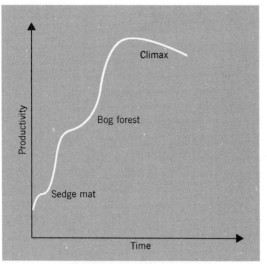

(A) Lindeman hypothesis (1942)

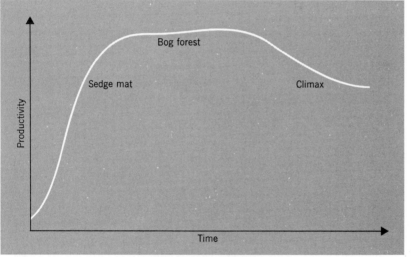

(B) Hypothesis of a 1970 ecologist

Figure 40.1

Hypotheses of changing productivity in succession. In 1942 Lindeman postulated that productivity should increase as succession proceeded, but later measurements were to show that this was not so. It is noteworthy that a 1970 ecologist who had read Walkers (1970) paper (Chapter 6) might even dispute the possibility of their being such an hydrarch succession in the first place. But if he drew a curve of productivity through time in a secondary succession he was sure of, it would be of the form of graph B.

The absence of many real data describing the fortunes of real successions in the field makes the hypothesis of Figure 40.1b very general indeed. It is likely that there should be small blips in a real curve, particularly as the age compositions of the different stands change, but we should expect these to be but minor perturbations on a general theme of maximum productivity quickly established early on in the succession and then being maintained until the final decline once the climax is established. We should bear in mind, however, that each succession might have its own unique development, that some might well involve changing environmental conditions which modify productivity as the succession proceeds. Perhaps the most obvious are involved in the physical changes brought about by xerarch successions on sand, like the classic at Lake Michigan. We should expect the productivity of the early stages of this succession to be low because plant growth is inhibited by shortage of water. So productivity may increase more slowly in the early stages of a xerarch succession, perhaps even requiring more than one successional stage before the maximum is reached. But this is due to the simple matter of water shortage, and is not a consideration of energetics. In the commonest successions of the old field type maximum productivity is apparently obtained very quickly, continues through the successive plant communities until climax forest is established, when productivity declines. We must ask ourselves: Why is a climax forest less productive than most of the stages which have preceded it in the succession?

From the outset it has seemed that this declining efficiency of climax forests must have something to do with aging. The climax forest is made up of mature trees, mostly of some considerable age, replaced at only rare intervals by the deaths of individuals. So the old do not produce as well as the young, but yet hold their space in the canopy by reason of their size. This seems all right, as far as it goes, and there are now a number of studies of aging stands of trees, particularly by Japanese workers, which encourage the view. Tatuo Kira and Tsunahide Shidei (1967) reviewed productivity measurements in forests by Japanese workers at various Pacific Ocean sites, much of the work their own. They had measured increments in growth of trees, which provided them with net production figures, and they also measured the respiration of whole trees to give them the other parameter of gross production. Elaborate census of stands in forest plantations of various ages then enabled them to calculate the gross primary productivity and to follow changes in productivity as stands aged. Kira and Shidei showed their general conclusions in a generalized productivity curve of the form shown in Figure 40.2. Productivity steadily increased as the plantation developed, reached a maximum, and then declined. In a plantation there is no succession, but we may well expect the growth phase at the left of their sketch to occur during early stages of a real succession. The development in a natural climax stage would be the decline in productivity shown at the right of the diagram, the same decline shown at the right of Figure 40.1b. Productivity of the final forest to occupy the site, whether planted or the result of succession, declines as the trees age. Which is, as I said above, all right as far as it goes. But why should not natural selection have replaced those senile trees with better adapted species that remain highly productive throughout their lives? Probably because the selection of trees happens when they are very young, when they are seedlings struggling together for light and space.

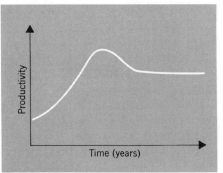

Figure 40.2

Productivity of an aging forest

If the bite of competition is felt by the young growing tree, then the strategy of the species must be to maximize growth in early youth. All young trees must be able to produce as fast as the physical circumstances of the site allow. But a mature tree of the climax forest is growing in quite different conditions to those it knew as a sapling. Strategies of organization which were necessary for its young days may not be quite so well adapted to the conditions of its maturity. Ideally, it should be able to reorganize its factories to suit its changed circumstances. But evidently the patterns of behavior are too fixed for natural selection to have been able to produce rival trees that can adapt both to that competitive crunch when young and to producing the maximum when old. There has been another of natural selection's politician's choices. Perhaps we should call this one an engineer's choice. The tree was designed for maximum efficiency when young, and the same design was not so efficient in the changed circumstances of the old tree.

This conception of the engineer's choice in designing a tree for efficient youth seems an adequate explanation for the falling productivity of the climax forest, one that shows why Lindeman's hypothesis proved to be false. But not all successions are like those on land which lead to forests and dominant trees. What of aquatic systems in which plants can never achieve dominance? Might not the climax of such successions maintain maximum productivity all the time? We know that populations of algae when grown alone can maintain maximum efficiency indefinitely (Chapter 10), so might we not expect an open water succession to proceed rapidly to maximum productivity and thence to stay there? In fact, we find that productivity declines in mature aquatic systems, just as it does in mature forests, although apparently for different reasons.

Successions involving typical small plants of the open water may readily be studied in laboratory microcosms, and many such studies have been performed. In a recent study of this kind, G. D. Cooke (1967) followed the fortunes of 18 replicate ecosystems in beakers containing 300 milliliters of water, each innoculated from a common-stock culture.

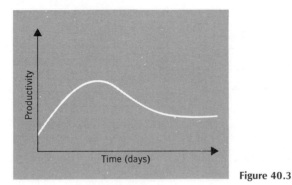

Figure 40.3

Productivity in aging microcosms.

The systems had propagules of a number of planktonic and benthic algae as well as planktonic animals. They were kept in growth chambers provided with artificial light for 12-hour days, the only other management being the addition of distilled water from time to time to replace that lost by evaporation. Productivity was measured by monitoring pH by day and night, a technique that is feasible because pH reflects the partial pressure of carbon dioxide dissolved in the water. There were visible plant successions in these systems, particularly involving early blooms of flagellate and open water plants and later mature stages when most of the plants were anchored or filamentous forms on the bottoms and sides. Changes of productivity during these successions are summarized by the diagram shown in Figure 40.3, which is in form closely comparable to that for forest trees given by Kira and Shidei.

Cooke attributed the decline in productivity at maturity to two processes whose effects combined: to the accumulation of ectocrine substances that depressed plant activity and to a locking-up of plant nutrients in the accumulated detritus as well as in the bodies of the larger anchored plants. The accumulation of ectocrines in such small closed systems was expected from the experience of workers like Woodruff (Chapter 8) and Gause (Chapter 23) with laboratory microcosms. A similar effect was the suppression of egg laying in flour beetles kept in old flour in which ethylquinone had accumulated (Chapter 22). And the effect had been previously demonstrated for green flagellate algae (Swanson, 1943). Such an effect might be expected to be particularly strong in artificially confined systems, but this is not to say that it does not occur to some extent in natural systems also. A lake may be bigger than a laboratory beaker, but then it has more organisms in it which might secrete the ectocrines. I know of no study that demonstrates the suppressive effects of such substances in lakes or the ocean, but there is abundant evidence that the second effect, the locking-up of nutrients, is important, and even crucial, in natural systems. The familiar spring blooms of temperate lakes and seas seem to take advantage of nutrients released into the water during the winter, and to terminate when free nutrients can no longer be found in quantity in the water layer in which the plants live. By the end of the blooms most nutrients seem to be present either in the bodies of living things or in the corpses that have been sent to the bottom, whence the nutrients may not be retrieved before the water-mixing of another winter.

The evidence seems clear that mature stages of both aquatic and terrestrial successions are less productive than some of the more youthful stages, contrary to the predictions of Lindeman. But it is also clear that this result is brought about by a number of different mechanisms, acting either singly or in combination. The persistence of old dominant individuals, the hoarding of nutrients in large structures, the accumulation of nutrients in detritus, the export of nutrients in corpses that fall through water columns, the accumulation of excretions that have physiological effects (the ectocrines); all may suppress productivity. The striking thing about this set of effects is that many of them may be thought of as side effects of adaptations of individuals to survive. By secreting ectocrines or hoarding nutrients, individuals are adapted to securing the largest possible energy supply; but an effect of this is that the energy supply to the whole ecosystem is reduced. Similarly, trees are adapted to securing the largest possible share of the available energy when young, but this ensures a senile old age which shall result in a reduced energy supply for the climax ecosystem. The individual is a device to secure the maximum energy supply; the ecosystem is not.

Under some conditions, at least, one of the most vital changes made during the course of succession is an enrichment of nutrients. A mature tropical rain forest, for instance, has a store of nutrients sufficient for its prodigious growth, but bare ground on the lowland tropics made by clearing the forest may be so deficient in nutrients that agriculture is next to impossible (Chapter 14). We postulate that the nutrient hoard in the rain-forest trees has been built up over time, notably during the succession that led to the establishment of the rain forest. Succession in this instance can be said to have been "directed" toward the accumulation, and efficient cycling, of nutrients.

A similar history of nutrient enrichment through succession was revealed when it was shown that leaving old fields to the pioneer plants for a year, leaving them fallow, resulted in enrichment of the soil with nitrates (Chapter 15). Wherever dense and complex vegetation is established on land, we find relatively efficient nutrient cycles with steady states apparently established. The vegetation acts to curb the rate at which nutrients leak from the soil to the water courses, reducing nutrient leak to no more than nutrient input. This allows another deterministic hypothesis: that successions proceed toward efficient use of cycling of nutrients.

The hypothesis casts the pioneer plants in the role of "accumulators," a term that appears in Odum's (1969) writings, since the period of their domain must see a nutrient hoard collected. We must then ask, "what

THE ROLES OF NUTRIENTS

special qualities have the pioneer plants to fit them for this role?" and a tempting answer can be found in their small size. Small size means a large surface to volume ratio. If nutrients are but diffusely spread through the substrate, it should be advantageous to have a large surface area over which they can be absorbed. Pioneer plants are both small and accumulators; so the circle of logic seems to be closed. But the logic is dubious, nevertheless. It is the same logic that has often been used to explain the smallness of the plants of the open sea, and which I have tried to show (Chapter 12) is unsound. Suitably large surface areas can be contrived without being small. In the sea, floating sargasso weed and anchored kelps are evidence enough for this. On land we may note that perhaps the most wonderful nutrient retrieval system of all is provided by the roots of tropical trees, in the underground and microscopic forest of root hairs and their associated miccrorhizza. It is better to think of pioneer plants as being small because their habitat is a place where there is no need to squander energy in maintaining bigness. They put their energy into dispersal devices that should let them occupy bare ground, rather than in achieving large size. Any effect this may have on their nutrient-getting powers (and it may be adverse, particularly on land) is part of the consequence of their size strategy, rather than a cause.

On dry land, the hypothesis that successions increase the nutrient supply seems no more than a statement of the empirical evidence. But there can be doubts, even there. Does the hypothesis hold good for an hydrarch succession leading to forest, for instance? Is the nutrient hoard of a mature forest growing on a filled lake basin indeed greater than that of the rich black mud with which the succession started? It seems rather unlikely. And when applied to successions of open water, the hypothesis fails completely. Consider the annual nutrient cycle in a temperate lake. The surface waters of early spring are rich in nutrients as a result of winter mixing; there is an algal bloom; the lake stratifies; the nutrients in the epilimnion become depleted as corpses carry them below the thermocline and to the bottom; the succession ends in nutrient poor maturity; and the surface water remains unproductive until winter mixing brings up the bottom water to start a fresh succession. This is a history of declining nutrient supply as the succession proceeds. It thus seems possible to say that the hypothesis of successions proceeding toward conservation and efficient use of nutrients, although sometimes true, has no general validity.

THE HYPOTHESIS THAT SUCCESSIONS ARE DIRECTED TOWARD MAXIMUM BIOMASS

But an explanation for the different fates of nutrients in terrestrial and aquatic systems is ready to hand: the terrestrial systems hoard nutrients by means of their great bulk whereas the enforced small size of planktonic plants (Chapter 12) denies them this chance. And the large size of forest trees not only provides a physical storage place for nutrients, it also allows them to develop that intricate network of roots which serves to retrieve nutrients leaking away in the soil water. The more striking differences in nutrient histories might thus be inevitable consequences of

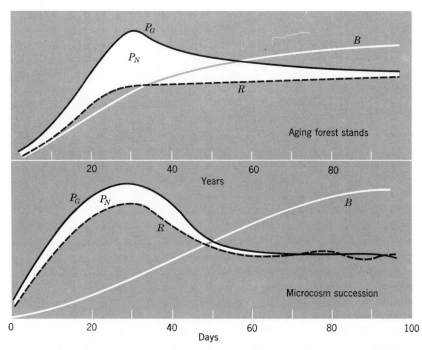

Figure 40.4

The increase in biomass during different successions. The upper diagram is compiled from data from forest stands of different ages. The lower diagram results from measurements of replicated laboratory microcosms, essentially of algae in water. Pg is gross production; Pn is net production; R is total community respiration; B is total biomass. The data suggest that as systems age they accumulate biomass (including detritus) until respiration balances production, when they may be called mature. Notice that gross productivity falls at maturity in both systems. (redrawn from Odum, 1969.)

the different sizes of the plants of land and water, leaving the way open for an alternative hypothesis that the real determining principle is a quest for size. Perhaps successions proceed toward the maximum biomass that the physical conditions of the habitat can be made to support. Such an hypothesis has the special interest that biomass is, of course, to be thought of as potential energy. The way is open for a fresh attempt to explain succession in terms of community energetics.

There seems no doubt at all that terrestrial successions lead to maximum biomass, at least, those that end in forest do. If you include as biomass the dead parts of the litter, which is fair since these represent potential energy of the ecosystem also, it seems obvious that bogs and prairies develop their greatest biomass at maturity also. Aquatic microcosms such as those studied by Cooke discussed earlier, also hold most living and dead biomass at maturity. Only floating planktonic communities may seem not to fit this pattern, because they lose biomass to the sediment in that rain of corpses; but if you include this exported biomass as part of the system, even planktonic communities may be said to achieve greatest biomass at maturity. It seems realistic to claim that successions proceed toward a maximum reserve of potential energy in the form of biomass.

E. P. Odum (1969) has summarized this view of succession with diagrams based on the work of Japanese foresters discussed above, and the microcosm studies of Cooke (Figure 40.4). In the early, highly productive stages of each succession, gross primary productivity is in excess of the community respiration, and biomass accumulates. Respiration continues to rise in the mature systems, even as productivity falls. This continual rise of respiration presumably represents the extra energy expended in maintaining structures of ever increasing size, although it may also represent more exploitation by animals, particularly the decomposers of the soil and litter. Eventually gross primary productivity and respiration are equal. The accumulation of biomass is at an end, and we may well say that the system is then definitely mature.

THE INCREASING COMPLEXITY OF STRUCTURE

This maturity represented by biomass is always accompanied by another apparent component of maturity, a complex vertical structure. The life of the pioneer stage of an old field succession lives essentially in one layer, or in two if you separate the zone of soil and roots from that of the aerial parts of the plants. But a climax forest usually has many layers. The green layer of producing structures is raised far above the ground in the upper canopy. There may be underneath this several layers of lesser producers working in the shade of those above. And the green carpet of pioneer plants which once covered the ground has given way to a brown leaf litter, grading more or less gradually through layers of humus into the mineral soil.

Layered physical structures serve to separate many of the essential functions of ecosystems in space. Production is mostly carried on in a layer high above the ground, but the decomposers work mainly on the ground below. Nutrients are raised from the soil to the production lines of the canopy in the sieve tubes and vessels of the trees, and they are returned to the soil surface by the force of gravity for decomposition. A similar layering of structure and function develops during successions in the sea and in lakes. At the onset of the spring bloom, all is stirred together and there is almost no structure. But then the surface waters warm and float as stable layers above the thermocline. Production is maintained in this surface layer; but gravity eventually carries the products down from the producing layer to be decomposed below. In the aquatic systems there are no tree trunks to return the raw materials to the producing layer above, and the producing parts of the ecosystem are dependent on the physical process of water-mixing by wind or current for the renewal of their supplies. But in spite of different mechanisms, it seems safe to say that all systems do develop vertical structures and spatial separations of functions, by one means or another and through the process of succession. This is nicely illustrated by two diagrams of Howard Odum's (Figures 40.5 and 40.6).

We thus find a set of linked physical changes always present in terrestrial successions, and which are, partly at least, present in aquatic successions. Nutrient reserves increase on land, and nutrient cycles are perfected. This seems most reasonably explained as a consequence of increasing size of the plants present, since nutrients can be both hoarded in their large structures and retrieved by their enveloping roots. Size, or at least mass, goes on increasing until production is no more than respiration. At this stage we may believe that maintenance costs of the ecosystem, represented by respiration, balance the productivity limit set by the physical condition of the habitat. But the large size then attained imposes a physical structure on the ecosystem. This structure always appears as a series of layers stacked on top of one another, a condition which is imposed by the fact that light always comes from above whereas gravity always carries corpses down below. The increasing vertical structure can therefore be explained as no more than the inevitable result of increasing size of plants in an ecosystem constrained by an energy source from above and gravity pulling from below. In aquatic systems the controlling influence of purely physical factors is even more obvious, because the structure is imposed by the physical structure of the lake, itself a product of the pull of light energy above and gravity below.

This analysis suggests that the only contribution of the living part of the ecosystem to vertical structure, and possibly to enrichment of nutrients also, comes about because individual plants grow larger throughout successions until limits are set by physical circumstances; on land by the energetics of maintaining large structures, in the water by the fluidity of the medium. There is nothing profound to be discovered in this steady increase in the size of plants. Successions typically start with no plant mass at all. On the way to achieving a physical limit to size, it is necessary to start small and to grow bigger.

In the early stages of all successions there is a progressive increase in diversity. This increase may not always continue through to the climax stage, probably as diversity is suppressed as part of the syndrome called ecological dominance. Climax deciduous forest has less species than the stages that came immediately before, reflecting the success with which the dominants shut out other species. But most of the time in most successions a progressively larger number of species come to occupy the site. An increase in the number of species means increasing complexity; a more mazelike food web. The ecosystem becomes more organized and more stable (Chapter 17). We may say that the "information" content of the system is increased during succession. But we have already noticed that biomass increases during successions also, and biomass is a measure of organized organic molecules, which constitute another measure of "information." Biomass is also a measure of part of the negentropy of the system. So is the organized vertical structure, which we know also increases as the system becomes mature. The concepts of "information" and "negentropy" are closely akin, and they are described by similar

THE HYPOTHESIS THAT
SUCCESSIONS PROCEED
TO EFFICIENTLY
MANAGE
INFORMATION

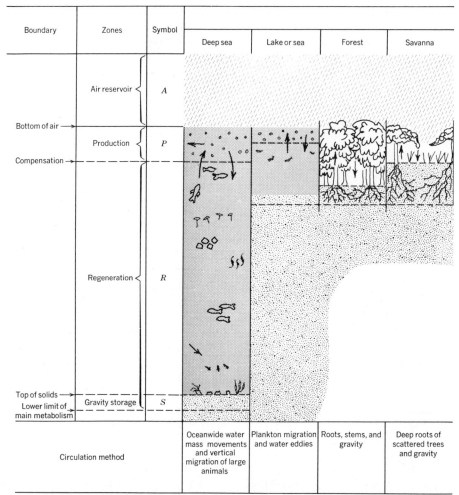

Boundary	Zones	Symbol	Deep sea	Lake or sea	Forest	Savanna
	Air reservoir	A				
Bottom of air →	Production	P				
Compensation →						
	Regeneration	R				
Top of solids →	Gravity storage	S				
Lower limit of main metabolism						
Circulation method			Oceanwide water mass movements and vertical migration of large animals	Plankton migration and water eddies	Roots, stems, and gravity	Deep roots of scattered trees and gravity

Figure 40.5

Vertical zones in land and water communities. The physical facts that light comes from above while gravity pulls from below cause a layering in all communities. This layering is reflected in a partial separation of functions so that production of organic matter by photosynthesis in surface layers is high, exceeding respiration. In lower layers respiration exceeds photosynthesis. There is a compensation layer where both are equal, and there is a physical transfer of organic products downwards, making necessary a physical return of raw materials upwards (From H. T. Odum: Environment Power and Society, John Wiley and Sons, New York, 1971).

mathematical equations. Both are measures of organization brought about by doing work. An ecosystem does this work during the course of succession, bringing about order, reducing entropy, increasing information. Suddenly we seem poised on a new understanding of that haunting determinism of the successional process; it is directed to achieve order, to maximize information. The information obtained by the accumulation of species, by the massing of organic molecules, by vertical structure, itself directs the accumulation of more information. The system feeds informa-

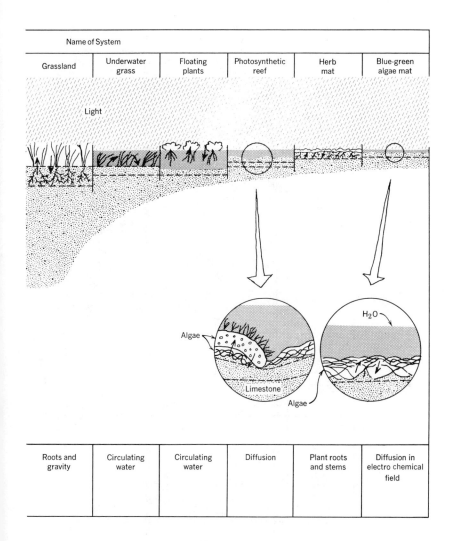

Name of System					
Grassland	Underwater grass	Floating plants	Photosynthetic reef	Herb mat	Blue-green algae mat
Roots and gravity	Circulating water	Circulating water	Diffusion	Plant roots and stems	Diffusion in electro chemical field

tion back on itself; it becomes self-regulating and self-determining. This is a concept to quicken the pulse. We seem to be explaining the grandest generalization of ecology in terms of the most fundamental and quantifiable physical concepts. It is easy to imagine the rising excitement in the Fisheries Research Institute in Barcelona when Ramon Margalef first thought these thoughts (1958, 1963, 1968).

A self-regulating self-determining system for the collection of information must have some limit set to its activities, a limit that should be manifest at maturity. But at maturity biomass is at a maximum, and respiration balances production. The greatest possible amount of structure is then maintained for the available energy flow. We can say that this represents the most efficient maintenance of structure. Margalef points out that ecosystems evolve through succession to achieve this most efficient

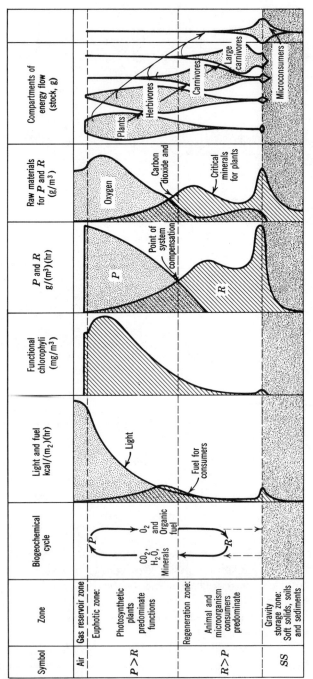

Symbol	Zone	Biogeochemical cycle	Light and fuel kcal/(m_2)(hr)	Functional chlorophyll (mg/m^3)	P and R g/(m^3)(hr)	Raw materials for P and R (g/m^3)	Compartments of energy flow (stock, g)
Air	Gas reservoir zone						
$P > R$	Euphotic zone: Photosynthetic plants predominate functions	P ↗ O_2 and Organic fuel ↙ CO_2, H_2O, Minerals	Light		P	Oxygen	Plants, Herbivores, Carnivores, Large carnivores
$R > P$	Regeneration zone: Animal and microorganism consumers predominate	R	Fuel for consumers		Point of system compensation, R	Carbon dioxide and, Critical minerals for plants	
SS	Gravity storage zone: Soft solids, soils and sediments						Microconsumers

Figure 40.6

Separations of functions resulting from the vertical zoning shown in Fig. 40.5 (From H. T. Odum: Environment Power and Society, John Wiley and Sons, New York, 1971).

maintenance of structure, this efficient management of information. This also is satisfying, because it always seems right that systems should be efficient. But it is as well to notice that efficiency as used by Margalef is not that same ecological efficiency of which Lindeman wrote. Lindeman postulated that ecosystems should develop to be efficiently productive; Margalef is postulating that they develop to efficiently maintain structure, which is something quite different. The Lindeman efficiency has been shown not to rise progressively through the stages of a succession, but there seems little doubt that the efficiency of maintaining structure of which Margalef talks does, indeed, increase with maturity.

Study of what was known of successions provided Margalef with many phenomena that seemed to fit into his general thesis. Climax systems are thought to have more symbiosis between species and perhaps longer food chains than earlier successional stages, properties that were certainly consistent with an hypothesis of increasing structure and information. The data here are perhaps subjective, however, and possibly even suspect. Think, for instance, of the successions that start with lichens on bare rock. A lichen is a symbiosis between algae and fungi, so that such xerarch successions commence with 100 percent symbiosis, a figure unlikely to be exceeded at maturity! But the general consensus of naturalists is probably that Margalef is right to consider symbiosis and longer food chains to be properties of mature ecosystems.

Of more significance to the hypothesis is the observation that biomass increases fastest in early succession stages, a fact brought out clearly in Odum's diagrams (Figure 40.4). This is apparently the same thing as saying that information and structure are increasing and being stored quickly. The young ecosystem is foregoing the use of present production to lay up stores of information (biomass) for the future: which seems to be self-regulation and self-determinism with a vengeance. Self-denial for the sake of the future! Clements' superorganism scarcely had more striking properties.

Not being content to store information for its own future, the evolving ecosystem seems able to export information to neighboring ecosystems. Margalef's own field studies were mostly of marine plankton communities. They never attained very great complexities or, as Margalef would say, they never attained great maturity, because of the fluidity of the medium. Individual size was small (Chapter 12), and physical organization like the spatial patterns of a forest was impossible. But, being kept in relative youth, the ecosystems of the marine plankton always had production (information) to spare. And they had a neighboring ecosystem on the sea floor beneath them; the benthos. They exported information to these benthic neighbors in the familiar rain of corpses. The benthic community, being able to organize itself against a solid substrate, was more mature, a fact no doubt mirrored in that remarkable diversity of the fauna of the abyss which Sanders noted (Chapter 39). So the immature system, the plankton, exports information for the use of the more mature community, the benthos.

There is another widespread form of export of biomass (information) familiar to all naturalists: migration. Typically, migrations involve breeding in some seasonally productive part of the world, and living between breeding seasons in more stable places which, reasonably enough, have more species in them. From Margalef's viewpoint, they may be said to do their growing (acquiring information) in immature systems, although they "belong" to more mature systems. The life histories of salmon illustrate this principle. They breed in the simple (immature) systems supported by inland streams, but they go down the river to spend much of their lives (and to do much of their growing) in the sea. On Margalef's hypothesis, this is an export of information from the simple system of their birth, which certainly is a place of surplus productivity, to the more mature ecosystems of their adult lives. A sceptic might like to know whether the return journey up river might not convey more information back to the immature system than was taken out, since salmon do much of their growing out at sea.

Margalef's (1963) most widely read paper summarizing these ideas was called, "On certain unifying principles in ecology." The unifying principles seemed peculiarly appealing because they rest on thermodynamics and can be expressed in formal mathematical terms through information theory. But, when looked at more closely, some may be found to lack profundity. The statement that information and negentropy develop during the course of succession, though elegant, is in reality no more than a statement of the original observations of succession themselves. A succession starting from a bare field begins with very little mass and starts, as it must, with a simple community. To produce a complex ecosystem much work has to be done. This work organizes molecules, thus reducing the entropy of the system. It produces structure. The structured system so produced can be conveniently described by information theory, but this description does not in itself throw new light on the process of succession. The "unifying" part of the Margalef hypothesis comes with the suggestion that information feedback in the evolving system becomes self-directing and self-organizing. The hypothesis then runs into grave theoretical and practical difficulties.

The hypothesis requires the foregoing of present production by the young ecosystem to build up future benefit in information passed on. But what in truth is done with the present production whose consumption is apparently foregone? Some goes to decomposers; some is used to enable large plants to hold space; some accumulates as litter. Is litter or the physical structure of trees to be thought of as "information," as an organizing principle? Rather than support the hypothesis, the fact that production exceeds respiration in the early days of successions seems more to suggest that the equating of organic molecules with "information" is a dubious process. Organic molecules represent a low state of entropy in the system, all right, but they only become information if they can be read. Litter and unused biomass represent energy passed on, but not information. This distinction is obvious when the export from the plankton to the benthos by gravity is considered; potential energy is passed on, and this is used by the benthos to produce negentropy and information, but the benthic organization itself cannot be said to have been passed down from the plankton.

But the greatest difficulty with the Margalef hypothesis is its treatment of the ecosystem as the functional biological unit. His ecosystem is a unit which is thought to develop by so organizing itself that information is conserved and managed. But an ecosystem is no more than a conceptual device. There are no ecosystems which we can classify like species; something that was found out the hard way by the labors of the old plant sociologists. Tansley's ecosystem was merely a convenient concept for empirical study. It is still no more. Animals, plants, and their physical habitat react on each other in ways that can be understood by systems theory and simulated by systems analysis. But ecosystems are built up by the immigrations and adjustments of the species that are their parts. They do not evolve during the process of succession. Margalef is led to think, and to write, about ecosystems being selected for. But natural selection does not choose between ecosystems, it chooses between individual living things. Group selection, as proposed by Wynne-Edwards to preserve group behavior, is hard enough to reconcile with the theory of natural selection (Chapter 33); Margalef seems to require that whole ecosystems, not just groups, be preserved by selection. I hesitate to say it, but there seems little in Margalef's information theory approach to succession which was not incorporated in Clements' superorganism. The two hypotheses describe the same wonderful organized beast, the one in forceful prose, the other in forceful mathematics.

A satisfactory explanation of succession must be compatible with the theory of evolution by natural selection. It must fully recognize that the individuals and species which make up the communities of the various stages exist, and are present in their regular place in the succession, because selection has preserved them as individuals. To drop into distorted metaphor, we may say that the forest is wonderful, but we must explain its wonder by looking at trees.

SUCCESSION EXPLAINED AS THE REPLACEMENT OF OPPORTUNISTS WITH EQUILIBRIUM SPECIES

To a field naturalist, the most evident and striking fact of succession is that species follow each other in regular and predictable order, first the familiar weeds of open fields, then the perennial herbs, the shrubs, and the successive communities of trees; or some equally familiar sequence of other successions. *Individuals* replace each other in regular array. Many of the adaptations of succeeding *individuals* to their appointed positions can be readily understood. The pioneers are opportunists, able to disperse well and to occupy open ground. They are replaced by more persistent plants, slower to move in, less adapted to the rigors of an open site, but more persistent when they do come. We may say that opportunists colonize, but equilibrium species exploit. The facts of succession are the displacement of opportunists by equilibrium species.

Opportunists are generalists, using their energy supplies to maintain plastic physiology, unspecialized behavior, and expensive methods of widespread dispersion. This strategy pays by allowing them to exploit the energy received in habitats laid bare by catastrophe. But the habitats which they exploit are so short-lived, and the strategy of dispersal which they follow is so expensive in energy, that they must be as highly productive as possible. Their energies cannot be diverted to such luxuries as woody stems and massive roots. Opportunists are thus, by their very strategies, not equipped to hold space or achieve lasting dominance.

Equilibrium species are adapted to living in relatively predictable environments. They may adopt specialized physiologies and stereotyped responses, strategies that avoid using energy to build structures and programs designed to meet emergencies which may never come. They need not be prepared to disperse rapidly and widely, either, so that their strategy does not require reserving so much energy for reproduction. They have energy available for what were luxuries to the opportunists, for building those massive structures and storage organs which give them the advantage in a long struggle. The extreme equilibrium species are those of the climax; plants which have won space and achieved lasting dominance, and specialized animals safe by reason that the parameters of their niches are unlikely to be changed.

There is a continuum of strategies possible between those of the extreme opportunists and those of extreme equilibrium species. We can see various intercepts on this continuum in the familiar stages of the old field successions. The extreme pioneers, the annual plants, are the extreme opportunists. They disperse magnificently, grow rapidly, and throw all their energies into seed production. Perennial herbs which replace them afford the luxury of underground storage organs. They must pay a cost in lesser dispersal powers, for which their arrival after the annual herbs is evidence, but they reap an energy return next spring when they occupy space before the annual plants may grow. Then come the successively increasing diversions of energy to holding space represented by the woody structures of shrubs and trees.

Successions occur essentially because the surface of the earth has many uncertain, unpredictable living places, but also because it has other living places whose tomorrows may be forecast from experience of their yesterdays. Natural selection has promoted different strategies for survival in these two sorts of places, the strategies of opportunist and equilibrium species. If some rare event, like a plague of men, displaces the equilibrium species of a place, opportunists may enjoy a few brief moments in the sun before the equilibrium species return. The predictability of successions in any country is a function of the species list that accidents of history, geography, and evolution have provided, and the relative dispersal powers of these species.

The standard phenomena of succession can thus be seen to follow from this central fact, *that opportunists must be displaced by equilibrium species.* The high productivity of early stages is a consequence of the strategy of opportunists in requiring large energy reserves for dispersal.

The increasing biomass of later stages reflects the luxuries of structure enjoyed by equilibrium species. The relative rise in respiration as succession proceeds is a product of the maintenance that these nonproducing structures require. The closing of the nutrient cycle is another consequence of the luxury of building unproductive structures, in this instance a complex root network.

But why does the number of species increase so much in the late stages of successions? The number must, of course, go up overall, because you start with none. But it does seem that the really large increase in numbers always happens late in successions, in the terminal phases that build the climax. Why should this be? The answer seems clear. There are more equilibrium species available. In Chapter 41 it was suggested that the places of greatest diversity were inhabited by equilibrium species, that these were numerous because the stability of their homes made extinction unlikely, letting their numbers collect. Climax stages become staffed by many species because there are many equilibrium species available to staff them. The early successional stages are staffed by opportunists, but opportunists are comparatively rare.

As immigration to the climax formation from the large pool of equilibrium species proceeds, so niches must be constrained, and so the parameters of niches represented by other living things become more important. There are more and more of those interactions that Margalef saw as information, and Clements saw as organismal properties. Far from being organizing principles, these complex interactions are the necessary consequence of crowding more and more species into the same space.

In the proper Darwinian view of ecology there is no organizing deterministic principle behind succession. Successions are not directed by some holistic process of the superorganism. Nor, and this is much more important to modern ecology, are they directed by negative feedbacks of ever-refining ecosystems. The ecosystem concept provides an understanding of natural events, allowing, when very rough approximations are made, examinations of nature by systems analysis. But the ecosystem itself has no biological identity. Complex ecosystems are the product of crowding many species into restricted spaces, of forcing them to live together and to adapt to each other's presence. The successional phenomena which lead to complex ecosystems are fully explainable as the result of both opportunist and equilibrium species being maintained in any country.

Men farm opportunist plants. They do so because opportunist plants grow well and swiftly on bare ground, and because such plants do not produce massive inedible structures like vast networks of roots and spreading woody branches. Such plants put as much of their energy as they can into reproductive structures like the grains of cereals, providing the farmer with the concentrated food energies that make farming pay. Or men farm the only slightly less opportunistic plants of postpioneer but still early succession stages, those that divert their energies into underground storage

SUCCESSION AND HUMAN FOOD PRODUCTION

for growth in the spring, plants like sugar beets and potatoes. The essential strategy of agriculture requires opportunist plants, since farmers have no use for plants that produce uneatable resistant structures with their energy reserves, and who then use these structures to maintain themselves. Even in his orchard crops, man is concentrating on relatively opportunistic plants. Fruits like apples and pears represent the massive investment in dispersal of shrub-trees of middle succession stages, trees that do not survive in the climax forest.

Everywhere, as man farms more and more of the earth, he drives successions back to early stages, removing the complex ecosystems of maturity as he goes. In doing so he acts as an agent of mass extinction, because he increasingly leaves no large undisturbed places behind where the plants and animals of the climax stages may persist. A world in which all successions are thrust back to immaturity is a world depauperate in species, because opportunistic species, all that are permitted to live in early successional stages, are rare. The plough is the most deadly agent of extinction ever devised; not even thermonuclear weapons pose such a threat to the beauty and diversity of life on earth. Not that encouraging early succession stages does not have aesthetic advantages, too. It certainly increases the plant production that can be used by animals, which is why one animal, man, encourages them. Many familiar beasts benefit from the early successional stages made by farmers. The extraordinary abundance of sparrows and robins around farms or of whitetail deer in the United States, are testimony to this. So, of course, are our abounding rats. But if 5 percent of every country—5 percent of the good land, not just the desert places—were spared the plough, we could probably maintain all the species we have inherited in local patches of climax, while driving the succession over the rest of it back to the highly productive pioneer stages.

But farming pioneer stages has some practical economic costs, too, the most unavoidable of which is probably the loss of nutrients. The nutrient cycles are not closed in plant communities that do not have massive expensive structures. Farmers have to expend energy (of fossil fuels nowadays) to carry nutrients back to the land. The climax formation would have used the energy of the sun to do the same thing.

The earth is made of patches; some hot, some cold; some high, some low; some wet, some dry; some seasonal; and some without seasons. In virtually all these earthly patches, life based on carbon chemistry is possible; but different strategies, different adaptations, are required in each. There is divergence of character in different countries as selection favors those out of large families who are most suited to the local environment. But the earthly patches shift with time, both with the slow change of geologic process and by the swifter progress of changing climate. And the animals and plants themselves move. Populations whose characters have diverged are brought together somewhere along the blending borders of habitats, where diverged characters are both fixed and further separated by the process of character displacement. The mosaic of earthly patches becomes mirrored in a mosaic of species.

On land, this initial richness or diversity of species caused by the earthly mosaic is physically enhanced by the complex structures of terrestrial plants. Mosaics are stacked on mosaics, providing new kinds of physical homes for animals, multiplying the initial diversity many times. But in the sea, except among coastal algal beds and coral reefs, there are no large living structures to multiply the diversity of animals or epiphytic plants. This must go some way toward explaining the comparative paucity of species in the open sea.

The diversity resulting from mosaics overlaid on mosaics is then enhanced yet again by the trophic structure of the community; it is multiplied by some fraction of the numbers of links in food chains (Hutchinson, 1959). But this multiplication is never likely to be very large. Food chains are short, being kept short by the stern realities of the second law of thermodynamics, so that the multiplier can usually only be three or four. And even this effect is muted because animals high on food chains have catholic tastes, feeding on many different kinds of the animals below them, and being much less diverse than their prey. So the additional diversity introduced by trophic dynamics is bound to be small.

So the patchiness of the earth takes us some way toward explaining the diversity of life; there is a mosaic of habitats, each habitat is made more physically diverse by the structures of plants, and the resulting diversity is then compounded by food chains. Some way; but not very far. The actual richness of species in the milder parts of the earth is such that any argument based on compounded environmental mosaics is unconvincing. A more complete explanation comes from reflecting on what actually happens when a new species does appear.

A new species arrives in a community as an immigrant, entering a place that is already well-used by its ancient inhabitants. Most of the resources that the newcomer will need have been appropriated for someone else's use, so that ownership must be disputed. In the ensuing competition, the newcomer (or perhaps the original owner) may be eliminated according to the Lotka-Volterra-Gause model. But it is much more likely that the newcomer will not find itself in such direct competition with anyone, but will rather be competing for the marginal resources of a number of the original inhabitants. There will be a set of character displacements, allowing the original inhabitants to relinquish their margins without a fight, at the same time drawing boundaries for the newcomer's niche. The diversity has been successfully increased by one species. And yet there has been no change in physical structure of the place, nor is it likely that the lengths of food chains have been altered. The increase in diversity has resulted from properties of the speciation process itself.

If one species can be added to a community in this way, so can others. The history could be repeated again and again, leading to an indefinite increase in the number of species, and suggesting that there should be an almost infinite number of species on all the pleasant places of this ancient earth. But there are not infinite numbers of species, merely very large numbers. Something must limit this process for automatically increasing diversity, and we must ask the question, why are there not more species still?

In some places the chances of becoming extinct may be so great that species may be lost to the community very quickly. Speciation will proceed, perhaps rapidly, but the number present at any one time is kept low. This should happen in places of unpredictable environmental hazard, places like estuaries, shallow lakes, arctic latitudes, and the tops of high mountains. These are all places known to support few species, for all that some of them may be fertile and teaming with life. In the equatorial lowlands, however, where predictable comfort is the order of existence, we find the greatest diversity of all. The numbers of species there may not be infinite, but they are apt to seem so to a visiting naturalist from the temperate North.

Yet there must be still other mechanisms operating to limit species number, because the diversities of some kinds of animals and plants are not nearly so influenced by environmental hazard as are others. The diversity of birds from place to place does not vary so dramatically as, for instance, the diversity of plants and insects. There are more kinds of birds in the tropics of course, but you can explain much of the difference by the more complex physical structure of tropical forests (MacArthur's foliage height diversity, Chapter 39). Even without making allowance for the size of tropical trees, the increased number of birds does not overwhelm the naturalist as does the diversity of other groups. Whittaker (1969) notes that whereas the diversity of birds may increase only modestly from a boreal forest through a hardwood forest to the tropical rain forest, there may be three species of tree in the boreal forest, 30 in the temperate hardwood forest, and 300 in the rain forest. Clearly bird species are subjected to limits that are not applied so harshly to trees, and these limits apply nearly as strongly in the unstable climates of the North as they do in tropical lowlands. An ecologist explains this by saying that birds compete strongly for food, and are able to press their competition far and wide through their ability to fly, so that the possibilities for speciation are limited to very few parameters of the niche (which makes the broken stick model so attractive). You can tell the distinctness of many a bird's niche by measuring its bill (Chapter 39). The restrictions on numbers of different sizes and shapes of bill that can exist in the tropics are still those that limit the numbers and shapes of bill size in the North, with the result that we have roughly similar numbers of bird species in each place. The actual number of species seems to be set by the mechanical possibilities of being a flying forager.

Similar restrictions on within-habitat diversity, though doubtless not so restraining, probably apply to most vertebrates, but there are also the possibilities for between-habitat diversity which less motile animals can perhaps better exploit. If the limits of niche size preclude more animals from sharing a habitat, some can often work their character displacements by specializing physically in a smaller part of the habitat. We should perhaps expect more regional differences in diversity among walking vertebrates than among birds.

The very great diversity of plants has perhaps long been one of the more remarkable oddities of nature. Plants are so fixed, so obviously living side by side that it has not been so easy to understand how they can be living in different ways. Why are there many kinds of grass in a pasture or many kinds of trees in a tropical forest? We seem to see the answer to this in the special ways which plants have been forced to adopt to defend themselves against being eaten. They are unable to move, apparently as vulnerable to an animal wanting a meal as the proverbial sitting duck is to a man with a gun. But plants have developed defense mechanisms all the same, obvious ones like thorns, more subtle ones like making seeds attractive to parrots so that some may be carried clear of the waiting insects to a place where they may germinate in safety (Chapter 34), and more subtle ones still, like making themselves taste nasty. Plants have long been known to contain varieties of curious compounds, often toxic, noxious compounds like the plant alkaloids on which medieval poisoners based their art, but which have no known physiological function. It seems likely that these compounds serve to make the plant uneatable to most animals, something for which the selective advantage is obvious. But selection has then favored animals that have been able to eat the plant in spite of the poison. Animals become specialists in eating particular plants; the plants, on an evolutionary time scale, respond with new poisons. New plant species are being made constantly as individuals occur which are inedible to the animals that ate the parental stock, and new species of herbivorous insects are evolved as fast to eat them. A similar game is played between dispersing seeds and the hunting skills of seed predators. Both processes should go on virtually indefinitely, being offset only by random extinction. In ancient equable tropics they lead to the bewildering numbers of species of both insects and plants which we see. The actual number is essentially a function of the time for which the tropical land has endured, and this is the result of geologic process.

So the answer given by ecologists to their discipline's greatest question, "Why does the world contain the number of kinds of plants and animals which it does?", is complex. It is that speciation will proceed indefinitely as organisms move about, diverge, and are then confronted with each other again. But some parts of the earth have such unpredictable climates, or such ephemeral states, that the inhabitants have hazardous lives and may easily become extinct. An equilibrium between speciation and extinction is obtained with relatively few species being present. In addition, there are ways of life, like those of birds, so restrictive that speciation by character displacement becomes difficult and slow. An equilibrium between extinction and speciation can then be attained with relatively low numbers even in the less hazardous equable places. But a third group of organisms, particularly including plants and insects, can maintain high rates of speciation even in communities that are already richly diverse. Perhaps these should be called **evasive** species, because evading persecutors is more important to their founding than avoiding competitors. Limits are set to the numbers of such evasive species only by chance natural upheavals that cause widespread catastrophe. The emergence of technological man is probably the most destructive of this kind of natural upheaval in all the history of life, enabling us to witness the greatest mass extinction that the geological records of the future will hold. An epoch or two must pass before our inherited richness is replaced.

A diverse community is thus built by a process of accretion. Species are forced by character displacement to refine their strategies, to accept limits, so that they avoid competition or evade persecutors. Yet all the individuals living in one place must share alike energy and raw materials, taking in turns their unique gulps at the flowing energy, but holding raw materials only briefly before passing them on to others. It follows that the animals and plants of a place must function together as a natural system, a splendid engine, doing work, driven by the unceasing radiations of the sun. It should have been possible to deduce the existence of ecosystems as soon as the distinctness of species and the theory of evolution by natural selection were widely understood. If only we had! We should then have had to our credit an intellectual *tour de force* the equivalent of Einstein's. But instead we had to discover the existence of ecosystems the hard and patient way; through geography and the analysis of communities. Even after we had clearly realized that natural communities functioned as systems, we still had to obfuscate the simple mechanical inevitability of them with quasi-philosophies that gave them lives of their own. Gifted ecologists have been led into intellectual heresies about superorganisms or self-organization. But an ecosystem is no more than the dynamic sum of its dynamic parts. When the parts are most numerous, an ecosystem should be most stable. But its stability is also affected by the stability of the physical world in which it is placed. These two things are enough by themselves to explain its wonderful properties.

The understanding we have thus built up of the working of ecosystems and the grand diversity of life of which they are properties comes ultimately from examining the strategies that animals and plants have adopted to win resources. If you avoid competition or evade a persecutor, you are adopting a strategy of life. Natural selection chooses between strategies, always giving the advantage to that strategy which leaves most viable offspring. Increasing diversity is a process of fitting together strategies. The result of this fitting together we understand by our concept of the ecosystem. It follows, then, that an ecologist setting out to learn the workings of some part of the natural world must study the strategies of individual species. The question he must ask himself is: What are the tricks used to turn resources into babies? This is the rewarding way to approach autecology. It is also the approach used by those who have made the great advances of ecology in the last few years.

The study of habit and habitat, the defined subject matter of ecology, is necessarily a study of fitness for a way of life. But fitness, in a Darwinian sense, is measured by the ability to produce young which themselves survive to reproduce, and success in this endeavor is dependent on adopting a successful strategy for obtaining resources. The outcome of this logic has led to an explanation of the rich diversity of life. Now also many perplexing details, particularly of behavior, are beginning to yield to it. Jantzen explained the diversity of rain-forest trees in terms of their strategies for getting seeds to where the waiting predators could not find them, and Harper showed that herbs and insects have developed strategies for their games of hide-and-seek (Chapter 34). Probably much of the behavior that Wynne-Edwards suggested might be the product of group selection (Chapter 33) will shortly be shown to reflect strategies of individuals. Consider the group action of penguins jumping into the water together; why this social custom? If there is a leopard seal in the water, the individual's chance of survival goes up if he is in a crowd, so it is better to jump with the crowd, even though you know there is a leopard seal below the rock. The social behavior has been evolved because the favored strategies of individuals are to act with the crowd.

Reflections on the strategies of individuals can also lead to new understandings of the ways in which numbers are limited. Compare again the strategies of opportunist and equilibrium species. An opportunist strategy for life in a temporary place or an hazardous place must have a contingency plan for dispersal. Babies must not only be numerous, as they must be for all kinds of life, but they must be wanderers, whether dangling under a pappus like a dandelion seed or flying like a grasshopper. Many must be lost, and so they must be small. The special qualities of the opportunist species which result are the vital ingredients in that striking phenomenon of ecological succession (Chapter 39). The same special qualities make possible extremely rapid population growth, leading to characteristic plagues of opportunistic species in marginal habitats. The mass death of such plague irruptions is always certain, a density-independent death of the kind which stimulated the philosophy of Andrewartha and Birch (Chapter 27). But, as my colleague Jerry Downhower points out, these very opportunist strategies may also be seen as promoting density-dependent controls. The strategy of opportunism means that you must be ready to take advantage of a good breeding opportunity with the maximum of young; but, if the resource fails, you must get through the most young possible at the time. The best way of adapting the number of young to this fluctuating resource is by some feedback mechanism, to whit, density-dependence. Large families and density-dependent control are the essence of the Lotka-Volterra equations, as they must also be for opportunist species. But, if you are an equilibrium species adopting a strategy suitable for a place of predictable resource, it is in your interest to evolve a family size preadapted to the size of the expected resource. A true feedback of information to the growing young, as applied by a density-dependent control, must always involve the loss of some individuals. This is wasteful. A strategy that instead puts all the resource into babies that are going to survive will leave the most surviving offspring, and may be expected to replace a strategy that depends on negative feedback after the young are born. This is presumably the explanation of Lack's (1968) observation that northern birds have larger clutches of eggs than southern birds. The northern birds are more opportunist, must rely more on density dependence to adapt their family size to the resources, and so must lay more eggs. But in the equable sourthern regions, natural selection has favored those individuals who lay just the number of eggs which the regular resource allows them to rear. These southern birds avoid the cost in lost babies of density-dependent restraints, leave more descendents to reproduce next year, and so are favored. Much of the controversy between the density-dependent and density-independent schools of thought would have been avoided if population size was always seen as the consequence of the history of resources and the strategies that this history made necessary. Populations must tend to be stable when resources are stable, but to fluctuate when resources fluctuate.

Ecology is reaching some sort of plateau in explaining the grander ecological phenomena; of successions, of biogeography, of the existence of species, and of the number of species; a plateau in its understanding of the distribution and abundance of life. Behind all the answers we give lies the realization that living things are the embodiment of strategies adopted to obtain resources. Darwin gave us our first awareness. Lotka, Volterra, and Gause passed on the torch. The natural order that was the product of individual strategies was revealed by analytical minds like those of Elton and Tansley. And now, with the main patterns of the natural world clearly understood, a spate of students is applying the science of strategy to bring the remaining perplexities tumbling before reason.

But in every ecologist's mind is the worry about the strategies for THE HUMAN SIDE collecting resources now being adopted by the most formidable animal of all. Man alone can change his niche without speciating, can adopt different strategies at will. An ecologist sees the very triumph of man over all other animals as being due to this ability to change his niche. From hunting and gathering, to herding, then farming, the city state and the great nation, an ecologist sees the human advance as being the fruits of a continual revision of the human niche. At every stage more energy has been won, more competing animals have been displaced, more resources have been made available, and it has been possible to breed more men to enjoy them. There has also come about a great broadening of the human niche, so that very many resources can be enjoyed by one man, giving wonderful possibilities for varied lives and self-fulfillment. Yet the promise of such lives for the men of the future is in jeopardy because of our inherited strategy for collecting resources only for the next meal. Like other animals we adopt the family size that the resources of the moment suggest can be fed, or that we can afford, and we are prepared to subvert whatever resources may be at hand so that we may continue this breeding strategy, regardless of the consequences for future generations. The time is already on us when our dreams of providing the widened way of life for all our people is being thwarted because there are not enough resources to go round, because the carrying capacity of our living space is not enough to provide a broadened niche for all the men who now exist. In every nation on the earth the birthrate exceeds the death rate and populations continue to rise. In the past this has not mattered because the very dynamism of our expansion has won both the resources for more people and a widened niche for at least some of them. But now the size of our inherited crowds is such that the space for more wide-ranging lives is not available. If we still let our numbers go up we must herd together in monotony, knowing that what is possible for today's men will not be possible for their grandchildren and the generations beyond.

The increased activities, and rising numbers of men, do not pose a threat of imminent catastrophe. The oxygen of the atmosphere is not at risk. Carbon dioxide and pollutants discharged into the air are unlikely to start a new Ice Age, and if they do it does not matter very much for the coming and going of Ice Ages are normal parts of the long-term human heritage. The contemporary worries about polluted lakes and smelly cities are trivial matters that can easily be put right. There is no danger that the toxic chemicals we use and waste can end life on earth, even though pesticides may well end forms of life we value. The prospect of more widespread famine, as we approach some theoretical upper limit to the world's food supply, is still some decades off; perhaps even a few generations away. It is true that many of the promises of plenty, which agriculturalists and others have put forward, are as false as the threats of catastrophe themselves. The oceans are not a rich productive resource waiting to be tapped but deserts that cannot be profitably farmed. Many a scheme for ploughing virgin lands will result in fresh deserts of the dust-bowl variety. Algae are no more productive than other plants, so that there can be no marvelous increase in the world's food supply through algal culture. Yet the world's food can probably be increased several times by efficiently farming all the land that can be farmed. Famine there will be, just as there is at the moment when millions probably die of mal-nutrition every year. There will be increasing millions dead from hunger as peoples crowd more and more in the decades ahead, but those in the better-off countries can continue to ignore them, as they have done up to now. It is thus not sudden disaster, or crushing hunger, which we must fear, but a steady and perhaps irreparable loss of our chances for a satis-fying life.

Our strategy of choosing the family size that each individual couple can afford can be continued for two or three generations more without sudden natural catastrophe, but only if governments are ruthless and strong, rationing resources, regimenting dissenters who do not like their rations, acting as if nothing matters but feeding families that are being freely born now. It will then soon come about that there are, say, 1000 million Americans, 100 million Englishmen, and peoples of other coun-tries in proportion. The game refuges will have gone, the last wild places will have been subdued, the cities will be compact and space saving, and the looks of all the lands will be dictated by the needs of the food plants. The human niche will have been changed once again as every individual accepts his simple, meatless diet, and his synthetic surroundings, denying himself the beauties of adventure and the wild places, adapting his spirit to the knowledge that a wider life cannot be lived by his children or those who will come after.

Those vast populations being prepared for us such a little time ahead will, of course, only be able to maintain even their spiritually impover-ished existence by finally accepting breeding restraints. If we, who have so much to win, cannot change our breeding strategy now, why should they, who will have lost almost everything of which men can be proud, be able to do it then? The thing has to be started soon. Fortunately, the crowding of men who have been taught to believe that a better life is pos-sible, will produce social upheavals under whose pressures historic human attitudes can change. There may be revolutions, tyranny, and war. In the politics that they bring about may lie our best hopes for altering the pattern of breeding that has served well enough in the past, but that will only serve us ill now. Man, the only animal to change his niche at will, must become the only animal to accept a voluntary restriction of his family size to something well below what he thinks he can afford. When an ecologist is asked by those who have received but garbled accounts of trouble to come, "What can I do to help?" the only useful answer he can give is, "Make sure you have no more than two children yourself and try to persuade your friends to do the same."

Allee, W. C., A. E. Emerson, O. Park, T. Park and K. P. Schmidt. 1949 "Principles of Animal Ecology," Philadelphia, W. B. Saunders, p. 837.

Allen J. A. 1871 "Mammals and Winter Birds of East Florida, and a Sketch of the Bird Faunae of Eastern North America," *Bulletin of the Museum of Comparative Zoology,* Cambridge, Massachusetts, 2:161–450, Pt V, pp. 375–450.

American Chemical Society. 1969 "Cleaning our Environment, the Chemical Basis for Action," Report by the Subcommittee on Environmental Improvement Committee on Chemistry and Public Affairs, Washington, D.C.

Andrewartha, H. G. and L. C. Birch. 1954 *The Distribution and Abundance of Animals,* Chicago: University of Chicago Press, p. 782.

Andrews, R. V. 1968 "Daily and Seasonal Variation in Adrenal Metabolism of the Brown Lemming," *Physiological Zoölogy,* 41:86–94.

Armstrong, J. 1960 *The Dynamics of Dugesia tigrina Populations and of Daphnia pulex Populations as Modified by Immigration.* Ph.D. Dissertation, Ann Arbor, p. 104.

Baylor, E. R. and W. H. Sutcliffe. 1963 "Dissolved Organic Matter in Seawater as a Source of Particulate Food," *Limnology and Oceanography,* 8:369–381.

Beard, J. S. 1955 "Tropical American Vegetation Types," *Ecology,* 36:89–100.

Benninghoff, W. S. 1966 "The Relevé Method of Describing Vegetation," *Michigan Botanist,* 5:109–114.

583

Billings, W. D. 1970 *Plants and the Ecosystem*. Belmont: Wadsworth p. 160.

Birch L. C. and H. G. Andrewartha. 1941 "The Influence of Weather on Grasshopper Plagues in South Australia," *Journal of the Department of Agriculture South Australia*, 1941, 45:95–100.

Birge, E. A. and C. Juday. 1934 "Particulate and Dissolved Organic Matter in Wisconsin Lakes," *Ecological Monographs*, 4:440–474.

Blumenstock, D. I. and C. W. Thornthwaite. 1941 "Climate and the World Pattern," *Climate and Man* (USDA Yearbook), pp. 98–127.

Bodenheimer, F. S. 1938 *Problems of Animal Ecology*, London: Oxford University Press.

Bonner, J. 1962 "The Upper Limit of Crop Yield," *Science*, 137:11–15.

Börner, H. 1960. "Liberation of Organic Substances from Higher Plants and Their Role in the Soil Sickness Problem," Botanical Review, 26:393–424.

Boyce, J. B. 1946 "The Influence of Fecundity and Egg Mortality on the Population Growth of *Tribolium confusum* Duval," *Ecology*, 27:290–302.

Braun-Blanquet, J. 1932 *Plant Sociology: The Study of Plant Communities* (Translated, revised, and edited by G. D. Fuller and H. S. Conard), New York: McGraw-Hill, p. 439.

Braun, E. Lucy. 1950 *Deciduous Forests of Eastern North America*, Philadelphia: Blakiston, p. 596.

Broecker, W. S. 1970 "Man's Oxygen Reserves," *Science*, 168: 1537–1538.

Brown, J. L. 1969 "Territorial Behavior and Population Regulation in Birds," *Wilson Bulletin*, 81:293–329.

Brown, L. L. and E. O. Wilson. 1956 "Character Displacement," *Systematic Zoology*, 5:49–64.

Bryson, R. A. 1966 "Air Masses, Streamlines, and the Boreal Forest," *Geographical Bulletin*, 8:228–269.

Burnett, T. 1958 "Dispersal of an Insect Parasite over a Small Plot," *Canadian Entomologist*, 90:279–283.

Cain, S. A. 1950 "Life-forms and Phytoclimate," *Botanical Review*, 41:1–32.

Cain, S. A. and G. M. de Castro. 1959 *Manual of Vegetation Analysis*. New York: Harper and Row, p. 325.

de Candolle, A. P. A. and C. de Candolle. 1824–1873 *Prodromus Systematis Naturalis Regni Vegetabilis*, 17 vols., Paris.

de Candolle, A. P. A. 1874 "Constitution dans le Règne Végétal de Groupes Physiologiques Applicables à la Géographie Ancienne et Moderne," *Archives des Sciences Physiques et Naturelles*, Geneva, May.

Carrick, R. 1963 "Ecological Significance of Territory in the Australian Magpie *Gymnorhina tibicen*," *Proc. XIII Intern. Ornithal. Cong.* 9:740–753.

Caughley, G. 1970 "Eruption of Ungulate Populations, with Emphasis on Himalayan Thar in New Zealand," *Ecology*, 51:53–72.

Chapman, R. N. 1928 "The Quantitative Analysis of Environmental Factors," *Ecology*, 9:111–122.

Chapman, R. N. 1931 *Animal Ecology with Especial Reference to Insects*, New York: McGraw Hill.

Chitty, D. 1958 "Self Regulation of Numbers through Changes in Viability," *Cold Spring Harbor Symposium of Quantitative Biology*, 1958, 22:277–280.

Christian J. J. 1950 "The Adreno-Pituitary System and Population Cycles in Mammals," *Journal of Mammology*, 31:247–259.

Christian, J. J., 1960 Vagn Flyger, and David E. Davis. "Factors in the Mass Mortality of a Herd of Sika Deer," *Cervus nippon, Chesapeake Science*, 1:79–95.

Christain, J. J. 1970 "Social Subordination, Population Density, and Mammalian Evolution," *Science*, 168:84–90.

Clements, F. E. 1916 "Plant Succession: An Analysis of the Development of Vegetation," *Carnegie Institution of Washington Publication 242*, facsimile reprint of Haffner.

Cohen, J. 1966 *A Model of Simple Competition, Cambridge:* Harvard University Press, 138 p.

Cohen, J. 1968 "Alternative Derivations of a Species-Abundance Relation," *American Naturalist*, 1968, 102:165–172.

Cole, LaMont C. 1970 "Playing Russian Roulette with Biogeochemical Cycles," *The Environmental Crisis* ed. by H. W. Helfrich, New Haven: Yale University Press, p. 187.

Colinvaux, P. A. 1963 "The Environment of the Bering Land Bridge," *Ecological Monographs*, 34:297–329.

Colinvaux, P. A. 1968 "Reconnaissance and Chemistry of the Lakes and Bogs of the Galapagos Islands," *Nature*, 219:590–594.

Connell, J. H. 1961 "The Influence of Interspecific Competition and Other Factors on the Distribution of the Barnacle *Chthamalus stellatus*," *Ecology*, 42:710–723.

Connell, J. H. and E. Orias. 1964 *The Ecological Regulation of Species Diversity*, American Naturalist, 98:387–414.

Coope, G. R. 1967 "The Value of Quaternary Insect Faunas in the Interpretation of Ancient Ecology and Climate," *Quaternary Paleoecology*, edited by E. J. Cushing and H. E. Wright, New Haven, Yale University Press.

Coupland, R. T., R. Y. Zacharuk and E. A. Paul. 1969 "Procedures for Study of Grassland Ecosystems," *The Ecosystem Concept in Natural Resource Management* ed. by Van Dyne, New York: Academic Press, pp. 25–47.

Cowles, H. C. 1899 "The Ecological Relations of the Vegetation on the Sand Dunes of Lake Michigan," *Botanical Gazette*, 27:95–117, 167–202, 281–308, 361–369.

Dansereau, Pierre. 1951 "Description and Recording of Vegetation on a Structural Basis," *Ecology*, 32:172–229.

Dansereau, Pierre. 1952 "The Varieties of Evolutionary Opportunity," *Review of Canadian Biology*, 1952, 11:305–388.

Dansereau, Pierre. 1957 *Biogeography, an Ecological Perspective*. New York: Ronald Press, p. 394.

Darlington, P. J. 1957 *Zoögeography: The Geographical Distribution of Animals*. New York: John Wiley, p. 675.

Darwin, Charles. 1859 *Origin of Species,* London: Murray, facsimile reprint, Cambridge: Harvard University Press, 1966.

Davidson, J. and H. G. Andrewartha. 1948 "The Influence of Rainfall, Evaporation, and Atmospheric Temperature on Fluctuations in the Size of a Natural Population of *Thrips imaginis* (Thysanoptera)," *Journal of Animal Ecology,* 17:193–199.

Davies, T. A. W. and P. W. Richards. 1933 "The Vegetation of Moraballi Creek, British Guiana," *Journal of Ecology,* 21:350–384.

Davis, D. E. and F. B. Gooley. 1963 *Principles Mammology,* New York: Reinhold.

Davis, M. B. 1963 "On the Theory of Pollen Analysis," *American Journal of Science,* 261:897–912.

Davis, R. B. 1967 "Pollen Studies on near-Surface Sediments in Maine Lakes," *Quaternary Paleoecology* ed. by E. J. Cushing and H. E. Wright, New Haven: Yale University Press, pp. 143–173.

Dawson, E. Yale. 1966 *Marine Botany: An Introduction,* New York: Holt, Rinehart, and Winston, p. 371.

De Bach P. (ed.) 1964 *Biological Control of Insect Pests and Weeds,* New York: Reinhold, p. 844.

Deevey, E. S. 1939 "Studies on Connecticut Lake Sediments 1. A Postglacial Climatic Chronology for Southern New England," *American Journal of Science,* 237:691–724.

Deevey, E. S. 1942 'Studies on Connecticut Lake Sediments III. The Biostratonomy of Linsley Pond," *American Journal of Science,* 240:235–264, 313–324.

Deevey, E. S. 1947 "Life Tables for Natural Populations of Animals," *Quarterly Review of Biology,* 22:283–314.

Deevey, E. S. 1959 "The Hare and the Haruspex: A Cautionary Tale," *Yale Review,* 49:161–179.

Deevey, E. S. 1969 "Coaxing History to Conduct Experiments," *Bioscience,* 19:40–43.

Deevey, E. S. 1970 "In Defense of Mud," *Bulletin of the Ecological Society of America,* 51:5–8.

Delwiche, C. C. 1965 "The Cycling of Carbon and Nitrogen in the Biosphere," pp. 29–58, "Microbiology and Soil Fertility" ed. by C. M. Gilmour and O. N. Allen, *Proceedings of 25th Annual Biology Colloquium,* Oregon State University, Corvallis.

Dice, L. R. 1952 *Natural Communities*. Ann Arbor: University of Michigan Press.

Dirks, C. O. 1937 "Biological Studies of Maine Moths by Light Trap Methods," *The Maine Agricultural Experimental Station Bulletin* 389, Orono, Maine.

Dodd, A. P. 1940 "The Biological Campaign Against Prickly Pear," *Commowealth Prickly Pear Board,* Bisbane, Australia, p. 117.

Dokuchaev, V. V. 1883 *Tchernozeme de la Russie d'Europe,* St. Petersburg.

Du Rietz, G. E. 1929 "The Fundamental Units of Vegetation," *Proceedings of the International Congress of Plant Science,* Ithaca, 1:623–627.

Du Rietz, G. E. 1930 "Classification and Nomenclature of Vegetation, *Svensk Botanisk Tidskrift,* 24:489–503.

Edmondson, W. T. 1945 "Ecological Studies of Sessile Rotatoria, Part II. Dynamics of Populations and Social Structures," *Ecological Monographs,* 15:141–172.

Edmondson, W. T. 1970 "Phosphorus, Nitrogen, and Algae in Lake Washington after Diversion of Sewage," *Science,* 169:690–691.

Elton, C. S. 1927 *Animal Ecology,* New York: Macmillian, p. 209.

Elton, C. S. 1942 *Voles, Mice, and Lemmings,* New York: Oxford University Press, p. 469.

Elton, C. S. 1958 *"The Ecology of Invasions by Plants and Animals,"* Metheun Landscape, London.

Elton, C. S. 1966 *"The Pattern of Animal Communities,"* Metheun Landscape, London p. 181.

Erdtman, G. 1969 *Handbook of Palynology,* New York: Hafner, p. 486.

Errington, P. L. 1946 "Predation and Vertebrate Populations," *Quarterly Review of Biology,* 21:144–177, 21:221–245.

Errington, P. L. 1963 *Muskrat Populations,* Iowa State University Press, p. 665.

Ewing, M. et al. 1968 "Shipboard Site Reports: Site 2," *Initial Reports of the Deep Sea Drilling Project,* Washington, D.C., Government Printing Office, 1:84–111.

Faegri, K. and J. Iversen. 1964 *Textbook of Pollen Analysis,* 2nd ed., New York: Hafner, p. 237.

Feller, W. 1951 *Probability Theory and its Applications,* New York: John Wiley.

Fischer, A. G. 1960 "Latitudinal Variations in Organic Diversity," *Evolution,* 14:64–81.

Fisher, R. A., A. S. Corbet, and C. B. Williams. 1943 "The Relation between the Number of Individuals and the Number of Species in a Random Sample of an Animal Population," *Journal of Animal Ecology,* 12:42–58.

Flint, R. F. 1971 *Glacial and Quaternary Geology.* New York: John Wiley.

Fontaine, A. R. and F. S. Chia. 1968 "Echinoderms: an Autoradiographic Study of Assimilation of Dissolved Organic Molecules," *Science,* 161:1153–1155.

Frey, D. G. 1953 "Regional Aspects of the Late-glacial and Post-glacial Pollen Succession of Southeastern North Carolina," *Ecological Monographs,* 23:289–313.

Gaarder, T. and H. H. Gran. 1927 "Investigations of the Production of Plankton in the Olso Fjord," *Rapp. et Proc.-Verb., Cons. Int. Explor. Mer.,* 42:1–48.

Gaastra, P. 1958 "Light Energy Conversion in Field Crops in Comparison with Photosynthetic Efficiency under Laboratory Conditions," *Medeleel. Landbouwhogeschool Wageningin*, 58(4):1–12.

Gates, D. M. 1962 *Energy Exchange in the Biosphere*, New York: Harper and Row, p. 160.

Gates, D. M. 1965 "Heat Transfer in Plants," *Scientific American*, 213(6):76–87.

Gates, D. M. 1968 "Energy Exchange between Organisms and Environment," *Australian Journal of Science*, 31:67–74.

Gates, D. M. 1968 "Energy Exchange between Organism and Environment," *Biometeorology, Proceedings of the 28th Annual Biology Colloquium* 1967, ed. By W. P. Lowry, Corvallis, Oregon: Oregon State University Press, pp. 1–22.

Gauld, G. T. 1951 "The Grazing Rate of Planktonic Copepods," *Journal of the Marine Biological Association of the United Kingdom*, 29:695–706..

Gause, G. F. 1934 *The Struggle for Existence*, Baltimore: Williams and Wilkins, p. 163.

Gause, G. F. 1936 "The Principles of Biocenology," *Quarterly Review of Biology*, 11:320–396.

Geist, V. 1966 "The Evolutionary Significance of Mountain Sheep Horns," *Evolution*, 20:558–566.

Gwynne, M. D. and R. H. V. Bell. 1968 "Selection of Vegetation Components by Grazing Ungulates in the Serengeti National Park," *Nature*, 220:390–392.

Gleason, H. A. 1917 "The Structure and Development of the Plant Association," *Bulletin of the Tory Botanical Club*, 43:463–481.

Gleason, H. A. 1926 "The Individualistic Concept of the Plant Association," *Bulletin of the Tory Botanical Club*, 53:7–26.

Godwin, H. 1956 *The History of the British Flora, a Factual Basis for Phytogeography*. Cambridge University Press, p. 385.

Good, R. 1953 *The Geography of Flowering Plants*, London: Longman Green.

Goulden, C. E. 1966 "La Aquada de Santa Ana Vieja: An Interpretive Study of the Cladoceran Microfossils," *Archiv fur Hydrobiologie*, 62:373–404.

Green, R. G., C. L. Larson, and J. F. Bell. 1939 "Shock Disease as the Cause of the Periodic Decimation of Snowshoe Hares," *American Journal of Hygiene*, Sect. B, 30:83–102.

Griffing, B. 1967 "Selection in Reference to Biological Groups, I. Individual and Group Selection Applied to Populations of Unordered Groups," *Australian Journal of Biological Science*, 20:127–139.

Hafsten, U. 1960 "Pollen-analytic Investigations in South Norway," *Geology of Norway* ed. by O. Holtedahl *Norges Geol. Unders*, pp. 208–434.

Hairston, N. G. 1959 "Species Abundance and Community Organization," *Ecology*, 40:404–416.

Hairston, N. G., F. E. Smith, and L. B. Slobodkin. 1960 "Community Structure, Population Control, and Competition," *American Naturalist*, 94:421–425.

Halbouty, M. T. 1967 *Salt Domes*, Houston: Gulf Publishing, p. 425.

Hardy. A. C. 1924 "The Herring in Relation to its Animate Environment," Part 1. *Fishery Investigations*, Series 2, Vol. 7, No. 3, Ministry of Agriculture and Fisheries.

Hardy, A. C. 1956 *The Open Sea and the World of Plankton*, London: Collins.

Harper, J. L. 1967 "A Darwinian Approach to Plant Ecology," *Journal of Ecology*, 55:247–270.

Harper, J. L. 1969 "The Role of Predation in Vegetational Diversity," *Brookhaven Symposium in Biology No. 22, Diversity and Stability in Ecological Systems*, pp. 48–62.

Harrison, H. L. et al. 1970 "Systems Studies of DDT Transport," *Science*, 170:503–508.

Harvey, H. W. 1950 "On the Production of Living Matter in the Sea off Plymouth," *Journal of Marine Biological Association of U.K.*, 29:97–137.

Henderson, L. J. 1913 *The Fitness of the Environment*, New York, MacMillan, p. 317.

Hensley, M. M. and J. R. Cope. 1951 "Further Data on Removal and Repopulation of the Breeding Birds in a Spruce-fir Forest Community," *Auk*, 68:483–493.

Hickey, J. J. (ed.) 1965 *Peregrine Falcon Populations, their Biology and Decline*, Madison: University of Wisconsin Press, p. 596.

Holling, C. S. 1959 "The Components of Predation as Revealed by a Study of Small Mammal Predation of the European Pine Sawfly," *Canadian Entomologist*, 91:293–320.

Holling, C. S. 1965 "The Functional Response of Predators to Prey Density and its Role in Mimicry and Population Regulation," *Memoirs of the Entomological Society of Canada* no. 45, p. 60.

Holling, C. S. 1968 "The Tactics of a Predator," *Symposia of the Royal Entomological Society of London*: Number 4 "Insect Abundance," ed. by T. R. E. Southward.

Holloway, J. K. 1964 "Projects in Biological Control of Weeds," *Biological Control of Insect Pests and Weeds* ed. by P. De Bach, New York: Reinhold, p. 844.

Hornocker, M. G. 1969 "Winter Territoriality in Mountain Lions," *Journal of Wildlife Management*, 33:457–464.

Hopkins, D. M. 1959 "History of Imuruk Lake, Seward Peninsula, Alaska," *Geological Society of American Bulletin*, 70:1033–1946.

Hopkins, D. M. 1959 "Some Characteristics of the Climate in Forest and Tundra Regions in Alaska," *Arctic*, 12:215–220.

Hopkins, D. M. (ed.) 1967 *The Bering Land Bridge*, Stanford: Stanford University Press, p. 495.

Howard, H. Eliot, 1920 *Territory in Bird Life*, New York: Dutton, 308 p.

Huffaker, C. B. 1958 "Experimental Studies on Predation: Dispersion Factors and Predator-prey Oscillations," *Hilgardia,* 27:343–383.

Humboldt, A. von. 1807 "Ideen zu einer Geographie der Pflanzen nebst einem Naturgemälde der Tropenlander," Tübingen, p. 182.

Hutchinson, G. E. 1949 "Circular Causal Systems in Ecology," *Annals of New York Academy of Sciences,* pp. 221–246.

Hutchinson, G. E. 1950 "The Biogeochemistry of Vertebrate Excretion," *Bulletin of the American Museum of Natural History,"* Vol. 96, p. 554.

Hutchinson, G. E. 1951 "Copepodology for the Ornithologist," *Ecology,* 32:571–577.

Hutchinson, G. E. 1953 "The Concept of Pattern in Ecology," *Proceedings of the Academy of Natural Sciences of Philadelphia,* 105:1–12.

Hutchinson, G. E. 1957 *A Treatise on Limnology,* Vol. 1, New York: John Wiley, p. 1015.

Hutchinson, G. E. 1957 "Concluding Remarks" *Cold Spring Harbor Symposia on Quantitative Biology, 22, Population Studies: Animal Ecology and Demography,* Cold Spring Harbor, pp. 415–427.

Hutchinson, G. E. 1959 "Homage to Santa Rosalia or Why are There So Many Kinds of Animals?" *American Naturalist,* 93:145–159.

Hutchinson, G. E. 1961 "The Paradox of the Plankton," *American Naturalist,* 95:137–145.

Hutchinson, G. E. 1965 *The Ecological Theater and the Evolutionary Play,* New Haven: Yale University Press, p. 139.

Hutchinson, G. E. 1966 *Treatise on Limnology,* Vol. 2., New York, John Wiley, p. 1115.

Huchinson, G. E. and V. T. Bowen. 1950 "Limnological Studies in Connecticut IX. A Quantitative Radiochemical Study of the Phosphorus Cycle in Linsley Pond," *Ecology,* 31:194–203.

Hutchinson, G. E. and A. Wollack. 1940 "Studies on Connecticut Lake Sediments II. Chemical Analyses of a Core from Linsley Pond, North Bradford," *American Journal of Science,* 238:493–517.

Ivlev, V. S. 1939 "Transformation of Energy by Aquatic Animals. Coefficient of Energy Consumption by *Tubifex tubifex* (Oligochaeta)" *International Revue Hydrobiol.,* 38:449–458.

Jantzen, D. H. 1970 "Herbivores and the Number of Tree Species in Tropical Forests," *American Naturalist,* 104:501–528.

Jantzen, D. H. 1970 "Reproductive Behavior in the Passifloraceae and Some of its Pollinators in Central America," *Behavior,* 32:33–48.

Jantzen, D. H. 1971 "Escape of Juvenile *Dioclea megacarp* (leguminosae) vines from Predators in a Deciduous Tropical Forest," *American Naturalist, 105:97–112.*

Keith, J. A. 1966 "Reproduction in a Population of Herring Gulls (Larus *argentatus*) Contaminated by DDT," *Journal of Applied Ecology* (supplement), 3:57–70.

Killick, H. J. 1959 "The Ecological Relationships of Certain Plants in the Forest and Savanna of Central Nigeria," *Journal of Ecology,*

47:115–127.

King, C. E. 1964 "Relative Abundance of Species and MacArthur's Model," *Ecology*, 45:716–727.

Kira, T. and T. Shidei. 1967 "Primary Production and Turnover of Organic Matter in Different Forest Ecosystems of the Western Pacific," *Japanese Journal of Ecology*, 17:70–87.

Klein, D. R. 1968 "The Introduction, Increase, and Crash of Reindeer on St. Matthew Island," *Journal of Wildlife Management*, 32:350–367.

Klopfer, P. H. 1969 *Habitats and Territories*, New York: Basic Books, p. 117.

Klopfer, P. H. and R. H. MacArthur. 1961 "On the Causes of Tropical Species Diversity: Niche Overlap," *American Naturalist*, 95:223–226.

Kluijver, H. N. 1951 "The Population Ecology of the Great Tit" *Parus m. major L. Ardea*, 39:1–135.

Kohn, A. J. 1959 "The Ecology of *Conus* in Hawaii," *Ecological Monographs*, 29:47–90.

Köppen, W. 1884 "Die Wärmezonen der Erde, nach der Dauer der Heissen, Gemässigten und Kalten Zeit, und nach der Wirkung der Wärme auf die Organische Welt betrachtet," *Meteorologische Zeitschrift*, 1:215–226.

Köppen W. 1900 "Versuch einer Klassifikation der Klimate, Vorzugsweise nach ihren Beziehungen zur Pflanzenwelt," *Geographische Zeitschrift*, 6:593–611.

Köppen W. 1918 "Klassifikation der Klimate nach Temperatur, Niederschlag, und Jahres lauf," *Petermann's Mitteilungen*, 64:193–203, 243–248.

Kozhov, M. 1963 *Lake Baikal and its Life*, Dr. W. Junk, p. 344.

Krebs, Charles J. 1964 "The Lemming Cycle at Baker Lake, Northwest Territories, during 1959–1962," Arctic Institute of North America, Technical Paper No. 15, p. 104.

Krögh, A. 1931 "Dissolved Substances as Food of Aquatic Organisms," *Biological Reviews of the Cambridge Philosophical Society*, 6:412–442.

Kubiëna, W. L. 1953 *The Soils of Europe*, Madrid, p. 318.

Lack, D. L. 1944 "Ecological Aspects of Species-formation in Passerine Birds," *Ibis*, pp. 260–286.

Lack, D. L. 1945 "Ecology of Closely Related Species with Special Reference to Cormorant (*Phalacrocorax carbo* and shag (*P. aristotelis*)," *Journal of Animal Ecolooy*, 14:12–16.

Lack, D. L. 1947 *Darwin's Finches*. New York: Cambridge University Press, p. 218.

Lack, D. L. 1954 *The Natural Regulation of Animal Numbers*, New York: Oxford University Press, p. 343.

Lack, D. L. 1966 *Population Studies of Birds*, Oxford University Press, 341 p.

Leighton, P. A. 1954 "Cloud Travel in Mountainous Terrain," *Quarterly Reports, Department of Chemistry, Stanford University,* (Defense Documentation Center AD Number 96571), pp. 111–113.

Leighton, P. A. 1966 "Geographical Aspects of Air Pollution," *Geographical Review* 56:151–174.

Leopold, Aldo. 1936 *Game Management,* New York: Charles Scribner, p. 481.

Leopold, Aldo. 1943 "Deer Irruptions," *Wisconsin Conservation Bulletin,* August (reprinted in Wisconsin Conservation Department Publication, 321:1–11).

Leshniowsk, W. O., P. R. Dugan, R. M. Pfister, J. I. Frea, and E. I. Randles. 1970 "Aldrin: Removal From Lake Water by Flocculent Bacteria," *Science,* 1970, 169:993–995.

Lewin, R. A. 1962 *Physiology and Biochemistry of Algae.* New York: Academic Press, p. 929.

Lindeman, R. L. 1941 "The Developmental History of Cedar Creek Bog, Minnesota," *American Midland Naturalist,* 25:101–112.

Lindeman, R. L. 1941 "Seasonal Food-cycle Dynamics in a Senescent Lake," *American Midland Naturalist,* 26:636–673.

Lindeman, R. L. 1942 "The Trophic Dynamic Aspects of Ecology," *Ecology,* 23:399–418.

Little, C. and B. J. Gupta. 1968 "Pogonophora: Uptake of Dissolved Nutrients," *Nature,* 218:873–874.

Livingstone, D. A. 1955 "Some Pollen Profiles from Arctic Alaska," *Ecology,* 36:587–600.

Livingstone, D. A. 1957 "On the Sigmoid Growth Phase in the History of Linsley Pond," *American Journal of Science,* 255:364–373.

Livingstone, D. A. 1963 "Data of Geochemistry, Chapter G. Chemical Composition of Rivers and Lakes," *Geological Survey Professional Paper,* 440-G, Washington, D.C. p. 63.

Livingstone, D. A. 1963 "The Sodium Cycle and the Age of the Ocean," *Geochimica et Cosmochimica Acta,* 27:1055–1069.

Livingstone, D. A. 1967 "Postglacial Vegetation of the Ruwenzori Mountains in Equatorial Africa," *Ecological Monographs,* 37:25–52.

Lloyd, M. and T. Park. 1962 "Mortality Resulting from Interactions between Adult Flour Beetles in Laboratory Cultures," *Physiological Zoölogy,* 35:330–347.

Longwell, C. R., R. F. Flint, J. E. Sanders. 1968 *Physical Geology,* New York: John Wiley, p. 685.

Lotka, A. J. 1925 *Elements of Physical Biology,* Baltimore: Williams and Wilkins, reprinted as *Elements of Mathematical Biology,* Dover Press, 1956, p. 465.

MacArthur, R. H. 1955 "Fluctuations of Animal Populations, and a Measure of Community Stability," *Ecology,* 36:533–536.

MacArthur, R. H. 1957 "On the Relative Abundance of Bird Species," *Proceeding of the National Academy of Sciences,* 43:293–295.

MacArthur, R. H. 1958 "Population Ecology of some Warblers of Northeastern Coniferous Forests," *Ecology,* 39:599–619.

MacArthur, R. H. 1960 "On the Relative Abundance of Species," *American Naturalist,* 94:25–36.

MacArthur, R. H. 1964 "Environmental Factors Affecting Bird Species Diversity," *American Naturalist,* 98:387–397.

MacArthur, R. H. 1965 "Patterns of Species Diversity," *Biological Reviews,* 40:510–533.

MacArthur, R. H. and J. MacArthur. 1961 "On Bird Species Diversity," *Ecology,* 42:594–598.

Machta, L. and E. Hughes. 1970 "Atmospheric Oxygen in 1967 to 1970," *Science,* 168:1582–1584.

MacFadyen, A. 1962 "Energy Flow in Ecosystems and Its Exploitation by Grazing," *Grazing in Terrestrial and Marine Environments,* ed. by D. J. Crisp, Oxford: Blackwell, pp. 3–20.

MacLulick, D. A. 1937 "Fluctuations in the Numbers of the Varying Hare (*Lepus americanus*)," *University of Toronto Studies Biology Series,* 43:1–136.

Margalef, D. R. 1957 "La Teoria de la information en ecologia," *Memorias de la Real Academia De Ciencias y Artes de Barcelona,* 23: No. 13, p. 79.

Margalef, D. R. 1958 "Information Theory in Ecology," *General Systems,* Vol. III.

Margalef, D. R. 1963 "On Certain Unifying Principles in Ecology," *American Naturalist,* 97–374.

Margalef, D. R. 1968 *Perspectives in Ecological Theory,* Chicago: University of Chicago Press, p. 111.

Mc Farland, W. N. and J. Prescott. 1959 "Standing Crop, Chlorophyll Content and in situ Metabolism of a Giant Kelp Community in Southern California," *Publications of the Institute of Marine Science,* 6:109–132.

Mech, L. D. 1966 "The Wolves of Isle Royale," *Fauna of the National Parks of the United States, Fauna Series 7,* Wash. U.S. Govt. Print. Office, p. 210.

Merriam, C. H. 1890 "Results of a Biological Survey of the San Francisco Mountain Region and Desert of the Little Colorado, Arizona," *North American Fauna,* 1–136.

Merriam, C. H. 1891 "The Geographic Distribution of Life in North America," *Proceedings of the Biological Society of Washington,* 1892 (reprint 1893 in Smithsonian Institute Annual Report, pp. 365–415, 7:1–64.

Merriam, C. H. 1899 "Zone Temperatures," *Science,* 9:116.

Mitchell, R. D. 1969 "A Model Accounting for Sympatry in Watermites," *American Naturalist,* 103:331–346.

Moore, H. B. 1934 "The Biology of *Balanus balanoides* I Growth Rate and its Relation to Size, Season, and Tide Level," *Journal of the Marine Biological Association of the United Kingdom,* New Series 19:851–868.

Morowitz, 1968 H. J. *Energy Flow in Biology,* New York: Academic Press, p. 179.

Mosby, H. S. 1969 "The Influence of Hunting on the Population Dynamics of a Woodlot Gray Squirrel Population," *Journal of Wildlife Management,* 33:59–73.

Mullen, D. A. 1969 "Reproduction in Brown Lemmings (*Lemmus trimucronatus*) and its Relevance to their Cycle of Abundance," *University of California Publications in Zoology,* Vol. 85, p. 24.

Muller, C. H., W. H. Muller, and B. L. Haines. 1964 "Volatile Growth Inhibitors Produced by Aromatic Shrubs," *Science,* 143:471–473.

Murie, A. 1934 "The Moose of Isle Royale," *Miscellaneous Publication No. 25 of the Museum of Zoology,* University of Michigan, Ann Arbor, p. 44.

Murie, A. 1944 "The Wolves of Mount McKinley," *Fauna of the National Parks of the United States Fauna Series 5,* Washington, Government Printing Office, p. 238.

Neave, D. J. and B. S. Wright. 1968 "Ruffed Grouse Adrenal Weights Related to Population Density," *Journal of Wildlife Management,* 32:633–635.

Nicholson, A. J. 1950 "Population Oscillations Caused by Competition for Food," *Nature,* 165:476–477.

Nicholson, A. J. 1954 "An Outline of the Dynamics of Animal Populations," *Australian Journal of Zoology,* 2:9–65.

Nicholson, A. J. and V. A. Bailey. 1935 "The Balance of Animal Populations," *Proceedings of the Zoological Society of London,* pp. 551–603.

Nutman, P. S. 1956 "The Influence of the Legume in Root-nodule Symbiosis," *Biological Reviews,* 31:109–151.

Nye, R. H. and D. J. Greenland. 1960 "The Soil Under Shifting Cultivation," *Commonwealth Bureau Soil Science (At. Brit.) Technical Communication 51.*

Odum, H. T. and E. P. Odum. 1955 "Trophic Structure and Productivity of a Windward Coral Reef Community on Eniwetok Atoll," *Ecological Monographs,* 25:291–320.

Odum, E. P. 1969 "The Strategy of Ecosystem Development," *Science,* 1969, 164:262–270.

Odum, E. P. 1971 *Fundamentals of Ecology,* 3rd ed., Philadelphia: W. B. Saunders, p. 574.

Odum, H. T. 1956 "Efficiencies, Size of Organisms, and Community Structure," *Ecology,* 37:592–597.

Odum, H. T. 1971 *Environment, Power, and Society,* New York: John Wiley, p. 331.

Odum, H. T., and E. P. Odum. 1959 In E. P. Odum, *Fundamentals of Ecology,* Philadelphia: Saunders, p. 75.

Ogden, J. G. III. 1969 Correlation of Contemporary and Late Pleistocene Pollen Records in the Reconstruction of Postglacial Environments in Northeastern North America, Mitt. Internat, Verein. Limnol., 17:64–77.

Oosting, H. J. 1956 *The Study of Plant Communities,* 2nd ed. San Francisco: Freeman, p. 440.

Overland, L. 1966 "The Role of Allelopathic Substances in the "Smother Crop" Barley," *American Journal of Botany,* 53:423–432.

Park, T., D. B. Mertz, W. Grodzinski and T. Prus. 1965 "Cannibalistic Predation in Populations of Flour Beetles," *Physiological Zoölogy,* 38:289–321.

Patrick, R. 1954 "A New Method for Determining the Pattern of Diatom Flora," *Notulae Natural of the Academy of Natural Sciences of Philadelphia,* 259:1–12.

Patrick, R. 1961 "A Study of the Numbers and Kinds of Species Found in Rivers in Eastern United States," *Proceedings of the Academy of Natural Sciences of Philadelphia,* 113:215–258.

Pearl, R. 1932 "The Influence of Density of Population upon Egg Production in *Drosophila melanogaster,*" *Journal of Experimental Zoology,* 63:57–84.

Pearl, R. and S. L. Parker. 1922 "Experimental Studies on the Duration of Life, IV: Data on the Influence of Density of Population on Duration of Life in *Drosophila,*" *American Naturalist,* 56:312–322.

Peterle, T. J. 1969 "Translocation of Pesticides in the Environment," Symposium on the Biological Impact of Pesticides in the Environment: Environmental Health Science Series, No. 1, J. W. Gillett, ed., pp. 11–16.

Peterle, T. J. 1969 "DDT in Antarctic Snow," *Nature,* 224:640.

Pielou, E. C. and A. N. Arnason. 1966 "Correction to One of MacArthur's Species-Abundance Formulas," *Science,* 151:592.

Pitelka, F. A. 1957 "Some Characteristics of Microtine Cycles in the Arctic," *Arctic Biology,* P. H. Hansen (ed.), 18th Annual Colloquium, Oregon State University, Corvallis.

Pitelka, F. A., P. Q. Tomich, and G. W. Treichel. 1955 "Ecological Relations of Jaegers and Owls as Lemming Predators Near Barrow, Alaska," *Ecological Monographs,* 25:85–117.

Preston, F. W. 1948 "The Commonness, and Rarity, of Species," *Ecology,* 29:254–283.

Preston, F. W. 1960 "Time and Space and the Variation of Species," *Ecology,* 41:785–790.

Provasoli, L. 1958 "Nutrition and Ecology of Protozoa and Algae," *Annual Review of Microbiology,* 12:279–308.

Rasmussen, D. I. 1941 "Biotic Communities of Kaibab Plateau, Arizona," *Ecological Monographs,* 11:229–275.

Raunkiaer, C. 1934 "The Life Forms of Plants and Statistical Plant Geography; Being the Collected Papers of C. Raunkiaer," Oxford p. 632.

Revelle, R. et al. 1965 "Atmospheric Carbon Dioxide," Appendix Y4. pp. 111–133, "Restoring the Quality of Our Environment: Report of the Environmental Pollution Panel, President's Science Advisory Committee," Washington, p. 317.

Rhyther, J. H., 1960.

Rhyther, J. H. 1970 "Is the World's Oxygen Supply Threatened?" *Nature*, 227:374–375.

Rich, E. R. 1956 "Egg Cannibalism and Fecundity in *Tribolium*," *Ecology*, 37:109–120.

Richards, P. W. 1952 *The Tropical Rain Forest.* Cambridge University Press.

Richman, S. 1958 "The Transformation of Energy by *Daphnia pulex*," *Ecological Monographs,* 28:273–291.

Riley, G. A. 1956 "Oceanography of Long Island Sound, 1952–54. IX Production and Utilization of Organic Matter," *Bulletin of Bingham Oceanography Coll.,* 15:324–344.

Robertson, F. W. and J. H. Sang. 1944. "The Ecological Determinants of Population Growth in a *Drosophila* Culture, II: Circumstances Affecting Egg Viability," Proceedings of the Royal Society of London (B), 132:258–277.

Royce, W. F., D. E. Bevan, J. A. Crutchfield, G. J. Paulik, and R. F. Fletcher. 1963 *Salmon Gear Limitation in Northern Washington Waters,* University of Washington Publications in Fisheries, new series, 2(1):1–123.

Russell, E. W. 1950 "Soil Conditions and Plant Growth," 8th ed., Chapter 7, London, Longman, Green, p. 635.

Ruttner, F. 1963 *Fundamentals of Limnology,* 3rd ed., translated by D. G. Frey and F. E. J. Fry, Toronto: University of Toronto Press.

Ryther, J. H. 1959 "Potential Productivity of the Sea," *Science,* 130:602–608.

Ryther, J. H. 1960 "Organic Productivity by Plankton Algae, and its Environmental Control," in C. A. Tryon and R. T. Hartman (eds.), "The Ecology of Algae "Special Publication No. 2, Pymatuning Laboratory of Field Biology, Pittsburgh, pp. 72–83.

Ryther, J. H. 1969 "Photosynthesis and Fish Production in the Sea," *Science,* 166:72–76.

Sanders, H. L. 1968 "Marine Benthic Diversity: a Comparative Study," *American Naturalist,* 102:243–282.

Saunders, A. A. 1936 "Ecology of the Birds of Quaker Run Valley, Allegany State Park, New York," *New York State Museum Handbook,* 16, Albany, New York.

Schaller, G. B. 1967 *The Deer and the Tiger,* Chicago: Chicago University Press, p. 370.

Schimper, A. F. W. 1903 *Plant-Geography upon a Physiological Basis,* Oxford: Clarendon Press.

Schjelderup-Ebbe, T. 1935 "Social Behavior of Birds," *A Handbook of Social Psychology,* ed. by Carl Murchison, Worcester: Clark University Press, pp. 947–972.

Schmidt-Nielsen, K. 1964 *Desert Animals,* Oxford University Press, p. 270.

Schofield, E. K. and P. A. Colinvaux. 1969 "Fossil *Azolla* from the Galapagos Islands," *Bulletin of the Torrey Botanical Club,* 96:623–628.

Schultz, A. M. 1969 "A Study of an Ecosystem: The Arctic Tundra," *The Ecosystem Concept in Natural Resource Management,* G. M. Van Dyne ed., New York: Academic Press, pp. 77–93.

Sclater, P. L. 1858 "On the General Geographical Distribution of the Members of the Class Aves," *Journal of Proceedings of Linnaean Society of London,* (Zoology), 2:130–145.

Selye, H. 1950 "The Physiology and Pathology of Exposure to Stress; a Treatise Based on the Concepts of the General-Adaptation-Syndrome and the Diseases of Adaptation," Montreal, *Acta,* p. 822.

Sernander, R. 1908 "On the evidences of Postglacial Changes of Climate Furnished by the Peat Mosses of Northern Europe," *Geologiska Föreningens Förhandlingar,* 30:465–473.

Shelford, V. E. 1908 "Life-histories and Larval Habits of the Tiger Beetles (*Cicindelidae*), *Zoology,* (published by the Linnean Society) 30:157–184.

Shelford, V. E. 1911 "Ecological Succession II: Pond Fishes," *Biological Bulleting,* 21:127–151.

Shelford, V. E. 1912 "Ecological Succession IV: Vegetation and the Control of Land Animal Communities," *Biological Bulletin,* 23:59–99.

Shelford, V. E. 1913 *Animal Communities in Temperate America,* Chicago: University of Chicago Press, p. 368.

Shelford, V. E. 1963 *The Ecology of North America,* University of Illinois Press, p. 603.

Slobodkin, L. B. 1962 "Energy in Animal Ecology," *Advances in Ecology,* 4:69–101.

Slobodkin, L. B. and H. L. Sanders. 1969 "On the Contribution of Environmental Predictability to Species Diversity," *Diversity and Stability in Ecological Systems,* Brookhaven Symposium in Biology No. 22, pp. 82–95.

Smith, A. G. and E. H. Willis. 1961 "Radiocarbon Dating of the Tallahogy Landman Phase," *Ulster Journal of Archaeology,* 24–25.

Smith, N. G. 1970 "On Change in Biological Communities," *Science,* 170:312–313.

Smith, M. W. 1948 "Preliminary Observations upon the Fertilization of Crecy Lake, New Brunswick," *Transactions of the American Fisheries Society,* 75:165–174.

Smuts, J. C. 1926 *Holism and Evolution,* New York: Macmillian, p. 362.

Solomon, M. E. 1949 "The Natural Control of Animal Populations," *Journal of Animal Ecology,* 18:1–35.

Southward, A. J. and E. C. Southward. 1968 "Uptake and Incorporation of Labelled Glycine by Pogonophores," *Nature,* 218:875–876.

Stanley, H. M. 1890 "The Great African Forest," *Through Darkest Africa,* Vol. II, New York: Scribners, Chapter 23:73–111.

Stehli, F. R., R. G. Douglas, and N. D. Newell. 1969 "Generation and Maintenance of Gradientes in Taxonomic Diversity," *Science,* 164:947–949.

Stewart, R. E. and J. W. Aldrich. 1951 "Removal and Population of Breeding Birds in a Spruce-fir Forest Community," *Auk* 68:471–482.

Summerhayes, V. S. and C. S. Elton. 1923 "Contributions to the Ecology of Spitsbergen and Bear Island," *Journal of Ecology,* 11:214–286.

Swanson, C. A. 1943 "The Effect of Culture Filtrates on Respiration in *Chlorella vulgaris,*" *American Journal of Botany,* 30:8–11.

Talbot, L. M. and M. H. Talbot. 1963 "The Wildebeest in Western Masailand, East Africa," *Wildlife Monographs* No. 12, p. 88.

Tansley, A. G. 1935 "The Use and Abuse of Vegetational Concepts and Terms," *Ecology,* 16:284–307.

Thompson, W. R. 1939 "Biological Control and the Theories of the Interactions of Populations," *Parasitology,* 31:299–388.

Transeau, E. N. 1926 "The Accumulation of Energy by Plants," *Ohio Journal of Science,* 26:1–10.

Tsukada, M. 1966 "Late Pleistocene Vegetation and Climate in Taiwan (Formosa)," *Proceedings of the National Academy of Sciences,* 55:543–548.

Tsukada, M. 1967 "Fossil Cladocera in Lake Nojiri and Ecological Order," *Quaternary Research (Japan),* 6:101–110.

Utida, S. 1950 "On the Equilibrium State of the Interacting Population of an Insect and its Parasite," *Ecology,* 31:165–175.

Utida, S. 1957 "Cyclic Fluctuations of Population Density Intrinsic to the Host-parasite System," *Ecology,* 38:442–449.

Varley, G. C. 1970 "The Concept of Energy Flow Applied to a Woodland Community," *Animal Populations in Relation to their Food Resources,* ed. by A. Watson, Oxford: Blackwell, pp. 389–406.

Vaurie, C. 1951 "Adaptive Differences Between Two Sympatric Species of Nuthatches (*Sitta*)" *Proceedings of the International Ornithological Congress,* 19:163–166.

Vestal, A. G. 1914 "Internal Relations of Terrestrial Associations," *American Naturalist,* 48:413–445.

Volterra, V. 1926 "Variations and Fluctuations of the Number of Individuals in Animal Species Living Together," in *Animal Ecology,* by R. N. Chapman, New York: McGraw-Hill, Appendix, pp. 409–448.

Walker, D. 1970 "Direction and Rate in Some British Postglacial Hydroseres," *The Vegetational History of the British Isles,* ed. by D. Walker and R. West, Cambridge: Cambridge University Press, pp. 117–139.

Wallace, A. R. 1876 *The Geographic Distribution of Animals,* 2 vols. London: Macmillan.

Wallace, A. R. 1878 *Tropical Nature and Other Essays,* London: Macmillan, p. 356.

Warming, E. 1909 *Oecology of Plants: An Introduction to the Study of Plant Communities,* English Translation by P. Groom and I. B. Balfour, Oxford, p. 422.

Wassink, E. C. 1959 "Efficiency of Light Energy Conversion in Plant Growth," *Plant Physiology,* 34:356–361.

Watt, K. E. F. 1964 "The Use of Mathematics and Computers to Determine Optimal Strategy and Tactics for a Given Pest Control Problem," *Canadian Entomologists,* 96:202–220.

Watt, K. E. F. 1968 *Ecology and Resource Management,* New York: McGraw-Hill, p. 450.

Went, F. W. 1970 "Plants and the Chemical Environment," *Chemical Ecology* ed. by E. Sondheimer and J. B. Simeone. New York: Academic Press, pp. 71–82.

West, R. G. 1968 *Pleistocene Geology and Biology,* New York: John Wiley p. 377.

Westlake, D. F. 1963 "Comparisons of Plant Productivity," *Biological Reviews,* 38:385–425.

White, C. M. N. 1951 "Weaver Birds at Lake Mweru," *Ibis,* 93:626–627.

Whitehead, D. R. 1964 "Fossil Pine Pollen and Full Glacial Vegetation in Southeastern North Carolina," *Ecology,* 44:403–406.

Whitehead, D. R. 1965 "Palynology and Pleistocene Phytogeography of Unglaciated Eastern North America," *The Quaternary of the United States,* ed. by H. E. Wright and D. G. Frey, Princeton: Princeton University Press.

Whittaker, R. H. 1956 "Vegetation of the Great Smoky Mountains," *Ecological Monographs,* 26:1–80.

Whittaker, R. H. 1960 "Vegetation of the Siskiyou Mountains, Oregon and California," *Ecological Monographs,* 30:279–338.

Whittaker, R. H. 1962 "Classification of Natural Communities," *Botanical Review,* 28:1–239.

Whittaker, R. H. 1965 "Dominance and Diversity in Land Plant Communities," *Science,* 147:250–260.

Whittaker, R. H. 1967 "Gradient Analysis of Vegetation," *Biological Reviews,* 42:207–264.

Whittaker, R. H. 1969 "Evolution of Diversity in Plant Communities" *Diversity and Stability in Ecological Systems,* Brookhaven Symposia in Biology, No. 22, pp. 178–260.

Whittaker, R. H. and P. P. Feeny. 1971 "Allelochemics: Chemical Interactions Between Species," *Science,* 171:757–770.

Whittaker, R. H. and W. A. Niering. 1967 "Vegetation of the Santa Catalina Mountains, Arizona: A Gradient Analysis of the South Slope," *Ecology,* 46:429–452.

Wiegert, R. G. and R. Mitchell. 1972 "Ecology of Yellowstone thermal effluent systems: intersect of blue-green algae, grazing flies (*Paracoenia,* Ephydridae) and water mites (*Partnuniella,* Hydrachnellae)," *Hydrobiologia,* in press.

Wiens, J. A. 1966 "On Group Selection and Wynne-Edward's Hypothesis," *American Scientist,* 54:273–287.

Williams, A. B. 1936 "The Composition and Dynamics of a Beech-Maple Climax Community," *Ecological Monographs,* 6:318–408.

Wilson, J. Warren. 1966 "An Analysis of Plant Growth and Its Control in Arctic Environments," *Annals of Botany,* N. S. 30:383–402.

Wood, B. J. 1971 "Development of Integrated Control Programmes for

Pests of Tropical Perennial Crops in Malaysia," AAAS Symposium on Biological Control, Boston, edited by C. B. Huffaker, New York, Plenum Press.

Woodruff, L. L. 1912 "Observations on the Origin and Sequence of the Protozoan Fauna of Hay Infusions," *Journal of Experimental Zoology*, 12:205–264.

Woodruff, L. L. 1913 "The Effect of Excretion Products of Infusoria on the Same and on Different Species, with Special Reference to the Protozoan Sequence in Infusions," *Journal of Experimental Zoology*, 14:575–582.

Woodwell, G. M. 1970 "Effects of Pollution on the Structure and Physiology of Ecosystems," *Science*, 168:429–433.

Woodwell, G. M. and R. H. Whittaker. 1968 "Primary Production in Terrestrial Ecosystems," *American Zoologist*, 8:19–30.

Wynn-Edwards, V. C. 1962 *"Animal Dispersion in Relation to Social Behavior,"* Haffner, p. 653.

Figure 2.1 From ANIMAL GEOGRAPHY by Marion Newbigin, Oxford, England, 1928.   **Fig. 2.2** R. H. Noailles   **Fig. 2.3** Ralph Crane/Black Star **Fig. 2.4** Mark Boulton/National Audubon Society   **Fig. 2.5A** Verna R. Johnston/Photo Researchers   **Fig. 2.5B** Courtesy French Government Tourist Office.   **Fig. 2.5C** Courtesy Victoria Forest Commission, Melbourne, Australia.   **Fig. 2.6** Marcella Gay/Black Star   **Fig. 2.25** Steve McCutcheon/Alaska Pictorial Service   **Fig. 2.27** Courtesy NASA   **Fig. 6.1** Helen Nestor/Monkmeyer   **Fig. 6.2** From THE ECOLOGY OF NORTH AMERICA by Victor E. Shelford, 1963. Courtesy University of Illinois Press. Urbana.   **Fig. 6.4A–F** Jack Dermid   **Fig. 7.10** Paul A. Colinvaux   **Fig. 8.3A** Leonard Lee Rue III/National Audubon Society **Fig. 8.3B** J. Alex Langley/DPI   **Fig. 8.3C** Leonard Lee Rue III/National Audubon Society   **Fig. 10.4** From A TROPICAL RAIN FOREST by Howard T. Odum, Book 3, 1970. Courtesy AEC.   **Fig. 11.2** Douglas P. Wilson   **Fig. 15.1A** Courtesy The Nitragin Company   **Fig. 15.1B** Jack Dermid   **Fig. 18.5** From A TREATISON LIMNOLOGY by G. E. Hutchinson, Vol. 1, John Wiley & Sons, Inc., 1957. Courtesy G. E. Hutchinson. **Fig. 18.8** Nancy Hays/Monkmeyer   **Fig. 19.1** Kenneth Murray/Nancy Palmer   **Fig. 19.2** Courtesy Brookhaven National Laboratory   **Fig. 20.4A** Donna N. Sprunt/National Audubon Society   **Fig. 20.4B** Beth Weston/ Rapho Guillumette   **Fig. 20.4C** Walter Dawn   **Fig. 20.8** Arthur Ambler/ National Audubon Society   **Fig. 21.1A&B** Courtesy C. H. Muller, University of California, Santa Barbara.   **Fig. 25.5** Haverschmidt/National

INDEX—
GLOSSARY

* The asterisk after a page number indicates pages on which there are color plates.

interspecific competition (Chapter 23), 330 ff
logistic (Chapter 22), 309 ff, 313
log-normal for relative abundance, 513 ff
predation, algebraic, 420
  computer simulation, 439 ff
  Holling, 432 ff
  Lotka-Volterra, 422
  Nicholson and Bailey, 430
  regulation of prey, 439
  seeds of forest trees, 479
rodent cycles, 494
thrips irruptions, 387
Monoculture, hazards of, 240
Montpelier School, *see* Zurich-Montpelier School
Moose, Isle Royale on, 378
  life table, 378
  predation by wolves, 405
Mores, 117
  compared with niche, 134
Mortality, 370
Mosaic, of diversity, 571
Mosby, H. S., life table for squirrels, 374
Mouse plagues (Chapter 35), 482
Mullen, D. A., on weather in lemming cycles, 489 ff
Murie, A., on Dall sheep, 373
Muskrats, 402 ff
Mycorrhizae
  Roots and fungi living symbiotically or commensally. Together they constitute the nutrient retrieval system.

Naturcomplex, 123
Nekton
  Animals of sea, such as fish, which can control their position by swimming: sea animals other than "plankton."
Neritic zone
  Shallow sea over continental shelf.
Neuston
  Organisms supported on water surface.
Niche, defined, 134
  diversity and geometry, 538
  herdsmen of, 141, 504
  hypervolume: formal definition, 522
  introduced, 133 ff
  "mores" foreshadows, 117
  overlapping or not, 540 ff
  partitioning by new species, 572
  peasants of, 142, 504
  possibilities for solute feeder, 190 ff

size and birds' beaks, 540
Nicholson, A. J., on population regulation, 361
  predation, model of, 430
Nigerian rain forest, 201
Nitrates, oxygen return from, 218
Nitrifying bacteria, 213
Nitrogen, cycle (Chapter 15), 212 ff, 217, (figure)
  fixation, 213 ff
Numerical response, 431
Nutrients, cycles (Chapter 14), 198 ff
  cycles closed by fossil fuels, 57
  farming for, 201
  lemming cycle hypothesis, 488
  limiting factors, 275
  nitrogen (Chapter 15), 211 ff
  phosphate from detergents, 260, 271
  phosphorus, 206 ff
  rain forests in, 199 (figure), 200
  sea, productivity of limited by, 162
  soil, retention in, 202
  succession conserves, 559
  sulfur, 220 ff
  temperate forests in, 203 (figure)
  water pollution, 256 ff
Nuthatches, character displacement, 350 (figure), 351

Oceans, *see* Sea
  farming, prospects, 164
    unrealistic, 187
Odum, E. P., on complexity-stability hypothesis, 239
  on coral reef productivity, 163
  on succession and biomass, 554 ff
Odum, H. T., on assimilation and production, 170
  on coral reef productivity, 163
  on hydraulic analogy, 230, 232
  on productivity of forests, 155–156
  on succession and complex structure, 560 ff
Ohms law, 311
Oligotrophic lake, 250
Ontogeny
  The development of an individual: especially the changes in the shapes of its parts as it grows to maturity.
Opportunist species, 392
  low diversity of, 545
  muskrats, 403
  relative abundance of, 517
  replaced in succession, 567